SOTHIS PRESS

Sothis Press aims to completely reimagine the public domain book with an emphasis on universal access to captivatingly designed print editions. We believe that technology will enable every single book ever published to be perpetually accessible in beautiful print form. Sothis is beauty in ubiquity.

www.sothis.press
1 Waterhouse Square, London, ECIN 2ST

A Text-Book of Ordnance and Gunnery

Prepared for the Use of Cadets
of the U. S. Military

Lawrence L. Bruff

SOTHIS PRESS

ISBN 978-1-0368-0300-1

A Text-Book of Ordnance and Gunnery
is in the public domain.

This edition is published by
Sothis Press © 2023. The company
is registered in England and Wales.
The Company Number is 14900389.

PREFACE.

THE present text-book has been compiled with the object of presenting as clearly as possible the elementary principles of the course in Ordnance and Gunnery as taught at the Military Academy, and of so arranging it that it can be readily used for recitations in the section-room. For this purpose it has been divided into separate subjects, each as well defined as possible. In its preparation I have followed the lines laid down by my predecessor Capt. Henry Metcalfe, U. S. A., retired, from whose labors in the same field I have derived the greatest assistance. I am also under many obligations to those who have kindly criticised and corrected my work in many important particulars, and especially to Colonel Buffington, Captains Smith, Blunt, Birnie, Mitcham and Crozier, Ordnance Department, and to Mr. W. R. Quinan of the California Powder Company.

Lieut. Babbitt, Ordnance Department, my assistant at the Military Academy, has made many valuable criticisms and suggestions which have added greatly to the clearness of the work.

WEST POINT, March 10, 1896.

TABLE OF CONTENTS.

TEXT-BOOK OF ORDNANCE AND GUNNERY.

CHAPTER I.

GUNPOWDER AND INTERIOR BALLISTICS.

GUNPOWDER.

1. Composition—Manufacture.

COMPOSITION.—Gunpowder is a mechanical mixture of nitre, charcoal, and sulphur, in the proportions of 75 parts nitre, 15 charcoal, and 10 sulphur.

The nitre furnishes the oxygen to burn the charcoal and sulphur. The charcoal is the principal combustible body, and the sulphur gives density to the grain and lowers its point of ignition.

The nitre is purified by solution in water and crystallization, the sulphur by distillation, and the charcoal is carefully prepared to make it as uniform as possible.

The distinguishing characteristic of charcoal is its color, being brown when prepared at a temperature up to 280° Cent., from this to 340° red, and beyond 340° black.

Brown charcoal is now generally used for powder.

MANUFACTURE.—The operations are:

1. Pulverizing, mixing, and incorporating the ingredients.

2. Compressing this mixture to give it a proper density.

3. Dividing the dense mass into grains.

4. Finishing the grains.

Pulverizing and Mixing.—The nitre is in fine crystals when received; the sulphur is rolled in an iron barrel with iron balls; and the charcoal also, or the latter may be ground in a mill.

The ingredients are mixed by hand or by machine.

Incorporating.—To make the mixture thorough, the above composition is moistened and incorporated in a wheel mill. This mill consists of a pair of heavy cast-iron cylindrical rollers running in a circular trough (Fig. 1). By their action they grind the products together, and give a thorough mixture.

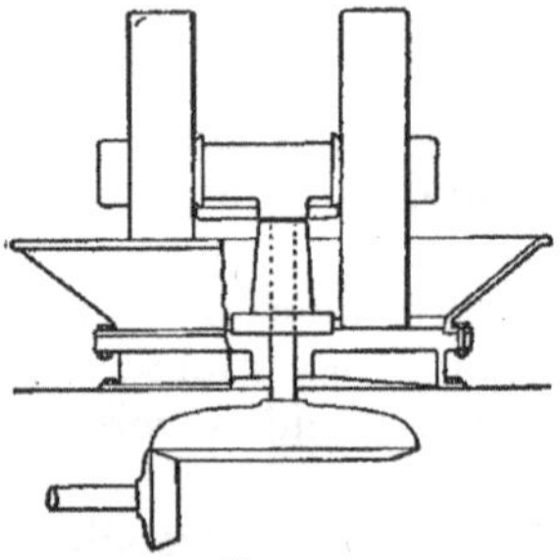

FIG. 1.

This is the most important operation in the manufacture, and if not well done no subsequent operation can remedy it.

Pressing.—The mill-cake which comes from the "wheel mill" is broken up, moistened, and placed, in layers about 2 inches thick, under a hydraulic press. The layers are reduced to a thickness of about 1 inch and become very dense and hard, and this is called "press-cake."

Graining.—In ordinary powders, the press-cake is broken up into grains of various sizes. The object of this is to increase the surface of combustion, and to regulate it according to the gun in which it is to be used. The graining is done by rollers acting on the press-cake, and the grains are afterwards assorted with sieves.

Glazing.—To remove the sharp angles, and give uniform density to the surface, the grains are placed in a wooden barrel revolving on its axis. They are thus made to rub against each other, and accomplish, by their mutual attrition, the objects mentioned.

Drying and Dusting.—The excess of moisture in the powder is now removed by a current of warm, dry air, and the dust which has been formed on the grains is removed by passing them through a revolving sieve.

Blending and Marking.—Different lots of the same kind of powder are mixed, to overcome, as much as possible, irregularities of manufacture. In our service the powder is packed in 100-lb. barrels and marked with certain letters, as I. K. B., E. V. X., etc., the first two letters denoting the kind of powder, or its use, and the third letter the lot.

In foreign services the letters indicate the use directly: as, R. F. G., rifle fine grain; P., pebble; etc.

Government powder is purchased by contract.

2. Specific Gravity and Gravimetric Density.

THE SPECIFIC GRAVITY, or actual density, of gunpowder, like that of any solid body, is the weight of a given volume referred to that of an equal volume of water as unity. Since water dissolves the nitre, mercury is used instead. The instrument employed is called a mercury densimeter, and consists of a glass globe *a*, Fig. 2, connected with an air-pump by a rubber tube *c*.

The globe is exhausted of air and its lower end immersed in mercury in the dish *d*.

The mercury is allowed to rise till it fills the globe and stands at a certain height in the glass tube *e*. The globe is then detached full of mercury and weighed. It is then emptied, and a given weight of powder placed in it, returned to its original position, the air again exhausted, and mercury allowed to enter till it stands at the same height as before; the globe with its mercury and powder again detached and weighed. The difference of the two weights of mercury gives the weight of the mercury whose volume is equal to that of the powder.

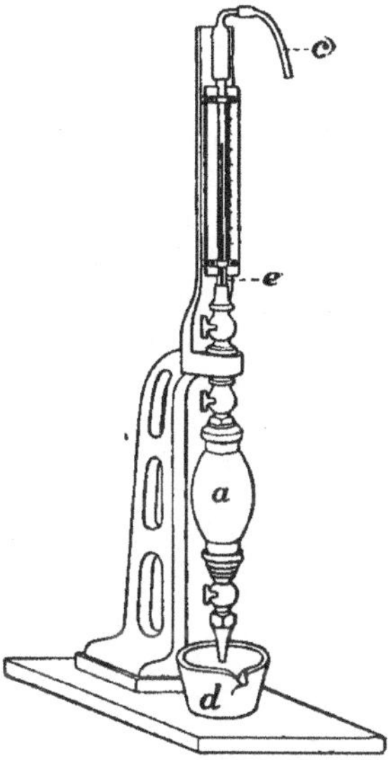

FIG. 2.

Let a = the weight of the powder;

P = the weight of the vessel and mercury;

P' = the weight of the vessel, mercury, and powder;

$S =$ the specific gravity of the mercury;
$\delta =$ the specific gravity of the powder.

Then $P' - a =$ the weight of the mercury and vessel when the latter is partially filled with powder;
$P - P' + a =$ the weight of the volume of mercury displaced by the powder.

Since the weights of equal volumes are proportional to the densities, we have

$$a : P - P' + a :: \delta : S,$$

or

$$\delta = \frac{aS}{P - P' + a} \quad . \quad . \quad . \quad . \quad . \quad . \quad . \quad (1)$$

The density varies between 1.68 and 1.85.

GRAVIMETRIC DENSITY is the name given to the density of powder when the spaces between the grains are considered. That is, it is the specific gravity of the powder in its natural form. Suppose we have a solid piece of powder weighing 1 lb. If we determine its specific gravity, we will obtain a value δ given by formula (1). Suppose the same powder broken up into grains. Its weight will not change, but its volume will be greater than before. If we determine its specific gravity under these circumstances, by comparing its weight with the weight of an equal volume of water, we have a particular value, called the "gravimetric density."

It is evident that the same powder will have only one value for δ, but may have many values for gravimetric density according to its granulation. If the shape of the grain is changed, the same weight of powder will occupy a greater or less space according as the spaces between the grains are greater or less.

Hence we say that gravimetric density measures the capacity of powder to pack, or measures the spaces between the grains.

A cubic foot of powder is usually taken in determining

gravimetric density. A cubic foot of water weighs 62.425 lbs. Hence we have

$$\gamma = \frac{w}{62.425}, \quad . \quad . \quad . \quad . \quad . \quad . \quad . \quad . \quad (2)$$

γ denoting the gravimetric density, and w the weight of a cubic foot of powder. γ varies between 0.875 and 1.00.

The space actually occupied by the solid powder in a given volume is determined as follows:

Let V = the total volume occupied by the powder;

v = the volume of the solid powder.

Since volumes are inversely as densities, we have

$$V : v :: \delta : \gamma;$$

$$v = V\frac{\gamma}{\delta} \quad . \quad . \quad . \quad . \quad . \quad . \quad . \quad . \quad . \quad (3)$$

If $\gamma = 1.00$ and $\delta = 1.8$ (the ordinary values), we have

$$v = .56V. \quad . \quad . \quad . \quad . \quad . \quad . \quad . \quad . \quad (4)$$

That is, the volume occupied by the solid powder in a charge is about .56 of the total volume of the charge.

3. Form of Grain.

Irregular Granulation.—The processes of manufacture are the same for all powders up to and including incorporation.

If the mill-cake be pressed into slabs, and these slabs broken up into irregular grains by rollers, we have powders of "irregular granulation." In our service the powders of irregular granulation are:

(*a*) *Small-arms powder*, used in the Springfield rifle and carbine, and in small arms generally, and also as a bursting-charge in field-shells.

(*b*) *Mortar-powder*, used in the 3.00-inch wrought-iron rifle, in the siege and sea-coast smooth-bore mortars, and in their shells.

(*c*) *Cannon-powder*, used in the old 8- and 10-inch smooth-bore guns.

(*d*) *Mammoth Powder*, used in the 15-inch Rodman guns.

(*e*) *I. K. Powder*, used in the 3.20-inch steel B. L. rifle, *model of* 1885, charge (3.50) lbs.

All these powders may be regarded as having grains which approach a sphere in shape, and whose mean radius is determined as follows:

Let N = the number of grains in one pound of powder;
r = the mean radius of the grain in inches;
δ = the specific gravity of the powder.

The volume of one grain is

$$V = \tfrac{4}{3}\pi r^3 \quad \ldots \ldots \ldots \quad (5)$$

Its weight is (Michie, eq. 1)

$$W = V\delta g', \quad \ldots \ldots \ldots \quad (6)$$

g' being the weight of one cubic inch of water, or

$$g' = \frac{62.425}{1728}.$$

Hence

$$W = \frac{\tfrac{4}{3}\pi r^3\delta \times 62.425}{1728}. \quad \ldots \ldots \quad (7)$$

and the weight of N grains is

$$NW = \frac{\tfrac{4}{3}\pi r^3\delta \times 62.425 \times N}{1728}. \quad \ldots \ldots \quad (8)$$

But $NW = 1$. Hence

$$1 = \frac{\tfrac{4}{3}\pi r^3\delta \times 62.425 \times N}{1728},$$

or

$$r = \frac{1.8766}{(\delta N)^{\frac{1}{3}}}. \quad \ldots \ldots \ldots \quad (9)$$

Using this method, we find

$2r$ = 0.04 small-arms powder;
$2r$ = 0.08 mortar powder;
$2r$ = 0.30 cannon powder;
$2r$ = 0.75 mammoth powder;
$2r$ = 0.24 I. K. powder.

Regular Granulation.—Powders of regular granulation are obtained by breaking up the mill-cake and pressing it between plates having depressions in them of regular shape, such as a sphere or a pyramid.

Under this head we have the Dupont powders, viz.;

(*a*) *The Hexagonal.*—The shape of the grains is that of two hexagonal pyramids joined base to base. The grains are connected by a thin cake, which is broken off, and leaves a rough surface at *a*, which facilitates ignition. Used in the 8-inch converted rifle, charge 35 lbs.

FIG. 3.—HEXAGONAL.

(*b*) *The Sphero-hexagonal*, Fig. 4, which is the same as the above, except that spheres are substituted for hexagonal pyramids.

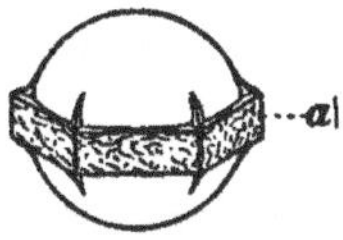

FIG. 4.—SPHERO-HEXAGONAL.

This powder is now used for all the field and siege guns and mortars. In the field service it has replaced the I. K. powder used with the earlier model guns.

Molded Powder.—This is made by reducing the mill-cake to powder and pressing it into any required form, each grain being made separately; or a number of grains of powder of irregular granulation may be compressed into a single large grain; the latter is also called concrete powder. Under this head we have prismatic or brown powder, which is a molded concrete powder made as above described. It is called brown or cocoa powder from its color, which is due to brown charcoal.

It is made in hexagonal prisms, Fig. 5, about 1 inch high and 1.375 inches between opposite faces. Each prism is pierced by a central hole parallel to the axis, and about 0.40 inch in diameter. The composition is generally given as

Nitre...........	81.5 per cent
Charcoal........	15.5 per cent
Sulphur........	3.0 per cent
	100.0

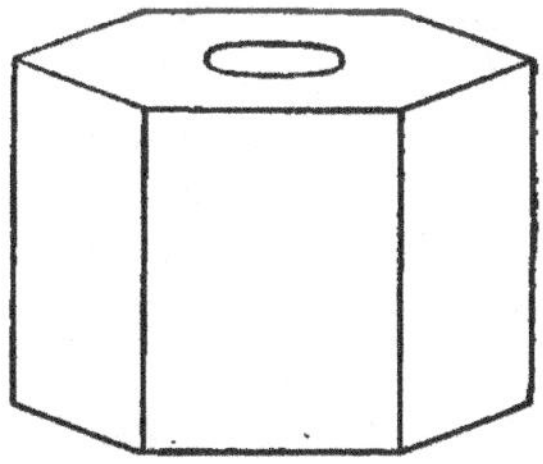
FIG. 5.—COCOA.

It is a slow-burning powder, and is used in modern high-power guns of large calibre.

4. Inspection and Proof of Powder—History.

INSPECTION.—The object is to see that the powder is properly manufactured, and that it has certain required qualities.

For small-arms powder 100 barrels are considered a lot, and from them five barrels are taken, one pound of powder from each barrel being selected for test. If the test is successful, the lot is accepted.

For Granulation.—All grains must pass through a sieve with a mesh of 0.06 inch and none through a mesh of 0.03 inch.

Specific Gravity between 1.75 and 1.80 and gravimetric density between 0.96 and 1.00.

Dust is detected by allowing a stream of powder to fall rapidly two or three feet in a strong light. There must be no dust.

Incorporation is tested by flashing 20 grains of powder on a copper plate. There should be little residue, and no globules of fuzed nitre on the plate.

Moisture is determined by exposing 1000 grains to a temperature of 100° F. for 24 hours. The loss in weight should be about 7 grains.

Capacity for absorbing moisture is tested by exposing 1000 grains to the vapor of water for 24 hours. The gain in weight should be about 6 grains.

Fouling is tested by firing rapidly 100 rounds of rifle-ball cartridges in series of 25 rounds each and weighing the rifle after each series. In a moderately dry atmosphere the weight of fouling from the 100 rounds should not exceed 15 grains.

For other powders the inspection will vary according to the terms of the contract.

PROOF.—The object of proving powder is to ascertain the initial velocity it will impart to the projectile, the corresponding pressure on the bore, and, for small arms, the accuracy of the projectile.

The powder is always proved in the gun in which it is

to be used, and with service charges. The velocities and pressures are measured with instruments to be described.

HISTORY.—Gunpowder was first used in Europe early in the fourteenth century, and in the form of powder or dust, whence the name. The guns in which it was used were weak, and the powder was suited to them, because it burned slowly and gave low pressures. In the form of dust, however, it was difficult to load at the muzzle, as cartridges were not used, and hence loading at the breech was introduced. This failed because no gas-check could be devised that would completely close the breech. As guns improved in strength, better results were obtained by graining the powder, but the grained powder became too strong for the guns, and large guns were not made. No marked change in powder was made until about 1860.

General Rodman, of the Ordnance Department, then proposed to vary the size of the grain with the calibre, using large-grained powder for large guns. He also advocated a perforated powder, which was not used. His mammoth powder, however, was adopted, and by an improvement in the process of gun-construction, together with that of the powder, he built smooth-bore guns up to 20 inches in calibre.

Following Rodman's plan, various forms of powder have been adopted by other nations, the general idea being to so modify the action that the gun will be strained less at the beginning of the motion of the projectile, and more uniformly throughout the bore, than with the old powders. The brown powder is the latest development of the old nitrate powders. It has, however, many objections, and at present smokeless powders are being developed, with the prospect that they will supersede the others.

5. Combustion in Air—Laws.

Explosion is the rapid conversion of gunpowder into gases and solids with evolution of heat. It may be divided into three parts, Ignition, Inflammation, and Combustion.

Ignition is the setting on fire of a part of the grain or charge, and for this purpose a temperature of 300° C. is required.

Gunpowder may be ignited by electricity, by contact with an ignited body, by friction, shock, or by chemical reagents.

A gradual heat will decompose the powder by subliming the sulphur, and the temperature of ignition will be raised thereby.

Flame, owing to its slight density, will not ignite powder readily. The time necessary for ignition will vary with the condition of the powder. Thus damp powder ignites less easily than dry; a smooth grain less easily than a rough one; a dense grain less easily than a light one; etc.

Powder is ordinarily ignited by a primer, by electricity, or by contact with an ignited body.

Inflammation is the spread of the ignition from point to point of the grain, or from grain to grain of the charge.

With small-grain powders, where the spaces between grains are small, the time of inflammation is large as compared with the time of combustion of a grain, but with modern large-grain powders, the facilities for the spread of ignition and the time of burning of the grain are so great, that the whole charge is supposed to be inflamed at the same instant, and the time of inflammation is not considered.

Combustion is the burning of the inflamed grain from the surface of ignition inward or outward or both, as the case may be.

LAWS.—Experiment shows that powder burns in the air according to the following laws:

1. In parallel layers, with uniform velocity, and the velocity is independent of the cross-section burning.

2. The velocity of combustion varies inversely with the density of the powder.

Hence if v denote the velocity of combustion, and δ the density of the powder, we have

$$v\delta = c, \quad . \quad . \quad . \quad . \quad . \quad . \quad . \quad . \quad (10)$$

the constant c depending on the nature of the powder.

3. The velocity decreases rapidly as the degree of moisture in the powder increases.

4. It increases with the amount of trituration of the ingredients, up to a certain limit.

5. For the same density, trituration, and moisture, the greatest velocity of combustion is obtained with a powder whose composition is 75 nitre, 15 charcoal, and 10 sulphur.

6. In air the actual velocity of combustion is from 0.4 to 0.6 inch per second.

7. The velocity varies with the pressure according to a law which is expressed by the following formula of Sarrau:

$$v = v_0\left(\frac{p}{p_0}\right)^{\frac{1}{2}}, \quad . \quad . \quad . \quad . \quad . \quad . \quad . \quad . \qquad (11)$$

in which v is the velocity of combustion for the pressure p, and v_0 the velocity in open air corresponding to the atmospheric pressure p_0.

According to this formula, a powder which burns 0.6 inch per second in air would burn about 29 inches per second in a gun under a pressure of 35,000 lbs. per square inch.

6. Formula for Burning in Air of Grains of Different Shapes.

To deduce a formula for the amount of powder burned at the time t, we proceed as follows:

The amount burned per unit weight or per unit volume at any time t will evidently be a function of t, and may be represented by $\phi(t)$. Suppose $\phi(t)$ to be developed according to the ascending powers of t, with constant coefficients, which are to be determined. We may then write

$$\phi(t) = \frac{at}{\tau}\left(1 - \lambda\frac{t}{\tau} + \mu\frac{t^2}{\tau^2} + \text{etc.}\right), \quad . \quad . \quad . \quad . \qquad (12)$$

in which a, λ, and μ are constants depending on the *form* of the grain, and τ is the total time of combustion of the grain, and hence depends on its *size*. The negative sign is used before λ because the sign of this term is found to be negative for all forms of grain in use. It is required to find values for a, λ, and μ for all service grains.

Spherical Grain.—Let r be the radius of the grain;
v the velocity of combustion;
τ the total time of combustion.

The original volume of the grain is $\frac{4}{3}\pi r^3$. At the end of the time t the radial distance burned over is vt, and the radius remaining unburned is $r - vt$. Hence the volume unburned at the end of t is $\frac{4}{3}\pi(r - vt)^3$.

The volume burned is then

$$\tfrac{4}{3}\pi r^3 - \tfrac{4}{3}\pi(r - vt)^3, \quad \ldots \ldots \quad (13)$$

or

$$\tfrac{4}{3}\pi r^3\left[1 - \left(1 - \frac{v}{r}t\right)^3\right]. \quad \ldots \ldots \quad (14)$$

But

$$\tau = \frac{r}{v}; \quad \ldots \ldots \ldots \quad (15)$$

and substituting for $\frac{v}{r}$ in (14) its value from (15), we have

$$\tfrac{4}{3}\pi r^3\left[1 - \left(1 - \frac{t}{\tau}\right)^3\right]. \quad \ldots \ldots \quad (16)$$

The expression (16) is the actual volume burned at the end of the time t. It is composed of two factors, the first being the original volume of the grain, $\frac{4}{3}\pi r^3$, and hence the second factor must be the proportional part burned; that is, the amount burned per unit of volume, or per unit of weight, in the time t, according as we consider volumes or weights, and is denoted by $\phi(t)$. Hence for this particular form of grain we have

$$\phi(t) = 1 - \left(1 - \frac{t}{\tau}\right)^3 = \frac{3t}{\tau} - \frac{3t^2}{\tau^2} + \frac{t^3}{\tau^3}; \quad \ldots \quad (17)$$

or

$$\phi(t) = \frac{3t}{\tau}\left(1 - \frac{t}{\tau} + \frac{1}{3}\frac{t^2}{\tau^2}\right). \quad \ldots \ldots \ldots \quad (18)$$

Comparing this with the general development in equation (12), we see that

$$a = 3, \quad \lambda = 1, \quad \mu = \tfrac{1}{3}.$$

Since all powders of irregular granulation may be considered as spheres whose mean radii can be calculated by equation (9), and since the hexagonal and sphero-hexagonal may also be so considered, these values of a, λ, and μ apply to all such powders.

Values of a, λ, and μ for Other Forms of Grain.—By a similar process we can find the values of a, λ, and μ for all service grains. These values are collected in the following table.

Form of Grain.	a	λ	μ
Spherical, Cubical, Irregular granulation	3	1	$\frac{1}{3}$
Parallelopipedon	$1+x+y$	$\frac{x+y+xy}{1+x+y}$	$\frac{xy}{1+x+y}$
Flat, $x=y$	$1+2x$	$\frac{2x+x^2}{1+2x}$	$\frac{x^2}{1+2x}$
Flat, $x=y=\frac{1}{2}$	2	$\frac{5}{8}$	$\frac{1}{8}$
Pierced cylinder, Pierced prism, Cocoa	$1+x$	$\frac{x}{1+x}$	0
Pierced cylinder; thickness of wall half of height	$\frac{3}{2}$	$\frac{1}{3}$	

In this table $x=\frac{\alpha}{\beta}$, $y=\frac{\alpha}{\gamma}$, in which for the parallelopipedon α, β, and γ are the lengths of the edges of the grain, α being the least dimension; also, for the pierced cylinder, $x=\frac{r-r'}{h}$, in which r is the exterior and r' the interior radius of the cylinder, and h its height.

By substituting for a, λ, and μ in the general formula (12), the values given in the above table, the amount of powder burned per unit weight, or per unit volume, for any of the above-shaped grains, can be determined; and this amount, multiplied by the total weight or volume, will give the total amount burned at any time t.

7. Velocity of Emission—Spherical Grain.

This is the rate of evolution of the gas of gunpowder; that is, the ratio of the part of the unit of weight of powder burned in a small interval of time to that time.

Considering unit weight, let

v denote the velocity of combustion;
S the total surface burning at any time t;
δ the density of the powder.

Then vdt is the space passed over by the burning surface in the time dt, $Svdt$ the volume burned, and $Sv\delta dt$ the weight, the value of g' being unity for French measures. Then, from the definition, we have for the velocity of emission

$$\eta = \frac{Sv\delta dt}{dt} = Sv\delta. \quad . \quad . \quad . \quad . \quad . \quad . \quad (19)$$

This equation shows that the velocity of emission depends upon—

1. The total burning surface S;
2. The velocity of combustion v;
3. The density of the powder δ.

But it has been found by experiment that in air

$v\delta = c$, (see equation (10).)

The velocity of emission in air depends, therefore, upon the surface burning, and this surface depends on the *form* and *size* of the grain.

We may obtain another expression for the velocity of emission which is more convenient for discussion, as follows:

The proportional part of powder burned per unit weight at any time t is, as we have seen, a function of t, and has been expressed by $\phi(t)$.

The proportional part burned in the time dt is $d[\phi(t)]$, and by definition we have

$$\eta = \frac{d[\phi)t)]}{dt}. \quad . \quad . \quad . \quad . \quad . \quad . \quad . \quad (20)$$

The value of $\phi(t)$ for different forms of grain can be determined just as for the spherical grain, and knowing these, we can determine the values of the velocity of emission for different forms of grain. It is evident that the form of grain whose velocity of emission is least at first will be most advantageous, since, the rate of emission being small, the gas will be given off gradually at first, and the pressure in the gun will increase slowly, and give time for the projectile to move before the gun is overstrained.

We can therefore determine what form of grain is best calculated to give the lowest pressure.

FOR SPHERICAL GRAIN.—Differentiating equation (12) with respect to t, we have

$$\eta = \frac{d[\phi(t)]}{dt} = \frac{a}{\tau}\left(1 - 2\lambda\frac{t}{\tau} + 3\mu\frac{t^2}{\tau^2}\right). \quad . \quad . \quad (20a)$$

Substituting for a, λ, and μ their values for the spherical grain, viz., $a = 3$, $\lambda = 1$, $\mu = \frac{1}{3}$, we have

$$\eta = \frac{d[\phi(t)]}{dt} = \frac{3}{\tau}\left(1 - \frac{t}{\tau}\right)^2.$$

At the beginning of combustion, when $t = 0$, we have

$$\eta = \frac{3}{\tau};$$

and at the end, when $t = \tau$,

$$\eta = 0.$$

8. Velocity of Emission for Parallelopipedon and for Pierced Cylinder.

FOR PARALLELOPIPEDON.—Substituting the values of a, λ, and μ for the parallelopipedon in (20a), we have

$$\eta = \frac{d[\phi(t)]}{dt} = \frac{1+x+y}{\tau}\left(1 - \frac{2(x+y+xy)}{1+x+y}\frac{t}{\tau} + \frac{3xy}{1+x+y}\frac{t^2}{\tau^2}\right).$$

When $t = 0$, we have

$$\eta = \frac{1+x+y}{\tau};$$

and when $t = \tau$,

$$\eta = \frac{(1-x)(1-y)}{\tau}.$$

For the flat grain whose thickness is one half its other dimensions, $x = y = \frac{1}{2}$. Hence for this grain we have at the beginning of combustion

$$\eta = \frac{2}{\tau},$$

and at the end

$$\eta = \frac{.25}{\tau};$$

or the velocity of emission is less at the beginning and greater at the end for a flat grain than for a spherical one.

FOR PIERCED CYLINDER—COCOA POWDER.—Substituting the values of a, λ, and μ for the pierced cylinder in (20a), we have

$$\eta = \frac{d[\phi(t)]}{dt} = \frac{1+x}{\tau}\left(1 - \frac{2x}{1+x}\frac{t}{\tau}\right).$$

When $t = 0$, we have

$$\eta = \frac{1}{\tau}(1+x);$$

and when $t = \tau$,

$$\eta = \frac{1}{\tau}(1-x).$$

For this grain $x = \frac{r - r'}{h}$. If $x = \frac{1}{2}$, as in the case of the flat grain, the thickness of the walls is one half the height, and we have, for the velocity of emission at the origin,

$$\eta = \frac{3}{2\tau} = \frac{1.5}{\tau};$$

and at the end,

$$\eta = \frac{.5}{\tau}.$$

Comparing this with the spherical and flat grains, we see that the velocity of emission for the pierced cylinder is less at the beginning and greater at the end than for any other form of grain. Hence, so far as velocity of emission is concerned, this is the best form of grain, and is the one now used in large guns.

The results are shown in the following table:

Form of Grain.	Velocity of Emission at Beginning.	Velocity of Emission at End.
Spherical / Irregular grain ..	$\frac{3}{\tau}$	[illegible]
Flat..............	$\frac{2}{\tau}$	$\frac{25}{\tau}$
Pierced cylinder ..	$\frac{1.5}{\tau}$	$\frac{.5}{\tau}$

9. Size of Grain—Density—Progressive Powders.

SIZE.—The velocity of emission depends on the surface in combustion, and this, as has been shown, depends on the *form* and *size* of grain. The effect of form has been discussed. To show the effect of size, suppose we have two charges of the same weight, composed of cubical grains.

Let a represent the edge of the grains in the first charge;
$2a$ the edge of the grains in the second charge;
N the number of grains in the first charge.

The surface of each grain in the first charge will be

$$a^2 \times 6,$$

and the total surface in the first charge

$$S = N \times 6a^2.$$

Since the weights are proportional to the cubes of the edges of the grains, each grain in the second charge will weigh

$$\frac{(2a)^3}{a^3} = 8$$

times as much as those of the first charge, and hence there will be only $\frac{N}{8}$ grains in the second charge.

The surface of each grain in the second charge will be

$$(2a)^2 \times 6,$$

and the total surface in the second charge

$$S' = \frac{N}{8} \times 6 \times (2a)^2 = \frac{N \times 6a^2}{2} = \frac{S}{2};$$

or the total surface in the second charge will be only one half that in the first; and since the velocity of emission depends on the surface, it will be in the beginning one half that of the first charge.

INFLUENCE OF DENSITY.—Considering a single grain, if we increase its density, the volume remaining constant, we decrease the velocity of combustion according to the formula

$$v\delta = c,$$

and hence do not change its velocity of emission, since $\eta = Sv\delta$. But considering a given weight of power, if we increase the density of the grains without changing their volume, we increase the weight of each grain, and hence decrease the number of grains contained in this weight. This decreases the initial surface of combustion in the given weight, and hence decreases the velocity of emission.

Ordinarily, for slow emission we increase at the same time both the size and the density of the grain. The limit of increase in size and density is reached when the grains cannot be consumed in the gun before the projectile leaves the bore.

PROGRESSIVE POWDERS.—For this reason progressive powders have been used. A progressive powder is one which burns slowly at first, and afterwards more rapidly.

The Italian Fossano powder is an example.

The larger grains are in the form of a cube, each composed of small dense grains united by a lighter powder.

The grain burns at first as a cube, with a small burning surface, but the light powder which unites the dense grains soon burns out, and the cube is then broken up into a number of dense grains, by which the burning surface and the velocity of emission are greatly increased.

The same progressive principle is found in nearly all powders. With molded powders and those of regular granulation the surface is more dense than the interior, owing to the method of manufacture, and hence this surface burns more slowly than the interior. In the molded powders especially, the ends of the prisms next the punches which mold it, are most dense. Even in ordinary powders of irregular granulation the same principle applies.

10. Combustion in a Close Vessel—Chemical Formula—Noble and Abel's Experiments.

To determine the chemical composition of the products of exploded gunpowder, and the various circumstances attending its combustion, it is necessary to burn it in a close vessel, and collect the products for examination and analysis.

CHEMICAL FORMULA.—It is generally admitted that no chemical formula will exactly represent the results of the combustion of gunpowder under all circumstances, since these results vary, for the same powder, with the conditions under which it is fired.

The formula generally adopted is

$$4KNO_3 + C_4 + S = K_2CO_3 + K_2SO_4 + N_4 + 2CO_2 + CO.$$

According to this formula we should have for the solids and gases of the exploded powder the following percentages by weight:

$$\left.\begin{array}{r} K_2CO_3 = 28.53 \\ K_2SO_4 = 35.96 \\ \hline 64.49 \end{array}\right\} \text{Solids.}$$

$$\left.\begin{array}{r} N_4 = 11.56 \\ 2CO_2 = 18.17 \\ CO = 5.78 \\ \hline 35.51 \end{array}\right\} \text{Gases.}$$

It will be seen later that the percentage of solids is less, and of gases greater, than the above, and also that the actual constituents of both solid and gaseous products are different.

NOBLE AND ABEL'S EXPERIMENTS. — *Apparatus and Methods.*—Numerous experiments have been made by Count Rumford (1792), Bunsen and Schischkoff (1859), and by

Rodman (1863), upon the composition of the products of fired gunpowder, the pressure produced by the gases, etc.; and while many of them are valuable, their results are not strictly accurate, owing to defective apparatus and other causes.

Captain Noble of the English Army, and Sir F. Abel, a chemist of the British War Department, made a series of experiments in 1874 and a second series in 1880, which are accepted as authoritative on this subject.

Apparatus.—The apparatus (see Fig. 6) was a strong steel vessel of the shape shown, capable of resisting very high

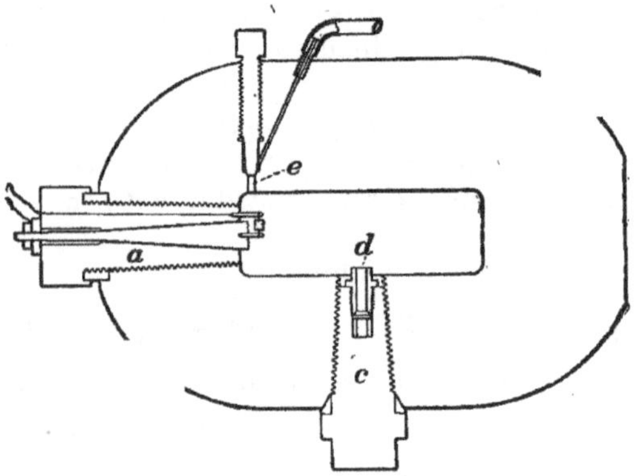

FIG. 6.

pressures. The charge of powder was introduced into the vessel through an opening *a*, which was then closed with a tapering screw-plug. Besides this screw-plug there was another, *c*, carrying a crusher-gauge *d* for measuring pressures, and a third opening, *e*, was for the purpose of drawing off the gases for analysis. The charge was fired by electricity.

Methods.—Different powders were used in these experiments, and for each kind of powder a series of experiments was made. The volume of the explosion-chamber being constant, the quantity of powder in each experiment was varied, starting with a very small charge and proceeding till the chamber was filled. The maximum charge was

2.2 pounds (1 kilogram). The results of all the experiments were compared in order to deduce the general laws pertaining to all the powders, and the variations due to particular kinds of powder, form of grain, density, etc.

11. Density of Loading—Object of Experiments.

DENSITY OF LOADING.—It is evident that the amount of powder fired in a given volume must greatly affect the resulting pressures.

It is necessary, therefore, to determine accurately the relation between this quantity and the space in which it is fired. If gunpowder were always of the same density, and of the same gravimetric density, we could compare the volume of the powder with that of the space containing it. But we know that both density and gravimetric density vary, and hence if a vessel were one half full of two different kinds of powder, while the volume of powder in the two cases would be the same, the actual weights would be different. By referring to gravimetric density, we see that the weights of equal volumes of powder and water are very nearly equal; hence we compare the weight of the powder fired with that of a volume of water which will fill the chamber in which the charge is fired.

This is called "density of loading," and is a very important ratio, which is constantly used in discussing the action of gunpowder in guns or in any closed vessel. It may be defined as "the ratio of the weight of the charge of powder to that weight of water, at its maximum density, which will completely fill the volume in which the charge is fired."

To determine an expression for it, let

Δ = the density of loading;
ϖ = the weight of the powder in pounds;
C = the volume in which the powder is fired in cubic inches.

One cubic foot of water weighs 62.425 lbs.: hence one pound of water occupies

$$\frac{1728}{62.425} = 27.68 \text{ cubic inches.}$$

The number of pounds of water that will fill the volume C is

$$\frac{C}{27.68},$$

and by definition

$$\varDelta = \frac{\bar{\omega}}{\frac{C}{27.68}} = \frac{27.68\bar{\omega}}{C}. \quad . \quad . \quad . \quad . \quad . \quad (21)$$

French Measure of Density of Loading.—In the metric system the weight of the charge is expressed in kilograms, and the volume of the chamber in litres or cubic decimetres.

Since a litre of water weighs one kilogram, the volume of the chamber in litres expresses at once the weight in kilograms of the water which would fill it, and hence the density of loading is obtained simply by dividing the weight of the charge by the volume of the chamber. Therefore

$$\varDelta = \frac{\bar{\omega}}{C}. \quad . \quad . \quad . \quad . \quad . \quad . \quad . \quad . \quad . \quad (22)$$

OBJECT OF EXPERIMENTS.—The object of the experiments was:

1. To determine the nature and composition of the products of combustion.

2. The effect of varying the size of the grain, and the density and composition of the powder.

3. The amount of heat generated.

4. The volume of the permanent gases.

5. The maximum pressure exerted by the gases, and the laws of its variation, and from the data thus obtained to calculate the effect in the bore of a gun.

12. Results—Nature and Composition of Products.

NATURE OF PRODUCTS.—In these experiments, for each kind of powder the density of loading was varied by varying the weight of the charge, starting with a density of 0.05, and increasing by constant increments up to a density 1.00. For the latter density the vessel was completely filled with the powders used. The products were found to consist of permanent gases and solids. Noble and Abel supposed the solid products to be liquid at the temperature of explosion, and to be diffused in a finely divided state throughout the gases. When the explosion chamber was opened after the combustion of the charge, the residue was found collected at the bottom in a solid form. The mass was compact and hard, and of an olive-green color, changing to black on exposure to the air. The volume of this residue *when cold* was about 0.3 the original volume of the powder.

To ascertain the condition of the residue immediately after the explosion, the following method was adopted.

One minute after the explosion, the vessel was inclined quickly at an angle of 45°. It remained in this position 45 seconds, and was then returned to its original position. When the vessel was opened, the solid products were found to be inclined to the walls at an angle of 45°.

From this it follows that one minute after the explosion the solid residue was in a liquid state, and 45 seconds after this it had become solid. Moreover, a slight crust adhering to the walls, and which had been partially broken by the liquid when it took up its inclined position, showed that the solidification had begun one minute after the explosion.

The effect of high temperature on the solid residue was tested as follows. It was exposed to a temperature of 1700° C. in a Siemens furnace. At first a slight effervescence appeared, which disappeared immediately. At the end of the experiment, a slight volatilization was visible. When the crucibles containing the residue were removed from the furnace, and allowed to cool, the increase in volume of the solid products, as determined by marks left on the walls of the crucibles, was about 78 per cent.

Composition.—This was determined by chemical analysis, and was found to be as follows for pebble powder. The results differed slightly for different powders, and are the percentages by weight.

$K_2CO_3 = 33$		
$K_2SO_4 = 7$		
$K_2S = 10$		
$S = 4$	Solids.	
Various = 2		
56		
$CO_2 = 27$		
$CO = 5$		
$N = 11$	Gases.	
Various = 1		
44		

That is, 100 pounds of pebble powder, when fired, will give 56 pounds of solid residue and 44 pounds of gases. A mean of the results from all the powders gave 57 pounds solids and 43 pounds gases. That is, the solid residue is 57 per cent and the gases 43 per cent of the original weight of the charge. The theoretical reaction gives 64.49 per cent solids, and 35.51 gases.

13. Results—Effect of Variations in Powder—Amount of Heat Generated—Volume of Permanent Gases.

VARIATIONS.—Slight variations in size and density of grain were found to have very little effect upon the composition of the products of combustion, and no effect whatever upon the pressures. Hence the pressure in a closed vessel is independent of the size, form, and density of the grain, and depends only on the density of loading or the quantity of powder. The case is very different for a gun, as will be seen later.

AMOUNT OF HEAT.—The amount of heat generated by the explosion was measured by immersing the steel explosion-vessel after discharge in a calorimeter containing a given quantity of water at a known temperature, and noting the rise of temperature of the water.

The mean result obtained was that 705 units of heat were given off per unit weight of powder burned. Now if the mean specific heat of the products of explosion were accurately known, the temperature of these products at explosion could be determined by the formula

$$T = \frac{Q}{c},$$

in which Q is the quantity of heat, and c the mean specific heat at constant volume. But this value of c is not known.

It is known that the specific heat of the *solid* products increases with the temperature, but the law of increase is unknown. Bunsen and Schischkoff found from their experiments a value for $c = 0.185$. From this we should have

$$T = \frac{705}{0.185} = 3811^\circ,$$

and

$$T_0 = T + 273 = 3811 + 273 = 4084^\circ \text{ C}.$$

Noble and Abel believed this value to be much too large, for the following reasons:

1. The specific heat of the solid products increases with the temperature, hence 0.185 is too small.

2. The heat measured by the calorimeter includes all the heat absorbed by the explosion-vessel, as well as that of the gaseous products. The heat absorbed by the vessel is taken from the products of explosion, and hence lowers the temperature; therefore 705 is practically too large.

3. A piece of platinum wire was enclosed in the explosion-chamber, and when the chamber was opened after explosion the platinum showed only slight signs of fusion. As this metal fuses completely at 2000° C., it was thought that if anything approaching a temperature of 4000° C. had been attained the metal would have been entirely fused. The temperature of explosion was therefore obtained by calculation, as will be explained.

VOLUME OF PERMANENT GASES.—This was determined by collecting the gases in a gasometer, and observing their volume at the ordinary atmospheric temperature and pressure, and afterwards reducing this volume to zero centigrade.

For the purpose of comparing the volumes of different

gases the "specific volume" of each is used. For ordinary gases this is the volume occupied by a *unit weight of the gas* at zero C. and under atmospheric pressure. For gunpowder, it is the volume occupied by the *gas from unit weight of powder* at the above temperature and pressure, and was found to be 280 times the original volume of the powder. That is, the gas from 1 kilogram of powder occupies at zero Centigrade and under atmospheric pressure a volume of 280 cubic decimetres.

14. Results—Pressure of Gases—Formula.

PRESSURE.—For each kind of powder the pressure was measured with the Noble crusher-gauge, for all densities of loading from 0.05 to 1.00. The results were plotted; the abscissas being the densities of loading, and the ordinates the corresponding pressures. The resulting curve gave the law connecting abscissas and ordinates. It was then necessary to deduce the equation of this curve so as to express analytically the law connecting density of loading and pressure, and afterwards to compare the results calculated by this formula with those obtained by experiment.

FORMULA.—To deduce the formula we proceed as follows: The experiments had shown that the products of explosion were partly solid and partly gaseous. Hence for a given volume of explosion-chamber, it is evident that the volume occupied by the gases at the moment of explosion is equal to the total volume of the chamber minus the volume occupied by the solid products. We can then calculate the pressure due to this volume by Mariotte's and Gay-Lussac's laws.

Let T_0 be the absolute temperature of the products at the moment of explosion;

T, the actual temperature of these products;

P, the pressure in kilograms per square decimetre on the walls of the chamber at the same instant;

p_0, the normal atmospheric pressure (103.33 kilograms per square decimetre);

V, the volume in cubic decimetres actually occupied by the gases at the moment of explosion, corresponding to the pressure P;

V_0, the volume in cubic decimetres occupied by the same gases at zero Centigrade and at p_0 pressure;
$\hat{\omega}$, the weight of the charge in kilograms;
v_0, the specific volume of the gases;
α, the volume occupied by the solid residue of 1 kilogram of powder at the temperature of explosion;
C, the volume of the explosion-chamber in cubic decimetres.

Then we have (Michie, equation 823), from Mariotte's and Gay-Lussac's laws,

$$PV = p_0 V_0 \left(\frac{273 + T}{273}\right). \quad . \quad . \quad . \quad . \quad (23)$$

But

$$273 + T = T_0;$$

hence

$$\frac{PV}{T_0} = \frac{p_0 V_0}{273}, \quad . \quad . \quad . \quad . \quad . \quad . \quad . \quad . \quad . \quad (24)$$

or

$$P = \frac{p_0 V_0 T_0}{273 V}. \quad . \quad . \quad . \quad . \quad . \quad . \quad . \quad . \quad (25)$$

Now

$$V_0 = \hat{\omega} v_0, \quad . \quad . \quad . \quad . \quad . \quad . \quad . \quad . \quad . \quad (26)$$

and

$$P = \frac{p_0 v_0 T_0}{273} \times \frac{\hat{\omega}}{V}, \quad . \quad . \quad . \quad . \quad . \quad (27)$$

but

$$\frac{p_0 v_0 T_0}{273} = \text{constant} = f. \quad . \quad . \quad . \quad . \quad . \quad (28)$$

Hence

$$P = f\frac{\hat{\omega}}{V}. \quad . \quad . \quad . \quad . \quad . \quad . \quad . \quad . \quad . \quad (29)$$

Now V, the volume actually occupied by the gases, is the difference between the volume of the chamber and that of the solid residue. The volume of the solid residue is

$$\alpha\hat{\omega},$$

and hence

$$V = C - \alpha\bar{\omega}. \quad . \quad . \quad . \quad . \quad . \quad . \quad . \quad . \quad (30)$$

The expression for density of loading is (see 22)

$$\Delta = \frac{\bar{\omega}}{C}; \quad \therefore C = \frac{\bar{\omega}}{\Delta}.$$

Substituting this value of C in (30), we have

$$V = \frac{\bar{\omega}}{\Delta}(1 - \alpha\Delta),$$

and this value of V in (29) gives

$$P = f\frac{\Delta}{1 - \alpha\Delta}, \quad . \quad . \quad . \quad . \quad . \quad . \quad . \quad . \quad (31)$$

the equation required.

15. Discussion of Formula (31).

To compare the results of this formula with those obtained by experiment it is necessary to know α and f. These can be calculated by equation (31) by taking from the experiments two values of Δ and the two corresponding values of P, and substituting in (31). We will thus have two equations containing the two unknown quantities α and f from which they may be determined. In this manner Noble and Abel found for these constants the following values:

$f = 291200$ kil. per sq. dee. $= 18.49$ tons per sq. inch.
$\alpha = 0.57$.

Taking Noble and Abel's numerical values, we have for the French units

$$P = \frac{291200\Delta}{1 - 0.57\Delta}, \quad . \quad . \quad . \quad . \quad . \quad . \quad . \quad (32)$$

or for English units

$$P = \frac{18.49\Delta}{1 - 0.57\Delta}. \quad . \quad . \quad . \quad . \quad . \quad . \quad . \quad (33)$$

This value of α means that when the volume of the charge is 1 cubic decimetre the volume of the solid products is 0.57 cubic decimetre.

Referring to equation (4), it is seen that the volume of the solid residue after explosion is nearly equal to that of the solid powder in the charge before explosion.

If the charge is in kilograms, the volume of the solid products is obtained by multiplying the number of kilograms by 0.57. If the charge is in pounds, the volume of the solid products is obtained by multiplying the number of pounds by 27.68 and this by 0.57, or

$$\text{Vol. of solid products} = \text{No. lbs.} \times 27.68 \times 0.57;$$
$$= \text{No. lbs.} \times 15.77.$$

When the chamber is full of powder the density of loading for the powders used by Noble and Abel is 1.00.

In this case, since no more powder can be introduced, we should get the greatest possible pressure which gunpowder will give. This is sometimes called the "absolute pressure." Its value is by (33), for $\varDelta = 1$,

$$P = 43 \text{ tons per square inch.}$$

16. "Force" of Powder—Temperature of Explosion.

FORCE.—Assume equation (31),

$$P = f\frac{\varDelta}{1 - \alpha\varDelta}.$$

If in this equation we make

$$\frac{\varDelta}{1 - \alpha\varDelta} = 1, \quad \dots \dots \dots \quad (34)$$

we have

$$P = f. \quad \dots \dots \dots \dots \quad (34a)$$

From (34) we have

$$\varDelta = \frac{1}{1 + \alpha}. \quad \dots \dots \dots \quad (35)$$

Comparing this with the general expression for density of loading,

$$\varDelta = \frac{\varpi}{C},$$

we see that when the weight of the powder is unity, and the volume of the chamber in which it is fired is $(1 + \alpha)$, we have, from (34*a*),

$$P = f.$$

The quantity represented by f in these equations is (see 28)

$$f = \frac{p_0 v_0 T_0}{273},$$

and is constant. This value of f is called the "force" of the powder, and from (34*a*) and (35) it may be defined as "the pressure per unit of surface exerted by the gases from unit weight of powder, the gases occupying at the temperature of explosion a volume equal to unity." The volume of the chamber is $1 + \alpha$, and α is the portion occupied by the solid products. The value of f as determined by Noble and Abel has been given.

This value of f is uncertain, and therefore it admits of being modified to account for various resistances in a gun which cannot be readily calculated.

It is found also that the values of f for different powders are nearly the same. This arises from the fact that the quantity of heat of a powder varies approximately inversely as the specific volume of the gas; and by (28) the product of these two quantities measures the force of the powder.

TEMPERATURE OF EXPLOSION.—Assume, (28),

$$f = \frac{p_0 v_0 T_0}{273};$$

from which

$$T_0 = \frac{f \times 273}{p_0 v_0}. \quad . \quad . \quad . \quad . \quad . \quad . \quad . \quad . \quad (36)$$

We have found

$$f = 291200$$
$$v_0 = 280;$$
$$p_0 = 103.33.$$

Substituting in (36), we have

$$T_0 = \frac{291200 \times 273}{103.33 \times 280} = 2748^\circ \text{ C.}$$

and

$$T = 2748 - 273 = 2475^\circ \text{ C.}$$

These values agree well with the melting-point of platinum, and could be accepted if there were no doubt about the value of f.

COMBUSTION IN A GUN—INTERIOR BALLISTICS.

17. Action of Gunpowder in a Gun.

Suppose we have a charge of powder which completely fills the chamber of a gun, the density of loading being unity. If this charge be completely burned before the projectile moves, we obtain, by equation (33),

$$P = 43 \text{ tons per square inch.}$$

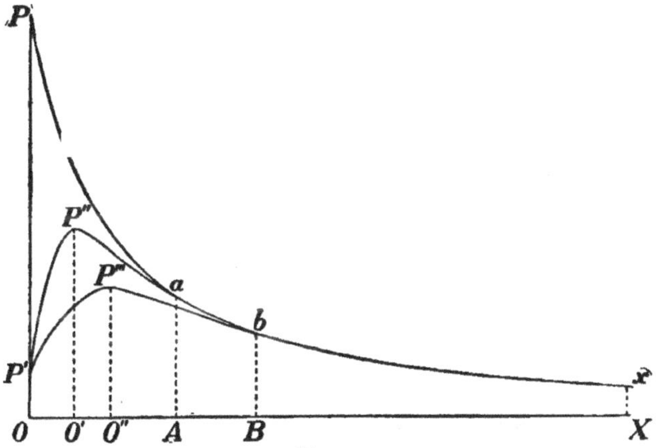

FIG. 7.

In Fig. 7 let O be the position of the base of the projectile before firing, OX the axis of the bore, and OP the axis of pressures. Lay off $OP = 43$ tons, and we have the pressure corresponding to the instantaneous combustion of the charge. From this point P the gas will expand according to the hypothesis adopted, and, acting on the projectile, will cause it to move rapidly down the bore, the ordinates of the curve Px representing the pressures at corresponding abscissas of travel. The equation of this curve will be deduced later.

Error in Supposition.—It is evident that the assumption that all the powder is consumed before the projectile moves cannot be true in practice. As soon as the pressure rises high enough to overcome the resistance of the projectile and gun to motion, they will both move in opposite directions; but for the present we will consider the motion of the projectile alone.

Quick-burning Powder.—Take a small-grained powder of cubical form. The time of combustion of this powder is small, and its velocity of emission at first great, as has been shown. Let OP' represent the pressure which is sufficient to start the projectile, and suppose the powder is all burned when the projectile reaches A. Under these circumstances the relation between the pressures and the travel of the projectile, or the "pressure curve," will be represented by a curve such as $OP'P''abx$.

Slow-burning Powder.—Take the *same weight of charge* of cocoa or slow-burning powder. The time of combustion is comparatively great, and its velocity of emission at first small. Let OP' represent, as before, the pressure required to start the projectile, and suppose the powder all burned when the projectile reaches B.

The pressure curve in this case will be $OP'P'''bx$; and from these curves we may deduce the following consequences:

a. The quick curve will rise above the slow one near the origin, because the volume of gas given off in the same time is greater with the quick powder.

b. The work done by the quick powder upon the projectile is greater than that done by the slow powder, because the area under the quick curve is greater than that under the slow one.

c. The quick powder strains the gun more than the slow

one, because the maximum pressure $O'P'' > O''P'''$, and this maximum pressure is what determines the maximum strain.

That part of each curve from P' to a and b, respectively, is called the combustion curve, because during this time the powder is still burning and giving off gas. The part from a and b to x is called the expansion curve, because from these points on the gas is expanding only.

18. Equation of Pressure Curve—Noble and Abel's Method.

The equation of the true pressure curve is very difficult to deduce, since at the origin, as we have seen, gas is being evolved while the projectile is moving, and this renders the problem very complex.

Noble and Abel deduced the equation of the curve $Pabx$, Fig. 7, under the following hypotheses:

1. That all the powder is burned before the projectile moves.

2. That the solid products of combustion give off heat to the gases during the expansion.

Let c_1 represent the specific heat of the solid products, supposed constant throughout the expansion, and dT_0 any small change of temperature of the products of combustion.

Then c_1dT_0 is the corresponding quantity of heat given to the gases by the solid products per unit of weight.

Let w_1 represent the number of units of weight of the solid residue· then the total quantity of heat given to the gases by the solid residue is $w_1c_1dT_0$. Let w_2 be the number of units of weight of the gases. These gases, by hypothesis, receive the heat above found, and hence they receive per unit of weight a quantity of heat dQ equal to

$$dQ = -\frac{w_1}{w_2}c_1dT_0 = -\beta c_1dT_0, \quad . \quad . \quad . \quad . \quad (37)$$

β being the ratio $\frac{w_1}{w_2}$, and the negative sign being used since T_0 is a decreasing function of Q.

When the volume, pressure, and quantity of heat of a gas change at the same time, we have a general law connecting them, which is expressed by the following equation (Michie, 832):

$$dQ = \frac{c_v vdp + c_p pdv}{R}, \quad . \quad . \quad . \quad . \quad . \quad . \quad (38)$$

in which dQ is the elementary quantity of heat imparted to the gases;

dp and dv the elementary changes of pressure and volume of the gas due to dQ;

c_p and c_v the specific heats at constant pressure and volume, respectively.

Substituting in equation (38) for dQ its value given by (37), we have

$$-\beta c_1 dT_0 = \frac{1}{R}(c_v vdp + c_p pdv). \quad . \quad . \quad . \quad . \quad (39)$$

This equation contains T_0 and R, while the equation of the pressure curve should contain only p, v, and constants, because the pressure curve is one showing the relation between p and v. To eliminate T_0 and R, assume the general equation (823, Michie) connecting the pressures, volumes, and temperatures of a gas.

$$pv = RT_0. \quad . \quad . \quad . \quad . \quad . \quad . \quad . \quad . \quad . \quad (40)$$

Differentiating, we have

$$RdT_0 = pdv + vdp. \quad . \quad . \quad . \quad . \quad . \quad . \quad (40')$$

from which

$$dT_0 = \frac{pdv + vdp}{R}.$$

Substituting this value of dT_0 in (39), we have

$$-(\beta c_1 + c_p)\frac{dv}{v} = (\beta c_1 + c_v)\frac{dp}{p}. \quad . \quad . \quad . \quad . \quad . \quad (40'')$$

For small changes of pressure and volume β, c_1, c_p, and c_v are constant. Hence

$$-(\beta c_1 + c_p)\int_{v_i}^{v}\frac{dv}{v} = (\beta c_1 + c_v)\int_{p_i}^{p}\frac{dp}{p};$$

p_i, p, v_i, and v being the initial and any subsequent values of p and v.

Integrating,

$$p = p_i\left[\frac{v_i}{v}\right]^{\left(\frac{\beta c_1 + c_p}{\beta c_1 + c_v}\right)}. \qquad (41)$$

Making

$$\frac{\beta c_1 + c_p}{\beta c_1 + c_v} = k', \qquad (42)$$

we have

$$p = p_i\left[\frac{v_i}{v}\right]^{k'}, \qquad (43)$$

the equation of the pressure curve.

19. Application of Formula (43)—Work of Gunpowder in a Gun.

APPLICATION.— Assume equation (43)

$$p = p_i\left[\frac{v_i}{v}\right]^{k'}.$$

In this equation v_i is the volume occupied by the gas at the moment of explosion, and p_i the corresponding pressure. To apply this equation to the case of a gun, the original volume v_i occupied by the gas, is the volume of the chamber in which the charge is fired, minus the volume occupied by the liquid products of the charge. The volume v occupied by the gases corresponding to the pressure p is the total volume to which the gas has expanded, including that of the chamber, minus the volume occupied by the liquid products.

Hence if v' denote the original volume of the chamber, *the density of loading being unity*, $\alpha v'$ is the volume occupied by the liquid residue, and $v_i = v' - \alpha v' = v'(1 - \alpha)$. Also, if v'' represent any subsequent volume of the bore, at which the pressure is p, the volume actually occupied by the gas is

$$v = v'' - \alpha v'.$$

Making these substitutions in (43), we have

$$p = p_i\left(\frac{v'(1 - \alpha)}{v'' - \alpha v'}\right)^{k'}. \qquad (44)$$

Taking from the experiments the values of the constants, we have

$$p_i = 43 \text{ tons per square inch};$$
$$\alpha = 0.57; \qquad \beta = 1.2957;$$
$$c_p = 0.2324; \quad c_v = 0.1762; \quad c_1 = 0.45;$$
$$k' = 1.074,$$

and substituting in (44), we have

$$p = 43\left(\frac{.43v'}{v'' - .57v'}\right)^{1.074} \quad . \quad . \quad . \quad . \quad . \quad (45)$$

for use in practice.

WORK OF GUNPOWDER.—The general expression for the work done by a gas expanding from a volume v_i to a volume v is

$$W = \int_{v_i}^{v_2} pdv. \quad . \quad . \quad . \quad . \quad . \quad . \quad . \quad . \quad (46)$$

Substituting for p its value from (44) and changing the limits to v'' and v', we have

$$W = \int_{v'}^{v''} p_i\left(\frac{v'(1 - \alpha)}{v'' - \alpha v'}\right)^{k'} dv'', \quad . \quad . \quad . \quad (47)$$

or

$$W = p_i v'^{k'}(1 - \alpha)^{k'} \int_{v'}^{v''} \frac{dv''}{(v'' - \alpha v')^{k'}}.$$

Integrating, we have

$$W = p_i v'^{k'} \frac{(1 - \alpha)^{k'}}{12 \times (k' - 1)}\left[\frac{1}{[v'(1 - \alpha)]^{k'-1}} - \frac{1}{(v'' - \alpha v')^{k'-1}}\right]. (48)$$

Multiply, and divide the second member by $[v'(1 - \alpha)]^{k'-1}$:

$$W = \frac{p_i v'(1 - \alpha)}{12 \times (k' - 1)}\left[1 - \left(\frac{v'(1 - \alpha)}{v'' - \alpha v'}\right)^{k'-1}\right]. \quad . \quad (49)$$

12 is used to reduce to foot-tons. This is Noble and Abel's formula for work.

Taking the volume corresponding to the muzzle of the gun, the corresponding work is obtained. If there were no loss of energy due to the friction, resistance of rifling, etc., the work thus calculated should be equal to the energy in foot-tons possessed by the projectile at the muzzle, which is

$$E = \frac{wV^2}{2g \times 2240},$$

w being the weight of the projectile in pounds, and V its muzzle velocity in feet per second. But E is always less than W, owing to the above causes; and the ratio

$$\frac{E}{W} = F$$

is called the "factor of effect" for the particular gun and charge.

Knowing this factor for any given gun we can find the muzzle velocity for a given charge by calculating W by (49), multiplying it by F, and we have

$$\frac{wV^2}{2g \times 2240} = FW;$$

from which

$$V = \sqrt{\frac{FW \times 2g \times 2240}{w}}. \quad . \quad . \quad . \quad . \quad (50)$$

Infinite Expansion.—When the length of bore becomes infinite, v'' in (49) is infinite, and we have

$$W = \frac{p_i v'(1 - \alpha)}{12(k' - 1)}. \quad . \quad . \quad . \quad . \quad . \quad . \quad . \quad (51)$$

Using the constants as given above, and substituting for v' its value, 27.68 cubic inches, the volume occupied by one pound of powder, we have

$$W = 576.35 \text{ ft.-tons} \quad . \quad . \quad . \quad . \quad . \quad . \quad . \quad (52)$$

for the work of one pound of powder expanded to infinity, under Noble and Abel's hypothesis.

20. Equation of Pressure Curve—Recent Hypothesis—Expression for Work under this Hypothesis.

EQUATION OF PRESSURE CURVE.—In recent discussions Noble and Abel's hypothesis is rejected, as it is believed that from the feeble absorbing power of gases generally, they receive only a very minute quantity of the heat radiated by the solid products. The equation of the pressure curve *Pabx*, Fig. 7, is therefore deduced under the following hypotheses:

1. That all the powder is burned before the projectile moves.

2. That the gases expand without receiving heat from or giving off heat to any external source, and that the work done on the projectile is due to their own heat; that is, the expansion is adiabatic.

Assume the equation expressing the general law connecting the heat, volume, and pressure of a gas, as before, (equation 832, Michie's Mechanics),

$$dQ = \frac{c_v v dp + c_p p dv}{R}.$$

Since no heat is gained or lost externally, $dQ = 0$ and we have

$$\frac{dp}{p} + \frac{c_p}{c_v}\left(\frac{dv}{v}\right) = 0.$$

Making $\frac{c_p}{c_v} = k$ and integrating between the limits p_i, p, v_i and v, we have

$$\log \frac{p_i}{p} = \log \left(\frac{v}{v_i}\right)^k, \quad \dots \dots \quad (53)$$

or

$$p = p_i \left[\frac{v_i}{v}\right]^k, \quad \dots \dots \quad (54)$$

which is the equation of the pressure curve, and differs from that deduced under Noble and Abel's hypothesis only in the value of the exponents k and k', k' being 1.074 and k, for powder gases, 1.30.

WORK.—To deduce the expression for work in this case, we have, as before, equation (46),

$$W = \int_{v_i}^{v} p dv.$$

Substitute for p its value from (54), integrate, and we have

$$W = -\frac{p_i v_i^k}{k-1} v^{1-k} + C. \quad . \quad . \quad . \quad . \quad . \quad (55)$$

When $v = v_i$ we have $W = 0$, and $C = \frac{p_i v_i}{k-1}$; hence

$$W = \frac{p_i v_i}{k-1}\left\{1 - \left(\frac{v_i}{v}\right)^{k-1}\right\}. \quad . \quad . \quad . \quad . \quad . \quad (56)$$

21. Work of Gunpowder in Terms of Force and Weight of Charge —Expression for it in Terms of Length of Travel of Projectile.

WORK IN TERMS OF FORCE AND WEIGHT.—If ϖ be the weight of the charge in kilograms, the initial volume occupied by the gas from each kilogram of powder, expressed in cubic decimetres, will be

$$v_1 = \frac{v_i}{\varpi};$$

hence

$$p_i v_i = p_i v_1 \varpi; \quad . \quad . \quad . \quad . \quad . \quad . \quad . \quad (57)$$

and from Mariotte's law we have

$$p_i v_1 = p'' v'' = C \text{ (a constant)},$$

see equation (814), Michie's Mechanics.

Now if we make $v'' = 1$ = one cubic decimetre, p'' becomes by definition the force of the powder, and hence

$$p_i v_1 = p'' = f;$$

and from (57),

$$p_i v_i = f\varpi.$$

Substituting in (56) for $p_i v_i$ this value, we have

$$W = \frac{f\varpi}{k-1}\left\{1 - \left(\frac{v_i}{v}\right)^{k-1}\right\}, \quad . \quad . \quad . \quad . \quad . \quad (58)$$

in which W is expressed in terms of the "force of the powder" and its weight.

In equation (58) the force of the powder is expressed in kilograms per square decimetre, and the volumes in cubic decimetres. Hence the work W will be expressed in kilogram-decimetres. It is usual, however, to express work in kilogram-metres, and this is done by dividing by 10; and we have

$$\frac{W}{10} = W' = \frac{f\hat{\omega}}{10(k - 1)}\left\{1 - \left(\frac{v_i}{v}\right)^{k-1}\right\} \quad . \quad . \quad . \quad (59)$$

WORK IN TERMS OF LENGTH OF TRAVEL OF PROJECTILE.—We can place this expression for work under a still more convenient form, as follows:

Reduced Length of Initial Air-space.—The initial air-space in the powder-chamber is equal to the total volume of the chamber minus the volume occupied by the solid powder; and the reduced length of this air-space is the length of a cylinder whose volume is that of the initial air-space, and whose area of cross-section is that of the bore proper.

To determine its value,

let Δ be the density of loading;
δ, the density of the powder;
ω, the area of cross-section of the bore;
z, the reduced length of the initial air-space.

We have

$$\Delta = \frac{\hat{\omega}}{C}; \quad \therefore C = \frac{\hat{\omega}}{\Delta}.$$

The volume occupied by the solid powder in cubic decimetres is

$$V = \frac{\hat{\omega}}{\delta};$$

hence the volume actually occupied by the gases, or the initial air-space, is

$$v_i = C - V = \frac{\hat{\omega}}{\Delta} - \frac{\hat{\omega}}{\delta}.$$

This volume divided by ω, the area of cross-section of the bore, gives z; hence

$$z = \frac{\hat{\omega}}{\omega}\left(\frac{1}{\Delta} - \frac{1}{\delta}\right) \quad . \quad . \quad . \quad . \quad . \quad . \quad . \quad . \quad (60)$$

Having this value for z, we have (see page 29)

$$v_i = \omega z; \quad v = \omega(z + x), \quad . \quad . \quad . \quad . \quad (60a)$$

x denoting the length of travel of the projectile.

Substituting these values in (59),

$$W' = \frac{f\hat{\omega}}{10(k-1)} \left\{ 1 - \left(\frac{z}{z+x} \right)^{k-1} \right\}. \quad . \quad . \quad (61)$$

Taking $k = 1.3$ and $f = 291,200$, we have the constants which enter (61), and W' can be calculated for any length of travel x of the projectile. When $x = \infty$, the bore becomes infinite in length, the powder expanded to infinity, and (61) becomes

$$W' = \frac{f\hat{\omega}}{10(k-1)}.$$

Making $\hat{\omega} = 1$, we have, for the work of one kilogram of powder expanded to infinity under the adiabatic hypothesis,

$$W' = 97066 \text{ kil.-metres per kilogram,}$$
$$= 142.2 \text{ ft.-tons per pound.}$$

Comparing this with Noble and Abel's value for the work of one pound, viz.,

$$W = 576.35 \text{ ft.-tons per pound,}$$

we see that the work is much less under the adiabatic hypothesis.

22. Division of Work of Gunpowder—Velocity of Recoil.

DIVISION OF WORK.—Having the value of the total work done by gunpowder, it is required to find how much of this work is done upon the gun and how much upon the projectile, and thence to deduce values for the velocity of recoil and of the projectile.

In this discussion we suppose:

1. That the gun recoils freely.

2. That gravity and resistance of the air can be neglected in comparison with the great pressures considered.

Then we have, from mechanics:

1. The total energy of the system is equal to the total work done by the powder upon it.

2. Since the gun and projectile move in opposite directions with equal momenta, the sum of the quantities of motion of the system is zero.

The energy of the system after the projectile has passed over a given path x is composed of three quantities: 1st, the energy of the projectile; 2d, that of the gun; 3d, that of the charge. The energy of the projectile is $\frac{mv^2}{2}$, m denoting its mass and v its velocity of translation. To this should be added the energy of rotation; but this is so small that it may be neglected. The energy of the gun is $\frac{Mv'^2}{2}$, M being the mass and v' the velocity.

The energy of the charge is unknown, since the velocity of its particles is unknown. The velocities of these particles vary from zero, near the bottom of the bore, to v, that of the projectile, for those in contact with the latter, not taking into account irregular motions which also exist. Hence the mean velocity of the particles is less than that of the projectile. If μ be the mass of the charge, $\frac{\mu v^2}{2}$ would be its energy if the velocities above mentioned were equal; as they are not, we represent the energy of the charge by $\frac{\mu v^2}{2} \times \theta$, θ being a coefficient whose value is between zero and unity.

We have, then, as a first equation,

$$2W = mv^2 + Mv'^2 + \theta\mu v^2. \quad . \quad . \quad . \quad . \quad (62)$$

The momenta of projectile and gun are mv and Mv'. As before, the quantity of motion of the charge is not known; but, reasoning as above, we may represent it by $\theta'\mu v$, and its sign will be $+$, because, as the centre of gravity of the system is fixed, the greater part of the gaseous mass moves in the same direction as the projectile. The second equation is then

$$mv + \theta'\mu v - Mv' = 0. \quad . \quad . \quad . \quad . \quad . \quad (63)$$

The values of θ and θ' are found by analytical methods to be

$$\theta = \tfrac{1}{3}, \quad \theta' = \tfrac{1}{2}.$$

VELOCITY OF RECOIL.—From (63) we have

$$v' = \frac{m + \theta'\mu}{M} v. \quad \ldots \ldots \quad (64)$$

Make $v = V$ the initial velocity, and we have $v' = V'$, the velocity of the gun at the moment the projectile leaves the bore. Making $\theta' = \frac{1}{2}$, and replacing masses by weights, we have

$$V' = \frac{p + \frac{\omega}{2}}{P} V. \quad \ldots \ldots \quad (65)$$

Experiment shows that this formula gives correct values for V' at the instant the projectile leaves the bore, supposing the gun to recoil freely. But this value of V' does not represent the *maximum* velocity of recoil; in fact, it gives only about three fourths of the maximum, since it applies only at the instant the projectile leaves the bore. For slow powders the velocity of recoil is increased, since the gas continues to act upon the piece after the projectile has left the bore.

The subject of recoil will be further discussed under Gun-carriages.

23. Velocity of Projectile—Passive Resistances—Limit of Length of Bore—Influence on Velocity and Maximum Pressures.

VELOCITY OF PROJECTILE.—Substitute the value of v' from (64) in (62), and we have

$$v^2 = \frac{2W}{m + \theta\mu + \frac{(m + \theta'\mu)^2}{M}}. \quad \ldots \quad (66)$$

This equation gives the velocity of the projectile as a function of the work of the powder. The third term in the denominator, being generally small, may be omitted, and we have

$$v^2 = \frac{2W}{m + \theta\mu}. \quad \ldots \ldots \quad (67)$$

If we have a quick powder, and suppose it all burned before the projectile moves, we may substitute for W in (67) its value from (61), which gives

$$v^2 = \frac{f\hat{\omega}}{5(k-1)(m+\theta\mu)}\left\{1-\left(\frac{z}{z+x}\right)^{k-1}\right\}.$$

Making $x = u$, the total length of travel of the projectile, v^2 becomes the initial velocity, and we have

$$V^2 = \frac{f\hat{\omega}}{5(k-1)\left(m+\frac{\mu}{3}\right)}\left\{1-\left(\frac{z}{z+u}\right)^{k-1}\right\}.$$

For a long gun u would be large and $\frac{z}{z+u}$ small, and V^2 would become

$$V^2 = \frac{f\hat{\omega}}{5(k-1)\left(m+\frac{\mu}{3}\right)}. \quad . \quad . \quad . \quad . \quad . \quad (68)$$

The value of f in this equation must be found by experiment, to compensate for the erroneous assumptions made in deducing it.

Passive Resistances are those due to the forcing of the band of the projectile into the grooves of the rifling, friction, etc.

Let ρ denote the work done against these resistances over the path x. In equation (62) this work is not accounted for, and it is therefore not exact. Introducing it into that equation, we have

$$2(W-\rho) = mv^2 + Mv'^2 + \theta\mu v^2; \quad . \quad . \quad . \quad (69)$$

and in equation (67) we have for the velocity

$$v^2 = \frac{2(W-\rho)}{m+\theta\mu}. \quad . \quad . \quad . \quad . \quad . \quad . \quad . \quad (70)$$

LENGTH OF BORE.—Although the value of ρ is unknown, we can use it as follows: Differentiate (70), v, W, and ρ being variable.

$$dv = \frac{1}{v} \frac{dW - d\rho}{m + \theta\mu}. \quad . \quad . \quad . \quad . \quad . \quad . \quad (71)$$

In Fig. 8 let OX be the axis of the bore and OP that

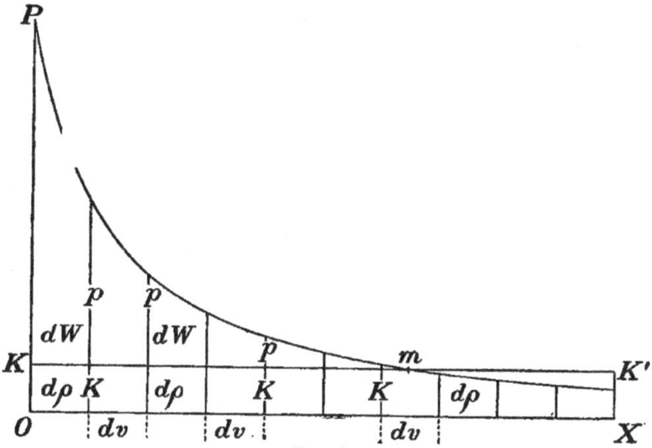

FIG. 8.

of pressures. Suppose the bore divided into elementary lengths dx. Then, since the length multiplied by the constant area of bore is the volume, we may replace dx by dv, the increment of volume, as in the figure. From equation (46),

$$dW = pdv;$$

and each of the small areas bounded by the pressure curve, the ordinates, and dv will be a value of dW. In the same way

$$d\rho = Kdv,$$

K being a constant and equal to the constant pressure between projectile and rifling multiplied by the coefficient of friction. Since K and dv are both constant, the values of $d\rho$ will all be equal, and they are bounded by the straight line KK', the ordinates, and the axis of X.

As the projectile moves from O toward X, the increment

of the work due to friction remains constant, while dW decreases and tends towards zero, since p constantly decreases.

There is some point, then, such as m, where

$$dW - d\rho = 0$$

or $dv = 0$. This value of X will be greater for a slow powder than for a quick one, because, as we have seen, for *equal charges* the slow powder gives a less maximum pressure, and hence we can use *larger charges* of slow powder without overstraining the gun; and as these large charges give off more gas, the pressure is kept up better along the bore than with the quick powder, or the values of p are greater along the bore, $d\rho$ being independent of the nature of the powder. When the point m is reached where $dv = 0$, or the velocity of the projectile ceases to increase, the limit of length is reached.

With small arms this limit is attained more nearly than with cannon; and the above reasoning shows that slow powder requires longer bores than quick powder.

INFLUENCE ON VELOCITY AND MAXIMUM PRESSURE.—Equation (70) shows that the passive resistances decrease the initial velocity of the projectile; but this is not always the case. Certain passive resistances, such as the resistance of the rifling, produce at first a more rapid combustion of the powder on account of the rise in pressure due to the delay of the projectile in moving off. Hence the powder acts as a quicker powder, the work done by it over a given path is increased, and this increase of work may more than compensate for the resistance. It follows as a consequence that the maximum pressure on the gun is increased. An accidental resistance, such as wedging of the projectile, may cause a great increase in pressure, and, if it cannot be overcome, may burst the gun.

24. Sarrau's Formulas—General Equation of Motion of Projectile in Bore.

The formulas deduced above furnish approximations to the initial velocity of the projectile, but are not exact for many reasons. Among these are:

1st. The powder is supposed to be all burned before the projectile moves. This is known to be incorrect.

2d. No account is taken of the passive resistances.

3d. The kind of powder and the calibre of the gun are not considered.

For these reasons it was customary to use empirical formulas. These formulas gave good results so long as the conditions under which they were deduced were not departed from, but they were limited in their applications and could not be generally used.

To obviate these difficulties, the subject has been discussed by M. Emile Sarrau, a distinguished French engineer of powders, etc. In his discussion the following hypotheses are adopted :

1. The inflammation of the charge is instantaneous.
2. The gases expand adiabatically.
3. The powder is not all burned before the projectile moves.

EQUATION OF MOTION.—At the end of any time t let

q be the weight of powder burned;
x, the length of bore passed over by the projectile;
p_1, the mean pressure;
v_1, the volume of the bore in rear of the base of the projectile, minus the volume occupied by the solid residue of the powder;
v, the velocity of the projectile.

The gas which is formed at the time t expands adiabatically, and we have, for any time t' after t (see equation 54),

$$p'v'^k = p_1 v_1^k, \quad . \quad . \quad . \quad . \quad . \quad . \quad . \quad . \quad (72)$$

p' and v' being the mean pressure and volume at the time t', and k the ratio of specific heats.

The total quantity of work done up to the time t' is, from (46),

$$W = \int p'dv'.$$

Substitute for p' its value from (72), integrate, and we have

$$W = -\frac{p_1 v_1^k}{k-1} v'^{1-k} + C. \quad . \quad . \quad . \quad . \quad . \quad (73)$$

To find the value of the constant: When $t' = t$, we have $v' = v_1$; and if we consider the energy of the projectile alone, $W = \frac{1}{2}mv^2$; hence

$$C = \frac{1}{2}mv^2 + \frac{p_1 v_1}{k-1},$$

and consequently

$$W = -\frac{p_1 v_1^k}{k-1} v'^{1-k} + \frac{1}{2}mv^2 + \frac{p_1 v_1}{k-1}. \quad . \quad . \quad (74)$$

If we now suppose the bore to be infinite in length, v' becomes infinite, and W becomes, from (61),

$$W = \frac{fq}{10(k-1)};$$

and since $k > 1$,

$$\frac{p_1 v_1}{k-1} v'^{1-k}$$

becomes zero. Hence

$$\frac{fq}{10(k-1)} = \frac{1}{2}mv^2 + \frac{p_1 v_1}{k-1}. \quad . \quad . \quad . \quad . \quad . \quad (75)$$

25. Transformation of Equation (75).—Errors in its Deduction and their Correction.

TRANSFORMATION.—Equation (75) is transformed as follows: Make

$$v = \frac{dx}{dt};$$

$$v_1 = \omega(z + x); \quad \text{(see equation 60}a\text{;)}$$

$$p_1 \omega = m\frac{d^2x}{dt^2}. \quad . \quad . \quad . \quad . \quad . \quad . \quad . \quad (75a)$$

This last equation expresses the fact that the total pressure on the projectile, $p_1\omega$, is equal to the accelerating force.

While this is not exactly true, it is sufficiently accurate, and can be corrected for, as will be explained.

Making these substitutions in equation (75) and including the constant 10 in f, which is equivalent to changing the unit in which the force of the powder is expressed, we have

$$\frac{fq}{k-1} = \frac{m}{2}\left(\frac{dx}{dt}\right)^2 + m\frac{d^2x}{dt^2}\,\frac{z+x}{k-1}. \quad . \quad . \quad . \quad (76)$$

ERRORS AND CORRECTIONS.—In deducing this equation certain errors have been made. These are:

1. The total work which a weight of powder is capable of doing, when expanded to infinity, is not equal to $\frac{fq}{k-1}$, because part of the heat is absorbed by the walls of the bore, and no allowance has been made for this loss.

On the contrary, we have supposed an adiabatic expansion without gain or loss of heat. Hence $\frac{fq}{k-1}$ is too large and must be diminished.

2. At any time t the work of expansion is not equal to $\frac{1}{2}mv^2$, but is equal to the total energy of the system, including gun, projectile, charge, and gun-carriage.

To correct for this we must increase the term $\frac{1}{2}m\left(\frac{dx}{dt}\right)^2$.

3. We have assumed $p_1\omega = m\frac{d^2x}{dt^2}$, or that the total pressure on the projectile is equal to the accelerating force.

This is not correct, because the force $p_1\omega$ not only produces acceleration of the projectile, but overcomes the passive resistances, such as forcing of the band, friction, etc. Hence, in order to make $p_1\omega = m\frac{d^2x}{dt^2}$, we must increase $m\frac{d^2x}{dt^2}$.

Instead of correcting each term of equation (76) as indicated, we can apply such a correction to the first member as will make it a true equation. The numerical value of f is uncertain, and we may therefore apply all the corrections to this quantity, and f then becomes a numerical coefficient whose value must be determined by experiment.

26. Deduction of Final Equation of Motion of Projectile in Bore.

Value of q.—The method of determining the value of f in equation (76) has been explained. It is necessary now to find the value of q, the quantity of powder burned at the end of any time t.

The proportional part of powder burned in air at any time t is given by the general expression, (12),

$$\phi(t) = \frac{at}{\tau}\left(1 - \lambda\frac{t}{\tau} + \mu\frac{t^2}{\tau^2} + \text{etc.}\right). \quad . \quad . \quad . \quad (77)$$

By multiplying this expression by the volume of the grain or charge we obtain the volume burned (see equation (16)); and multiplying the same expression by the weight, we have the weight burned at any time.

Hence, $\bar{\omega}$ being the weight of the charge, we have

$$q = \bar{\omega}\,\phi(t). \quad . \quad . \quad . \quad . \quad . \quad . \quad . \quad . \quad (78)$$

But the expression (77) applies to the burning of a grain or charge in air. In a gun, the pressure varies and is much greater than in air, and hence the velocity of combustion varies and is much greater, as has been shown; and this variation of velocity is expressed by Sarrau's formula already given (equation 11),

$$v = v_0\left(\frac{p}{p_0}\right)^{\frac{1}{2}}; \quad . \quad . \quad . \quad . \quad . \quad . \quad . \quad . \quad (79)$$

and the expression for $\phi(t)$ must be modified accordingly.

Spherical Grain.—Take the simplest case, that of a spherical grain. When burning in air, we have found for the volume burned at any time t, equation (13),

$$\tfrac{4}{3}\pi r^3 - \tfrac{4}{3}\pi(r - vt)^3.$$

Since v is no longer constant, owing to the variation of pressure in the gun, the space burned in the time dt is vdt;

and in the time t it is $\int_0^t vdt$, instead of vt, as in the case of constant pressure. The above expression for the volume of powder burned in the case of a spherical grain under varying pressure becomes, then,

$$\tfrac{4}{3}\pi r^3 - \tfrac{4}{3}\pi\left(r - \int_0^t vdt\right)^3; \quad . \quad . \quad . \quad . \quad . \quad (80)$$

and the value of $\phi(t)$ in equation (14) becomes

$$\phi(t) = 1 - \left(1 - \frac{1}{r}\int_0^t vdt\right)^3. \quad . \quad . \quad . \quad . \quad . \quad (81)$$

If in this equation we substitute for v its value from (79), we have

$$\phi(t) = 1 - \left[1 - \frac{v_0}{r}\int_0^t \left(\frac{p}{p_0}\right)^{\frac{1}{3}} dt\right]^3. \quad . \quad . \quad . \quad (82)$$

Now

$$\tau = \frac{r}{v_0},$$

the time of combustion of the grain in air under the normal pressure, and

$$\frac{v_0}{r} = \frac{1}{\tau}.$$

Substituting in (82), we have

$$\phi(t) = 1 - \left[1 - \frac{1}{\tau}\int_0^t \left(\frac{p}{p_0}\right)^{\frac{1}{3}} dt\right]^3. \quad . \quad . \quad . \quad (83)$$

Comparing this value of $\phi(t)$ with that for uniform pressure, which is, equation (17),

$$\phi(t) = 1 - \left(1 - \frac{t}{\tau}\right)^3, \quad . \quad . \quad . \quad . \quad . \quad . \quad . \quad . \quad (84)$$

we see that the only difference is the substitution of

$$\int_0^t \left(\frac{p}{p_0}\right)^{\frac{1}{3}} dt$$

for t.

Following the same method for other forms of grain, the same results will be obtained. Hence we conclude that if the combustion of powder under constant pressure is represented by

$$\phi(t) = \frac{at}{\tau}\left(1 - \lambda\frac{t}{\tau} + \mu\frac{t^2}{\tau^2} + \text{etc.}\right),$$

the combustion under variable pressure will be represented by

$$\phi(t) = \frac{a}{\tau}\int_0^t\left(\frac{p}{p_0}\right)^{\frac{1}{2}}dt\left[1 - \frac{\lambda}{\tau}\int_0^t\left(\frac{p}{p_0}\right)^{\frac{1}{2}}dt + \text{etc.}\right]. \quad . \quad (85)$$

FINAL EQUATION.—In equation (85) make

$p = p_1$ the mean pressure at the time t.

$p_1\omega = m\dfrac{d^2x}{dt^2}$ (see equation (75*a*)).

Whence

$$p_1 = \frac{m}{\omega}\frac{d^2x}{dt^2};$$

and we have for the value of the term $\int_0^t\left(\frac{p}{p_0}\right)^{\frac{1}{2}}dt$,

$$\int_0^t\left(\frac{p}{p_0}\right)^{\frac{1}{2}}dt = \left(\frac{m}{\omega p_0}\right)^{\frac{1}{2}}\int_0^t\left(\frac{d^2x}{dt^2}\right)^{\frac{1}{2}}dt; \quad . \quad . \quad . \quad (86)$$

and substituting this value in (85), we have

$$\phi(t) = \frac{a}{\tau}\left(\frac{m}{\omega p_0}\right)^{\frac{1}{2}}\int_0^t\left(\frac{d^2x}{dt^2}\right)^{\frac{1}{2}}dt\left[1 - \frac{\lambda}{\tau}\left(\frac{m}{\omega p_0}\right)^{\frac{1}{2}}\int_0^t\left(\frac{d^2x}{dt^2}\right)^{\frac{1}{2}}dt + \text{etc.}\right]; \quad (87)$$

and for the value of q,

$$q = \hat{\omega}\phi(t) = \hat{\omega} \times \text{2d member of equation (87)}. \quad . \quad (88)$$

In equation (76) make

$$\theta = \frac{k-1}{2},$$

and it becomes

$$(z + x)\frac{d^2x}{dt^2} + \theta\left(\frac{dx}{dt}\right)^2 = \frac{fq}{m}. \quad . \quad . \quad . \quad . \quad (89)$$

Substituting in this equation the value of q from (88), we have the final equation desired, which is

$$(z+x)\frac{d^2x}{dt^2}+\theta\left(\frac{dx}{dt}\right)^2=\frac{fa\hat{\omega}}{m\tau}\left(\frac{m}{\omega p_0}\right)^{\frac{1}{2}}\int_0^t\left(\frac{d^2x}{dt^2}\right)^{\frac{1}{2}}dt$$

$$\times\left\{1-\frac{\lambda}{\tau}\left(\frac{m}{\omega p_0}\right)^{\frac{1}{2}}\int_0^t\left(\frac{d^2x}{dt^2}\right)^{\frac{1}{2}}dt+\frac{\mu}{\tau^2}\left(\frac{m}{\omega p_0}\right)\left[\int_0^t\left(\frac{d^2x}{dt^2}\right)^{\frac{1}{2}}dt\right]^2+\text{etc.}\right\}. \quad (90)$$

27. Integration of Equation (90)—Practical Formulas for Velocity and Pressure—Values of *A* and *B*—Characteristics *α* and *β*.

INTEGRATION.—Equation (90) must be integrated before it can be used practically. Sarrau has done this by the use of auxiliary functions which are numerical and independent of the variable elements of fire. As a final result of his process, the general values for velocity and pressure are expressed in the form of definite series, which are very convergent. On this account it is necessary to consider only two terms of the series in the expression for the velocity, and only one term in the expression for the pressure.

BINOMIAL FORMULA FOR VELOCITY.—Considering the two terms of the series, Sarrau's formula for velocity is expressed as follows. It is called the binomial formula for velocities, and is the result of the integration of equation (90):

$$v=A\alpha(\hat{\omega}u)^{\frac{1}{2}}\left(\frac{\Delta}{pc}\right)^{\frac{1}{4}}\left[1-B\beta\frac{(pu)^{\frac{1}{2}}}{c}\right]; \quad . \quad . \quad . \quad (91)$$

in which v is the velocity of the projectile at the point u, and becomes the muzzle velocity when u is equal to the total length of travel of the projectile in the bore;

u, the length of bore passed over by the base of the projectile in inches, measured from the position occupied by it before firing. In equation (90) this length is denoted by x, and is changed to u in formula (91);

c, the calibre or diameter of the bore in inches;

p, the weight of the projectile in pounds;

$\hat{\omega}$, the weight of the charge of powder in pounds;

Δ, the density of loading;

α and β, two coefficients depending on the nature of the powder used, and called the "characteristics" of the powder;

A and B, two numerical coefficients which are independent of the elements of fire.

The values of α and β are

$$\alpha = \left(\frac{fa}{\tau}\right)^{\frac{1}{2}}, \quad \beta = \frac{\lambda}{\tau}, \quad . \quad . \quad . \quad . \quad . \quad . \quad (92)$$

in which f is the force of the powder;

a and λ, coefficients depending on the form of the grain of powder, and whose values for the different forms of grain have been given;

τ, the total time of burning of the grain in air.

FORMULA FOR MAXIMUM PRESSURE ON BASE OF PROJECTILE.—In the same way, as the result of integration, the expression for the maximum pressure on the base of the projectile is

$$P = K\alpha^2 \Delta \frac{(p\tilde{\omega})^{\frac{1}{2}}}{c^2}. \quad . \quad . \quad . \quad . \quad . \quad . \quad (93)$$

in which the quantities are the same as before, and K is a constant whose value is to be determined as will be explained.

VALUES OF A AND B.—If for a given powder we know the values of α and β, we can fire this powder from two different guns, and measure the resulting initial velocities. This will give v. The values of c, u, and p are known for the two guns, and the values of $\tilde{\omega}$ and Δ also, and, having two equations containing A and B, they can be calculated. But it is difficult to find exact values for α and β, since they depend on f and τ, equation (92), and the values of these quantities are uncertain. To avoid this difficulty, Sarrau adopts a particular powder which he calls a type powder, and assumes for it the values $f = 1$, $\tau = 1$.

The values of a and λ for the type powder are calculated as explained previously. Hence all the quantities in (91) are known except A and B, and they can now be calculated.

Since A and B are constants to be determined experimentally, whatever errors we make in assuming the values of f and τ will be corrected for in the values of A and B as found by experiment; and since these values of A and B are independent of all the elements of fire, they will be true for all powders and all guns. A and B are found by experiment because they depend on v, which is determined by experiment.

In this way Sarrau found the values of A and B to be

$$\log A = 2.56635;$$
$$\log B = \bar{2}.30964.$$

CHARACTERISTICS α AND β.—Having the values of A and B in (91), we must know the characteristics α and β for any powder before we can apply the formula to this powder. The values of α and β depend on f, a, λ, and τ (see (92)). a and λ can be calculated, as before explained, for any grain of ordinary shape, and their values for most service forms have been given. The value of f is uncertain, and therefore for simplicity Sarrau assumes $f = 1$ for all powders, the same as for the standard powder.

We have seen that f is practically constant for all powders, and hence the above assumption may be made. The value of τ cannot be accurately determined except by the use of a formula not yet deduced, and hence the method of determining it will be explained later.

28. Maximum Pressure on Breech of Gun—Value of K.

MAXIMUM PRESSURE.—Equation (93) gives the maximum pressure on the base of the projectile, and in order to use it K must be known. If we could measure accurately the pressure on the base of the projectile, the value of K could be found by firing a shot from a gun, since all the quantities except K in (93) would be known, and hence it could be determined. But this pressure cannot be accurately measured, and hence we determine first the maximum pressure on the breech.

To do this assume equation (64):

$$v' = \frac{m + \theta'\mu}{M}v.$$

Differentiate with respect to t:

$$\frac{dv'}{dt} = \frac{m + \theta'\mu}{M}\frac{dv}{dt}. \quad . \quad . \quad . \quad . \quad . \quad . \quad (94)$$

Multiply both members by $\frac{M}{\omega}$, ω being the area of cross-section of the bore:

$$\frac{M}{\omega}\frac{dv'}{dt} = \frac{m}{\omega}\frac{dv}{dt}\left(1 + \theta'\frac{\mu}{m}\right). \quad . \quad . \quad . \quad . \quad (95)$$

Now denoting by P_0 the maximum pressure per unit of surface on the bottom of the bore, and by P the corresponding pressure on the base of the projectile, we have

$$\left.\begin{aligned} P_0\omega &= M\frac{dv'}{dt}; \\ P\omega &= m\frac{dv}{dt}; \end{aligned}\right\} \quad . \quad . \quad . \quad . \quad . \quad . \quad . \quad . \quad (96)$$

and substituting in (95),

$$P_0 = P\left(1 + \theta'\frac{\mu}{m}\right). \quad . \quad . \quad . \quad . \quad . \quad . \quad (97)$$

Substituting weights for masses, and for θ' its value $\frac{1}{2}$,

$$P_0 = P\left(1 + \frac{\varpi}{2p}\right). \quad . \quad . \quad . \quad . \quad . \quad . \quad . \quad (98)$$

This equation does not give true values for P_0, since in placing the total pressure equal to the accelerating force, equation (96), we have evidently neglected the force necessary to overcome the passive resistances. Sarrau has therefore adopted, as more nearly agreeing with experiment, the following formula:

$$P_0 = P\left(1 + \frac{3}{2}\frac{\varpi}{p}\right). \quad . \quad . \quad . \quad . \quad . \quad . \quad . \quad (99)$$

Substituting for P in (99) its value from (93), we have

$$P_0 = K\left(1 + \frac{3}{2}\frac{\hat{\omega}}{p}\right)\alpha^2 \varDelta \frac{(p\hat{\omega})^{\frac{1}{2}}}{c^2}. \quad . \quad . \quad . \quad (100)$$

Making

$$\left(1 + \frac{3}{2}\frac{\hat{\omega}}{p}\right) = K'\left(\frac{\hat{\omega}}{p}\right)^{\gamma}, \quad . \quad . \quad . \quad . \quad . \quad (101)$$

we have

$$P_0 = KK'\left(\frac{\hat{\omega}}{p}\right)^{\gamma}\alpha^2 \varDelta \frac{(p\hat{\omega})^{\frac{1}{2}}}{c^2}. \quad . \quad . \quad . \quad . \quad (102)$$

Making

$$KK' = K_0$$

and

$$\gamma = \tfrac{1}{4},$$

which is justified by experiment, we have finally, for the pressure on the breech,

$$P_0 = K_0\left(\frac{\hat{\omega}}{p}\right)^{\frac{1}{4}}\alpha^2 \varDelta \frac{(p\hat{\omega})^{\frac{1}{2}}}{c^2};$$

or reducing,

$$P_0 = K_0 \alpha^2 \varDelta \frac{\hat{\omega}^{\frac{3}{4}} p^{\frac{1}{4}}}{c^2}. \quad . \quad . \quad . \quad . \quad . \quad . \quad . \quad . \quad (103)$$

Since we can measure very accurately the pressure on the breech of a gun, we fire with a given powder, measure this pressure, substitute it for P_0 in (103), and, as all the other quantities except K_0 are known, we thus obtain its value, which is

$$\log K_0 = 4.25092.$$

VALUE OF K IN (93).—Having the value of P_0, we substitute it in (99) and find the corresponding value of P. This value of P in (93), together with the known values of the other quantities, will give K, whose value is

$$\log K = 3.96197.$$

Collecting the pressure formulas for convenience, we have

$$P = K\alpha^2 \Delta \frac{(p\omega)^{\frac{1}{2}}}{c^2}, \text{ on base of projectile}$$

$$P_0 = K_0 \alpha^2 \Delta \frac{p^{\frac{1}{2}}\omega^{\frac{1}{2}}}{c^2}, \text{ on breech of gun.}$$

$$\log K = 3.96197;$$

$$\log K_0 = 4.25092.$$

29. Theoretical Maximum Velocity—Time of Burning Corresponding to the Maximum Velocity.

MAXIMUM VELOCITY.—Assuming the binomial formula for velocity, and replacing α and β in it by their values, equation (92),

$$\alpha = \left(\frac{fa}{\tau}\right)^{\frac{1}{2}}, \quad \beta = \frac{\lambda}{\tau},$$

we have

$$v = A\left(\frac{fa}{\tau}\right)^{\frac{1}{2}}(\omega u)^{\frac{1}{2}}\left(\frac{\Delta}{pc}\right)^{\frac{1}{2}}\left[1 - B\frac{\lambda}{\tau}\frac{(pu)^{\frac{1}{2}}}{c}\right]. \quad . \quad . \quad (104)$$

If in this equation we make v and τ the only variables, it can be shown by the usual rules of calculus that v will have a maximum value for a particular value of τ. That is, as τ decreases in value, v will increase till it reaches a maximum.

But this ought not to be the case, because, theoretically, as τ decreases, or the powder becomes more quick, v should increase; and this increase should continue up to the limit where the combustion is practically instantaneous and $\tau = 0$.

Formula (104) gives this maximum value for v because it is not absolutely correct, but only approximate. It will be remembered that in its deduction it was stated that the value for v was expressed in the form of a series, of which the first two terms only were retained. But the function represented by this series may go on increasing when τ decreases below the value which makes the sum of the first two terms a maximum, provided the sum of the other terms goes on increasing.

The value of τ corresponding to the maximum of the first two terms is nevertheless important, because it marks the limit below which a decrease in τ gives only a slight increase in the velocity. It is not advisable to pass below this value of τ in practice, because $\alpha = \left(\frac{fa}{\tau}\right)^{\frac{1}{2}}$ enters the pressure formulas (93) and (103) to the second power, and τ being in the denominator of the value of α, a small decrease in τ causes a rapid increase in the value of α, and hence in that of the maximum pressure, while the gain in velocity, as shown by the previous discussion, is very small.

VALUE OF τ CORRESPONDING TO THE MAXIMUM VALUE OF v.—The particular value of τ which corresponds to this maximum value of v is called the "time of the maximum."

Differentiating equation (104) with respect to τ, placing $\frac{dv}{d\tau} = 0$, and solving for τ, we have, calling the resulting value τ_1,

$$\tau_1 = 3B\lambda\frac{(pu)^{\frac{1}{2}}}{c}. \quad . \quad . \quad . \quad . \quad . \quad . \quad (105)$$

This shows that for a given form of grain, the value of τ_1 or the time of the maximum, depends on the calibre, weight of projectile, and length of travel, and is independent of the charge of powder and density of loading.

For the same powder, the weight of the projectile p is proportional to the cube of the calibre, the length of travel u to the first power of the calibre, and hence, since $3B\lambda$ is constant, we may write, from (105),

$$\tau_1 = f\frac{(c^3c)^{\frac{1}{2}}}{c} = f(c), \quad . \quad . \quad . \quad . \quad . \quad (106)$$

or the time of the maximum is proportional to the calibre of the gun; that is, to obtain the greatest velocity, the time of burning should increase as the calibre of the gun increases, or large-grain powder should be used in large guns, which proves the principle enunciated by Rodman.

30. Modulus of Quickness—Value of Modulus—Velocity and Pressure as Functions of this Modulus.

MODULUS OF QUICKNESS.—Powders are called "quick" or "slow" depending upon their action in a given gun. A given powder may be "quick" when used in one gun and "slow" when used in another. For example, the I. K. powder which is used in the 3.20-inch field-guns is "quick" when used in the 8-inch rifle, and "slow" when used in the Springfield rifle. From equation (105) we can calculate the value of τ_1, or the time of the maximum, for any gun and powder.

A powder whose time of combustion is much greater than this is called a slow powder for this gun, and one whose time of combustion is nearly equal to this is called a quick powder for the same gun.

Also two powders fired in different guns are considered equal, as regards their quickness, if their times of combustion are proportional to the "times of the maximum" of the two guns considered. Hence, if we make

$$x = \frac{\tau_1}{\tau}, \quad . \quad . \quad . \quad . \quad . \quad . \quad . \quad . \quad (107)$$

and call this ratio x the "modulus of quickness" of the powder, we can say that the quickness of a powder is measured by its modulus.

On this basis Sarrau classifies powders as follows:

$x = 1.0$, very quick powder;
$x = 0.9$, quick powder;
$x = 0.8$, medium powder;
$x = 0.7$, slow powder;
$x = 0.6$, very slow powder.

VALUE FOR MODULUS.—We have, from (107),

$$x = \frac{\tau_1}{\tau}.$$

Substitute for τ, its value from (105), and we have

$$x = 3B\frac{\lambda}{\tau}\frac{(pu)^{\frac{1}{2}}}{c}; \quad . \quad . \quad . \quad . \quad . \quad . \quad (107a)$$

and since from (92)

$$\beta = \frac{\lambda}{\tau},$$

we have

$$x = 3B\beta\frac{(pu)^{\frac{1}{2}}}{c}. \quad . \quad . \quad . \quad . \quad . \quad . \quad . \quad (108)$$

VELOCITY AS A FUNCTION OF THE MODULUS x.—To express the velocity as a function of the modulus, we have for the subtractive term of the binomial formula (104) from (107a),

$$B\frac{\lambda}{\tau}\frac{(pu)^{\frac{1}{2}}}{c} = \frac{x}{3};$$

and for τ from the same equation,

$$\tau = \frac{3B\lambda(pu)^{\frac{1}{2}}}{xc}.$$

Substituting the value $\frac{x}{3}$ for the subtractive term, and for τ its value above, in the factor $\left(\frac{fa}{\tau}\right)^{\frac{1}{2}}$ in the binomial formula, we have

$$v = \tfrac{1}{3}A(3B)^{-\frac{1}{2}}\left(\frac{fa}{\lambda}\right)^{\frac{1}{2}}\varpi^{\frac{1}{2}}\Delta^{\frac{1}{4}}c^{\frac{1}{4}}u^{\frac{1}{4}}p^{-\frac{3}{4}}x^{\frac{1}{2}}[3 - x]. \quad . \quad (109)$$

Making

$$\tfrac{1}{2}x^{\frac{1}{2}}[3 - x] = f(x), \quad . \quad . \quad . \quad . \quad . \quad (109a)$$

we have

$$v = \tfrac{2}{3}A(3B)^{-\frac{1}{2}}\left(\frac{fa}{\lambda}\right)^{\frac{1}{2}}\varpi^{\frac{1}{2}}\Delta^{\frac{1}{4}}c^{\frac{1}{4}}u^{\frac{1}{4}}p^{-\frac{3}{4}}f(x). \quad . \quad . \quad . \quad (110)$$

MAXIMUM PRESSURE AS A FUNCTION OF THE MODULUS x.—By a similar process the pressures on the base of the projectile and on the breech may be expressed in terms of x.

For the maximum pressure on the base of the projectile we have

$$P = K(3B)^{-1}\frac{fa}{\lambda}\frac{\Delta\hat{\omega}^{\frac{1}{2}}}{cu^{\frac{1}{2}}}x, \quad . \quad . \quad . \quad . \quad . \quad (111)$$

and for that on the breech

$$P_0 = K_0(3B)^{-1}\frac{fa}{\lambda}\left(\frac{\hat{\omega}}{p}\right)^{\frac{1}{2}}\frac{\Delta\hat{\omega}^{\frac{1}{4}}}{cu^{\frac{1}{2}}}x. \quad . \quad . \quad . \quad (112)$$

31. Limit of Use of Binomial Formula.

It has been shown that v in formula (91) becomes a maximum when τ decreases to a particular value τ_1, and also that v should not be a maximum for this particular value of τ, but should increase continuously as τ decreases.

It is also evident from (107) that when τ becomes τ_1, x, the modulus, becomes unity, or $x = 1$. That is, the velocity is a maximum by formula (91) when $x = 1$.

The value of x from (108) is

$$x = 3B\beta\frac{(pu)^{\frac{1}{2}}}{c}, \quad . \quad . \quad . \quad . \quad . \quad . \quad (113)$$

and the subtractive term of the binomial formula is

$$B\beta\frac{(pu)^{\frac{1}{2}}}{c}, \quad . \quad . \quad . \quad . \quad . \quad . \quad . \quad (114)$$

which is $\frac{1}{3}$ of x. Hence when v becomes a maximum in (91), x becomes unity, and the subtractive term of (91) becomes $\frac{1}{3}$.

This value for the subtractive term would then mark the limit of the use of the binomial formula, were it not for the fact that as a function approaches its maximum it changes its value very slowly, and hence before we reach the value $x = 1$, the binomial formula will cease to give correct results. For this reason Sarrau adopts the value $x = \frac{9}{11}$ for the particular value of the modulus at which it is best to cease the use of formula (91).

When $x = \frac{9}{11}$, the subtractive term, being $\frac{1}{3}$ of x, will be $\frac{1}{3}$ of $\frac{9}{11} = 0.273$. Hence we have for determining the limit of the use of the binomial formula the rule: Calculate the value of the subtractive term in the binomial formula; if it is greater than 0.273, the binomial formula is not applicable; if less than 0.273, it is applicable; or

$$\left.\begin{array}{l} B\beta\dfrac{(pu)^{\frac{1}{3}}}{c} > 0.273, \text{ do not use binomial formula;} \\ B\beta\dfrac{(pu)^{\frac{1}{3}}}{c} < 0.273, \text{ use binomial formula.} \end{array}\right\} \quad (115)$$

32. Monomial Formula for Velocity.

It is necessary, from what precedes, to have a formula for velocity that can be used when the binomial formula ceases to apply.

It is deduced as follows: The values of the modulus for all powders in use vary between narrow limits (0.6 to 1.0).

Hence, assuming the equation (109*a*),

$$f(x) = \tfrac{1}{2}x^{\frac{1}{2}}[3 - x],$$

we may place

$$f(x) = Nx^n, \quad . \quad . \quad . \quad . \quad . \quad . \quad . \quad . \quad . \quad (116)$$

since when a variable changes its value within narrow limits, the function is proportional to some power of the variable properly chosen. It is necessary now to find the proper value of n in (116).

Differentiating (116), we have

$$\frac{df(x)}{dx} = f'(x) = nNx^{n-1};$$

$$n = \frac{f'(x)}{Nx^{n-1}} = x\frac{f'(x)}{f(x)}. \quad \text{[See (116).]} \quad . \quad . \quad . \quad (116a)$$

Substituting for $f'(x)$ and $f(x)$ their values from (109*a*), we have

$$n = x\frac{f'(x)}{f(x)} = \frac{3}{2} \cdot \frac{1 - x}{3 - x} \quad . \quad . \quad . \quad . \quad . \quad (117)$$

for the value of n required.

Assume equation (110); and substitute in it for $f(x)$ its value (116), and we have

$$v = \frac{2}{3}A(3B)^{-\frac{1}{4}}N\left(\frac{fa}{\lambda}\right)^{\frac{1}{4}}\frac{\bar{\omega}^{\frac{3}{8}}\Delta^{\frac{1}{2}}c^{\frac{1}{4}}u^{\frac{1}{8}}}{p^{\frac{1}{2}}}x^{n}. \quad . \quad . \quad (118)$$

Substitute for x its value (107a), and make

$$M = \frac{2}{3}A(3B)^{n-\frac{1}{4}}N, \quad . \quad . \quad . \quad . \quad . \quad (119)$$

and we have

$$v = M\left(\frac{fa}{\tau}\right)^{\frac{1}{4}}\left(\frac{\tau}{\lambda}\right)^{\frac{1}{4}-n}\frac{\bar{\omega}^{\frac{3}{8}}\Delta^{\frac{1}{2}}c^{\frac{1}{4}-n}u^{\frac{1}{8}+\frac{n}{2}}}{p^{\frac{1}{2}-\frac{n}{2}}}. \quad . \quad . \quad . \quad (120)$$

Formula (120) is a general form of the binomial formula (104), and will give the same values for v as the binomial formula, if the proper values for n be substituted. For the particular case when $x = \frac{9}{11}$ and $n = \frac{1}{8}$ (120) becomes

$$v = M\left(\frac{fa}{\tau}\right)^{\frac{1}{4}}\left(\frac{\tau}{\lambda}\right)^{\frac{1}{8}}\frac{\bar{\omega}^{\frac{3}{8}}\Delta^{\frac{1}{2}}c^{\frac{1}{8}}u^{\frac{3}{16}}}{p^{\frac{7}{16}}}. \quad . \quad . \quad . \quad . \quad (121)$$

This equation (121) is strictly applicable only to the particular case for which it was deduced; that is, for $x = \frac{9}{11}$ and $n = \frac{1}{8}$; but by examining it we see that v increases continuously as τ decreases, which should be the case, while in the binomial formula, as already shown, v ceases to increase as τ decreases.

Hence if we use equation (121) for all values of x equal to or greater than $\frac{9}{11}$, we will obtain a value for v which is correct for the value $x = \frac{9}{11}$, and for all values of x greater than $\frac{9}{11}$, values for v which will be more nearly correct than those given by the binomial formula.

This is called the "monomial formula" for velocity, and we say that it is used whenever the subtractive term in the binomial formula is greater than 0.273; since when that is the case the binomial formula is no longer applicable.

When the subtractive term is nearly equal to 0.273, either formula can be used.

The monomial formula is usually written

$$v = M\alpha\beta^{-\frac{1}{4}}\frac{\varpi^{\frac{3}{8}}\Delta^{\frac{1}{4}}c^{\frac{1}{8}}u^{\frac{3}{16}}}{p^{\frac{7}{16}}}, \quad . \quad . \quad . \quad . \quad . \quad . \qquad (122)$$

by substituting α and β for their values, equation (92).

To find M, Sarrau assumes a type powder as before, making $f = 1$, $\tau = 1$, and thus determines α and β. The powder is then fired in a given gun, v measured, and thus everything is known in (122) except M, which may be calculated.

Its value thus determined is

$$\log M = 2.84571.$$

33. Calculation of the Value of τ.

For the type powder $\tau = 1$, and under this supposition the values of A and B are deduced. The values of τ for all other powders must therefore be expressed in terms of the type powder as unity. That is, the value of τ for any powder is the ratio of its true time of burning to that of the time of burning of the type powder.

The force of all nitrate powders is practically constant, as has been shown; and since $f = 1$ for the type powders, it may be assumed as unity for all powders as an approximate value. Making $f = 1$ in the binomial formula (104), we have

$$v = A\left(\frac{a}{\tau}\right)^{\frac{1}{2}}(\varpi u)^{\frac{1}{4}}\left(\frac{\Delta}{pc}\right)^{\frac{1}{2}}\left[1 - B\frac{\lambda}{\tau}\frac{(pu)^{\frac{1}{2}}}{c}\right]. \quad . \quad . \quad . \qquad (123)$$

For any particular powder to which this formula is applicable, we could measure v and determine τ, since all the other quantities are known, if the equation could be solved for τ. But it is found that this solution is impossible.

If the monomial formula applies to this particular powder, we have, making $f = 1$,

$$v = M\left(\frac{a}{\tau}\right)^{\frac{1}{2}}\left(\frac{\tau}{\lambda}\right)^{\frac{1}{4}}\frac{\hat{\omega}^{\frac{1}{2}}\Delta^{\frac{1}{4}}c^{\frac{1}{4}}u^{\frac{3}{16}}}{p^{\frac{7}{16}}}. \quad . \quad . \quad . \quad . \quad (125)$$

In (125), placing

$$X = \frac{M\hat{\omega}^{\frac{1}{2}}\Delta^{\frac{1}{4}}c^{\frac{1}{4}}u^{\frac{3}{16}}}{vp^{\frac{7}{16}}}, \quad . \quad . \quad . \quad . \quad . \quad . \quad . \quad . \quad (126)$$

we have

$$\tau = \frac{a^4X^8}{\lambda^2}, \quad . \quad . \quad . \quad . \quad . \quad . \quad . \quad . \quad . \quad . \quad (127)$$

and this may be easily solved and the value of τ obtained. For any powder, however, we do not know beforehand whether the binomial or the monomial formula is applicable, since τ must be known to determine this point. Again, while the value of f is very nearly unity for all powders, it is not exactly unity for any except the type powder, and hence the value of τ determined as above by the monomial formula would not be correct.

Under these circumstances we proceed as follows: The value of τ determined by (127) is approximate, but the approximation is sufficiently correct to show which formula is to be used. Substitute the value of τ from (127) in the subtractive term of the binomial formula

$$\gamma = B\frac{\lambda}{\tau}\frac{(pu)^{\frac{1}{2}}}{c}. \quad . \quad . \quad . \quad . \quad . \quad . \quad . \quad (128)$$

If the result obtained is greater than 0.273, the monomial formula applies; if less than 0.273, the binomial.

Then calculate α by the pressure formula (103). Substitute this value of α in either the monomial or the binomial formula according as the former or the latter applies, as determined by the test, and solve for β. This value of β substituted in the formula

$$\beta = \frac{\lambda}{\tau}$$

will give τ. The value of τ thus obtained, substituted in the formula

$$\alpha = \left(\frac{fa}{\tau}\right)^{\frac{1}{3}},$$

together with the correct values of a and α, will give f.

34. Determination of the Characteristics α and β.

1st Method.—We have, equation (92),

$$\alpha = \left(\frac{fa}{\tau}\right)^{\frac{1}{3}}; \quad \beta = \frac{\lambda}{\tau}.$$

Assuming $f = 1$ for all powders, we can calculate a and λ for any service form of grain by the methods already explained, and illustrated in the case of the spherical grain; τ can be calculated by the method explained above, in terms of the type powder as unity, and hence we find α and β.

2d Method.—In the second method we find the values of α and β directly, without determining f, a, λ, and τ, as follows:

The characteristics α and β enter the binomial and monomial formulas (91) and (122), and α enters the pressure formula (103). Hence, for a given piece, powder, and projectile, if we measure accurately the pressure, and substitute it in (103), we can find α, and substituting the measured velocity and the value of α, just found, in either the monomial or the binomial formula, whichever is applicable, we can find β.

3d Method.—In the third method two different guns are used, and the velocities accurately measured under two different conditions of firing. The results being substituted in the binomial formula, or in the monomial and binomial formulas together, will give two equations, containing the two unknown quantities α and β, from which they may be obtained.

The following table gives data with reference to American guns and powders, and was calculated from data furnished from the Ordnance Proving Ground, Sandy Hook, N. J.

Gun.	Model.	Gun Constants.				Powder.									
		c	p	u	C	Kind.	$\bar{\omega}$	Δ	a	λ	τ	f	Characteristics.		
		Ins.	Lbs.	Ins.	Cu. Ins.		Lbs.						log α	log α^2	log β
Hotchkiss 3″ B. L. R..		3	12	36.729	27.12	Granular......	0.875	0.9	3	1	0.5363	0.8674	0.34297	0.68594	0.27058
3″.2 B. L. Rifle........	1885	3.2	13.5	72.0	109.7	Sphero-hexag..	3.5	0.8832	3	1	0.6923	1.0714	0.33338	0.66676	0.15971
3″.2 B. L. Rifle........	1890	3.2	13.5	69.792	108.9	Sphero-hexag..	3.5	0.8897	3	1	0.7066	1.0855	0.33179	0.66358	0.15083
3″.6 B. L. Rifle........	1891	3.6	20	72.288	149.2	Sphero-hexag..	4.1875	0.7769	3	1	0.6429	1.1340	0.36183	0.72365	0.19187
3″.6 B. L. Mortar......	1890	3.6	20	16.092	33.2	Sphero-hexag..	0.937	0.8339	3	1	0.3890	0.8983	0.42031	0.84062	0.41008
5″ B. L. Rifle..........	1890	5	45	119.8	395.6	Sphero-hexag..	12.5	0.8746	3	1	1.0269	1.1429	0.26180	0.52360	$\bar{1}$.98848
7″ B. L. Howitzer....	1890	7	105	81.27	315.7	Granular......	10	0.8768	3	1	0.5535	0.9072	0.34585	0.69170	0.25688
7″ B. L. Mortar.......	1892	7	125	44.66	182.8	....	5.5	0.8328	3	1	0.4436	0.8131	0.37014	0.74027	0.35299
8″ B. L. Rifle..........	1888M	8	300	205.28	3597	Brown Pris....	125	0.9619	1.5	$\frac{1}{8}$	1.7820	1.0549	$\bar{1}$.97419	$\bar{1}$.94837	$\bar{1}$.27195
10″ B. L. Rifle........	1888M	10	575	275.1	7064	Brown Pris....	250	0.9797	1.5	$\frac{1}{8}$	2.0456	0.9387	$\bar{1}$.91891	$\bar{1}$.83781	$\bar{1}$.21205
12″ B. L. Rifle........	1888M	12	1000	330.96	12114	Brown Pris....	450	1.0285	1.5	$\frac{1}{8}$	2.7099	0.9558	$\bar{1}$.86176	$\bar{1}$.72352	$\bar{1}$.08993
12″ B. L. Rifle........	1891	12	1000	364.92	13582	Brown Pris....	520	1.0598	1.5	$\frac{1}{8}$	3.7413	1.1491	$\bar{1}$.83171	$\bar{1}$.66341	$\bar{2}$.94986
16″ B. L. Rifle........	1892	16	2370	453.92	29341	Brown Pris....	1060	1.0000	1.5	$\frac{1}{8}$	3.2464	0.8875	$\bar{1}$.80643	$\bar{1}$.61285	$\bar{1}$.01148
12″ B. L. Mortar, C. I.		12	800*	91.92	1990	Brown Pris....	80	1.1128	1.5	$\frac{1}{8}$	0.7650	0.7158	0.07363	0.14726	$\bar{1}$.63924
12″ B. L. Mortar (steel)		12	800*	98.70	2636	Brown Pris....	105	1.1026	1.5	$\frac{1}{8}$	0.9407	0.7904	0.05024	0.10048	$\bar{1}$.54944

* A 1000-lb. shell may be used for ranges up to 2½ miles with reduced charges.

35. Effect of Variation of Elements of Loading upon Velocity, and Maximum Pressure.

The variable elements of loading for any gun are:

1. The weight of the charge of powder, $\hat{\omega}$
2. The density of loading, Δ.
3. The time of combustion of the grain, τ.

The fixed elements are:

1. The calibre, c.
2. The travel of the projectile in the bore, u.
3. The weight of the projectile, p.

For the same force of the powder and the same form of grain, having c, u, and p constant, we may vary $\hat{\omega}$, Δ, and τ so as to obtain the same muzzle velocity with a different maximum pressure on the breech, or the same maximum pressure with a different muzzle velocity. In this discussion the maximum pressure on the breech is alone considered, since it is always greater than that on the base of the projectile. There are an infinite number of sets of values of $\hat{\omega}$, Δ, and τ which will satisfy these conditions, and the question is to determine what set to use.

Assume equations (110) and (112), and regarding $\hat{\omega}$, Δ, and τ as variables, take the Napierian logarithms of both members of each, differentiate, and we have

$$\frac{dv}{v} = \frac{3}{8}\frac{d\hat{\omega}}{\hat{\omega}} + \frac{1}{4}\frac{d\Delta}{\Delta} + \frac{f'(x)}{f(x)}dx; \quad . \quad . \quad . \quad . \quad (129)$$

$$\frac{dP_0}{P_0} = \frac{3}{4}\frac{d\hat{\omega}}{\hat{\omega}} + \frac{d\Delta}{\Delta} + \frac{dx}{x}. \quad . \quad . \quad . \quad . \quad . \quad . \quad (130)$$

Now

$$x = \frac{\tau_1}{\tau}.$$

Hence

$$dx = -\frac{\tau_1 d\tau}{\tau^2} = -\frac{\tau_1}{\tau}\frac{d\tau}{\tau} = -x\frac{d\tau}{\tau}. \quad . \quad . \quad . \quad (131)$$

We have also, from (116*a*),

$$n = x\frac{f'(x)}{f(x)}.$$

Substituting these values in (129) and (130), we have

$$\frac{dv}{v} = \frac{3}{8}\frac{d\hat{\omega}}{\hat{\omega}} + \frac{1}{4}\frac{d\Delta}{\Delta} - n\frac{d\tau}{\tau}; \quad . \quad . \quad . \quad . \quad . \quad (132)$$

$$\frac{dP_0}{P_0} = \frac{3}{4}\frac{d\hat{\omega}}{\hat{\omega}} + \frac{d\Delta}{\Delta} - \frac{d\tau}{\tau}. \quad . \quad . \quad . \quad . \quad . \quad . \quad . \quad (133)$$

These equations show that when $\hat{\omega}$ and Δ increase, v and P_0 increase, and when τ increases, v and P_0 decrease; and this should evidently be the case.

36. Change of Velocity when Maximum Pressure remains Constant—Fixed Powder-chamber.

VARIATION OF VELOCITY.—A gun, like any other structure, is built to stand a certain fixed maximum pressure, and this pressure must not be exceeded. Therefore the most important consideration is to find how the velocity will vary for such changes in $\hat{\omega}$, Δ, and τ as will keep the maximum pressure constant and within limits.

To do this we will consider the three variables $\hat{\omega}$, Δ, and τ in order, keeping one constant and varying the other two, and find the effect upon the velocity, the pressure being always constant and a maximum.

1st. *$\hat{\omega}$ constant, Δ and τ variable.*—Since P_0 and $\hat{\omega}$ are constant, we have

$$dP_0 = 0; \quad d\hat{\omega} = 0;$$

and hence, from (133),

$$\frac{d\Delta}{\Delta} = \frac{d\tau}{\tau}. \quad . \quad . \quad . \quad . \quad . \quad . \quad . \quad . \quad . \quad (133a)$$

This condition must hold in order that P_0 be constant.

Substituting the above value of $\frac{d\tau}{\tau}$ in (132), we have

$$\frac{dv}{v} = (\tfrac{1}{4} - n)\frac{d\Delta}{\Delta}. \qquad \qquad (134)$$

The value of n is, from (117),

$$n = \frac{3}{2} \cdot \frac{1 - x}{3 - x}.$$

When the modulus $x > 0.6$, which corresponds to a very slow powder, $n < \frac{1}{4}$, and hence $\frac{dv}{v}$ is positive and increases with Δ. Therefore, when the weight of the charge is constant, we see from (134) that we may increase the velocity by increasing the density of loading; but in order to keep the pressure constant, (133*a*) shows that we must use a slower powder.

2d. *Δ constant, $\hat{\omega}$ and τ variable.*—P_0 and Δ being constant, we have

$$dP_0 = 0;$$

$$d\Delta = 0;$$

and from (133),

$$\frac{3}{4}\frac{d\hat{\omega}}{\hat{\omega}} = \frac{d\tau}{\tau}, \qquad \qquad (134a)$$

for the condition that P_0 shall be constant.

Substituting this value of $\frac{d\tau}{\tau}$ in (132), we have

$$\frac{dv}{v} = \frac{3}{4}\left(\frac{1}{2} - n\right)\frac{d\hat{\omega}}{\hat{\omega}}. \qquad \qquad (135)$$

The value of n is

$$n = \frac{3}{2}\frac{1 - x}{3 - x}. \qquad \qquad (135a)$$

When $x = 0$, $n = \frac{1}{2}$. It is less than $\frac{1}{2}$ for every other value of x. Hence $(\frac{1}{2} - n)$ is always positive, and $\frac{dv}{v}$ increases with $\hat{\omega}$.

Therefore when the density of loading is constant, (135) shows that we may increase the velocity by increasing the weight of charge, but (134*a*) shows that in order to keep the pressure constant we must also use a slower powder.

This means that we can obtain an increase of velocity with a constant maximum pressure, by increasing the size of the chamber and using a larger charge of slower powder, and this is the general method employed at present.

3d. *τ Constant, $\hat{\omega}$ and Δ Variable.*—P_0 and τ being constant, we have

$$dP_0 = 0;$$
$$d\tau = 0;$$

and from (133),

$$\frac{3}{4}\frac{d\hat{\omega}}{\hat{\omega}} = -\frac{d\Delta}{\Delta}, \quad \ldots \ldots \quad (135b)$$

for the condition that P_0 shall be constant.

Substituting this value of $\frac{d\Delta}{\Delta}$ in (132), we have

$$\frac{dv}{v} = \frac{3}{16}\frac{d\hat{\omega}}{\hat{\omega}}. \quad \ldots \ldots \quad (136)$$

Since this is always positive, (136) shows that for the same kind of powder we may increase the velocity by increasing the weight of charge, but (135*b*) shows that in order to keep the pressure constant we must also decrease the density of loading.

FIXED POWDER-CHAMBER.—In the preceding discussion it has been supposed that we could vary the size of the powder-chamber. But in the ordinary case, with a gun already built, the powder-chamber will be fixed and its volume constant. In this case we have

$$\Delta = \frac{\hat{\omega}}{C}.$$

Taking the Napierian logarithms of both members and differentiating, we have

$$\frac{d\Delta}{\Delta} = \frac{d\hat{\omega}}{\hat{\omega}}. \qquad (137)$$

Substituting this value of $\frac{d\Delta}{\Delta}$ in (132) and (133), we have

$$\frac{dv}{v} = \frac{5}{8}\frac{d\hat{\omega}}{\hat{\omega}} - n\frac{d\tau}{\tau}; \qquad (138)$$

$$\frac{dP_0}{P_0} = \frac{7}{4}\frac{d\hat{\omega}}{\hat{\omega}} - \frac{d\tau}{\tau}. \qquad (139)$$

The variables are thus reduced to two. Now if we suppose $\hat{\omega}$ and τ to vary so as to keep P_0 constant, we have $dP_0 = 0$, and from (139)

$$\frac{d\tau}{\tau} = \frac{7}{4}\frac{d\hat{\omega}}{\hat{\omega}}, \qquad (140)$$

for the condition that P_0 shall be constant.

Substituting this value of $\frac{d\tau}{\tau}$ in (138),

$$\frac{dv}{v} = \frac{7}{4}\left(\frac{5}{14} - n\right)\frac{d\hat{\omega}}{\hat{\omega}}. \qquad (141)$$

When $x = 0.6$, $n = \frac{1}{4}$ (see 135*a*); and for larger values of x, n becomes smaller; therefore for all cases in practice the second member of (141) is positive, and v increases with $\hat{\omega}$.

Hence (141) shows that when the powder-chamber is fixed, we may increase the velocity by increasing the weight of the charge, but (140) shows that in order to keep the pressure constant we must use a slower powder. That is, we use a larger charge of slower powder.

37. Relative Variation of Velocity and Time of Combustion—Of Velocity and Maximum Pressure—Limits of Modulus—Useful Practical Formulas.

VELOCITY AND TIME OF COMBUSTION.—To determine the relative change in velocity for a given change in the time of combustion, suppose τ to be the only variable in equation (132). Then

$$d\hat{\varpi} = 0; \quad d\Delta = 0;$$

and the equation becomes

$$\frac{dv}{v} = -n\frac{d\tau}{\tau}. \quad \ldots \ldots \quad (142)$$

As powder becoms slower x decreases. But as x decreases n increases (see 135a). In fact, n may be called the "modulus of slowness," since it increases as the powder becomes more slow, while x, or the "modulus of quickness," increases as the powder becomes quicker. From (142) it is evident that for the same relative change in the time of burning the effect upon the velocity will be greater as the powder becomes more slow, since n becomes greater.

This is one of the principal objections to using very slow powder, because small irregularities of manufacture, which are always apt to occur, affect τ, the time of burning, and cause irregularities in velocity.

VELOCITY AND MAXIMUM PRESSURE.—In the same way, to determine the relative change of velocity and maximum pressure, suppose τ the only variable in equation (133).

Then

$$d\hat{\varpi} = 0; \quad d\Delta = 0;$$

and the equation becomes

$$\frac{dP_0}{P_0} = -\frac{d\tau}{\tau}. \quad \ldots \ldots \quad (143)$$

Substitute this value of $\frac{d\tau}{\tau}$ in (142) and we have

$$\frac{dv}{v} = n\frac{dP_0}{P_0}. \quad \ldots \ldots \quad (144)$$

From this we see, generally, that with quick powders, since n is small, a given increase of pressure gives only a very small increase of velocity, and for slow powders, since n is large, a given increase of pressure gives a considerable increase of velocity. Hence it is advantageous on this account to use slow powders.

LIMITS OF THE MODULUS.—The preceding considerations may be applied in fixing the inferior limit of the modulus as follows:

Equation (144) shows that as x decreases, or n increases, a given increase in the pressure will give a considerable increase in the velocity, and hence it appears to be advantageous to use a slow powder, for which n is large. But (142) shows that for large values of n we have great irregularities in v, as previously explained, and Sarrau has fixed upon 0.6 as the value of x below which it is not expedient to go in practice in order to avoid these irregularities.

For the superior limit, when $x = \frac{9}{11}$ or $n = \frac{1}{8}$, equation (144) shows that the relative increase of velocity is only one eighth that of the maximum pressure; and since the monomial formula was deduced for this value, it was formerly regarded as the superior limit of the modulus. Other considerations, however, have led to an increase of this value up to 1.2 for some powders.

USEFUL PRACTICAL FORMULAS.—In practice it is frequently required to find what change in velocity and pressure a given change in weight of charge will produce in the same gun. For this purpose assume the monomial formula for velocity

$$v = M\alpha,\beta^{-\frac{1}{4}} \frac{\hat{\varpi}^{\frac{3}{8}} \Delta^{\frac{1}{4}} c^{\frac{1}{8}} u^{\frac{3}{16}}}{p^{\frac{7}{16}}}.$$

For any other charge of the same powder whose weight is $\hat{\varpi}'$, all the quantities in the formula will remain constant except v, $\hat{\varpi}$, and Δ. The value of Δ is

$$\Delta = \frac{27.68\hat{\varpi}}{C} = K\hat{\varpi},$$

K being a constant. Raising both members to the one-fourth power and multiplying by $\hat{\omega}^{\frac{1}{4}}$,

$$\omega^{\frac{1}{4}}\Delta^{\frac{1}{4}} = K^{\frac{1}{4}}\hat{\omega}^{\frac{1}{4}}.$$

Dividing the value of v by that of v', we have

$$\frac{v}{v'} = \frac{\hat{\omega}^{\frac{3}{4}}}{\hat{\omega}'^{\frac{3}{4}}}. \quad . \quad . \quad . \quad . \quad . \quad . \quad . \quad . \qquad (145)$$

Similarly, for the pressures we have

$$P_0 = K_0 \alpha^2 \Delta p^{\frac{1}{2}} \hat{\omega}^{\frac{3}{4}} c^{-2};$$
$$\Delta \hat{\omega}^{\frac{1}{4}} = K \hat{\omega}^{\frac{3}{4}}.$$

Dividing as before, we have

$$\frac{P_0}{P_0'} = \frac{\hat{\omega}^{\frac{7}{4}},}{\hat{\omega}'^{\frac{7}{4}}}. \quad . \quad . \quad . \quad . \quad . \quad . \quad . \quad . \qquad (146)$$

These formulas are correct for quick powders and approximately correct for slow ones. The velocity formula is useful where it is necessary to find the charge required to give a certain velocity to a projectile at a target at reduced range, as in armor-plate experiments.

38. Pressure Curves in Guns—Noble and Abel's Method—Mayevski's Method.

It is necessary in designing a gun to know the pressures at different points along the bore, as the projectile moves through it, under the action of the powder-gas, in order that the gun may be given sufficient strength to withstand these pressures.

The accurate solution of this problem is attended with great difficulties, and can hardly yet be said to have been successfully accomplished. Enough is known, however, to furnish safe working limits in designing the strength of the gun at different points.

NOBLE AND ABEL'S METHOD.—They assumed an expression of the form

$$x = at^{\alpha + \beta t + \gamma t^2}, \quad . \quad . \quad . \quad . \quad . \quad . \quad . \quad (147)$$

in which x is the distance travelled by the projectile, t the

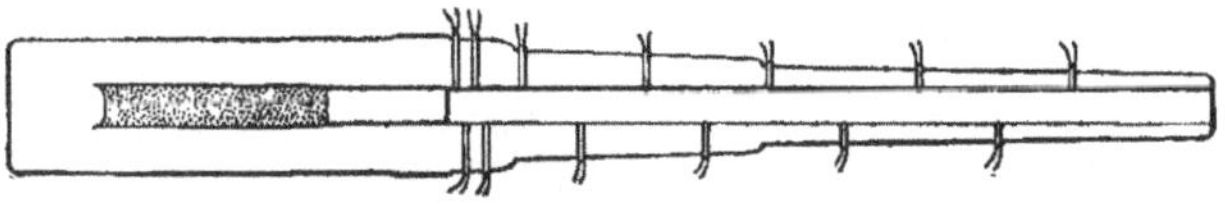

FIG. 9.

corresponding time, and a, α, β, and γ constants to be determined by experiment.

Wires were inserted into a gun through holes drilled at short intervals, as shown in Fig. 9. These wires carried currents of electricity, which were broken by the projectile in its passage through the bore, and these breaks were recorded on the Noble chronoscope, which is an instrument for measuring very small intervals of time.

The distance between the holes gave x, and the record of the chronoscope t, and substituting the values thus obtained in (147), the most probable values of the constants were determined by the method of least squares.

Differentiating (147) with respect to t gives

$$\frac{dx}{dt} = v, \quad . \quad . \quad . \quad . \quad . \quad . \quad . \quad . \quad . \quad (148)$$

and differentiating this with respect to t gives

$$\frac{d^2x}{dt^2} = \frac{dv}{dt}, \quad . \quad . \quad . \quad . \quad . \quad . \quad . \quad . \quad (149)$$

the acceleration.

If W be the weight of the projectile and P the total pressure on its base, we have

$$P = \frac{W}{g}\frac{d^2x}{dt^2}. \quad . \quad . \quad . \quad . \quad . \quad . \quad . \quad (150)$$

MAYEVSKI'S METHOD.—General Mayevski assumed

$$x = At + Bt^2 + Ct^3 + Dt^4 + \text{etc.}, \quad . \quad . \quad . \quad (151)$$

and by experiment determined t, having x given by the nature of the experiments (see Fig. 10). His general plan resembles that of Noble and Abel, but differs in the method of conducting the experiments.

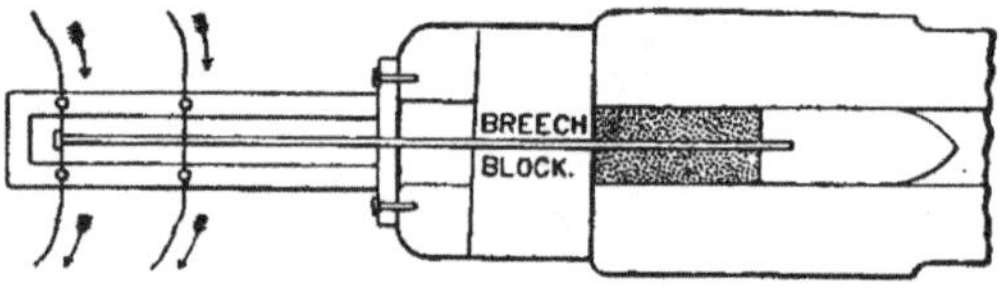

FIG. 10.

From these results values of A, B, C, and D were determined by the method of least squares.

Differentiating (151) with respect to t, we have

$$v = \frac{dx}{dt} = A + 2Bt + 3Ct^2 + 4Dt^3 + \text{etc.}, \quad . \quad (152)$$

for the velocity at any point.

From (152) we have

$$\frac{dv}{dt} = \frac{d^2x}{dt^2} = 2B + 6Ct + 12Dt^2 + \text{etc.}, \quad . \quad . \quad . \quad (153)$$

for the acceleration; and for the point where this is a maximum we have, from (153),

$$\frac{d^3x}{dt^3} = 6C + 24Dt + \text{etc.} = 0. \quad . \quad . \quad . \quad . \quad (154)$$

39. Longridge's Method.

Mr. Longridge, an English engineer, uses a combination of Noble and Abel's and of Sarrau's formulas as follows:

Noble and Abel's formula for the pressure curve is, see (45),

$$p = 43\left(\frac{.43v'}{v'' - .57v'}\right)^{1.074}, \quad . \quad . \quad . \quad . \quad (155)$$

in which p is the pressure corresponding to the volume v'', v' is the volume of the powder-chamber supposed to be entirely filled with powder, and v'' any other volume of expansion.

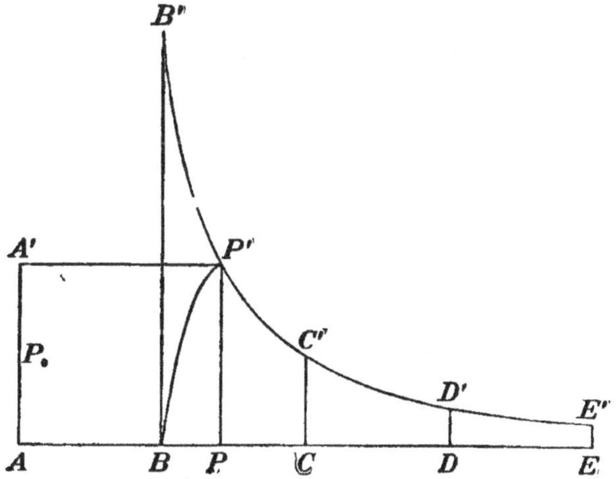

FIG. 11.

Let AB, Fig. 11, represent the reduced length of the powder-chamber, that is, the length of a cylinder whose diameter is that of the bore and whose volume is that of the powder-chamber.

Suppose, according to Noble and Abel's hypothesis, that all the powder is burned before the projectile moves from its position B. Make $v'' = v'$ in (155), and we have $p = 43$ tons.

This is the pressure that would exist in the chamber if the powder were all burned before the projectile moved.

Lay off $BB' = 43$ tons. Assume different values for v'', corresponding to C, D, E, etc. Calculate the corresponding ordinates from (155), and erect them at the corresponding points.

The resulting curve $B'C'D'E'$ will be Noble and Abel's pressure curve.

Now it is known that the powder is not all burned before

the projectile moves, and hence the pressure $BB' = 43$ tons cannot exist in a gun.

The maximum pressure that does exist is given by Sarrau's formula for the maximum pressure on the breech.

$$P_0 = K_0 \alpha^2 \Delta p^{\frac{1}{2}} \varpi^{\frac{1}{2}} c^{-2}.$$

Calculate this pressure, lay off AA' at the breech equal to it, and assume this pressure to be uniform from the breech to the point of maximum pressure in the bore. Substitute this value of P_0 for p in (155) and find the corresponding value of v''. This value of v'' will give the point P in the bore at which the maximum pressure occurs. The line $A'P'$ will be parallel to AP, and the curve of pressures will be $A'P'C'D'E'$.

Let $x_1 = AB$, the reduced length of the powder-chamber; $x =$ any other length measured from A. Then

$$v' = \pi r^2 x_1;$$

$$v'' = \pi r^2 x;$$

and equation (155) becomes

$$p = 43 \left\{ \frac{.43}{\frac{x}{x_1} - .57} \right\}^{1.074}, \quad \ldots \ldots \quad (156)$$

which is more convenient for use.

The pressure at B is originally zero and rises to a maximum at P'. Hence the actual pressure curve has the form $BP'C'D'E'$.

The form of the curve from B to P' is not important, as its maximum ordinate only is required.

40. Pressure Curve by Sarrau's Formulas.

The pressure curve in a gun may also be obtained from Sarrau's velocity formulas, as follows:

For a slow powder we have

$$v = A\alpha\left(\varpi u\right)^{\frac{1}{2}}\left(\frac{\Delta}{pc}\right)^{\frac{1}{4}}\left[1 - B\beta\frac{(pu)^{\frac{1}{2}}}{c}\right]. \quad \ldots \quad (156a)$$

In (156*a*) u is expressed in inches. In (161), following, v and g are in feet, and u must be expressed in feet in order that, when du is substituted in (161), all the quantities may have the same unit.

Hence in (156*a*) we write

$$u^{\frac{1}{4}} \text{ inches} = 12^{\frac{1}{4}} u^{\frac{1}{4}} \text{ feet};$$

and in the subtractive term

$$u^{\frac{1}{2}} \text{ inches} = 12^{\frac{1}{2}} u^{\frac{1}{2}} \text{ feet}.$$

Place

$$v = au^{\frac{1}{4}}(1 - bu^{\frac{1}{2}}), \quad . \quad . \quad . \quad . \quad . \quad . \quad . \quad (157)$$

in which u is in feet, and

$$a = \frac{A\alpha\widehat{\varpi}^{\frac{1}{4}}\Delta^{\frac{1}{2}} \times 12^{\frac{1}{4}}}{(pc)^{\frac{1}{2}}}, \quad . \quad . \quad . \quad . \quad . \quad . \quad (158)$$

$$b = \frac{B\beta p^{\frac{1}{2}} \times 12^{\frac{1}{2}}}{c}. \quad . \quad . \quad . \quad . \quad . \quad . \quad . \quad (159)$$

From mechanics we have

$$P' = m\frac{dv}{dt};$$

$$du = vdt; \qquad \therefore dt = \frac{du}{v}.$$

Substituting in the value of P', we have

$$P' = m\frac{vdv}{du} = \frac{p}{g}\frac{vdv}{du}, \quad . \quad . \quad . \quad . \quad . \quad . \quad (160)$$

p being the weight of the projectile, and m its mass; and if ω denote the area of the base of the projectile in square inches, we have for the pressure in tons per square inch on its base,

$$P'' = \frac{p}{2240\omega g}\frac{vdv}{du}. \qquad (161)$$

Differentiating (157), we have

$$\frac{dv}{du} = \frac{3}{8}au^{-\frac{5}{8}} - \frac{7}{8}abu^{-\frac{1}{8}},$$

and

$$\frac{vdv}{du} = \frac{3}{8}a^2u^{-\frac{1}{4}} - \frac{5}{4}a^2bu^{\frac{1}{4}} + \frac{7}{8}a^2b^2u^{\frac{3}{4}}. \qquad (162)$$

Hence

$$P'' = \frac{p}{2240 \times \omega \times g}\left[\frac{3}{8}a^2u^{-\frac{1}{4}} - \frac{5}{4}a^2bu^{\frac{1}{4}} + \frac{7}{8}a^2b^2u^{\frac{3}{4}}\right]. \qquad (163)$$

For quick powders we have

$$v = \frac{M\alpha\beta^{-\frac{1}{4}}\bar{\omega}^{\frac{3}{8}}\Delta^{\frac{1}{4}}c^{\frac{1}{4}}u^{\frac{3}{16}}}{p^{\frac{7}{16}}};$$

which may be placed in the form

$$v = a'u^{\frac{3}{16}}, \qquad (164)$$

u being in feet as before explained, and

$$a' = \frac{M\alpha\beta^{-\frac{1}{4}}\bar{\omega}^{\frac{3}{8}}\Delta^{\frac{1}{4}}c^{\frac{1}{4}} \times 12^{\frac{3}{16}}}{p^{\frac{7}{16}}}. \qquad (165)$$

Differentiating (164), we have

$$\frac{dv}{du} = \frac{3}{16}a'u^{-\frac{13}{16}}, \qquad (166)$$

and

$$\frac{vdv}{du} = \frac{3}{16}a''^2u^{-\frac{1}{4}}. \quad . \quad . \quad . \quad . \quad . \quad . \quad (167)$$

Hence

$$P'' = \frac{p}{2240 \times \omega \times g}\left(\frac{3}{16}a''^2u^{-\frac{1}{4}}\right). \quad . \quad . \quad . \quad (168)$$

Since these curves are obtained on the adiabatic hypothesis, they may be considered as marking the inferior limit of pressures, and the true pressure curve probably lies between the latter and those obtained by Longridge's method.

It must be remembered that the pressures given by these equations (163) and (168) are those producing motion of the projectile, and do not represent the total pressure on the base.

41. Determination of Velocity by Experiment—General Principles—Targets for Cannon—For Small Arms.

In order to verify the formulas for velocity and pressure previously deduced, it is necessary to determine accurately by experiment the velocity of the projectile, and the pressure in the gun, due to a given charge of powder, under given conditions of loading.

VELOCITY.—The velocity of a projectile is determined by measuring the time of its passage over a given distance.

Let A and B be two points whose distance apart is s, and t the time of passage of the projectile over this distance. Then, since

$$v = \frac{s}{t},$$

v will be the mean velocity of the projectile over the distance s, or its velocity at the middle point between A and B. In order that this may be true, the space s must be so small that the motion of the projectile may be considered uniform and in a right line. As neither of these conditions holds in practice, v will not be the velocity at the middle point between A and B, but it will be sufficiently correct for all practical purposes to assume that it is.

GENERAL PRINCIPLES.—In order that we may know exactly the time of passage of the projectile over the distance s, we must have first an accurate scale of time, and second, we must mark on this scale the instant that the projectile passes the two points A and B. The difference between the times of passage of the points A and B will then give the time of passage over the distance s, and knowing this time we can find the mean velocity. The passage of the projectile over the points A and B is marked on the time-scale as follows:

Two targets are set up, one at A and the other at B. Electric currents circulate through these targets, and also through the instrument which furnishes the scale of time.

When the projectile passes the target at A it breaks the circuit, and this break is registered by the instrument. When it passes B the same thing occurs. The difference between these breaks measured on the scale of time, and corrected for errors, gives the time of passage required.

TARGETS FOR CANNON.—The functions of the targets are then to mark the points in the path of the projectile between which its velocity is to be measured, and to support the wires carrying the electric currents which are to be broken by the passage of the projectile. For cannon, the first target is placed at such a distance from the muzzle that it will not be injured by the blast. Call this distance x_1, and the distance from the muzzle to the middle point between the targets x; then

$$x = x_1 + \frac{s}{2}.$$

The velocity found by experiment is at the point $\frac{s}{2}$; that which we wish to find is at the muzzle, or the initial velocity. By formulas in "Exterior Ballistics" we can find the latter, when the former, at the distance x, is known.

Each target for cannon generally consists of a frame of wood carrying a number of small parallel copper wires.

The wires are so arranged that the current entering one side of the target must traverse all of them before passing out at the other side, so that the breaking of any wire will break the current. The wires are drawn as tight as possible in order that the break may be abrupt, and the distance between them depends on the diameter of the projectile, as it must be impossible for it to pass through without breaking at least one wire. The breaks are repaired after each fire.

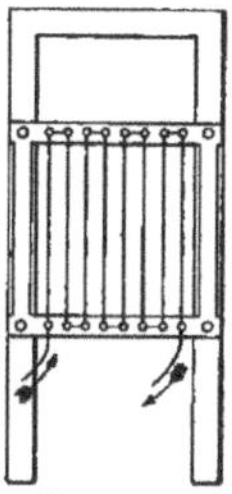

FIG. 12.

The target is shown in Fig. 12.

SMALL-ARM TARGETS.—As there is practically no blast with small arms, the first target is placed at the muzzle, and consists of a single wire drawn tightly across it. To avoid repairing the second target, it consists of a steel plate to stop the bullets. On its rear face is secured a spring insulated from the plate (Fig. 13). This spring, *s*, is fixed at one end to an insulating substance, such as a block of wood, *w*, and the other end rests on a metallic pin, *p*. The current passes through the spring and pin. When the bullet strikes the steel plate, the shock causes the spring to leave the pin, and thus the current is broken. The elasticity of the spring causes it instantly to resume its former contact with the pin, and thus renders any repairs unnecessary.

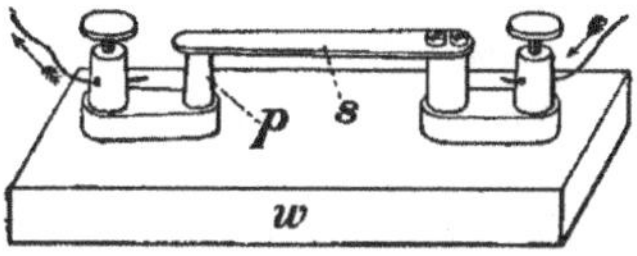

FIG. 13.

42. The Ballistic Instruments—Description of the Le Boulengé Chronograph.

The functions of the ballistic instruments are to furnish an accurate scale of time, and to record on that scale the rupture of the targets by the passage of the projectile.

LE BOULENGÉ CHRONOGRAPH.—The instrument generally used for this purpose was invented by Captain Le Boulengé of the Belgian Artillery, and is called the Le Boulengé Chronograph.

Scale of Time.—Its scale of time is furnished as follows:

Two rods are suspended vertically from electro-magnets, and the currents which pass through the magnets pass also through the targets. Each electro-magnet has its own current and its own target, and is independent of the other.

When the first target is broken, one of the magnets is demagnetized and its rod falls. When the second target is broken, its magnet is demagnetized and its rod falls. The fall of this second rod makes a mark on the first rod while it is falling, and the distance h' from the origin of fall to this mark is measured. Then we have, from the laws of falling bodies,

$$h' = \tfrac{1}{2}gT^2; \quad \therefore\ T = \sqrt{\frac{2h'}{g}},$$

which furnishes the scale of time.

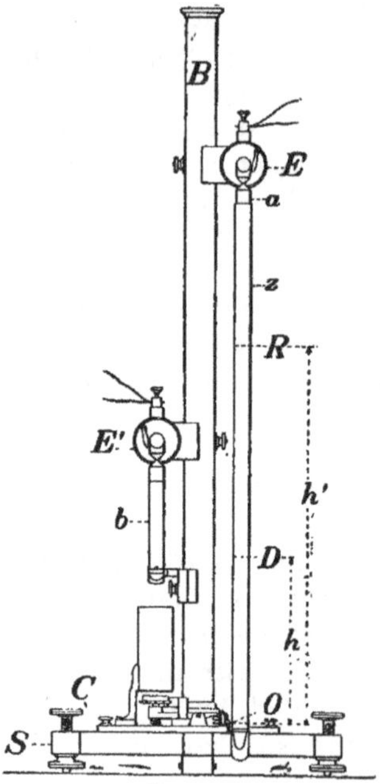

FIG. 14.

Record of Breaking of Target—Description of Instrument.—The method of making the record will appear from a description of the instrument, Fig. 14.

Its principal parts are a vertical column of brass, B, which is supported by a triangular bed-plate, C, and this bed-plate rests upon a support or stand, S. To the brass column are attached two electro-magnets, EE'. The magnet E supports the long rod a of the instrument, called the *chronometer*. This rod when in use is enveloped by a zinc or copper tube, z, called the *recorder*, upon which the mark is made. The magnet E' supports the short rod b, called the *registrar*.

Fig. 15 shows the details of the marker, or part of the

instrument which makes the record of the breaking of the target. It consists of a circular knife, *m*, on the end of a spring, *s*, which causes it to move to the right in the figure. The trigger, *t*, is supported in its fulcrum on the bed-plate. Its right end terminates in a catch, which engages in a corresponding one on the knife, and prevents the latter from moving to the right, under the action of the spring, till the catch is freed. The left end of the lever is acted on by a spring *s'*, which presses it upwards, and keeps the catch engaged with the knife. The piece marked *b* is a disk which screws into the left-hand end of the lever, and which may be raised or lowered by means of the screw.

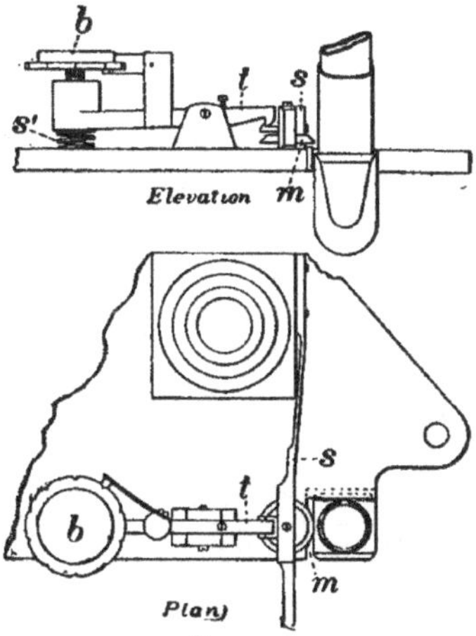

FIG. 15.

Above the disk is a tube or cup which retains the short rod *b* after its fall. The record is made by the short rod falling on the disk *b*, depressing it, and releasing the knife *m* from the catch on the trigger. The knife then moves to the right and, striking the long rod in its fall, makes the required record.

43. Arrangement of Wires—Working of Instrument—Disjunction.

ARRANGEMENT OF WIRES.—The arrangement of the wires depends upon whether the time to be measured is comparatively great or small. When great, the wires are arranged as follows, Fig. 16.

The chronometer is supported by the upper magnet. The first current comes from the battery to the upper magnet, *E*; from the magnet *E* to the disjunctor, whose functions will be explained later; from the disjunctor to the first target, and from the first target to the battery. The course of the second current is similar and can be readily followed.

The instrument thus arranged is called a megagraph.

WORKING.—When the first target is broken, the chronometer falls; when the second target is broken, the registrar

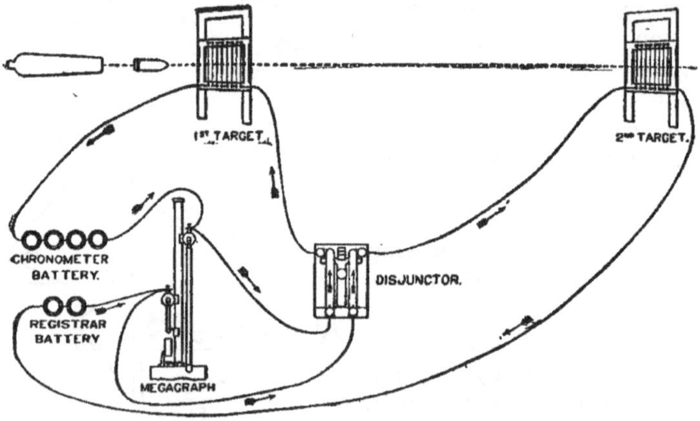

FIG. 16.

falls, and, striking the disk of the trigger, makes the record on the chronometer, as at R, Fig. 14.

The point on the chronometer from which all heights are measured is the mark O, Fig. 14, made on this rod by the knife when the chronometer is suspended by its magnet. Denoting the height OR by h', the corresponding time is

$$T = \sqrt{\frac{2h'}{g}},$$

and is the time which elapses from the fall of the chronometer till the record is made. It is not, however, the time of passage of the projectile between the targets, because—

1. There is a certain time required for the demagnetization of the magnet E. Hence the chronometer does not fall at the instant the first target is broken, and the time is too short by this amount, which we call t_1.

Instead of making the record the instant the second target is broken, there is a delay caused by—

2. The time required to demagnetize $E' = t_2$.

3. The time required for the registrar to fall to the disk of the trigger $= t_3$.

4. The time necessary to disengage the trigger, and for the knife to move forward and make the record on the chronometer $= t_4$.

During these last three intervals the long rod is falling, and hence the height of fall, and consequently the time, is too great by their sum. Hence the true time is

$$t = T - (t_2 + t_3 + t_4) + t_1,$$

and to find the true value of t it is necessary to find the values of these times, since T is known.

To do this it is not necessary to find the value of each single interval, since the total time can be readily obtained. If we break both currents at the same instant, it is evident that all the delays mentioned will still exist. The delay in falling of the chronometer, and that of making the record by the registrar, will be marked on the long rod as it falls, and will be found at a certain height above O, as at D, Fig. 14. This height is called "the disjunction," and the time corresponding to this height is the algebraic sum of all the times before named.

Let

$$\theta = (t_2 + t_3 + t_4) - t_1.$$

Then

$$\theta = \sqrt{\frac{2h}{g}},$$

in which h is the height OD. Hence

$$t = T - \theta = \sqrt{\frac{2h'}{g}} - \sqrt{\frac{2h}{g}}.$$

It must be remembered that difference of *times* and not difference of *heights* is to be taken.

FIXED DISJUNCTION.—For the velocity at the middle point between targets we have

$$v = \frac{s}{t}.$$

Substituting for t its value, we have

$$v = \frac{s}{\sqrt{\frac{2h'}{g}} - \sqrt{\frac{2h}{g}}}.$$

From this equation we see that if the values of s, and of $\sqrt{\frac{2h}{g}}$ or the disjunction, be fixed, the values of v can be calculated and tabulated for all values of h' within the limits of practice. This has been done for the values $s = 100$ feet and $\sqrt{\frac{2h}{g}} = 0.15$ second, and this value of $\sqrt{\frac{2h}{g}}$ is called the fixed disjunction. If the above table is not at hand, this fixed value of the disjunction avoids the labor of calculating θ or $\sqrt{\frac{2h}{g}}$ for each shot.

Hence in this case

$$t = T - 0.15 \text{ sec.} = \sqrt{\frac{2h'}{g}} - 0.15.$$

To fix the disjunction, the disk b on the trigger t may be raised or lowered to regulate the height of fall of the registrar till $\theta = 0.15$ sec.

44. Arrangement of Wires for Small Times—Disjunctor—Measuring-rule.

ARRANGEMENT OF WIRES.—Under ordinary conditions, the distance between targets is so great, that the chronometer acquires considerable velocity in falling, before the record is made by the registrar. As the distance between targets decreases there will be less interval between the breaking of the two currents, and consequently between the fall of the two rods. Hence the record will be made before the chronometer has acquired much velocity, and small differences in reading will correspond to considerable differences in time. A small error in reading, therefore, will correspond to a large error in time. As the distance between targets decreases, the record will approach the disjunction circle, and will fall on that circle when the distance is zero, or when both currents are broken simultaneously. To measure these short intervals of time accurately it is necessary to allow the chronometer to acquire considerable velocity before the record is made upon it.

This necessitates a new arrangement of wires and magnets, as in Fig. 17. The magnet which supports the registrar is changed from below to above that which supports the chronometer, and the first current runs from the battery to the registrar magnet, thence to the disjunctor and to the first target, so that the registrar will fall first. With this arrangement, if both currents be broken simultaneously, the "disjunction" will be made near the top of the chronometer at *D*, when its velocity is greatest. When the registrar falls first, as it does in actual use in determining velocity, it is evident that the record will be made at some point below the disjunction, as at *R*. The same method is followed in determining the time as before, except that the time corresponding to the record must be subtracted from that corresponding to the disjunction, for the time of passage between targets. The instrument thus arranged is called a micrograph.

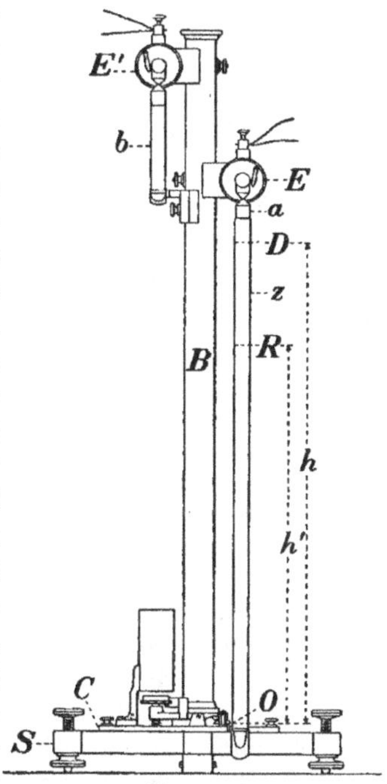

FIG. 17.

DISJUNCTOR.—This instrument is used for breaking both

currents simultaneously in order to determine the algebraic sum of all the times t_1, t_2, t_3, etc.

It consists (Fig. 18) of two steel blades, *n n'*, mounted on

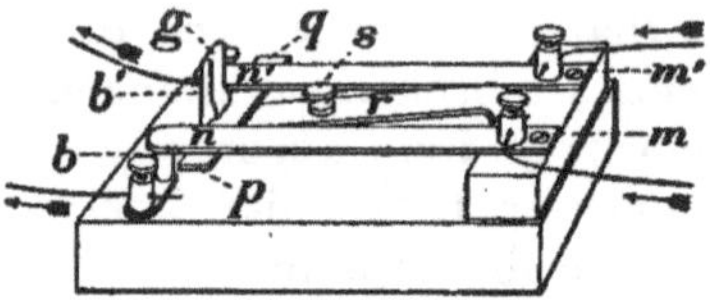

FIG. 18.

a block of wood. These blades are attached at one end, *m m'*, to the block, and carry binding-screws at this end. At the other end they rest on two brass pins, *b b'*, and these pins are connected with the binding-screws shown. Between these blades is a strong spring, *r*, with a knob, *s*, and a spring-catch, *g*. At right angles to, and attached to this spring *r* is a cross-piece, *pq*. The action is as follows:

When the spring *r* is pressed down by pushing on the knob *s*, it is caught and held under the spring-catch *g*, and the cross-piece is not in contact with the blades *n n'*.

Under these circumstances the blades rest on their pins *b b'*, and the current from each battery enters its own blade by the binding-screws and posts, and passes to its target.

But when the trigger or catch *g* is pulled back quickly, the spring *r* is released, and, rising, its cross-piece *pq* strikes both blades *n n'* at the same instant, lifts them from the pins *b b'*, and breaks both circuits.

MEASURING-RULE. — To facilitate measurements, the heights corresponding to all velocities within the ordinary limits of experiment are inscribed on a metal rule furnished with a sliding index. The heights are in millimetres, and must be reduced to feet for use with English measures. A table of times corresponding to heights in millimetres has been calculated, and by its use the above reduction may be avoided. The sliding index has a knife-edge attached to it, and, to obtain the reading, this knife-edge is placed on the mark made by the marker on the chronometer, a pin on the lower part of the scale having been inserted in a hole in the chronometer at the lower end to bring the zero point of the scale opposite the origin of fall. The height can then be read off.

45. Adjustments—Use—Objections to the Instrument—Bréger's Improvements.

ADJUSTMENTS.—The instrument must be properly mounted on a stand at such a distance from the gun that it will not be affected by the shock of discharge, and connected with the batteries and targets, and be then adjusted for use.

The adjustments are three:

1. *Levelling.*—The object of this adjustment is to make the bed of the instrument level, and consequently the brass column or standard vertical. The chronometer is used for this purpose. The enveloping tube or recorder is first put on, and when in place must rest closely against the bob. Having cocked the knife, suspend the chronometer and recorder from its magnet, and move the levelling-screws which pass through the bed-plate, till the bob of the chronometer rests in a square notch in the bed-plate. The standard is now vertical,

2. *Regulating the Magnets.*—To regulate the strength of the magnets, each of the rods is provided with a weight which is one tenth that of its rod. Place the proper weight on the chronometer, and suspend it with this weight from its magnet, the core of which is movable, and draw out this core till the rod and weight fall. The strength of the magnet is by this means regulated. Do the same for the short rod, or registrar, and its magnet.

3. *Fixing the Disjunction Reading.*—For the megagraph this reading is at a fixed height, corresponding to 0.15 second. To make the adjustment, place the sliding index on the rule at the mark "disjunction," and clamp it.

Place the pin of the rule in the hole in the bob of the chronometer, bring the knife-edge of the rule to bear against the copper tube on the chronometer, and turn this tube around the chronometer. The knife-edge will describe a circle on this tube, called the "disjunction circle," and the disjunction reading must fall on this circle. To test it, suspend both rods from their magnets, and break both currents by means of the disjunctor. If the mark made by the knife falls on the circle, no adjustment is necessary; if above the circle, the fall of the registrar is too great; and if below, the fall is too small. The height of fall of the registrar is diminished or increased by raising or lowering the disk on the left-hand end of the trigger *t*. The instrument is now ready for use.

USE.—In using it, first cock the knife or marker, suspend the long and short rods, and take a disjunction reading. If the disjunction is not exact, correct as above. If exact, cock the knife again, suspend the rods, fire the piece, and read the height with the rule. Find the time corresponding to this reading, subtract from it the time corresponding to the disjunction, which is 0.15 second when the instrument is used as a megagraph; or if used as a micrograph, subtract the time corresponding to this height from that corresponding to the disjunction, as previously determined, and the remainder will give the time of passage of the projectile between targets. Divide the distance between the targets in feet by this time in seconds, and the quotient will be the velocity of the projectile at a point midway between the targets.

OBJECTIONS TO THE INSTRUMENT.—The principal source of error in the Le Boulengé arises from the fact that the circuits are not broken similarly by the disjunctor and by the projectile.

When the circuits are broken by the projectile, the retardation of demagnetization is modified, and unequally so for the two magnets, because they sustain different weights and are consequently of different strength.

BRÉGER'S IMPROVEMENTS.—This has led to modifications of the instrument by Captain Bréger (Fig. 19). The principal of these are, the two rods are made of exactly the same weight, and consequently the electro-magnets are of the same strength. Their axes are vertical instead of horizontal. The parts generally are heavier and more firmly supported.

The height of fall of the registrar is regulated by raising

or lowering its magnet, E', and the disk of the trigger on which the registrar strikes is fixed with reference to the lever. The knife is square instead of circular. The disjunctor has been modified so as to insure the simultaneous rupture of the two circuits, and the strength of the currents is regulated by resistance-coils. These improvements render the instrument much more accurate than the old form.

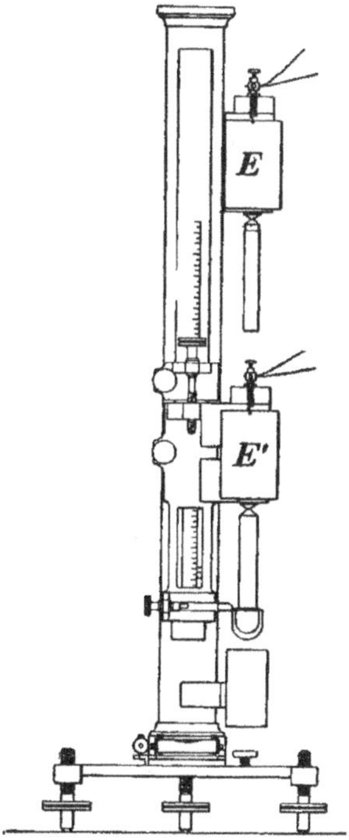

FIG. 19.

46. Schultz Chronoscope — Marcel-Deprez Registers—Bashforth Targets.

SCHULTZ CHRONOSCOPE.—The Le Boulengé Chronograph measures velocity at one point only. If the velocity of a projectile is to be measured at several points, a separate instrument is required for each point, and this arrangement would be troublesome, besides having other objections.

It is frequently necessary to measure the velocity of the same projectile at different points, as in determining the laws of the resistance of the air to its motion, and also it is sometimes required to determine its velocity at different points in the bore. For such purposes an instrument must be used which will give a scale of time of such an extent that all the phenomena may be registered upon it.

There are several instruments of this class, and as a type of them the Schultz chronoscope, one of the best known, will be briefly described (Fig. 20).

Scale of Time.—In this instrument a cylinder *a* revolves by means of clockwork, and this cylinder has also, in the older form of machines, a motion of translation parallel to its axis. In the most recent form the cylinder rotates only, while the point *b*, which describes the scale, has the motion of translation.

The point *b* is a quill attached to one branch of a tuning-fork, *c*. This point may be made to rest lightly against the surface of the cylinder, or may be withdrawn from contact with it. On each side of the tuning-fork is an electro-magnet, *d*.

The object of the magnets is to start the fork vibrating, to keep up this vibration during the experiment, and to

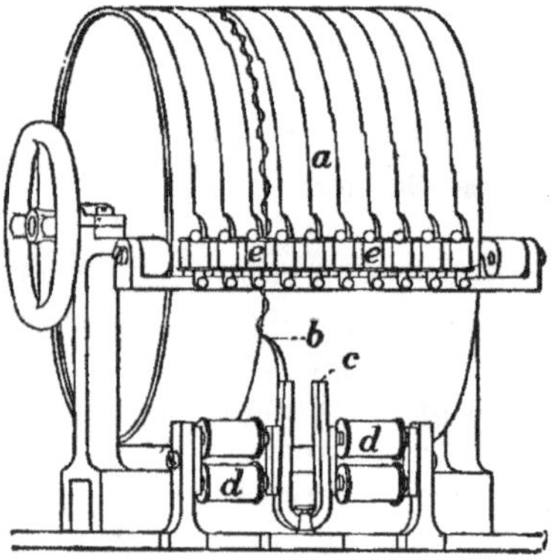

FIG. 20.

equalize the amplitude. The surface of the cylinder is covered with lampblack before using. When the quill is placed in contact with the coated cylinder, the latter is set to rotating and the fork to vibrating. The quill-point will then trace on the cylinder a sinusoidal curve, and, the number of vibrations of the fork per second being known, we have an accurate scale of time. If the time to be measured is greater than can be registered on one revolution of the cylinder, the cylinder or fork is given a motion of translation along the axis, and the sinusoidal curve then becomes a helix, and the whole length of the cylinder can be used.

MARCEL DEPREZ REGISTERS.—The record is made as follows: Small electro-magnets, *ee*, Fig. 21, are placed in front of the cylinder, above the time-register. They are provided with very light armatures, *f*, acted on by a spring, *g*, which is almost in equilibrio with the magnetic attraction. A point, *h*, connected with the armature, rests against the surface of the cylinder. When the current at a target is broken, the corresponding armature *f* yields to the action of the spring *g* and is drawn aside quickly, the point *h* recording the motion on the cylinder by the side of the time-scale. The number of vibrations of the fork between any two breaks divided by the number per second gives the corresponding time. To assist in counting vibrations, the quill *b*, Fig. 20, is first allowed to trace a simple helix before the fork is put in vibration. The quill-point is then returned to its starting-point. This line is called the mean helix. If the targets are at such a distance apart that the current which is broken at one point may be restored before the projectile reaches the next, one register and one circuit will be sufficient. If the targets are too close together for this restoration of current, each target must have its own current and register. The registers have a motion of translation in common with that of the tuning-fork.

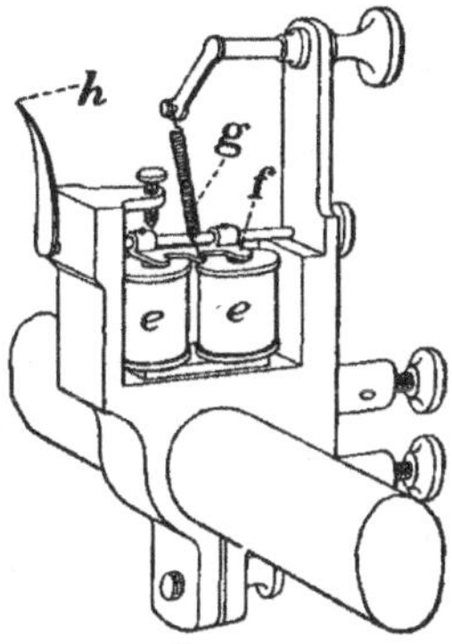

FIG. 21.

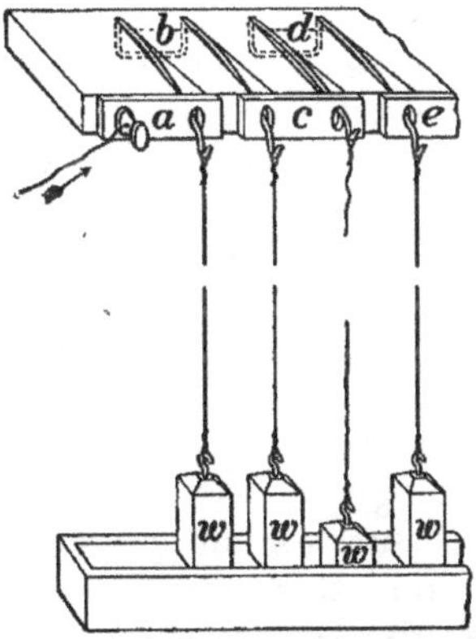

FIG. 22.

BASHFORTH TARGETS. — For restoring the current as above described, the simplest device is the Bashforth target, invented and used by the Rev. Francis Bashforth in his celebrated experiments on the resistance of the air to the motion of projectiles.

This target (Fig. 22) consists of a series of wire springs, *bd*, inserted in a board. On the front of this board are brass plates, *a c e*, having oblong holes in them through which the springs pass.

The springs are held down in contact with the lower side of the holes by weights, *w w*, attached to them by strings. The current entering the plate *a*, will pass through the wire spring *b* to the plate *c*, and so on. When one of the strings is cut by a projectile, the corresponding spring will fly up to the upper side of the hole in the brass plate *c*, and the current will be broken during the passage of the spring from bottom to top of hole, and will be made again as soon as the spring strikes the top.

47. Determination of Pressures by Experiment—Static Method—Discussion—Conclusions.

There are two methods of measuring a force or pressure:

1. *The Static Method*, in which the unknown force is balanced by a known resistance;

2. *The Dynamic Method*, in which the unknown force is determined by the acceleration which it communicates to a given mass. Its measure is, from mechanics,

$$P = m\frac{dv}{dt} = m\frac{d^2s}{dt^2}.$$

STATIC METHOD.—*General Principles.*—The general method adopted in this case is to balance the unknown force by the resistance which a body offers to deformation. If we have a cylinder of metal of known length and diameter, and uniform in quality, and apply to it a known force in the direction of its length, the cylinder will be decreased in length by a certain amount. We measure accurately this decrease in length and note the force producing it. Proceeding in this manner we can form a table one column of which will contain the decrease in length of the cylinder, and the other the corrresponding pressure for all pressures within the limits of experiment. From this table a curve may be constructed whose abscissas give the pressures, and the ordinates the corresponding compressions.

Such a curve is called the "tarage" of the cylinder.

If now a cylinder of the same material and dimensions be subjected to the force to be measured, and this force be applied in the same manner as that producing the "tarage," it is only necessary to measure the compression produced by the unknown force, and find from the "tarage," or from the table, the corresponding pressure.

DISCUSSION.—The pressure we wish to measure is that of the powder-gas. This gas acts upon the cylinder to be compressed, through the medium of a piston whose area is exactly known.

This piston moves in a cylindrical channel, and its head rests against the cylinder to be compressed, the gas acting upon the opposite end of the piston (see Noble crusher-gauge). In order that the results of the compression may agree with those of the "tarage," the mass of the piston and its velocity must be as small as possible. To show this: At any instant let

P be the intensity of the force to be measured;
R, the resistance to deformation offered by the cylinder;
m, the mass of the piston;
v, its velocity;
x, the length of path passed over by it.

The work of the pressure on the piston over the path x is

$$\int_0^x Pdx,$$

and that of the resistance over the same path

$$\int_0^x Rdx,$$

and the difference between these is the energy of the piston; hence

$$\tfrac{1}{2}mv^2 = \int_0^x Pdx - \int_0^x Rdx.$$

In order that $P = R$, which is the condition sought, we must have at all times

$$\tfrac{1}{2}mv^2 = 0.$$

Now m cannot be zero, but it must be as small as possible, and v must also be small. The former condition is attained by making the piston small, and the latter by compressing the cylinder before firing by a force nearly equal to the value of P anticipated.

CONCLUSIONS.—From numerous experiments Sarrau concludes:

1. Gunpowder is the only explosive which under ordinary conditions produces compressions agreeing with the "tarage."

2. This conclusion is true only when the gauge is in rear of the projectile. In the powder-chamber the pressure rises from zero to a maximum in a short time, but the time is appreciable. Hence the application of the pressure resembles in some degree that of the force producing the tarage. When, however, the gauge is situated in front of the base of the projectile, the gas suddenly strikes it, upon the passage of the projectile, and we have a case similar to that of the high explosives, and the same rule applies as with them. (See 3.)

3. For the high explosives, the rate of application of the force is so great that as a general rule the maximum pressure is measured by the "tarage" corresponding to one half the compression of the cylinder.

48. Rodman and Noble Gauges—Advantages of Noble.

RODMAN GAUGE.—The Rodman Pressure-gauge, Fig. 23, consists of a body or housing, H, which is a receptacle for all the working parts. A copper disk, C, is placed in the housing, and a knife, K, rests against it. The knife is attached to the piston P, which fits accurately in the cylindrical hole in the housing. The housing is closed by a screw-plug, J.

The gas acts on the end P' of the piston, and presses the knife into the copper disk, causing it to make a cut, whose length measures the pressure. A small copper cup, c, is placed at the outer end of the piston to act as a gas-check, and prevent the entrance of gas into the housing. This gauge, when used, is placed in the centre of the bottom of the cartridge-bag and tied to it with a string around the groove *g*. When in the gun, it must rest against the bottom

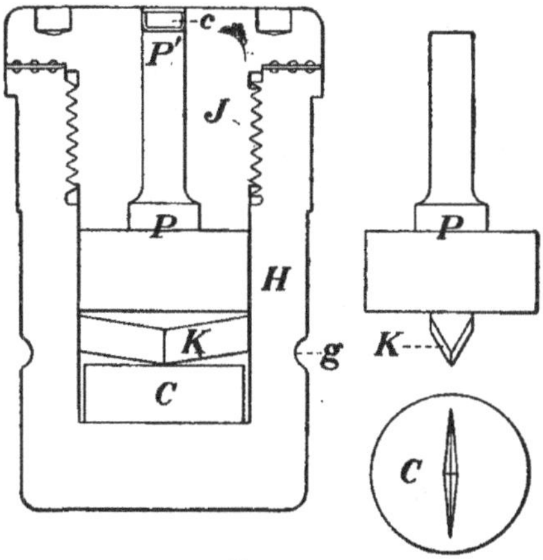

FIG. 23.

of the bore. The gauge may also be screwed into the breech-block, or walls of the bore, in which case it is threaded on the exterior.

NOBLE CRUSHER GAUGE.—This has replaced the Rodman gauge generally, for reasons which will appear later. It was used in Noble and Abel's experiments. It consists (Fig. 24) of a housing, *H*, closed by a screw-plug, *J*, and forming a receptacle for the working parts.

These consist of a piston, *P*, moving in a cylindrical channel as shown, and a copper cylinder, *C*, to be compressed, which is in contact with the piston. The cylinder is central, and kept in the axis of the housing by the spring *S*.

A copper cup, *c*, is used as a gas-check as in the Rodman, and another method for the same purpose, called "air-packing,"

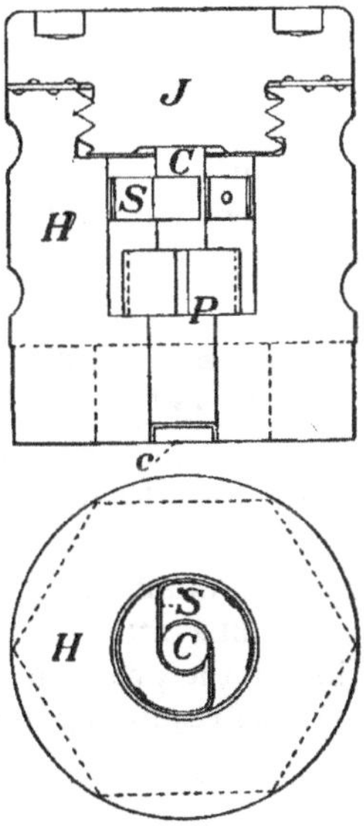

FIG. 24.

is also employed. A series of grooves, *a* (Fig. 25), are made around the piston. If gas enters between the piston and its channel, it escapes into the first groove, and by expanding, its tension is diminished. It may also escape into the second groove, and so on, and by each expansion its tension is still further reduced till it is unable to penetrate into the body of the housing. The action of the gauge is evident. In using it, the piston must always be in contact with the copper cylinder.

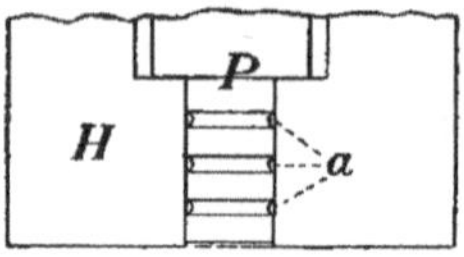

FIG. 25.

ADVANTAGES OF NOBLE GAUGE.—1. It is smaller than the Rodman, since the copper cylinder is smaller than the disk. It therefore takes up less room in the gun. The mass of the piston is also less than that of the knife and piston in the Rodman.

The advantage of this has been shown.

2. The knife of the Rodman is difficult to reproduce if broken, while the piston of the Noble can always be duplicated.

3. The copper disk offers very little resistance to motion at first, while that offered by the cylinder is more nearly uniform.

4. The cylinder can be given a preliminary compression, but a preliminary cut cannot be given to the copper disk.

49. Determination of Pressures by the Dynamic Method—Noble and Abel's Method—Letard's Apparatus—Sebert's Velocimeter.

In this method the pressure is determined by the acceleration of a known mass. The mass may be either the projectile, the gun, or a piston lodged in the walls of the bore, and communicating with it by a radial channel.

NOBLE AND ABEL'S METHOD.—In this method the motion of the projectile is used, as already explained, page 77, and the result is given by equation (150),

$$P = \frac{W}{g}\frac{d^2x}{dt^2};$$

the value of $\frac{d^2x}{dt^2}$ being determined by calculation from data obtained by the experiment.

LETARD'S APPARATUS.—To avoid piercing the walls of the bore, as in Noble and Abel's method, this apparatus is

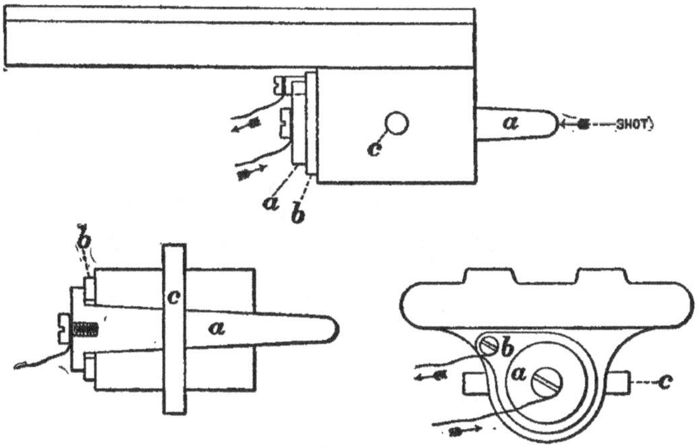

FIG. 26.

employed. It consists (Fig. 26) of a body of wood, on the front of which is a metallic ring, *b*. A metal bolt, *a*, passes through the wood body and projects to the rear, its head being in contact with the ring *b*. A pin, *c*, which is easily broken, holds the bolt *a* in place. When in this condition the current passes through the ring and bolt.

The wood body is attached to a second piece of wood, and the whole is placed in the bore of the gun, and secured against the wall with resin or cement. When the projectile strikes the projecting end of the bolt *a*, the pin *c* is broken, and the bolt driven out, thus breaking the circuit.

SEBERT'S VELOCIMETER.—With this instrument the motion of the gun, or of the projectile, or of both, may be used. The general principles are as follows: A ribbon of steel *S* (Fig. 27) is attached to the trunnion of the gun by the rod *T* and the gun mounted so as to recoil with very little friction. As recoil takes place, the ribbon has the same motion as that of the gun. A tuning-fork, *A*, whose rate of vibration is known, is fixed, with reference to the gun, above the ribbon, and carries a quill-point, *b*. The fork is made to vibrate by electro-magnets, *c*, as in the Schultz Chronoscope, and during recoil the quill-point traces on the blackened surface of the ribbon a sinusoidal curve which is the scale of time. In rear of the tuning-fork are placed several Marcel-Deprez registers, *R*, connected with Letard interrupters in the gun. When the projectile passes a point at which one of the inter-

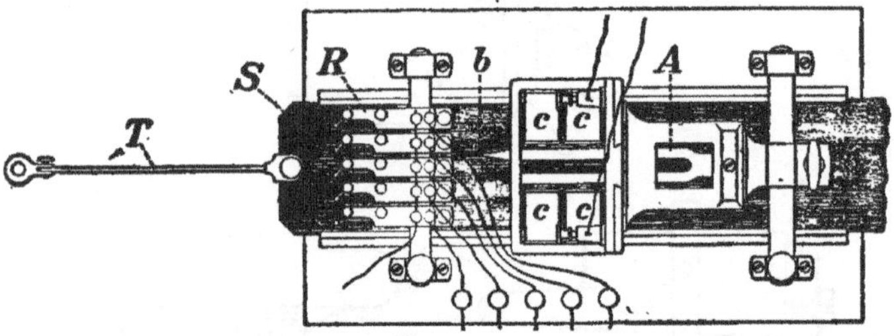

FIG. 27.

rupters is situated, the break is registered on the steel ribbon beside the scale of time, and so for each successive break. The number of vibrations between breaks, divided by the number of vibrations per second of the fork, gives the time of passage of the projectile over the distance between interrupters, and from this we can determine the velocity. From these velocities we can determine the accelerations, and hence the pressures, using the mass of the projectile.

Since the ribbon contains a complete record of the motion of recoil of the gun, we can also determine velocities and accelerations, and hence the pressures, using the mass of the gun.

CHAPTER II.

HIGH EXPLOSIVES AND SMOKELESS POWDERS.

HIGH EXPLOSIVES.

50. Definitions and Classification.

An Explosive is a substance which is capable of a sudden change from a solid or liquid to a gaseous state, with evolution of great heat.

A High Explosive is one in which this change is very rapid, and is accompanied by a crushing or shattering effect.

A Low Explosive is one in which the change is relatively slow, and accompanied by a propelling or pushing effect.

CLASSIFICATION.—Explosives may be classed into

1. Explosive mixtures;
2. Explosive compounds.

Explosive Mixtures are intimate mixtures of certain substances which are in themselves inexplosive, and which undergo no chemical change till the moment of explosion. They consist generally of a combustible body, such as carbon, and an oxidizing agent, such as potassium nitrate. The best example is gunpowder, which has already been discussed.

Explosive Compounds are chemical compounds, the molecules of which are explosive in themselves. They contain one or more combustible elements, such as carbon and hydrogen, together with the oxygen necessary to oxidize these elements.

The constitution of the molecule is more or less unstable, and when heated to a certain degree, the molecule breaks up with the formation of the gaseous products of oxidation.

The most important explosive compounds are the organic nitrates or nitric ethers, whose composition may be represented by R–O–NO_2, and the nitro-substitution compounds, represented by R–NO_2, R in each case representing the hydrocarbon radical.

Both are derived from organic substances by the action of nitric acid,—the former from complex alcohols, such as glycerine, etc.; the latter from certain hydrocarbons,—by the substitution in each case of NO_2 of the acid for H of the alcohol or hydrocarbon.

Usually from each substance a series of explosive compounds can be made, depending upon the number of atoms of H replaced by NO_2.

In the explosive mixtures, relatively great distances exist between the atoms which are to combine, while in the compounds each molecule constitutes a complete explosive, and hence the transformation is much more rapid with the latter.

51. Orders of Explosion—Berthelot's Theory—Detonators.

ORDERS OF EXPLOSION.—When gunpowder is fired in the ordinary manner we have an explosion of the second order; when it is mixed with nitro-glycerine and fired, we may have an explosion of the first order, or a detonation.

The difference consists in the time necessary to produce the chemical change. In the case of the explosion of the second order, the time is appreciable; in the case of detonation the change is practically instantaneous throughout the whole mass of the body.

BERTHELOT'S THEORY.—Berthelot, the great French authority, accounts for the difference in these orders as follows: Every explosion is caused by heating some part of the substance to the temperature of decomposition, and this temperature is transmitted successively to all parts of the body.

In the case of explosions of the second order, the portion of the substance first heated explodes; but if the gases have space in which to expand, they are cooled to a certain extent, and heat only a small additional portion of the explosive body to the temperature of explosion. This new portion then explodes, and the cooling again takes place; and so on, the explosion being propagated successively from layer to layer. This is the ordinary case with gunpowder.

Suppose, now, that a violent shock is given to any part of the explosive body, and that the pressures resulting from this shock are too great to be transmitted throughout the mass of the explosive.

The energy of this shock will be transformed into heat, and this heat will affect the first layers of the explosive body and cause them to be suddenly converted into gas, or will produce detonation. This gas being suddenly produced, the body causing the shock will not have time to be displaced, and therefore the expansion of the gas thus produced will cause a new shock, more violent than the first, to the layers below.

The energy of this shock will be transformed into heat, and will cause the second layer to detonate, and so on.

Hence we have an alternate conversion of energy into heat and of heat into energy, and this conversion resembles the propagation of a sound-wave in a given medium, except that its rate of travel is much greater. We may also have a combination of these orders of explosion, so that the distinction between the two cannot be sharply defined.

Every explosive seems capable of producing the two different orders of explosion, according to the manner in which the initial heating or shock is given. The high explosives give ordinarily the first order of explosion, the low explosives the second order.

DETONATORS.—The order of explosion generally depends on the intensity of the initial shock. If this is not great enough, the explosive may burn quietly, or give an explosion of a lower order. To produce this initial shock, a small quantity of some violent explosive, called a detonator, is required.

The principal detonating agent in use is fulminate of mercury, which, on account of its great force, gives rise to a high temperature when the initial shock is converted into heat.

52. Modes of Producing Explosion—Fuzes—Detonation by Influence. An explosion of the second order may be produced by shock, friction, the direct application of heat, by electricity or by an ordinary primer, and by certain chemical or physical changes; but to produce detonation a special fuze, called a detonating fuze, is generally employed. The material used in these fuzes is ordinarily merçuric fulminate, and one form is shown in Fig. 28. *A* is a copper shell; *B*, the chamber filled with mercuric fulminate; *C*, the electric wires; *D*, the ends of these wires; *E*, the platinum bridge which is heated by the current; *F*, the sulphur cement holding the wires and fulminate in place. This fuze is placed in the mass of the explosive, as its effect is weakened if a layer of air is interposed. Other varieties of fuzes are used. Those fired by electricity are classed as high and low tension, according to the kind of current used with them.

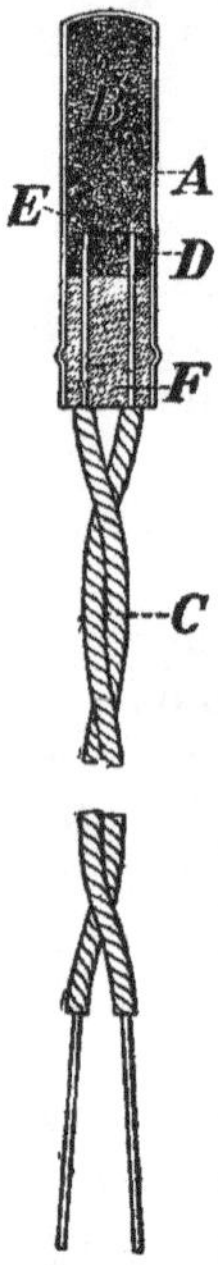

FIG 28.

Mass of Fuze. — The mass of the detonator should bear a certain proportion to that of the explosive. If it is too weak, it produces a low order of explosion; if too strong, it may scatter the explosive.

The exception to the rule is nitro-glycerine, which detonates equally well with a small or a large primer; but it holds for gun-cotton, and for those explosives which have been rendered less sensitive by various means.

DETONATION BY INFLUENCE.—If a series of cartridges of dynamite or gun-cotton be placed at certain distances apart, and one of them be detonated by a fulminate primer, the others will also detonate. This is called "detonation by influence" or "sympathetic" detonation. It appears to be governed by the following laws:

1 The distance apart at which detonation occurs depends on the envelope of the cartridges, and the nature of the material on which the cartridges rest. If the initial cartridge is enveloped in a non-resisting material, such as a paper envelope, the influence extends much further than

with a resisting envelope. If the cartridges rest on a resisting material, as an iron rail, the effect is propagated to a greater distance than if they rest upon the ground.

2. The envelope of the secondary charges should be as thin and elastic as possible, in order to oppose the minimum resistance to the shock.

3. An explosion thus propagated will become weaker from cartridge to cartridge, and may even change its order.

4. Similar effects are observed under water.

5. The shock is better transmitted by a liquid than by a gas.

6. The density of the secondary charges should be as great as possible, in order that the effect may not be reduced by motion among the particles.

53. Strength of an Explosive — Potential — Force — Rapidity of Reaction.

STRENGTH.—An explosive must be considered as exerting pressure and having potential energy, and in order to estimate its strength, and its value for different purposes, we must be able to determine the relative values of the pressure and potential for each explosive.

Water in freezing exerts great pressure if confined, and may burst the walls of the containing vessel. The fragments, however, will not be projected to any distance.

In this case we have a great pressure, but no potential energy or capacity to do work. If instead of water a high explosive be confined in the envelope and detonated, two effects will be observed: the walls will be ruptured, and the fragments thrown violently in all directions. In this case we have the pressure required to rupture the walls, and the potential energy necessary to project the fragments.

Again, if we compare equal weights of large and small grained powder of the same composition, exploded in a closed vessel, it is evident that the potential is the same for both, since the products and the quantity of heat disengaged are the same; the force or pressure is also the same, since this depends solely on the density of loading. The effect of these powders is, however, very different.

The small-grained powder, if exploded in a shell, will burst it into many fragments and project them to great distances, while the large-grained powder will give few fragments and small propelling force. Similar effects are observed with different classes of high explosives, and depend on the rapidity of the reaction by which they are converted into gas.

The strength of an explosive, then, depends on—

1. Its force or pressure;
2. Its potential;
3. The rapidity of its reaction or conversion.

FORCE OR PRESSURE.—The force of an explosive, as already defined in the case of gunpowder, is the pressure exerted by its gaseous products per unit of surface, when unit weight of these products is confined in unit volume. The expression for it in the case of gunpowder is, equation (28),

$$f = \frac{p_0 v_0 T_0}{273};$$

and the same expression measures the force of any explosive, v_0 being the specific volume, p_0 the atmospheric pressure, and T_0 the absolute temperature.

POTENTIAL.—The potential energy of an explosive is the total work it can do, when the products are indefinitely expanded without loss of heat, all the heat being expended in the performance of work.

Let E be the potential energy of unit weight of the explosive, J the mechanical equivalent of a heat-unit, T_0 the absolute temperature of explosion, and K the mean specific heat of the products. Then

$$E = KT_0J.$$

RAPIDITY OF REACTION.—This depends on the rapidity with which the chemical transformation is propagated throughout the mass of the explosive. Certain explosives, such as nitro-glycerine and gun-cotton, have very great velocities of conversion, and they may be regarded as undergoing an instantaneous change.

This change being so rapid, the heat is employed almost entirely in expanding the gases and performing mechanical work. Hence these substances are violent explosives, and shatter everything in their path.

The transformation is made to take place less rapidly by removing the particles to a greater distance from each other, as in the case of gunpowder, which then decomposes comparatively slowly and exerts a pushing effect rather than that of a blow.

It is evident that the choice of an explosive depends upon the relative values of these three elements. If an explosive is required for a shell, we need one having the greatest possible potential to scatter the fragments, a relatively small force, so as not to break it into very small fragments, and great rapidity of reaction in order that all the gases may be formed before the shell breaks. For mining, we require moderate force, small potential, and moderate rapidity of reaction, etc.

54. Principal Explosives—Gun-cotton.

PRINCIPAL EXPLOSIVES.—Gunpowder may be regarded as a type of the explosive mixtures, and as its properties are possessed to a greater or less extent by all these mixtures, the high explosives only will be considered in what follows.

The principal ones in use for military purposes are:

1. Gun-cotton;
2. Nitro-glycer ine;
3. Dynamites;
4. Picric acid and picrates;
5. Fulminates;
6. Sprengel safety mixtures;
7. Smokeless powders.

GUN-COTTON.—Its chemical formula is $C_6H_7O_2(ONO_2)_3$, and it is formed from cotton wool by the action of strong nitric acid. The reaction is

$$\underset{\text{Cellulose.}}{C_6H_7O_2(OH)_3} + \underset{\text{Nitric Acid.}}{3HNO_3} = C_6H_7O_2(ONO_2)_3 + 3H_2O.$$

Sulphuric acid is added to the nitric to take up the water and prevent the dilution of the later acid, which would give the lower orders of nitration, such as collodion cotton. The method of preparing it is described in chemistry.

In the earlier processes of manufacture the long fibres of cotton were used. These became filled with the acids, and being capillary tubes, it was found impossible to wash them out, and hence the product was unstable and liable to spontaneous decomposition. Abel, however, improved the process of manufacture by selecting the cotton waste, and cleaning it with alkaline washing, and especially by cutting up, or pulping the gun-cotton after it had been partially freed from the acids employed in its manufacture. By this operation the long fibres were reduced to very short tubes which could be thoroughly washed. A final washing in alkaline water completed the neutralization of free acids.

Forms.—Gun-cotton occurs ordinarily—

1. In the form of wool, like the original cotton;
2. In compressed cylinders and slabs.

The first form is that in which the cotton was used up to the time of Abel's improvement. It was sometimes twisted into strands, and woven, to regulate its rate of burning.

The compressed cylinders are made by the action of a hydraulic press upon the pulped and washed gun-cotton.

Properties—Density.—Its density is about 0.2 for the wool form and 1.1 for the dry compressed.

Solubility.—It is insoluble in water, alcohol, or sulphuric ether, but is soluble in acetone and acetic ether. This insolubility in alcohol and sulphuric ether distinguishes it from the lower orders of nitrated cellulose, which are soluble, giving collodion.

Effect of Foreign Substances.—The addition of *water* decreases the sensitiveness to explosion. The addition of *paraffine* has a similar effect, with the advantage that it does not evaporate. The reason for this decrease in sensitiveness is that the water or paraffine gives a certain elasticity and solidity to the gun-cotton, so that the initial shock of the detonator is propagated through a much greater mass, and consequently its local energy is diminished.

Nitre is sometimes added to increase the supply of oxygen, which is deficient.

Free Acids.—These cause spontaneous decomposition, with elevation of temperature and increased sensitiveness, so that explosion frequently results.

The presence of these acids formerly caused many accidents.

Effect of Heat and Cold.—As a general rule, the sensitiveness of a high explosive increases with the temperature. If wet, the application of heat will cause the evaporation of the water and thus increase the sensitiveness of the gun-cotton.

Cold will cause the freezing of the wet compressed gun-cotton and its consequent flaking and disintegration.

Ignition.—A temperature of about 180° C. is required. If the cotton is dry and unconfined, the application of flame will cause it to burn quickly. If the mass is large, an explosion may occur, but it will be ordinarily of a low order. If wet, the cotton will burn when unconfined, only in successive layers as they become dry.

Detonation.—Dry gun-cotton is detonated by a fulminate fuze.

Wet or paraffined gun-cotton requires a large detonator or an initial priming charge of dry cotton with a fulminate fuze.

Reaction.—The reaction on explosion is

$$2C_6H_7O_2(ONO_2)_3 = 7H_2O + 3CO_2 + 9CO + 6N.$$

There is evidently a deficiency of oxygen, and hence of potential energy, and therefore nitre is sometimes added as before stated.

Use in Blasting.—From the CO given off it is disadvantageous in blasting unless a nitrate be added. It has also the disadvantage of being of comparatively low density and solid, and hence it cannot be introduced so readily into bore-holes, nor in such large quantities as other explosives which are plastic or have higher densities. It is, however, very safe.

Use for Military Purposes.—It has been used for charging torpedoes, for bursting charges for shell, and for destroying obstacles such as walls, palisades, guns and carriages, etc.

Storage.—It is best stored wet, as under these conditions it is perfectly safe. It should not, however, be exposed in this state to a freezing temperature on account of disintegration.

55. Nitro-glycerine.

Its chemical formula is $C_3H_5(ONO_2)_3$, and it is formed by the action of strong nitric acid upon glycerine.

The reaction is

$$\underset{\text{Glycerine}}{C_3H_5(OH)_3} + \underset{\text{Nitric acid}}{3HNO_3} = C_3H_5(ONO_2)_3 + 3H_2O.$$

Sulphuric acid is added to the nitric, as in the case of gun-cotton, to take up the water formed in the reaction.

The method of preparing it is to add the glycerine slowly to the mixture of acids; to keep the mixture cool by cooling coils in the vessel, and by passing a current of air through it, which also insures a thorough mixture; and to wash the product thoroughly with water to which a small quantity of alkali is added, to insure the removal of free acid.

Form.—Nitroglycerine when first made is in the form of an opaque, white, oily liquid, becoming colorless with time.

Properties—Density.—Its density is 1.6.

Solubility.—It is very slightly soluble in a large quantity of cold water. It is freely soluble in alcohol, ether, chloroform, and slightly in glycerine.

It has a sweet, pungent, aromatic taste; is poisonous, and causes headache.

Heat and Cold.—It freezes at 4°.4 C. to a white crystalline solid, and is almost inexplosive in this condition by ordinary means, unless a small mass be acted upon by a shock.

When frozen, it is thawed at a temperature of 37°.7 C. by placing the vessel containing it in another of water at this temperature.

Free Acids.—These cause its decomposition, as in the case of gun-cotton, and render it more sensitive to friction and percussion, and hence they must be carefully removed.

Ignition.—Its temperature of ignition is about the same as that of gun-cotton, 180° C.

If unconfined and subjected to a blow, the particle struck will explode, and scatter the remainder.

If confined and struck, it will detonate; when unconfined, in small masses, the application of flame causes it to burn rapidly without explosion.

Detonation.—It is detonated by mercuric fulminate, and the detonator should be placed in the liquid. If frozen, it may also be detonated, but the action is generally less violent, owing to incomplete conversion.

Reaction.—The reaction on explosion is

$$2C_3H_5(ONO_2)_3 = 6CO_2 + 6N + 5H_2O + O.$$

Here we have an excess of oxygen, and the reaction, following a general law, is always stable.

Use in Blasting.—Nitro-glycerine is one of the strongest of the high explosives, possessing great force, potential, and rapidity of reaction. Owing to its liquid form it can be poured into holes of any shape, provided they do not communicate with fissures, and from its great rapidity of reaction, the depth of the hole may be decreased, and no tamping except water is required. It is therefore a valuable agent for blasting, but owing to its liquid form it is very unsafe in handling, as it is liable to leak, and thin films of it may be easily exploded. For this reason, except for special purposes, it is now generally replaced by dynamite.

Use for Military Purposes.—The same remarks apply in this case as for blasting.

Storage.—If possible nitro-glycerine should be kept frozen, and should be transported and handled in this state, being thawed before using.

Test for Purity.—Free acid may be detected by using blue litmus-paper. The acid will redden it.

56. Dynamite—With Inert Base.

Owing to the dangers involved in the transportation, handling, and storage of nitro-glycerine as previously noted, efforts were made to find an absorbent for it, so that it could be given a solid form. The addition of these absorbents has given rise to dynamite and various other derivatives of nitro-glycerine.

Absorbents Classified.—These may be classified into: 1. Inert; 2. Chemically active.

INERT ABSORBENTS—DYNAMITE NO. 1.—The most important of the inert bases is kieselgühr, a siliceous infusorial earth which is porous, and will absorb and retain about three times its weight of nitro-glycerine.

When it absorbs about 75 per cent of nitro-glycerine, the product is called dynamite No. 1.

Form.—It is either granular, or in compressed cylinders, wrapped in paraffined paper.

Properties—Density.—Its density is about 1.5 to 1.6.

Heat and Cold.—Dynamite freezes at 4°.4 C., and in this condition is detonated with great difficulty when solid, but when loose it may be detonated, the explosion being less violent. If frozen, it must be thawed before exploding, and this should be done very carefully, as in common with all preparations of nitro-glycerine it becomes more sensitive as it is heated. At all high temperatures the nitro-glycerine exudes, and hence the dynamite becomes dangerous.

Free Acids.—These are very dangerous, and the same remarks apply as to nitro-glycerine.

Detonation.—A fuze of fulminate of mercury is used, which must be placed in the mass of the cartridge.

Reaction.—This is the same as for nitroglycerine.

Use in Blasting.—It has been found very useful in blasting on account of safety in handling. The potential is diminished by the presence of the silica, and hence its action is less violent, and its effects more distributed. By regulating the percentage of nitro-glycerine present, this effect may be still further modified.

Use for Military Purposes.—In the U. S. service, Dynamite No. 1 is used for charging torpedoes, and may be regarded as the standard high explosive for this purpose. General Abbott of the U. S. Engineers has made a series of researches upon this subject, and has deduced formulas for the intensity of different explosives when used under water. (See "Professional Papers, Corps of Engineers," No. 23—1881.)

Storage.—Dynamite No. 1 is stored in boxes of wood, in magazines free from dampness, and no fulminate caps or primers should be stored with it.

57. Dynamite with Chemically Active Bases.

Instead of an inert base, a combustible one may be used,

which is capable of combining with the excess of oxygen of the nitro-glycerine and thus increasing the potential.

Various substances have been used for this purpose.

90 per cent of nitro-glycerine and 10 per cent of charcoal form carbo-dynamite. Sawdust, treated with superheated steam, becomes a jelly, and is capable of absorbing a large quantity of nitro-glycerine. Other compounds of the same class are also found.

On the other hand, by using a nitrate or chlorate mixture as a base, additional effect is obtained by inducing a higher order of explosion in the base. When gunpowder is used as an absorbent, the detonation of the nitro-glycerine causes the detonation of the powder. Potassium-chlorate mixtures are also used for this purpose, but are generally regarded as dangerous.

High Explosive Bases.—The most important of these compounds is explosive gelatine or "gum-dynamite."

It has been shown that when gun-cotton detonates, there is a deficiency of oxygen, and in the case of nitro-glycerine there is an excess of it.

If these two explosives are mixed in such proportions as to have the excess of oxygen in the one, neutralize the deficiency in the other, we have a considerable increase of potential. The result is best realized in a substance called explosive gelatine or gum-dynamite, which was invented by the Swedish engineer Nobel.

It is made by dissolving 7 parts of a special grade of soluble gun-cotton in 93 parts of nitro-glycerine, by the aid of heat.

For military purposes about 4 per cent of camphor is added to decrease its sensitiveness.

Form.—It is a translucent jelly of a yellowish or dark brown color, which may become in time hard and opaque.

Properties—Density.—Its density is 1.6.

Solubility.—It is insoluble in water and is unaffected by it.

Heat and Cold.—If heated slowly to 204° C., it explodes; and it freezes at low temperatures.

Ignition.—When ignited unconfined it burns readily, but

does not explode; but if confined, explosion occurs. It is not affected by shock, and bullets have been fired through it without producing explosion.

Detonation.—For detonation a special primer is required, and the strength of the primer must be increased as the sensitiveness of the gelatine is decreased.

Use for Military Purposes.—It has been tried as a bursting charge for shells; but as it requires a large primer, the advantages of the decrease in sensitiveness of the explosive are lost by the increase of sensitiveness of the primer.

Storage.—Some doubt exists as to the stability of this compound, and the effect upon its sensitiveness of the evaporation of the camphor.

58. Picric Acid and Picrates—Fulminates.

The chemical formula for picric acid or tri-nitro-phenol is $C_6H_2(NO_2)_3OH$, and it is formed by the action of nitric acid on carbolic acid. The reaction is

$$C_6H_5OH + 3HNO_3 = C_6H_2(NO_2)_3OH + 3H_2O.$$

Form.—It occurs in yellow crystals which are slightly soluble in water. It explodes when heated rapidly, but is ordinarily not used as an explosive by itself, and is only of importance from its compounds.

Potassium Picrate.—This salt is a violent explosive. Mixed with nitre and charcoal and grained, it forms Designolle's powder, which has been used for small arms and cannon, and also for torpedoes, with good results, but it is expensive, and some cases of premature explosions have been noted.

Ammonium Picrate.—This is less sensitive than the potassa salt, and burns without explosion in the air. Mixed with nearly equal parts of nitre, it forms Brugere's powder, which has about twice the strength of ordinary gunpowder, but is expensive, somewhat hygroscopic, and too violent for small arms.

The picrates form the bases for certain smokeless powders.

Emmens Acid.—This acid is said to be formed by the action of nitric acid upon picric acid.

Emmensite is a mixture of emmens acid and sodium or ammonium nitrate. It is yellow and crystalline in appearance, and is used in mining and as a substitute for gunpowder, both as a propelling agent and for charging shells. It is much stronger than gunpowder, is smokeless, and almost insensitive to shock.

It is hygroscopic, and its stability after long storage is not yet well settled. It was invented by Dr. Emmens, and is still undergoing trial.

Melinite.—This French explosive is generally supposed to be a mixture of gun-cotton with picric and cresylic acids dissolved in ether.

FULMINATES.—The most important is mercury fulminate, the chemical formula for which is $HgC_2N_2O_2$. It is formed by the action of alcohol upon mercury nitrate. The reaction is rather complex, and may be found in the chemistry.

Form.—It is in fine gray crystals.

Properties—Density.—Its density is 4.42.

Solubility.—It is insoluble in water, not affected by the air, and is poisonous.

Water.—When saturated with water it is inexplosive, and hence it is always kept under water for safety.

Detonation.—When dry it is very sensitive to a blow and detonates with violence, and also when heated to 182° C. or when subjected to friction, or to contact with any ignited body, or to the action of the electric spark.

Reaction.—The reaction on explosion is

$$HgC_2N_2O_2 = Hg + 2CO + 2N.$$

Use.—The great value of this explosive is as a detonator for the other high explosives. Its effects are due to its great force, since the volume of gas given off is very great; and also to its high density, in consequence of which a large mass is contained in a small volume. The gases also are not subject to dissociation, and hence impart all their energy to the explosive to be detonated. It is said to have ten times the force of gunpowder. Being comparatively low in potential, an oxydizing agent is sometimes added when the primer is at a distance from the charge.

Storage.—It must be kept under water for safety, and must not be allowed to come in contact with a metallic surface, as it then tends to decompose. Hence percussion-caps are varnished before it is placed in them. It must not be stored with high explosives.

59. Nitro-Benzines—Sprengel Safety Mixtures.

NITRO-BENZINES.—These are formed by the action of nitric acid on benzine, and we have the mono-, di-, and tri-nitro-benzines resulting.

They are not explosive, but are used in the manufacture of a class of explosives called Sprengel safety mixtures.

SPRENGEL SAFETY MIXTURES.—These were invented by Dr. Sprengel; the idea being to mix an oxydizing with a combustible agent at the time it is to be used, the constituents being each non-explosive before mixture, and therefore safe to handle and transport.

Rack-a-rock is a Sprengel mixture of liquid mono-nitro-benzine and potassium chlorate. If the cartridges are kept awhile, their sensitiveness to friction or percussion increases. This explosive was used at Hell Gate in 1885; 240,000 lbs. of it being exploded together with 42,000 lbs. of dynamite.

Hellhoffite is a mixture of di-nitro-benzine and nitric acid. It has been used as a bursting charge for shells, by placing the components in separate vessels in the shell, and causing their mixture automatically, either during its flight or on impact.

Bellite is a mixture of tri-nitro-benzine with ammonium nitrate. It is not sensitive to blows or friction, is chemically stable, and can be stored and transported without change or danger.

Another class of explosives of the same kind are the flameless explosives, which when confined and detonated, evolve gases which quench any flame. They are especially useful in mines where fire-damp is prevalent.

Roburite is one of the class of flameless explosives, made by mixing ammonium nitrate and chlorinated di-nitro-benzine, and is a yellowish powder. It is flameless because the ingredients are so proportioned as to cause complete oxidation, and the products of combustion are carbon dioxide, water, nitrogen, and HCl; the gases given off quenching any flame that may be produced.

Many other explosives of this class are made, and their composition may be found in "The Dictionary of Explosives," by Major J. P. Cundill.

SMOKELESS POWDERS.

60. Changes in Black Powders — Early History of Smokeless Powders.

CHANGES IN BLACK POWDERS.—A general idea has been given previously of the history of gunpowder, and the changes made in it. It was found to be too strong even for modern guns in the small-grained form, and hence Rodman conceived the idea of suiting the grain to the calibre.

His perforated powder was also designed with the idea of burning on an increasing surface, and thus decreasing the volume of the gas emitted at first, and hence the maximum pressure.

No change had been made in the components of the powder. Later still these changes in form were combined with changes in the nature of the materials, and their proportions. In the cocoa-powder the nature of the charcoal was changed, as were also the proportions of nitre, sulphur, and charcoal, and certain carbo-hydrates were introduced.

While these changes made the powder slower, they necessitated larger charges. This increased the cost, occupied a greater volume of bore, and thus reduced the path over which the gases worked, and necessitated long bores. and also gave great volumes of smoke.

With small arms, when the calibre was reduced to 0.30, the length of the bullet remaining the same, it became necessary, in order to obtain an increase in velocity, to increase the mean pressure per unit of area of the projectile, and hence to adopt some agent having better ballistic qualities than the old powders. Increased charges of compressed black powders were first tried, but they gave high and irregular pressures and relatively lower velocities than with the old charges.

EARLY HISTORY OF SMOKELESS POWDERS.—To obviate these defects a new explosive was sought, which would increase the velocities, without increasing the pressures beyond safe limits. Naturally the high explosives were tried, and of these the most promising was gun-cotton. It was known that this gave no smoke, that it burned freely when unconfined, but that it detonated when confined. Attempts were therefore made to regulate its burning, by twisting it into strands and winding these on the exterior of a hollow wood cylinder, so that the cartridge thus made would fit the chamber of the gun. It was supposed that when sufficient pressure was developed, the cylinder would crush, and thus give a very much larger volume for the gases to expand in, and hence prevent detonation. It was found, however, that detonation did occur, with destruction of the gun, and attempts in this direction were abandoned.

Another method was to mix gun-cotton with ordinary cotton. The two were after mixture subjected to a strong compression, but it was difficult to obtain a homogeneous mixture, the velocities were not increased, and the gun-cotton still detonated.

An attempt was also made to place a charge of black powder in front of the charge of gun-cotton. The projectile was started by the burning of the black powder, and then the gun-cotton was inflamed.

This plan gave excessive pressures in practice, and was soon abandoned.

Abel's compressed cotton was also tried with the same results, and gun-cotton was then abandoned as a propelling agent. This was about 1884.

61. Modification of Gun-cotton—Effect of Calibre.

MODIFICATION OF GUN-COTTON.—In all the early trials of gun-cotton no essential modification of its physical condition was attempted.

The fibres of the gun-cotton were not compact, and on being subjected to the action of a highly-heated gas, the flame readily penetrated all parts of the mass, raising it to the temperature of explosion, and detonation followed.

In 1884 it was proposed to dissolve the gun-cotton in some solvent, which could afterwards be evaporated, leaving a compact horny substance, which would resist the penetration of flame, and burn regularly. This was the first step in the successful manufacture of smokeless powder.

EFFECT OF CALIBRE.—With black or nitrate powders, as has been shown, the size of the grain must increase with the calibre for all large guns, but for all small arms the same powder (small-arms) may be used with good results.

With smokeless powders, however, each change in calibre of the small arm requires a change in the powder used. This may be explained as follows: As the calibre decreases, the length of the bullet remaining constant, while its weight decreases, it is necessary to increase the initial velocity of the projectile to obtain superior ballistic results, and this increase of velocity can only be obtained by an increase of pressure per square inch of the powder-gas.

To obtain this increase the physical qualities of the smokeless powder are modified so as to obtain a quicker powder, and this is accomplished by stopping the solution of the cotton at the proper point, and by decreasing the thickness of the grain. For the larger calibres the solution of the cotton is more complete and the grains thicker.

Those physical qualities, therefore, which principally affect the velocities and pressures given by the powder are:

1. Its degree of solution or density;

2. Its thickness, or least dimension of grain;

both of which regulate its burning.

In order to increase the pressure per unit of area, a proper combination of these qualities is required for each particular calibre, and this requires a special powder for each gun.

62. Operations in the Manufacture of Smokeless Powder—Solution.

OPERATIONS.—The principal operations in the manufacture of a smokeless powder of the nitro-cellulose class are:

1. Preparation of the nitro-cellulose;

2. Solution of the nitro-cellulose in a proper solvent;

3. Compression of the material after evaporation of the solvent;

4. Rolling into sheets or pressing into rods or tubes;

5. Cutting up the sheets, rods, or tubes into grains;

6. Drying the grains.

SOLUTION.—The principal precautions to be taken in this operation are, to avoid the formation of lumps or undissolved particles of nitro-cellulose; to have the solvent act regularly; and to prevent the cotton from collecting in masses, so that the solvent cannot readily penetrate it. If the powder is to be quick, the cotton must not be completely dissolved, and hence the operation must be stopped at the proper time, which requires great delicacy in manipulation.

In the operation, the gun-cotton must not be plunged in the solvent, but the latter must be poured over the cotton. For this purpose the cotton, which is finely divided, is placed in layers of the proper thickness, in ebonite pans of slight depth. These are enclosed in a glass vessel, and the solvent added in the form of a spray. The gun-cotton gradually dissolves, or rather becomes gelatinized; the supply of the solvent is then stopped, and the solution allowed to proceed to the proper degree. A current of warm air is then passed over the gelatinized gun-cotton, carrying off the solvent in a state of vapor, which is afterwards condensed in a cool vessel. In this manner the drying of the powder is assisted, and the solvent which is removed can be used again. The cost of these powders depends principally upon the solvent, and hence it is important to collect as much of it as possible.

63. Compression and Rolling—Cutting Up—Drying.

COMPRESSION.—During the evaporation of the solvent, bubbles are formed, the effect of which is to render the mass more or less porous in places. This causes irregular density, and hence in the same sheet some parts would burn more quickly than others. It is necessary, therefore, to get rid of these bubbles.

The thickness of the sheet, after the solvent has evaporated, is not uniform, as it is impossible to spread the cotton regularly before it is acted on by the solvent. This thickness is very important, as affecting the ballistic properties of the powder.

To get rid of the bubbles and at the same time regulate the thickness, the sheet is subjected to strong pressure, which is kept up for some time. This pressure has also the effect of completing the solution of certain parts which were not completely dissolved.

ROLLING.—The sheet is then passed between two rolls of polished bronze. The upper roll must be so arranged that its weight will not rest upon the sheet. In this way when the sheet has reached its proper thickness it will not be reduced further.

The reduction to the required thickness is gradual, so as not to tear the surfaces. In general three or four successive passes through the rolls are necessary.

CUTTING UP.—Black powder is grained, but this operation is impossible with smokeless powder, which is tough and flexible and cannot be broken. It is therefore cut into the required form by special machines, or pressed, while still pasty, through holes in a die, thus forming strings or cords.

DRYING.—The sheets are cut while still saturated with the solvent, and the drying accomplished after the powder is reduced to grains.

This operation should take place slowly, and at a relatively low temperature. Without this precaution there is danger of evaporating the remaining solvent too rapidly, which would cause disintegration of the material and increase the porosity.

Smokeless powders are difficult to dry, especially when thick, and hence when considerable thickness is required, as with cannon-powder, several thin sheets previously dried are placed on each other, and the whole compressed to the required thickness by hydraulic pressure.

The drying should not be complete, a certain quantity of the solvent being left; as in the case of black powders, a certain quantity of moisture is retained. This tends to diminish the pressure and to keep up the normal velocity.

The great difficulty in the manufacture of smokeless powder has been to make it in large quantities by machinery, so that it shall give uniform results as to velocity and pressure. It is not difficult to make small quantities in a laboratory, but very difficult to reproduce them on a large scale.

64. Classification of Smokeless Powders—Classes 1, 2, and 3—Wetteren Powder.

The manufacture described above applies only to the nitro-cellulose classes of smokeless powders; but as the others are generally mixtures of different substances, in the state of powder, their preparation resembles that of gunpowder and requires no special description.

CLASSIFICATION.—Smokeless powders are generally classed as:

1. Those derived from picric acid and the picrates;
2. Those derived from ammonium nitrate;
3. Those derived from nitro-cotton or from mixtures of nitro-glycerine and nitro-cotton, with the addition of certain agents which act to modify the rate of burning.

Class 1. *Picric-acid Powders.*—Of these Designolle's and Brugère's powders have already been described. This class is no longer of much importance, as it has been abandoned for powders of class 3.

Class 2. *Ammonium-nitrate Powders.*—The objection to this class of powders is that they are all highly hygroscopic, and they are no longer used.

Class 3. There are a few well-known powders of this class. Very little is known about their actual composition, and hence only a general description of them can be given.

WETTEREN POWDER.—This is made in Belgium. It is said to be composed of nitro-cotton dissolved in acetic ether, with the addition of nitrate of baryta. Another composition given is nitro-cotton dissolved in acetic ether, with the addition of nitro-mannite, which is formed by the action of nitric acid on manna-sugar. The grains are hard, square in form, and of a slate color.

To protect it from moisture the grains are varnished with a special collodion.

The defects of this powder are, it is expensive; the acetic ether does not thoroughly dissolve the nitro-cellulose; and after the solvent is evaporated, white scales are formed on the surface. In time this powder loses its compact structure, and under the influence of shock reduces to dust in the cartridge-case, and this dust will cause excessive pressures. Also the acetic ether is difficult to evaporate, and hence different parts of the powder may contain different amounts of the solvent, and this gives rise to irregular ballistic qualities. This powder is now being tried in the U. S. cal. 30 rifle, charge 37 grains, muzzle velocity 2000 ft.-secs.

65. Powder B. N. F.—Ballistite—Cordite.

POWDER B. N. F.—This powder is used in France in the Lebel rifle. Its composition is unknown, but it is supposed to be a mixture of cottons of different degrees of nitration, gelatinized by suitable solvents. It is first formed into plates, these are rolled into sheets, which are cut up into grains for small arms or into strips for large guns. It has a grayish or yellowish color, is difficult to ignite, and is said to be very regular in its action and not affected by change of climate.

BALLISTITE.—This powder is used in Germany and Italy, and is the invention of Alfred Nobel. It was the first successful smokeless powder made by uniting nitro-glycerine and nitro-cellulose. By acting upon a soluble gun-cotton with nitro-glycerine in the proportions previously given, Nobel produced explosive gelatine. By increasing the proportions of the nitro-cellulose to 30 or 40 per cent and reducing the nitro-glycerine to 60 or 70 per cent, the resulting mixture becomes a horny compact mass capable of definite granulation. About 7 per cent of camphor dissolved in the nitro-glycerine is found to assist the process. In the manufacture, benzole is mixed with the nitro-glycerine, to render the nitro-cellulose temporarily insoluble in order to facilitate its equal distribution and absorption. The benzole is then evaporated and the material repeatedly passed through steam-heated rolls and made into sheets. These are afterwards cut up into cubical grains which are dark brown in color and are horny and translucent when cut. It has about three times the ballistic force of black powder, and its effects are very regular.

CORDITE.—This powder is used in England and is the invention of Sir F. Abel and Professor Dewar. It is very similar to ballistite, except that a highly nitrated gun-cotton is used.

As this is insoluble in nitro-glycerine, to obtain a stable union with the latter it is necessary to dissolve the gun-cotton in a solvent. Acetone is used. Various slowing agents have been tried. Tannin is called for in the patent, but at the present time about 10 or 15 per cent of vaseline is preferred.

These powders are perfectly smokeless, and give high velocities with safe pressures. They are said to deteriorate rapidly, the manufacture is dangerous, and in some samples the nitro-glycerine exudes, rendering the powder sensitive. They give great heat on explosion, and this may, it is thought, injuriously affect the bores of guns. They are also difficult to explode.

66. Leonard Powder—Peyton Powder.

LEONARD POWDER.—This is an American powder whose composition is gun-cotton dissolved in acetone, a large percentage of nitro-glycerine, and a slowing agent.

In shape it resembles cordite for large guns, the diameter of the cord increasing with the calibre of the gun. For small arms the cords are very small, and are cut in short pieces, so that the powder is granular in appearance.

This powder has given good results in proof, and it is now undergoing trial. The grains are rather soft.

PEYTON POWDER.—This is also an American powder, manufactured by the California Powder Works. It is a gelatinized mixture of nitro-glycerine 38 per cent and gun-cotton 40 per cent, acted on by acetone, with certain other substances added.

The mixture is incorporated in a small wheel-mill, covered to prevent loss of solvent. After incorporation the plastic mass is pressed by hydraulic pressure through a hole, in the centre of which is a rod. This forms the mass into a hollow cylinder or pipe. As the cylinder passes out of the hole, it is cut open longitudinally by a cutter, and is spread out into a flat sheet, about 10 inches wide and ⅛ inch thick. This sheet is run through a set of rollers with transverse grooves and ridges. By this means the sheets are cut into a series of strips, but the strips are not entirely separated, as they are still united by a thin film of the material, so that the sheet can be handled as a whole. These sheets are then passed under a cutter, acting at right angles to the strips, by which they are cut into grains. The length and width of the grains are about equal. By this operation most of the grains will be separated from each other. Those that stick are rubbed on a sieve or rolled in a barrel. The grains are then dried at a temperature of about 51°.6 C., to drive off the solvent, which is collected, and are finally polished and glazed. The size of the grain increases with the calibre of the gun, and must be determined by experiment for any particular gun.

The above particulars were furnished by Mr. W. R. Quinan of the California Powder Company.

This powder is also being tried by the United States, a lot of 5000 lbs. having been purchased for the cal.-30 rifle.

Mr. Longridge gives the following as the relative energies in foot-tons developed by equal weights of the three powders given below, the energy of brown powder being unity:

Cordite	4.16
Ballistite	3.44
Poudre B. N	2.48

67. Conditions to be Fulfilled by Smokeless Powders—Smokelessness—Velocities and Pressures—Stability.

CONDITIONS.—Smokeless powders should fulfil the following conditions:

1. They should be approximately smokeless.

2. They must give high and uniform velocities, with safe and regular pressures.

3. They must be chemically and physically stable, under varying conditions of moisture, temperature, and age.

4. They must not cause excessive fouling, or excessive heating of the gun.

5. They must not be sensitive to friction or shock.

6. The manufacture should not be difficult or dangerous, or the ingredients very expensive.

7. The products of combustion should not be noxious, and should not corrode the gun.

8. There must be no chemical action upon the cartridge-case.

9. They should give the required ballistic results with reduced weight of charge, and the charge should not occupy a large volume, and should be so grained as to be loaded in the ordinary loading-machine.

SMOKELESSNESS.—Nearly all the powders introduced satisfy this condition. Those of class 3, however, are the only ones which are in general truly smokeless, but the smoke from the others is rapidly dissipated. In most of them a slight mist is visible, since the water formed in the explosion is condensed by the air, and the priming or lubricant or the slowing agent also produces visible smoke.

VELOCITIES AND PRESSURES.—Very few of the powders fulfil this condition.

Experiment shows that, especially for small arms, a very slight variation of size of grain, weight of charge, or density of loading gives a great variation of pressure. For instance, a variation of weight of one grain in the small-calibre rifle increased the pressure from 44,740 to 51,620 lbs. per square inch, while the velocity was increased only 88 ft.-secs.

CHEMICAL AND PHYSICAL STABILITY.—This is another point about which there is great doubt. Most of these powders are of such recent date that sufficient time has not elapsed to test their stability. They are generally made as wanted for purposes of experiment, and used in a short time. Tests are being made, however, upon this subject by all countries.

Cordite has been tested as to changes of climate in India and Canada with good results.

Ballistite has been soaked in water, dried and fired, with very little change in ballistic qualities.

68. Fouling — Sensitiveness — Safety and Cost of Manufacture — Character of Products of Explosion — Chemical Action — Weight of Charge and Specific Gravity.

FOULING.—As a rule there is very little fouling with smokeless powders. The fact that they are smokeless indicates at once the absence of the residue of solid particles which causes fouling with ordinary black powders. This absence of fouling, however, has proved to be a disadvantage in small arms, owing to the increased friction between the bullet and the bore, which the fouling prevented, by acting as a lubricant. This friction has sometimes been so great as to strip the covering off the bullet and leave it in the bore.

Various lubricants have been tried to overcome this defect, but none of them have proved satisfactory, and the defect has been overcome by using a proper covering for the bullet, either of copper, nickled steel, or German silver.

SENSITIVENESS.—All the powders are safe in this respect, and have been pretty thoroughly tested. The difficulty lies in the opposite direction, as they are so insensitive that they are difficult to explode, and for most of them, special primers, or more powerful ones, are required.

Nitro-cellulose powders are specially insensitive.

SAFETY AND COST OF MANUFACTURE.—The principal danger arises in the manufacture of the ingredients, nitro-glycerine and gun-cotton, and in handling the former. While explosions sometimes occur, the manufacture can hardly be considered more dangerous than that of gunpowder. The question of cost is subordinate to that of efficiency, and would only enter in deciding between two or more powders of equally good ballistic properties.

CHARACTER OF PRODUCTS.—The character of the gaseous products differs very little from those of gunpowder, being principally CO, CO_2, H_2O, and N, and hence no danger is to be apprehended from them, and they should not corrode the gun. Corrosive effects have been noticed, but they are due to the great heat of the gases.

CHEMICAL ACTION.—So far as is known, these powders do not act chemically upon the cartridge-cases.

WEIGHT OF CHARGE AND SPECIFIC GRAVITY.—These

powders have much higher ballistic qualities than the old nitrate powders, and hence a much smaller charge will give greater velocities. The weight of the charge, for the .30-cal. rifle is 37 grains, and that of the projectile 220 grains. That of the old cal.-45 rifle was, powder 70 grains, bullet 500 grains.

This reduction in weight of cartridge has an important bearing upon the number of rounds carried by the soldier. It is evident that with a magazine arm the number of cartridges used will be greatly increased, and the reduction in weight enables them to be carried with ease.

The specific gravity and gravimetric density of the new powders are less than those of the old, and hence they occupy a greater volume for the same weight; but as the weight necessary to give the same or better results is less for each charge, the decrease in density occasions no difficulty.

69. Cause of Ballistic Superiority of Smokeless Powders.

The superior effects of the smokeless powders may be explained by considering their potential, force, and rapidity of reaction.

1. *Potential.*—This is much greater than with the old nitrate powders, as the quantity of heat evolved in the combustion of gun-cotton and nitro-glycerine is very much greater than that of ordinary powder. This heat measures the quantity of work which the gases can do upon the projectile, and hence the energy of the latter is much greater, and we have higher velocities.

2. *Force.*—This is also greater than with the old powders, as the specific volume of the gases and their temperature are higher.

The specific volume is greater because all the powder is converted into gas. This force tends to increase the pressure exerted by the gases upon the gun at the origin, and hence this pressure would be very great if it were not that—

3. *The Rapidity of Reaction* is very much decreased, so that the gas is given off slowly, and allows the projectile

to start from its original position before this pressure has reached too great a value.

These powders burn very slowly in air, but, like the nitrate powders, their rate of burning increases very rapidly with the pressure, and probably if this pressure were very high they would detonate.

Another circumstance concurs to prevent this, however, and that is that the powder has no solid residue, and hence *all* the space in the powder-chamber and in rear of the projectile is occupied by the gas. In ordinary powders about .57 of this space, or more than half, is occupied by solid residue. Hence the pressure is kept down at first, and, owing to the high temperature and great volume of the gas, it is maintained better along the bore than with the old powders.

We have therefore in the new powders a propelling agent which for less weight gives safe pressure at first, more gas, more heat, and more sustained pressure than the old powders, and hence their ballistic superiority.

TABLE OF HIGH EXPLOSIVES AND SMOKELESS POWDERS.

(From Cundill's "Dictionary of Explosives.")

I. Nitrate Mixtures.

Name.	Composition.	
Amide Powder	Nitre,	101 parts
	Ammonium nitrate,	80 "
	Charcoal,	40 "

II. Chlorate Mixtures.

Rack-a-rock	Chlorate potash,	3 parts
	Mono-nitro-benzine,	1 part

III. Nitro Compounds.

Abel's Glyoxiline	Gun-cotton
	Nitro-glycerine
Also	Gun-cotton
	Potassium nitrate
	Nitro-glycerine

Ætna Powder	Nitro-glycerine,	15 to 65 parts
	Sodium nitrate	
	Wood-pulp	
Atlas Powder (A)	Sodium nitrate,	2 parts
	Wood fibre,	21 “
	Magnesium carbonate,	2 “
	Nitro-glycerine	75 ‘
Bellite	Ammonium nitrate	83 “
	Tri-nitro-benzine	17 “
Blasting Gelatine	Di-nitro-cellulose dissolved in nitro-glycerine	
Borland Powder / Carbo-dynamite	Nitro-glycerine,	90 parts
	Charcoal,	10 “
Dittmar Powder / Dualine	Nitro-glycerine,	50 “
	Sawdust,	30 ‘
	Potassium nitrate	20 “
Dynamite No. 1	Nitro-glycerine,	75 ‘
	Kieselguhr,	25
Giant Powder	(See Dynamite.)	
Judson Powder	Gunpowder coated with nitro-glycerine	
Lithofracteur	(See Rendrock.)	
Nitro-mannite	Sugar of manna	
	Nitric acid	
	Sulphuric acid	
Rendrock	Potassium nitrate,	40 parts
	Nitro-glycerine,	40 “
	Wood-fibre,	13 “
	Paraffine,	7
Roburite	Ammonium nitrate	
	Chlorinated di-nitro-benzine	
Shultze Powder (a sporting powder)	Nitro-lignin	
	Potassium nitrate	
	Barium nitrate	
	Sawdust	
	Paraffine	
Tonite (blasting-powder)	Gun-cotton	
	Barium nitrate	

IV. Picric Powders.

Brugere's Powder	Ammonium picrate,	54 parts
	Potassium nitrate,	46 “

Designolle's Powder.....	Potassium picrate,	9 parts.
	Potassium nitrate,	80 "
	Charcoal,	11 "
Melinite..............	Gun-cotton dissolved in ether	
	Picric acid	
	Cresylic acid	

V. Sprengel Mixtures.

Hellhoffite............	Di-nitro-benzine
	Nitric acid
Rack-a-rock..............	(See *ante.*)

VI. Miscellaneous.

Caps for Toy Pistols....	Chlorate potash	
	Amorphous phosphorus	
Percussion-caps.........	Fulminate mercury,	6 parts
	Potassium chlorate,	6 "
	Antimony sulphide,	4 "
	Ground glass,	2
Mercury Fulminate	Mercury	
	Nitric acid	
	Alcohol	
Railroad Fog-signals.....	Tin cases filled with powder and having cones with ordinary percussion-caps.	

CHAPTER III.

GUNS.

GUN-STEEL.

70. Definition of Gun-steel—Chemical Composition—Different Constituents and their Effect.

DEFINITION.—Steel is an alloy of iron and carbon, the percentage of the latter being from 0.10 to 2.5. This percentage, however, does not always serve to classify steel, as it runs into wrought iron on the one hand, and into cast-iron on the other. It is distinguished from cast iron by its quality of becoming hard when heated to a certain temperature and cooled quickly, and of having this hardness reduced by a process called tempering.

It is distinguished from wrought iron by this same quality, and also by being cast into molds or ingots, which is not possible with wrought iron, the latter not being fluid except at very high temperatures. In gun-steel the proportion of carbon is low, not exceeding 0.5 per cent as a rule.

CHEMICAL COMPOSITION.—Upon this point there is great difference of opinion. Steel is called an alloy of iron and carbon, but the exact condition of the carbon is not known.

It is sometimes called dissolved carbon for the harder steels, and undissolved for the softer; also "hardening" and "cement" carbon, for the harder and softer steels respectively. More recently it is called "fixed" carbon for the hard, and "free" carbon for the soft steels. These two carbons may be changed from the one to the other, as the result of special treatment.

OTHER CONSTITUENTS.—Besides the carbon, there are always other substances present, some of which are beneficial and others injurious to its quality. Among the principal of these substances are:

1. *Sulphur.*—This is injurious to the steel, as it makes it difficult to forge, producing "hot-shortness," or brittleness when hot.

2. *Phosphorus.*—This is also injurious, as it has the effect of making steel brittle when cold, or "cold-short."

3. *Manganese.*—This when added in proper proportions improves the quality of the steel, rendering it hard and tough.

4. *Silicon.*—Is valuable, as it forms a fusible slag with the iron oxide in manufacture, and prevents the formation of gas, and consequently of blow holes in the steel. If in too great quantity, it causes brittleness.

5. *Chromium.*—Gives great hardness to steel without brittleness, and hence the best forged steel projectiles are made of chrome-steel.

6. *Nickel.*—This gives great toughness to steel, so that armor-plates made of nickel-steel resist racking very well.

Nickel-steel is also being experimented with for guns, at present.

71. Physical Qualities — Hardness — Toughness — Elastic Limit— Hooke's Law—Tensile Strength.

HARDNESS.—Gun-steel should be sufficiently hard to resist deformation from blows, and also the action of the projectile and the powder-gases; but hardness is generally accompanied by an undesirable quality, brittleness, and hence a modification called *toughness* is sought in this metal.

TOUGHNESS is the quality which enables a metal to undergo considerable change of form under the action of a force, without rupture, and with great resistance to that change.

ELASTICITY AND ELASTIC LIMIT.—When a tensile stress or force is applied to a piece of steel, it will elongate a certain amount.

This total elongation, divided by the original length, will give the elongation per unit of length. When the stress or force ceases to act, the steel will recover its original length, provided the stress is not too great. If an additional force be applied, a similar effect will be obtained, the elongation per unit of length being greater in this case, and the metal returning again to its original length when the stress ceases to act. The same effects will be observed till the stress reaches a certain amount, when the metal will not return to its original length, but will acquire a permanent *set*. If the stress next below the one which produces the permanent set be measured, and be divided by the area of cross-section of the metal, it will give the elastic limit of the metal; and if the elastic limit be divided by the corresponding elongation per unit of length, the result will be the modulus or coefficient of elasticity of the metal.

Let a be the area of cross-section of the metal;
l, its length;
K, the total elongation;
W, the stress acting at the elastic limit;
E, the modulus of elasticity;
θ, the elastic limit.

Then

$$\theta = \frac{W}{a};$$

$$E = \frac{\theta}{\frac{K}{l}} = \frac{Wl}{aK}.$$

Denoting by λ the elongation per unit length, we have

$$\lambda = \frac{K}{l},$$

and

$$E = \frac{\theta}{\lambda}; \quad \therefore \lambda = \frac{\theta}{E}. \quad . \quad . \quad . \quad . \quad (169)$$

HOOKE'S LAW.—The ratio of stress to elongation remains constant up to the elastic limit, and this constant ratio is the modulus of elasticity E. This is expressed as follows:

Within the elastic limit of a metal, the stress is proportional to the strain. This is called Hooke's law.

If we compare two kinds of steel, one having a high percentage of carbon, and the other a low percentage, it will be found that the steel high in carbon has a high elastic limit, and that low in carbon a low limit. Since the modulus of elasticity for all steel is nearly constant, and equal to about 30,000,000 lbs. per square inch, the high steel will elongate more at the elastic limit than the low (equation 169)). From this alone it would appear that the high steel is best for gun-construction, since it enables the metal to yield more to the stresses of the powder-gas, and to recover its original form without permanent set.

The reason why high or hard steel is not used is, that it is liable to flaws, strains, or incipient cracks, produced in manufacture, especially in large pieces. A hard steel is also dangerous, because after passing its elastic limit, it has very little remaining strength, and breaks easily, and with little warning, while the soft steel yields considerably without fracture, after passing the elastic limit, exhibiting the quality of toughness, previously defined.

TENSILE STRENGTH.—By this is commonly understood the stress per unit area required to rupture the metal. It is not of great importance in gun-steel, although limits are prescribed for it in the tests, since we consider the elastic limit only in gun-construction.

For clearness, a tensile stress only has been considered.

The same relations hold, however, for compression or torsional stress, and each has its corresponding elastic limit and modulus.

72. Structure of Steel—Defects—Blow-holes—Pipes.

STRUCTURE.—Steel is always a crystalline metal, and has no fibrous structure like wrought iron. These crystals are generally small, and vary in size and appearance with the treatment the metal receives after casting. They are very small in the best steels, and may be so small that the fracture will lose its crystalline appearance.

DEFECTS.—These are common to all cast metals, and steel has some in addition peculiar to itself.

Slow Cooling in large masses gives large crystals, and consequently a weak steel. It also causes a lack of uniformity in the steel. This being an alloy of iron and carbon, and the carbon being combined in different proportions throughout the fluid mass, the alloys highest in carbon are the lightest, and will rise to the top. Hence the hardest steel is found here. For the same reason, the softest will be at the bottom of the ingot. The middle portion of the length will therefore give the best steel. As the fusibility of the steel increases with the percentage of carbon, those portions low in carbon will solidify first, and hence, in the same cross-section, the part high in carbon will be near the centre, where the mass remains fluid longer.

BLOW-HOLES.—This defect is peculiar to steel, and is due to the gases in the melted metal, which, being unable to escape, are imprisoned in the casting, and form holes. These blow-holes are causes of weakness in steel, as it is impossible to discover them, and forging or compression only changes their form, but does not remove them. Various theories have been advanced to account for their presence, and attempts made to get rid of them. They are more prevalent in the Bessemer than in the open-hearth steels, by which latter process gun-steel is made.

The lower the temperature at which the steel is cast, the more apt are these blow-holes to occur, because the metal hardens before the gas has time to escape.

PIPES.—These are cavities formed in the axis of the ingot, due to internal strains from cooling. They generally occur when the metal is cast too hot. Thus on the one hand too low a temperature causes blow-holes, and too high a temperature pipes.

To avoid these defects in gun-steel, 6 per cent of the total weight of the cast ingot is cut from the bottom and $33\frac{1}{3}$ per cent from the top, the remainder being used for the forging. The piping and weak metal in the centre of the ingot are removed by boring or cutting out the central part of the ingot. Blow-holes can be prevented only by careful treatment in casting, and their presence cannot be detected except by subsequent working, and not then if they are beyond the reach of the tools employed.

73. Working Qualities of Steel — Fusibility — Malleability and Ductility—Welding—Annealing.

FUSIBILITY.—This quality enables steel to be cast into various shapes, and into ingots for gun-forgings. It requires, however, a relatively high temperature, and has caused the introduction of various special processes for obtaining this temperature. In the Bessemer process, the heat necessary is obtained, by blowing air through a melted mass of cast iron, by which the carbon and silicon are oxidized, and a high temperature produced. In the open-hearth process, the high temperature is obtained by the use of gaseous fuel, and by storing up the waste heat of the furnace in chambers of fire-brick, through which the gaseous fuel passes, and by which it is raised to a high temperature.

MALLEABILITY AND DUCTILITY.—Steel, when heated to a red heat, possesses the property of malleability, and it is due to this fact that it can be forged into any shape. When cold, owing to its ductility, it can be drawn into wire, which is used in wire guns and for various other purposes.

WELDING.—Ordinarily steel cannot be welded except when very low in carbon, and approaching wrought iron. Lately, however, the process of electric welding has been introduced, and by it the welding can be readily accomplished.

ANNEALING.—This is a very valuable property possessed by steel. By heating it to a certain temperature and allowing it to cool slowly, a piece of hard steel will become soft, so that it can be readily worked in the lathe. After working, it can be returned to its former hard condition, by heating it again, and cooling it quickly. After forging or working steel, it generally has internal strains due to these processes, and these strains may be removed by annealing. By cooling in oil, the tensile strength and elastic limit of steel are greatly increased, and these qualities, especially elasticity, are very valuable in gun-construction.

74. Manufacture of Gun-steel—Open-hearth Process—Gas-producer and Regenerators.

OPEN-HEARTH PROCESS.—All gun-steel at the present day is made by this process, which derives its name from the fact that the receptacle in which the steel is melted is open at the top, and exposed to the flame of the fuel which plays over the surface, and performs a principal part in the formation of the steel. It is also called Siemens or Siemens-Martin steel, according to the ingredients used to form the steel.

APPARATUS.—The furnace used is that invented by Dr. Siemens, and a general description of it is given. It con-consists of the following essential parts:

1. The gas-producer;
2. The regenerators;
3. The furnace proper.

THE GAS-PRODUCER.—The fuel used in the Siemens furnace is gaseous, and is obtained from ordinary fuel, by subjecting the latter to a preliminary process in the gas-producer. This apparatus, Fig. 29, consists of a rectangular

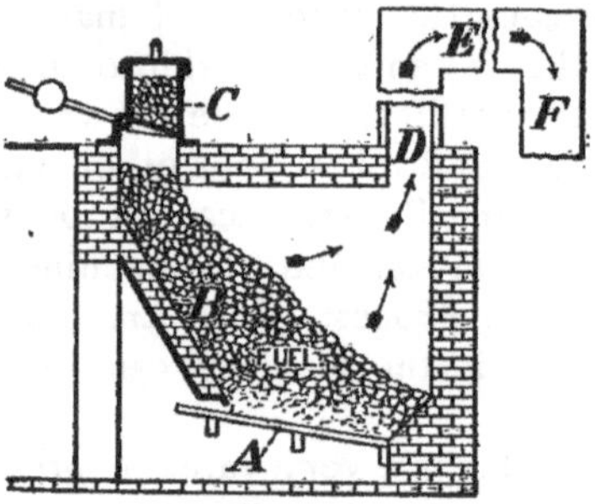

FIG. 29.

chamber of fire-brick, one side, *B*, being inclined at an angle of 45° to 60°. *A* is the grate. The fuel, which may be of any kind, is fed into the producer through the hopper *C*. As the fuel slowly burns, the CO_2 rises through the mass above it, and absorbs an additional portion of C, becoming converted into 2CO. This gas passes out of the opening *D*, into a flue. In order to cause it to flow towards the furnace, it is led through a long pipe, *E*, where it is partially cooled, and it then descends the pipe *F* leading to the furnace. The gas in *F* being cooler than that in *E* and *D*, a constant flow of gas from producer to furnace is maintained.

THE REGENERATORS.—The gas entering the furnace is, as has been stated, CO. To burn it to CO_2, air must be mixed with it. This mixture is made in the furnace proper, the CO and air being kept separate till they reach the point where they are to burn. The CO is cooled to some extent, as shown, before being admitted to the furnace.

To heat both air and CO before they are mixed and burned, and to accomplish this economically, and raise them to a high temperature, the waste heat of the furnace is employed. This is the object of the regenerators, Fig. 30.

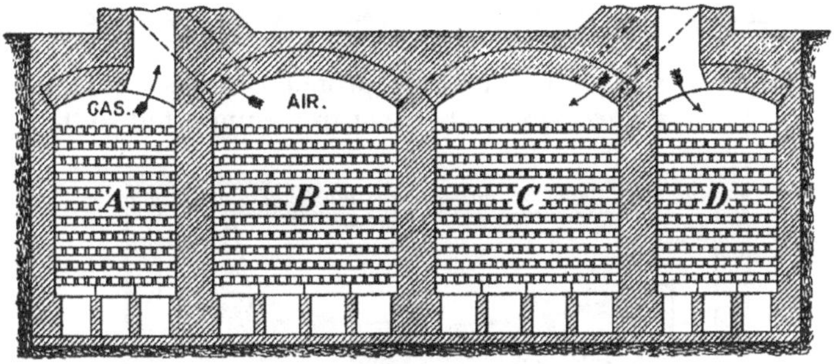

FIG. 30.

They consist of four large chambers below the furnace, filled with fire-brick, piled so that there are intervals between the bricks to allow the gas and air to pass through. Their action is as follows: When the furnace is started, CO is admitted through *A* and air through *B*, both *A* and *B* being cold. These pass up through the fire-bricks in *A* and *B* and through flues at the top, and flow into the furnace proper, where they are lighted. The products of combustion are caused to pass through *C* and *D*, which are similar chambers. In doing so these products heat the fire-bricks in *C* and *D*. After some time,—about one hour generally,—by the action of valves controlled by the workmen, the CO and air are caused to enter the furnace through *C* and *D* respectively, and the products of combustion to pass out through *A* and *B*. In this case the CO and air entering the heated chambers *C* and *D* are raised to a high temperature before ignition, and the temperature of the furnace thereby greatly increased. It is also evident that *A* and *B* will be more highly heated than *C* and *D* were, and hence when the next change is made the gas and air passing through *A* and *B* will be more highly heated than when passing through *C* and *D*, and so on.

The action of the furnace is therefore *cumulative*, and its only limit in temperature is the refractoriness of the material. By regulating the proportions of gas and air, which is readily done, the temperature may be kept constant.

75. Manufacture of Steel — The Furnace — Operation — Crucible Process.

THE FURNACE.—The furnace proper consists (see Fig. 31) of a dish-shaped vessel *D* of cast iron, supported so that

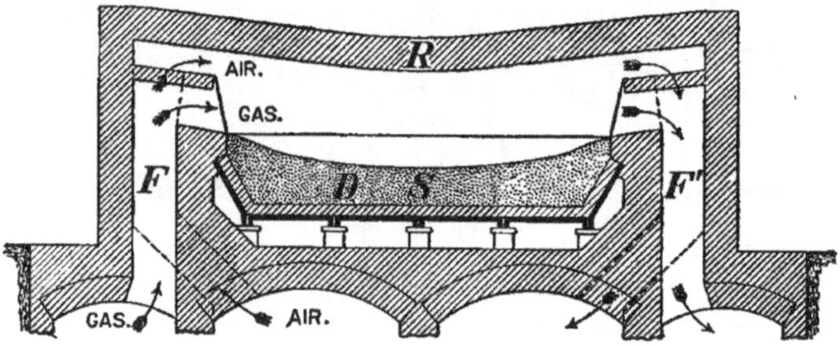

FIG. 31.

the air can circulate freely around it and keep it from melting. This is lined with refractory sand *S*; and in order to repair it when necessary, the pan *D* is generally arranged so that it can be run out of the furnace. This allows it to cool quickly. The pan is placed over the regenerators, and the gaseous fuel and air enter by the flues *F*, and the products of combustion escape by the flues *F'*, or the reverse, according to the position of the regulating-valves.

The arrows show the direction of these currents. The

roof *R* is lined with fire-brick, and by its shape deflects the flame over the metal in the hearth. At opposite ends of the furnace are a charging-door for admission of the metal, and a tap-hole for drawing off the finished steel. These are not shown in the drawing.

OPERATION.—The principle of the process is that when wrought-iron or steel scrap is added to melted cast iron, the percentage of carbon is thereby reduced till it reaches that required for steel. The charge consists of pig-iron heated red-hot in a separate furnace, and then placed on the hearth of the Siemens furnace. By the action of the furnace this pig-iron is soon melted. Scrap wrought iron or steel is then added in suitable proportions, till the percentage of carbon is low. When it has reached the proper point, the percentage is made exact by adding a pig iron containing a known percentage of carbon, such as Spiegeleisen or ferro-manganese, or by the addition of ore. The percentage of carbon is judged of during the process by taking samples from the melted metal, cooling them, observing their fracture on breaking, and by dissolving portions of the specimen in HNO_3 and comparing the color with that of standard solutions of steel in HNO_3 containing different percentages of carbon. In this way the composition of the steel can be exactly regulated, as the metal can be kept in a melted state without damage for a considerable time, and the character of the flame made oxidizing or reducing at will, according to the relative amounts of air and CO admitted.

The operation ordinarily lasts about eight hours for each charge.

When the steel has attained its proper composition, the furnace is tapped and the metal cast into ingots, ready for the succeeding operations.

CRUCIBLE PROCESS.—This is used by Krupp. The ingredients of the steel are melted in crucibles, and the resulting steel from the crucibles is poured into a common reservoir from which the ingots are cast.

The Bessemer process, though important and producing large quantities of steel, is not as yet used in making gun-steel.

76. Casting Ingots—Ladle—Crane—Ingot-mold—Pouring—Sinking Head.

After the proper percentage of carbon is obtained, the steel is cast in ingots.

LADLE.—The first step is to tap the furnace and draw off the steel into a ladle. This ladle is made of boiler-iron lined with refractory sand. It has two trunnions on the exterior which support it, and around which it revolves when tipped, to pour the metal into the mold; or it may have a tap-hole at the bottom closed with a plug of fire-clay, which is lifted by an iron rod covered with refractory material.

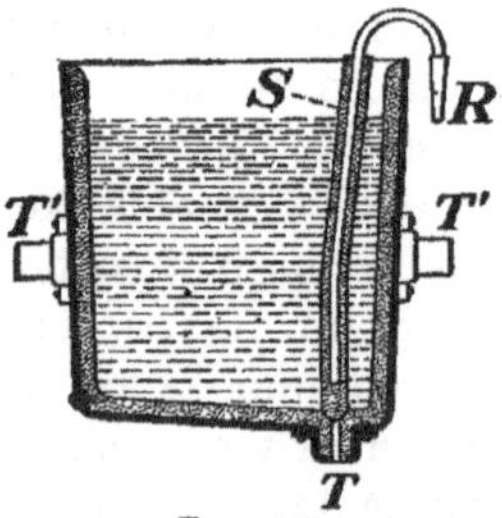

FIG. 32.

In Fig. 32, *T* is the tap-hole, *T'* the trunnions, *R* the rod, and *S* its casing. The advantage of tipping is that it is quicker, and of the tap-hole, that it gets rid of scoria and impurities on the surface of the melted steel, and keeps them out of the mold.

CRANE.—This is used to convey the ladle to the molds, or, more generally, for handling the ingots and molds after the casting. It is very often found more convenient to run the ingot-molds on cars under the ladle, or under a spout attached to the furnace.

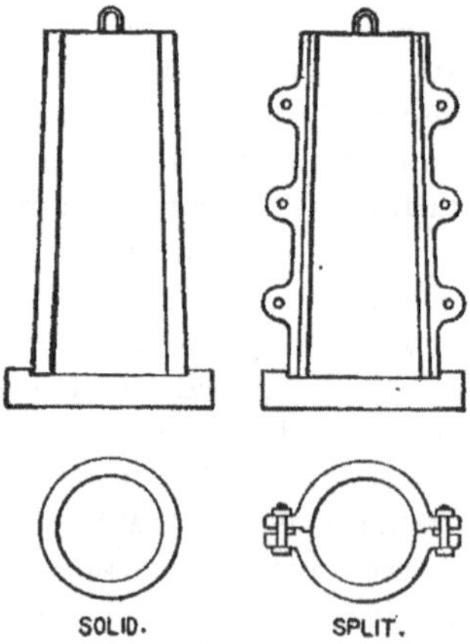

FIG. 33.

INGOT-MOLDS.—These are generally made of cast iron, and are circular in cross-section, to insure uniform cooling. They are in one piece, and slightly conical on the interior, so that the ingot, after casting, may be readily withdrawn. They may also be made in halves, parting on an axial plane; but in this case they are liable to open at the joint, due to warping.

The interior surface is protected by a wash of clay or plumbago. Melted steel poured into an ingot-mold will not adhere to the sides, while melted cast iron will adhere. The reason is that the steel chills and contracts away from the mold, while the iron cools more slowly and fuses the mold. The general shape of the ingot-molds is shown in Fig. 33.

POURING.—If the steel is very hot, it must be poured slowly into the molds in a thin stream. This allows the gases time to escape. If at a lower temperature, it may be poured more quickly.

The ingot-molds may be warmed before casting to prevent undue cooling and consequent strains, and also the formation of pipes.

After the steel is cast, the molds must be covered to exclude air and cause slow cooling.

SINKING-HEAD.—In all castings, whether of iron, steel, or other metal, an excess of metal, called the sinking-head, is left at the top of the mold. This column of metal acts by its weight to give greater density to the lower portions of the ingot; it also serves to collect the scoria and impurities which rise to the top, and it fills any cracks or cavities that may form in the cooling of the ingot. It is necessary to keep this sinking-head fluid as long as possible, and hence it is generally cast in a sand-mold for gun-ingots.

77. Whitworth's Process of Fluid Compression.

This process was invented by Sir Joseph Whitworth of England, and gives by hydraulic pressure, the same effect as that due to the sinking-head. It may be regarded as an artificial sinking-head of great height. The process consists in forcing the piston of a hydraulic ram down upon the melted steel in the mold, and maintaining the pressure till the steel solidifies. The ingot-mold used in this process must be very strong to withstand the great pressure, and it has a peculiar arrangement by which the gases driven out by the pressure are allowed to escape. Fig. 34 gives the general arrangement.

The mold consists of a strong cast-steel cylinder, *A*,

with its bottom, *B*. This cylinder is lined with rectangular bars of wrought iron, *C*, which have grooves, *D*, cut at intervals along their faces in a radial direction. Their rear edges are also cut off longitudinally so that when placed side by side and forming a lining for the cylinder *A* they have continuous longitudinal channels, *E*, parallel to the elements of the cylinder. The grooves *D* communicate with the channels *E*, and thus allow the gas to escape at the top and bottom of the mold. The interior of the mold is lined with refractory sand.

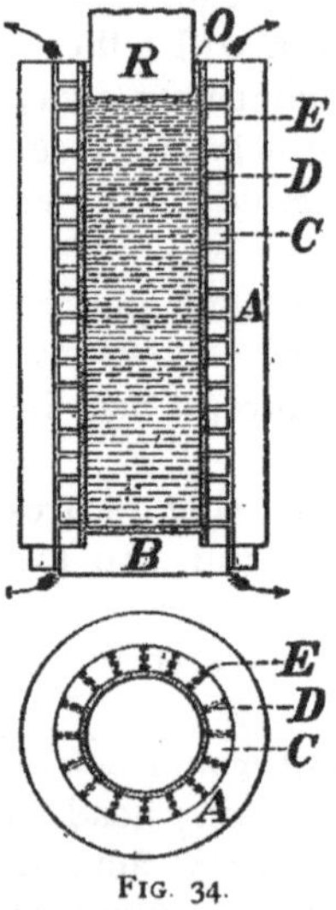

FIG. 34.

Action.—When the melted steel is poured into the mold and the ram *R* forced down upon it, the hot metal is at first forced through the openings *O* between ram and mold. But the metal quickly cools and forms a solid mass, completely closing these openings *O*. The gas is forced out through the channels as shown, and the effect of the pressure and shrinkage is to shorten the ingot about 1½ inches for each foot of length.

Theory.—It appears at first that the metal, as well as the gases, would be expelled through the channels, and also that, since fluid pressure is equal in all directions, there is no reason why this pressure should force the gas out of the melted metal. Dr. Siemens suggests as an explanation that the steel cools first on the exterior where it is in contact with the mold, and offers a greater resistance here to the motion of the ram. It is broken up in consequence by the pressure, and becomes porous. The interior of the mold remaining fluid, and offering less resistance than the outside, receives consequently more compression, and hence the result will be to force the gas outward through the porous exterior. This porous exterior, while allowing the gas to escape, retains the fluid metal. It is also claimed that the pressure increases the solvent action of the metal upon the gases. Krupp maintains, however, that this process of fluid compression simply closes up the cavities but does not expel the gas.

78. Treatment after Casting — Testing — Reheating — Forging— Cranes—Hammer.

TESTING.—After the ingot is cast and cooled, specimens of it are tested chemically to determine its composition, and also in the testing-machine to determine its physical qualities. The ingot is graded according to these tests, and a hole is then bored through it parallel to its axis, removing the central part of the ingot. This hole is for purposes of forging, as will be explained.

If the ingot is short, this hole may be punched; and for small tubes or any solid forgings the hole is not necessary.

REHEATING.—For forging, the ingot is then reheated in a furnace which is a modification of the Siemens. In the reheating, care must be taken to apply the heat slowly and regularly, so as to avoid overheating the exterior before the interior is brought to the proper temperature. If the heat is applied too quickly, the ingot is liable to crack from unequal expansion, and the exterior to be overheated or "burned."

CRANES.—The ingots when heated to the proper temper-

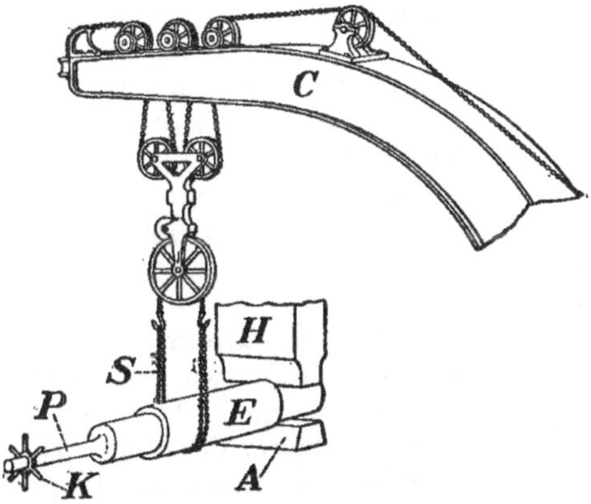

FIG. 35.

-ature are handled by heavy cranes, which remove them from the furnace and carry them to the hammer or press. They

are slung from the crane by a chain called the sling-chain, and are balanced by the addition of extra weights at the cool end, so that they may be readily swung by the workmen, and turned axially under the hammer. Fig. 35 shows the general arrangement for forging an ingot under a hammer. *H* is the hammer, *A* the anvil, *C* the crane, *E* the ingot, *S* the sling-chain, *P* the porter-bar; the handles *K* are used for rotating the ingot under the hammer.

HAMMER.—The steam-hammer is used in forging ingots, except where hydraulic pressure is preferred. These hammers consist of a heavy head or tup attached to the piston of a steam-cylinder. The cylinder is vertical and is supported by two legs, and the hammer thus formed is called an A hammer, from its general appearance. When the steam raises the hammer, and is then exhausted, and the tup allowed to fall by its own weight, we have a single-acting hammer; when the steam acts also to drive the tup down, we have a double-acting hammer. The foundation for the anvil is separate from the hammer to diminish the effect of the blow upon the latter.

Since the same energy may be obtained from a light hammer moving quickly or from a heavy hammer moving slowly, the latter is preferred for heavy masses, as the effect of the blow is better distributed through the mass. Armor-plates are generally made by hammering, as the effect of the blow is felt more on the face and less in the interior, and it is important to have a good quality of face. With gun-forgings both hammer and hydraulic pressure are used with excellent results, the hydraulic press, however, being preferred, as it is more slow in its action and distributes the effect throughout the mass.

79. Whitworth's Hydraulic Forging—The Press—Mandrels.

In this process the ingot is drawn into shape by the pressure of a powerful hydraulic ram. As the action is slow, it is claimed that the effect is better distributed throughout the mass, as before stated, and consequently produces a better effect upon the metal as a whole.

THE PRESS.—This is a large hydraulic ram so arranged

that it may be quickly adjusted to any size of ingot. The general arrangement is represented in Figs. 36 and 37, although the ram actually has many arrangements for adjustment, etc., not shown.

THE MANDRELS.—With this press a secondary or movable anvil, called a mandrel, is used. It is shown in Figs. 36 and 37, and its use is as follows:

If the ingot is to be drawn out into a long forging such as a gun-tube, it is first bored out on the interior. It is then heated, and the mandrel passed through the bore. The ingot is now placed under the forging-press, resting on the fixed anvil as shown in Fig. 36. When pressure is applied

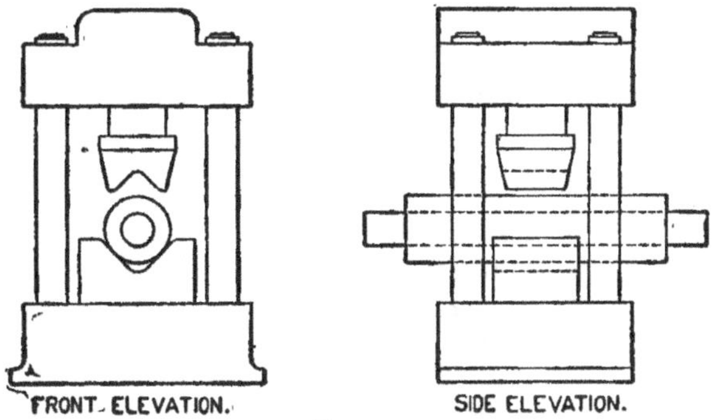

FIG. 36.

under these circumstances, the effect will be to lengthen the ingot, keeping its interior diameter unchanged. On the other hand, if the ingot is to be forged into a hoop, it is bored as before, heated, and the mandrel passed through the bore; but in this case the ends of the mandrel are supported as shown in Fig. 37, the ingot being allowed to swing on the mandrel.

When the pressure is applied under these conditions, it is evident that the walls of the ingot will be compressed, the interior diameter increased, and the length of the ingot will remain practically unchanged.

In the first case we have a fixed mandrel, and in the second a swinging mandrel. A current of water sometimes circulates through the centre of the mandrels to keep them cool, and in the case of the fixed mandrel, it is withdrawn from the forging by a second hydraulic press acting on it.

Comparing the hammer and press, it is claimed for the hammer that its effects are more local, and therefore that it

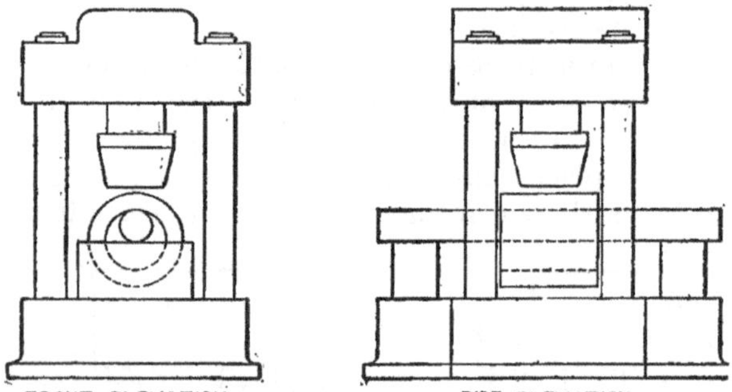

FIG. 37.

is better for armor-plates; that the effect of its blows is to heat the metal, and therefore the temperature may be lower in forging; and that it uncovers defects in the metal, while the press conceals them.

80. Gun-forgings—Treatment after Forging—Annealing—Boring and Turning—Oil-tempering—Re-annealing—Tests.

GUN FORGINGS.—The principal gun-forgings are the tube, the jacket, and the hoops.

The forging of the tube and hoops has been explained, and that of the jacket is exactly similar.

ANNEALING.—After forging, the hammer or press leaves certain strains in the metal, and they must be removed. This is done by annealing. This process consists in heating the forging carefully to a certain temperature, which is determined by experience, and allowing it to cool slowly, in the furnace, the latter being allowed to cool naturally. By this process the steel becomes soft, and all strains are removed.

BORING AND TURNING.—The forging is now placed in a

lathe and bored and turned to near its finished size. Pieces are also taken off the ends as specimens, and tested, to determine the qualities of the metal and as a guide to subsequent treatment.

OIL-TEMPERING.—The object of this process, to which the forging is now subjected, is to give the peculiar property called "toughness" to steel. The practical effect is that it increases the elastic limit and tensile strength, and reduces the elongation before rupture.

Process.—The forging is slowly heated and carefully inspected, till all the parts have acquired the same temperature, which is judged by the color. A long forging is generally heated vertically to avoid warping. When at the proper temperature, it is raised by a crane and lowered vertically into a tank of oil, a current of which is caused to flow through the bore. The oil is surrounded by a water-jacket to keep down the temperature.

Being a poor conductor of heat, the oil allows the steel to cool correspondingly slowly, and thus gives the particles time to adjust themselves, and the result is a considerable increase in elasticity and tenacity, and it acquires the property of toughness already defined.

RE-ANNEALING.—The process of oil-tempering causes internal strains in the metal, and these are removed by reannealing as before. This annealing process reduces slightly the elastic limit and tensile strength and increases the elongation before rupture.

TESTS.—The physical qualities of the metal are now tested. For this purpose specimens are cut from the breech and muzzle ends of each tube, jacket, or hoop forging, and the results of these tests compared. No great difference in quality must exist between breech and muzzle specimens, as this would indicate a variation in quality of the metal from breech to muzzle. In the U. S. service the requirements of the Ordnance Department are about as follows:

Elastic limit, 46,000 to 50,000 pounds per square inch;

Tensile strength, 86,000 to 93,000 pounds per square inch;

Elongation at rupture, 15 to 17 per cent in a length of 3 inches.

81. Brinell's Experiments.

To determine the effects of heating and cooling on the change of structure and the hardening of steel, Mr. J. A. Brinell, a Swedish engineer, made a series of experiments.

Taking a certain kind of steel which contained about the same percentage of carbon as gun-steel, he heated bars of it to different temperatures and cooled them at different rates. After heating and cooling, the bars were broken and the fracture carefully examined, and chemical tests were made to determine the condition of the carbon. He found that there were two states of the carbon—one which he called free carbon and which was associated with soft steel, and the other, fixed carbon, associated with hard steel. In general the soft steel had a crystalline structure and the hard steel an amorphous structure, or one in which the crystals were so small as to lose their crystalline appearance.

His conclusions were as follows:

1. For each steel, hard and soft, there is a certain temperature, called the critical temperature, to which if the steel be heated, and be suddenly cooled, all the carbon will become fixed, and the structure will be amorphous. This is the hardest condition of steel; and hence, to harden it, it is heated to this temperature and cooled suddenly.

2. Hard steel, if heated to this critical temperature and cooled slowly, will acquire the crystalline structure, and all the carbon will become free. Soft steel heated to this temperature and cooled slowly undergoes no change. This is the softest condition of steel, and hence, to anneal it, it is heated to this critical temperature and cooled slowly.

3. If hardened steel, or steel which has been subjected to the first process, be heated to any temperature below the critical temperature, it becomes softer as the temperature increases. That is, with hard steel, as the critical temperature is approached, more and more of the fixed carbon becomes free, and if the steel be cooled either slowly or

quickly after having been heated to any temperature below the critical one, the hardness of the steel is diminished. This process is called tempering, and by it the degree of hardness can be regulated to any extent. It is the process commonly employed by the blacksmith in tool-making. The less the steel is heated the harder it will be.

4. When steel is heated to the critical temperature and cooled very quickly, as by immersion in mercury or acidulated water, it becomes harder than if cooled by immersion in ordinary water; and on the other hand, if cooled more slowly, as in oil, it acquires less hardness but more elasticity.

MACHINES USED IN GUN-MANUFACTURE.

82. General Principles of Machines—Definition—How Motion is Transmitted and Modified.

In order to understand the operations in the manufacture of a modern gun, some knowledge of the general principles of machines is necessary, since all the operations upon the gun after the forgings are received, are conducted in a machine-shop, and the success of the modern gun as a machine for propelling projectiles, depends upon the accuracy with which the machine-work is done in building it.

DEFINITION.—A machine is any instrument or device designed to receive energy from some source, and to overcome certain resistances in transferring this energy to other bodies. (Michie, p. 246.)

The mechanical principles of machines are discussed in the Mechanics, pages 246–281, and it is intended here to give the practical application of these principles as seen in the shops.

Another definition of a machine is, an assemblage of moving parts for the purpose of transmitting and modifying motion and energy.

HOW MOTION IS TRANSMITTED AND MODIFIED.—A machine receives its motion from some source of energy such as the steam-engine, water-wheel, etc., and transmits it through a series of wheels, sliding surfaces, etc., to the point where the work is done.

The source of motion is called the driving-point or prime mover; the parts through which the motion is transmitted, the train; and the point where the work is done, the working-point.

Motion may be transmitted and modified by:

1. Rolling contact of two or more surfaces;
2. Sliding contact, as in gear-wheels, screws, etc.
3. Belts or bands;
4. Linkwork;
5. Cords or ropes;
6. Hydraulic connection.

83. Rolling Contact—Different Forms of Pieces in Rolling Contact.

ROLLING CONTACT.—Let *A* and *B*, Fig. 38, represent two wheels whose axes are parallel. When motion is communicated to *B* it will impart this motion to *A* by the friction of the two surfaces in contact at the point *a*. These circles in contact at *a* are called the pitch-circles, or pitch-lines; the point of contact *a*, the pitch-point. The line *cd* joining the centres of the wheels is called the line of connection, and is the line along which the velocity of the moving pieces is zero.

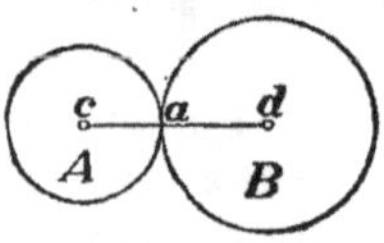

FIG. 38.

The general principle which governs the motion of the pieces in rolling contact is that each pair of points in the pitch-lines which are in contact at any instant, must at that instant be moving in the same direction, and with the same velocity.

This principle leads to the following results:

Since each pair of points in contact must move in the same direction at the same instant, the axes of the wheels and their points of contact must lie in the same plane, because the motion of the points of contact is at right angles to the axis of each wheel; and since the velocities of their points of contact must be equal, the angular velocities of the wheels must be inversely as their radii.

DIFFERENT FORMS OF PIECES IN ROLLING CONTACT.—Besides the circular wheels, we may have—

1. A wheel, *A*, and rack, *B*, Fig. 39,

2. Two wheels with intersecting axes, Fig. 40;

3. Two wheels with axes which are neither parallel nor intersecting. This case will not be considered.

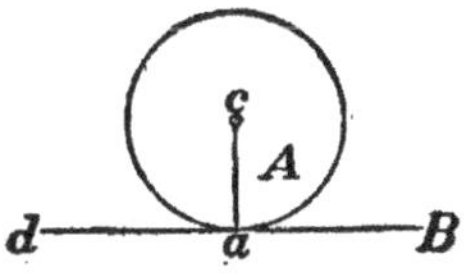

FIG. 39.

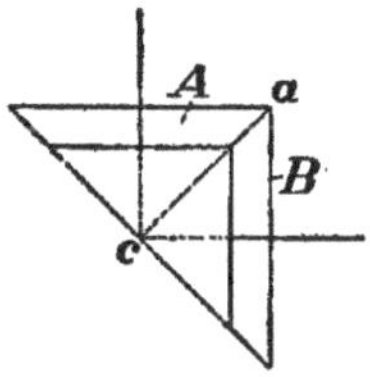

FIG. 40.

For the wheel and rack, since all points in the wheel move at right angles to its axis, while all points of the rack move parallel to itself, or at right angles to the axis of the wheel, the general principle that the points of contact shall be moving in the same direction requires that the axis of the wheel and all points of contact must lie in a plane perpendicular to the motion of the rack, and that, since the points of contact must have the same velocity, the actual velocity of the rack must be equal to the product of the angular velocity of the wheel by its radius. For the two wheels with intersecting axes, if the line *ac* joining the point of contact *a* with the intersection of the axes be regarded as the line of contact of two cones, whose axes are those of the wheels, it is evident that, as the surfaces of the cones come into contact along this line, each pair of points in contact will be moving at that instant in the same direction and with the same velocity. Hence the surfaces of two wheels whose axes intersect, are frusta of cones, whose element of contact passes through the point of intersection of the axes.

84. Sliding Contact—Principles of Teeth—Figures of Teeth—Action.

SLIDING CONTACT.—In the method of communicating motion by rolling contact, it is evident that no great force can be transmitted without danger of the slipping of one wheel on the other. If this happens, the velocity ratio of the two wheels is not constant, and hence this method will not answer for accurate work. Where an exact ratio is to be maintained between the velocities transmitted by two wheels, these wheels must be so connected that one cannot move without the other.

This connection is usually made by means of projections on each wheel called teeth.

PRINCIPLES OF TEETH.—Their construction and operation depend on the following general principles:

Let *A* and *B*, Fig. 41, represent two wheels whose axes are *a* and *b*, and suppose these wheels in contact at *c*.

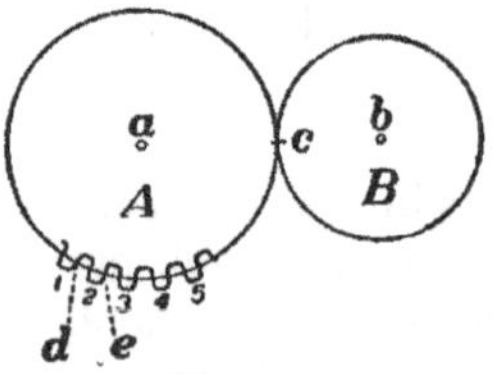

FIG. 41.

Then the circumferences in contact are the pitch-circles, as before explained. Let 1, 2, 3, etc., represent teeth formed upon the wheel *A*. Then the pitch of the teeth is the distance *de* along the pitch-circle from the front of one tooth to the front of the next. Hence—

1. In wheels which rotate continuously for one revolution or more, the pitch must be some aliquot part of the pitch-circle, in order that it may be contained in that circle an even number of times. For a rack, or a wheel which does not perform a complete revolution, this condition is not necessary.

2. In order that two wheels, or a wheel and rack, may work correctly together, the pitch must be the same in each.

3. Hence, in a pair of wheels which work together, the number of teeth in each wheel is directly as the circumference or radius, and therefore inversely as the number of revolutions in a given time.

FIGURES OF TEETH.—These are regulated by the principle that the velocity ratio given by the teeth sliding on each other shall be the same as that given by the pitch-circles rolling on each other.

ACTION OF TEETH.—To give a general idea of this action, let Fig. 42 represent the teeth of two wheels in contact.

The tooth *a* of the lower wheel first touches *b* of the upper at the point *c*. These teeth then slide towards each other till the point *d* is reached, when they slide away from each other and finally lose contact at *e*. This process continues for all the teeth, the arc *cd* being the arc of approach, and *de* the arc of recess, the whole curve *cde* representing the various positions occupied by the point of contact during the action of the teeth.

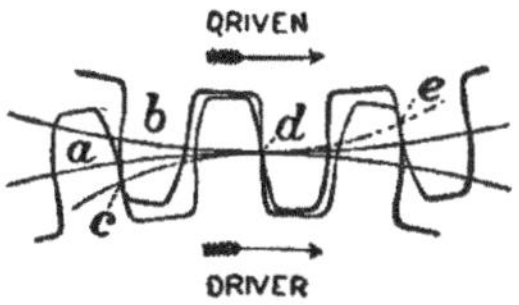

FIG. 42.

The method of describing the figures of teeth is too extensive for discussion here.

85. Belts or Bands—Rounded or Crowning Pulleys—Speed-Cones—Starting and Stopping.

BELTS.—When teeth are used to communicate motion, they possess the great advantage of preserving always the same velocity ratio between two wheels. They have, however, the disadvantage of being a rigid connection, so that they do not allow for starting or stopping, or sudden changes of speed. Hence for the transmission of energy from the engine or other prime mover to the different machines in a shop, belts are almost universally employed. After the energy has been received at any machine, the parts of that machine are connected by gearing, or teeth, if accurate velocity ratios are required.

Belts are generally made of leather or gutta percha, and are broad and flat, and hence require correspondingly shaped pulleys.

The velocity ratio of two pulleys connected by a belt follows the same principle as in the case of rolling or sliding

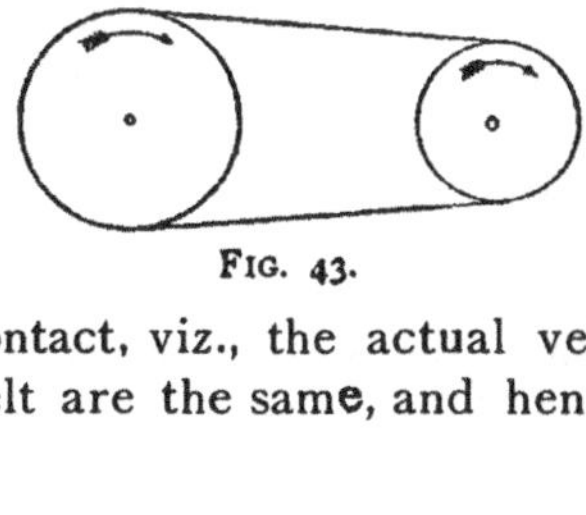
FIG. 43.

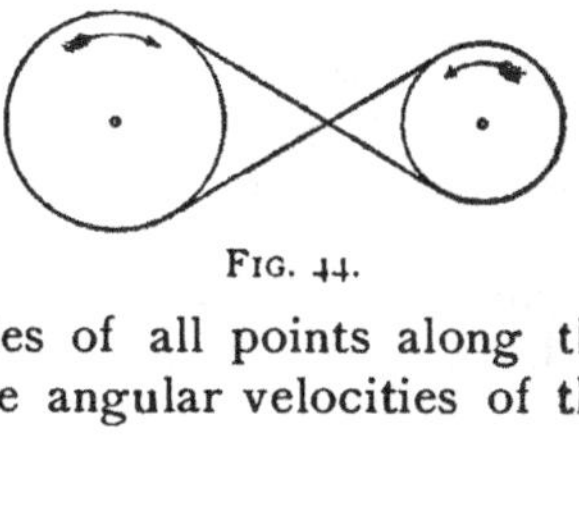
FIG. 44.

contact, viz., the actual velocities of all points along the belt are the same, and hence the angular velocities of the pulleys are inversely as the radii. If the pulleys are to move in the same direction, the belt must be open, Fig. 43; if in opposite directions, the belt must be crossed, Fig. 44.

ROUNDED PULLEY.—To prevent the belt from leaving the pulley, the latter is made crowning or rounded, Fig. 45.

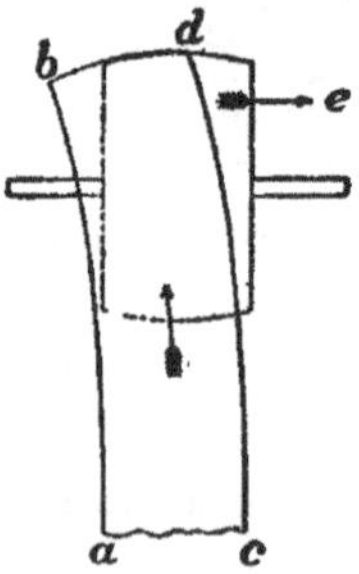

FIG. 45.

A belt always moves toward that part of the pulley whose radius is greatest, and the reason is as follows: When the belt moves to one side of the pulley, the side *ab* of the belt becomes compressed, Fig. 45. The resistance of the side *ab* to this compression produces a force in the direction of the arrow *e*, which straightens the belt and causes it to move to the highest part of the pulley.

SPEED-CONES.—To vary the velocity ratio communicated between a pair of parallel pulleys or shafts by a belt, without stopping the motion of the machinery, speed-cones are used. These may be either continuous cones, Fig. 46, or stepped cones, Fig. 47. In the first case we can obtain a gradual variation of speed, and in the second, certain fixed variations only.

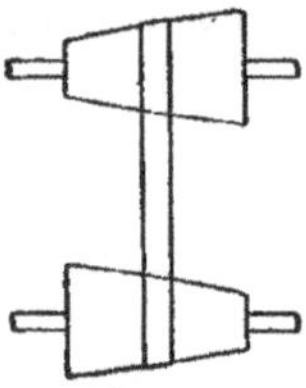
FIG. 46.

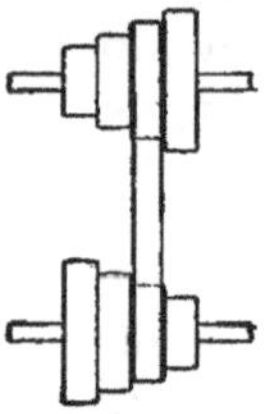
FIG. 47.

The second method is generally used.

STARTING AND STOPPING.—As individual machines require to be started or stopped without interfering with the source of power, each machine is in general provided with two pulleys.

These pulleys are mounted on an independent shaft called a counter-shaft, and one of them is fixed to this shaft, while the other turns freely upon it. When the machine is to be stopped, the belt is shifted to the "loose pulley," as it is called; and when started, to the fixed pulley.

86. Linkwork—Cords and Ropes—Hydraulic Connection.

LINKWORK.—When two rotating pieces are connected

by a rigid bar, as the driving-wheels of a locomotive, this bar is called a link. It may also connect a rotating piece and a sliding piece, as the piston-rod and crank of a steam-engine, which are connected by a link. In the case of link-work, the velocity of all points of the link being the same at any instant, the angular velocities of the rotating pieces are inversely as the perpendiculars let fall from the axes of rotation to the link.

In the case of a rotating and a sliding piece, as in Fig. 48, every point of the sliding piece is moving at a given instant perpendicular to the line *ab*, and at the same instant the point *c* is moving perpendicular to *bc*. Hence a line through the point *b*, perpendicular to the plane of the paper at the intersection of these two lines, is at this instant the instantaneous axis about which the two points *a* and *c* are moving.

Their velocities are therefore directly as their distances from this axis, or

$$v : v' :: ab : cb.$$

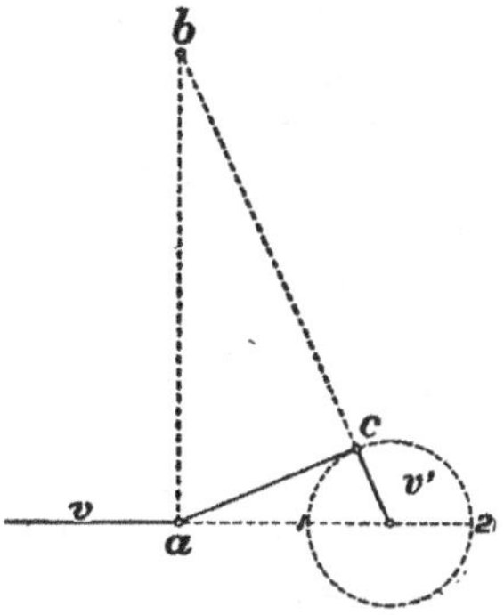

FIG. 48.

The same principle may be applied to linkwork in general. The actual velocity of the point *a* becomes zero when the point *c* reaches the positions 1 and 2, and these points are called "dead-points." In the steam-engine the stored-up energy of the fly-wheel carries the point *c* over the dead-points.

CONNECTION BY CORDS.—This connection is principally made between blocks, forming a block and fall, or block and tackle.

Although very useful, it is not employed to any extent in machine construction. Wire ropes are sometimes used instead of belts to transmit power, in which case grooved pulleys are required to keep the rope from slipping off.

HYDRAULIC CONNECTION.—This is of great importance in modern machinery, as the gun-steel is forged by a hydraulic press, and hydraulic cranes are employed for lifting the heavy weights. The general principle of these machines is explained in mechanics, and depends on the fact that if two cylinders fitted with pistons are in hydraulic communication, the volume of liquid forced out of one is equal to that forced into the other. As this volume is the product of the length of the cylinder by its area of cross-section, it follows that the velocities of the pistons are inversely as their areas. From this principle we can obtain a slow motion and great power, as in the hydraulic press, or a quick motion and less power, as in the hydraulic crane, by regulating properly the size of the cylinder.

87. General Arrangement of Machine-shops — Distribution of Energy.

GENERAL ARRANGEMENT.—All machine-shops are arranged upon the same general principles, though differing greatly in details, depending on the work to be done.

In general, there is first a source of energy, as a steam-engine or water-wheel. This source of energy may be regarded as the reservoir from which energy is drawn as required; and as different amounts of energy are needed at different times, according as different machines are working or not, some arrangement must be made to regulate the amount of energy. Without this regulation, if several machines are suddenly stopped, the energy will be in excess, and the remaining machines will increase in speed. This is injurious to the work and to the machines. The reverse will happen when machines are suddenly started. To regulate the energy of the prime mover, a fly-wheel and governor are used. The fly-wheel stores up energy and gives it out when it is suddenly required, and prevents sudden changes in speed, and the governor regulates the supply of steam, etc., to the engine.

The principles are explained in mechanics.

DISTRIBUTION OF ENERGY.—To distribute the energy from the prime mover to the various machines, any one of the methods previously described may be used. Belts are generally preferred.

The pulleys which carry these belts run upon lines of

shafting. Extending lengthwise through the shop, there is a "main shaft," *a*, Fig. 49.

The motion is communicated directly from the prime mover *b* to this main shaft, by a belt. The shaft is supported by hangers, *c*, bolted to the beams or walls. At intervals along the main shaft are pulleys, *d*, each of which carries the belt for a particular machine.

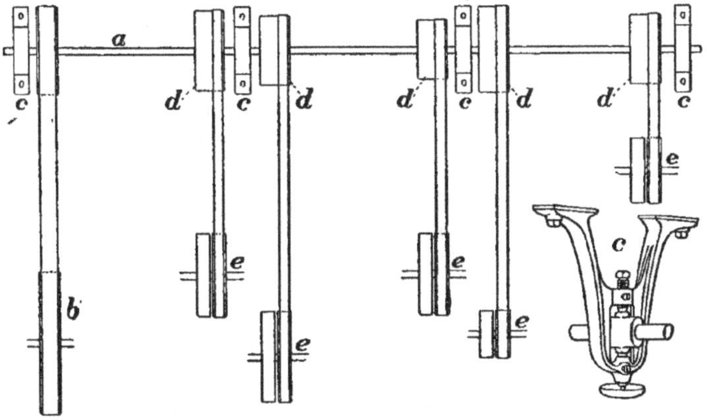

FIG. 49.

Over each machine is a short shaft, *e*, called a counter-shaft.

This carries at least three pulleys, the first running loose upon the shaft, the second fixed to it, and the third also fixed, and driving the machine. Their use has been explained.

In addition to affording a means of starting and stopping any machine without interfering with the main shaft, the counter-shaft affords a means of increasing or decreasing the speed of any machine, by decreasing or increasing the size of the pulleys as compared with those on the main shaft which transmit the power.

88. Machine-tools—Shearing—Cutting—Scraping—Drills—Reamers and Milling-cutters.

In every machine the working point, or part by which the work is actually done, is called a tool. Machine-tools may be classified according to the manner in which they act, as—

1. Shearing-tools;
2. Cutting or paring-tools;
3. Scraping-tools.

SHEARING-TOOLS.—These tools act to divide a plate or bar of the material operated on, by causing the parts to separate from each other by sliding or shearing. This class includes also punches and dies. They are not used to any extent in gun-construction.

CUTTING-TOOLS.—These cut a thin chip or shaving from the surface of the work and thus produce a new surface.

SCRAPING-TOOLS.—These tools scrape off small particles from the surface of the work, and correct any irregularities that may have been left by the cutting-tool.

ACTION OF CUTTING AND OF SCRAPING-TOOLS.—The general method of the action of these tools is shown in Figs. 50 and 51.

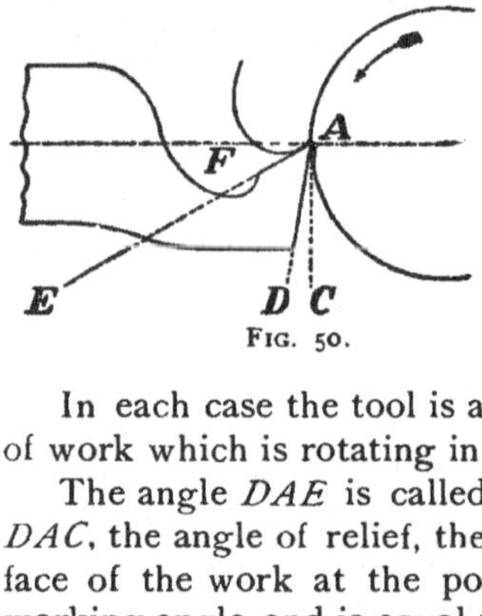

FIG. 50.

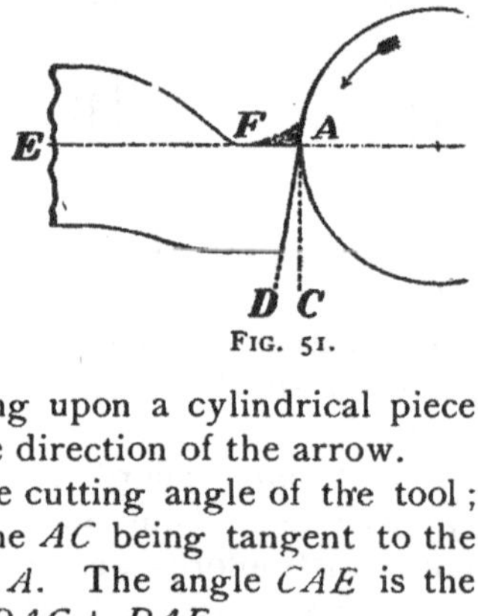

FIG. 51.

In each case the tool is acting upon a cylindrical piece of work which is rotating in the direction of the arrow.

The angle *DAE* is called the cutting angle of the tool; *DAC*, the angle of relief, the line *AC* being tangent to the face of the work at the point *A*. The angle *CAE* is the working angle, and is equal to *DAC* + *DAE*.

In cutting-tools the angle *CAE* is always acute; in scraping-tools the angle *CAE* is either right or obtuse: and the tools are thus distinguished by their working angles.

The hook *F* is given to the tools in order that the cutting edge *A* shall not be above the axis or centre line of the tool. If this were the case, any springing of the cutting edge caused by excessive resistance of the material, would move the edge *A* further into the work, or cause it to "dig into" it, while as arranged the cutting edge will spring away from the work. In plan, the tool may be of various shapes, as shown in Fig. 52, these shapes depending on the nature of the work.

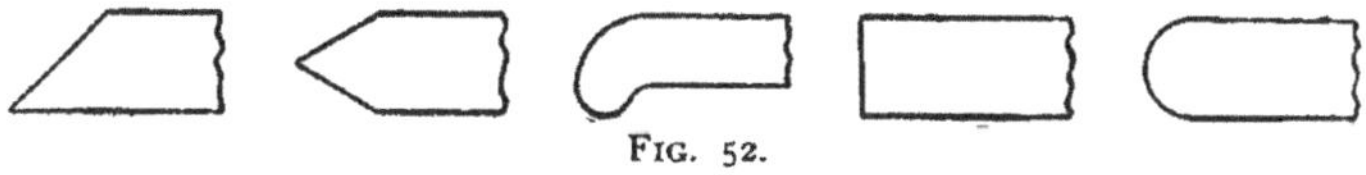

FIG. 52.

DRILLS AND REAMERS.—For making cylindrical holes, drills and reamers are used. The ordinary drill is shown in Fig. 53, the cutting edge being *adb*; the reamer in Fig. 54.

The reamer consists of a number of parallel cutters forming a cylinder, and is used to finish a cylindrical hole that is required to be very true and smooth. Drills and reamers rotate about the vertical axis *cd* and have generally a motion in the direction of this axis.

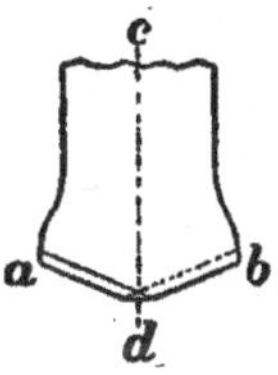

FIG. 53.

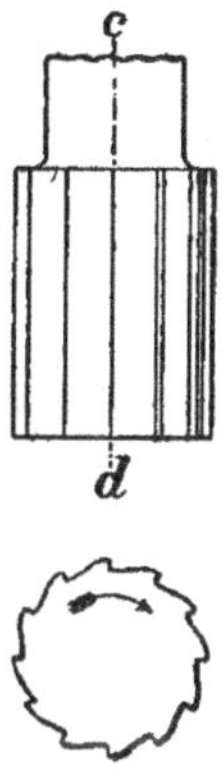

FIG. 54.

MILLING-CUTTERS.—These may be used to form surfaces of almost any shape, and they vary greatly in form. The general method of their operation is indicated in Fig. 55, in which the irregular surface *abcd* is cut by the milling-cutter *A* rotating on the axis *B*.

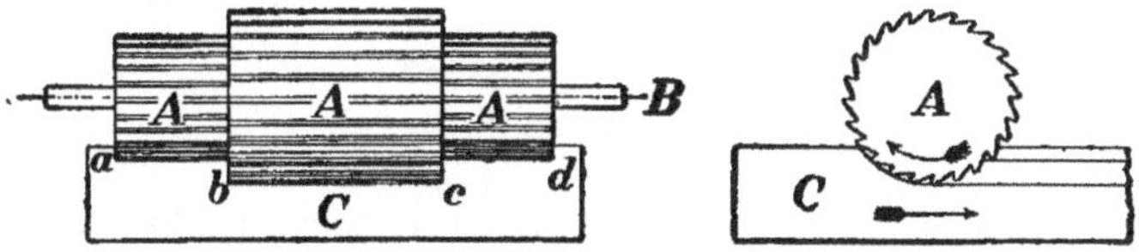

FIG. 55.

The work *C* moves along a plane director at right angles to the axis *B*.

89. Machines in General Use—The Lathe—Parts.

MACHINES IN GENERAL USE.—The machines in general use are

The lathe;
The planer;
The shaper;
The drill-press;
The milling-machine.

THE LATHE.—This machine is intended principally to produce accurate surfaces of revolution. Its general arrangement is as follows:

A piece of metal or wood is caused to revolve about one of its lines as an axis. A cutting- or scraping-tool is made to bear against the metal or wood. As the latter, which is called the "work," revolves, the tool is caused to move either parallel or perpendicular to the axis of the work, or

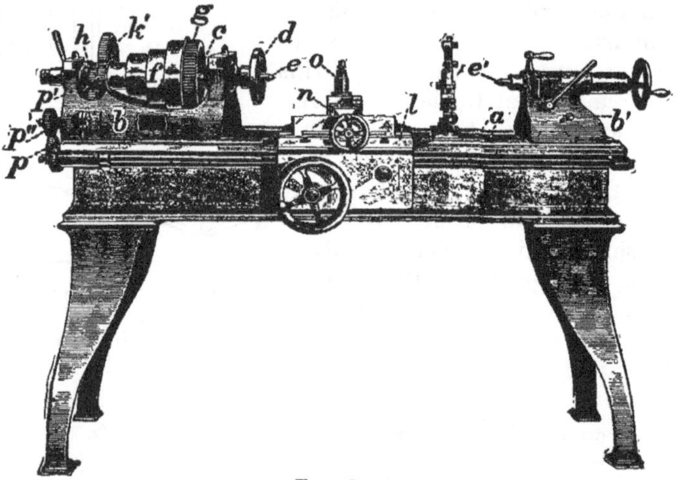

FIG. 56.

in a direction which is compounded of these two motions. The tool cuts a chip or shaving from the surface of the work, and by a continuation of its action produces either a cylinder, a plane surface, a cone, or any other surface of revolution which may be formed by combining the two motions at right angles to each other.

PARTS.—The principal parts of the lathe, Fig. 56, are the bed, consisting of two parallel ways or guides, *a*, of a Λ-shaped cross-section.

On these guides slides the support for the tool, which is thus made to travel parallel to the axis of the work. At one end of the ways is fixed a heavy block of metal, *b*, called the head-stock. This forms a support for the spindle *c*. To this spindle (see Fig. 57) is attached the face-plate *d*, by means of a screw, *d'*, on the end of the spindle. This spindle is hollow at one end, and in this hollow fits a conical piece of metal, *e*, called the live-centre. The spindle also carries a speed-cone, *f*, and a gear-wheel, *g*. The gear-wheel is fixed to the spindle, while the cone revolves freely upon it. The gear-wheel *g* and cone *f* may be connected by a bolt, *i*, passing through *g*. The small end of the speed-cone terminates in a gear-wheel, *h*, which is a part of the cone, and hence runs free on the spindle, but revolves with the same angular velocity as the cone. Parallel to the lathe-spindle *c* is another axis, *k*, Fig. 57, carrying two toothed wheels, *k'* and *k''*.

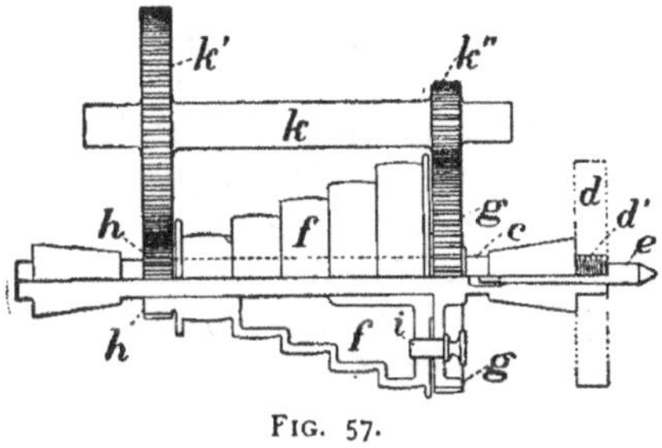

FIG. 57.

This axis *k* is mounted in eccentric bearings, and may be moved so that its wheels, *k'k''*, will engage or disengage with those on the lathe-spindle *c*. The arrangement of the axis *k* and wheels *k'* and *k''* is called the back gear. At the opposite end of the lathe-bed is a second block of metal, *b'*, resting on the ways, called the tail-stock. It also supports a spindle, called the dead-spindle, and this spindle terminates in a conical piece of metal, *e'*, called the dead-centre or back centre. The tail-stock may be moved to any position along the ways, and clamped there, and the dead-spindle has a sliding motion parallel to the axis of the lathe, which enables the distances between the centres *e* and *e'* to be very accurately adjusted. These centres *e* and, *e'*, form the axis of revolution for any work in the lathe; and if they are removed, the prolongation of the axis of the live and dead spindles forms this axis.

90. The Lathe—Slide-rest—Feed—Action.

SLIDE-REST.—This forms the support for the cutting-tool, and through it motion is given to the tool in any direction. It consists (Fig. 58) of a slide or bed, *l*, resting upon the ways, *a*, of the lathe, and capable of travelling along them by the action of the feed-screw *m*. Upon this slide rests a second or cross slide, *n*, which moves at right angles to the first slide, and hence at right angles to the axis of the lathe. This cross-slide carries a tool-holder, *o*.

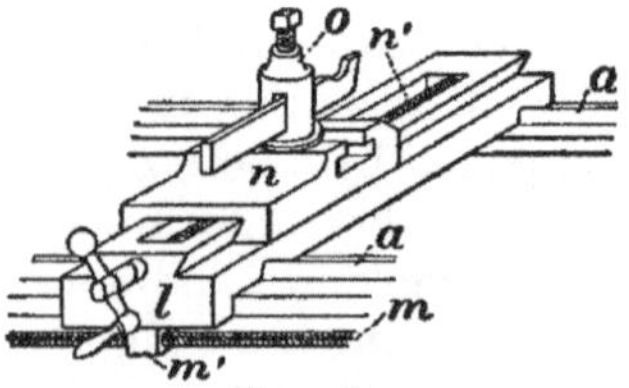

FIG. 58.

FEED.—The screw *m* is called the feed-screw. It passes through a nut, *m'*, on the slide-rest, and this nut is made in halves which can be separated, thus freeing the nut from the feed-screw, and stopping the longitudinal travel of the slide-rest. The cross-feed is given by hand or automatically by gearing, by means of the screw *n'*. On one end of the feed-screw *m* is fixed the gear-wheel *p* (Fig. 56). Attached to the lathe-spindle is a second gear-wheel, *p'*, and upon an axis fixed to the head-stock or some convenient part of the lathe-bed is a third gear-wheel, *p''*. This arrangement may be varied according to circumstances, and is intended to regulate the velocity ratio of the lathe-spindle and that of the feed-screw. Suppose, for example, it is required to cut a screw having ten threads to the inch, and that the feed-screw of the lathe has this number. Then it is evident that the work must turn ten times while the tool moves one inch, and also that, in order to move the tool one inch, the feed-screw must turn ten times. In other words, the velocity ratio of the feed screw and of the work is that of equality. Hence, from what has been stated under Toothed Wheels, it follows that p and p' must have the same number of teeth. The number of teeth upon p'' will not affect the velocity ratio, since, being a lever of equal arms, it receives and transmits the motion from p to p' without change.

ACTION OF LATHE.—Motion is imparted to the lathe from the belt running on the speed-cone f. By placing the belt on the different steps of this cone considerable variation of speed may be obtained. If a slower speed than that given by the cone is desired, the back gear is used. The action of the back gear is as follows:

When the back gear is in gear with the lathe, the cone-pulley is detached from the large gear g by removing the bolt, i, Fig. 57. The cone-pulley then rotates, and its small gear h drives the large wheel k' of the back gear. The speed of the back-gear shaft is thus reduced in the inverse ratio of the numbers of teeth of h and k', and with this reduced speed the gear k'' drives g, which in turn drives the lathe-spindle. The speed is again reduced here in the inverse ratio of the numbers of the teeth of g and k''.

The action of the feed-screw is evident. By throwing the feed-screw out of action and causing the cutting-tool to move at right angles to the axis of the lathe by the screw n' a plane surface will be formed, and by combining the longitudinal and transverse motions in various ways any surface of revolution may be produced.

91. The Planer—Parts—Action.

The object of this machine is to make a flat surface, as nearly plane as possible. Its general principles are as follows:

A large table is made to slide along two parallel plane surfaces. Upon this table is fixed the work. Above the table the cutting-tool is firmly supported. As the table slides, the tool bears against the work, and cuts a chip or shaving, leaving a surface which is an exact copy of the parallel plane guiding surfaces of the table. The table and work then slide back, and at the end of this motion the cutting-tool is moved sidewise an amount equal to the width of the cut. This side motion is called the feed. The table with the work again moves forward, and the tool makes a second cut, and these operations are repeated till the work is finished.

PARTS.—The machine (Fig. 59) consists of a bed, *a*, which is essentially two parallel beams or cheeks having on the

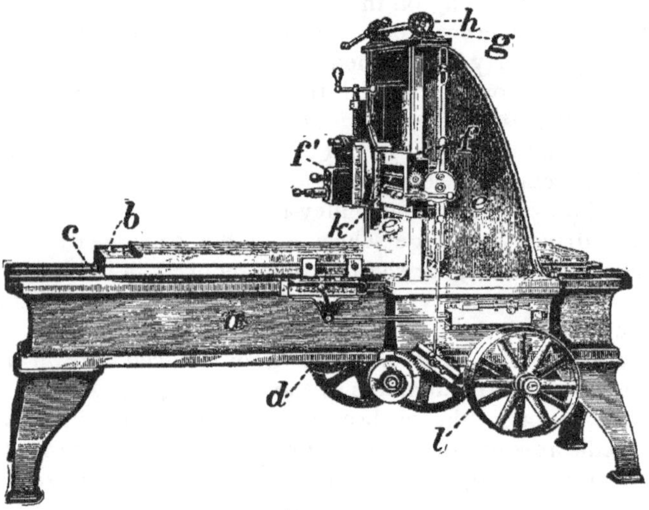

FIG. 59.

upper surfaces two V-shaped grooves, which are the guide-grooves. The table *b* has two corresponding projections on its under side which fit into these guide-grooves. Along the middle of the under side of the table is a rack, *c*, into which gears a toothed wheel, *d*, by which the table is driven. Two vertical standards, *e*, support a cross-slide, *f*, and this cross-slide carries the tool-holder *f'* and tool. The cross-slide can be raised or lowered upon the standards by the screws *g*, acted on by the bevel gears *h*. A feed-screw, *k*, runs through the cross-slide, and gives the feed motion already spoken of to the tool.

ACTION.—The machine is driven by two belts passing over pulleys, *l*. As the motion is reversed at every stroke, one of the belts is open and the other crossed, as previously explained.

The gearing is also arranged so that the backward movement of the table after the cut is much quicker than the forward motion, when the tool is working. This is to save time. The action of the machine is automatic both as to motion of table and feed, and can be set to any length of stroke. At the end of the forward travel of the table a projecting arm on it moves a lever, and this shifts the belts on the pulleys, bringing the reversing-belt into action. At the end of the backward motion of the table the feed is brought into action, and the tool prepared for its next cut. To prevent breaking the cutting edge of the tool by dragging it over the cut in the backward motion of the table, the tool-holder is hinged so that it allows the tool to rotate in the direction of the return stroke, but holds it firmly against rotation in the opposite direction. The same principle will be found later in the rifling-tool.

92. The Shaper—Parts—Action.

There are certain objections to the planer which have led to the introduction of a modified form of the machine called the shaper. For small work, or for short strokes of the tool, power is wasted in moving the heavy bed of the planer, and when it is necessary to stop the stroke of the tool at some definite point, as at a shoulder, it is difficult to do this with the planer on account of the delay in shifting the belts. The shaper remedies these defects.

PARTS.—It consists of a bed, *a* (Fig. 60), along which slides a head, *b*. This head carries a ram, *c*, upon which the cutting-tool is fixed. This ram moves backward and forward at right angles to the bed, and this transverse motion is given by a link, *d*, attached at one end to the ram and at the other to an arm, *e*, upon the toothed wheel *f*. The work is supported upon the tables *g* or *h*, as the sliding-head may be moved to any position along the bed; or if the work is too long for either table, it is supported at each end by them. *i* is an arbor or shaft attached to the bed, and intended for cylindrical surfaces. The tables *g* and *h*

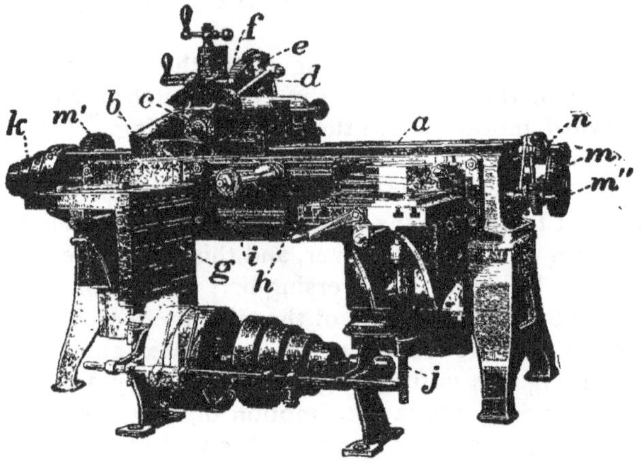

FIG. 60.

can be adjusted vertically by screws, one of them being shown at *j*.

ACTION.—The machine is driven by a belt on the speed-cone *k*. Motion is communicated from this cone to a back shaft *m* through the gear-wheel *m'*. On this back shaft is a small pinion splined to the shaft, so that it will slide freely along the latter and yet turn with it. The toothed wheel *f* is driven by this pinion, and this gives motion to the arm *e*, and this to the link *d* and ram *c*. A feed-screw, *n*, is connected with the sliding head *b*, and is driven by the toothed wheel *m''* on the back shaft *m*. This gears into a pinion on the feed-screw, and by means of proper gears any feed can be given to the sliding head.

By this arrangement the sliding head is fed along the bed *a* a certain distance, just before the beginning of each stroke. By changing the point of attachment of the link *d* nearer to or further from the centre of *e*, the length of stroke of the ram may be decreased or increased, and by changing its point of attachment to the ram the position of the tool may be regulated. The speed is varied by the cones. There is also a very ingenious mechanical device invented by Sir Joseph Whitworth to cause a slow forward motion of the tool while cutting, and a quick backward motion.

93. The Drill-press—Parts—Action.

This machine is used for making cylindrical holes of comparatively small size. For large sizes, such as the interior of tubes, gun-hoops, etc., a boring-mill, or boring lathe, is used.

PARTS.—The principal parts, Fig. 61, are the frame *a*, which supports all the parts; the table *b*, upon which the work is held; the speed-cone *c*, which gives motion to the drill and the other parts of the machine; the spindle *d*, which holds the tool *e*; the feed-screw *f*, which gives a ver-

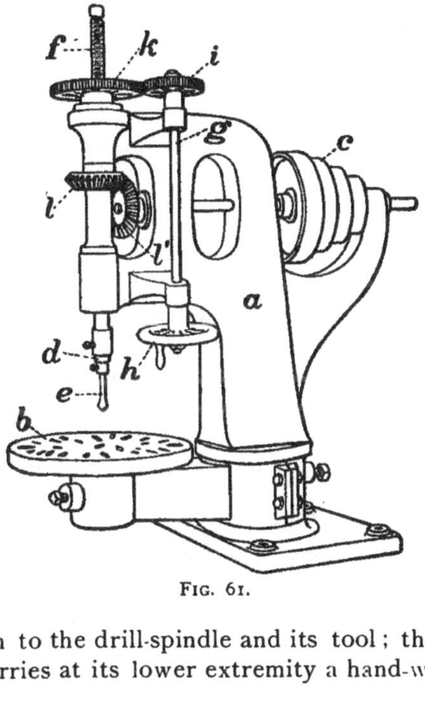

FIG. 61.

tical motion to the drill-spindle and its tool; the feed-shaft *g*, which carries at its lower extremity a hand-wheel, *h*, and at its upper end a pinion, *i*; this pinion gears into a toothed wheel, *k*, whose hub or centre forms a nut through which the feed-screw *f* passes; this wheel and nut are held in a collar, so that it can rotate freely but cannot change its position vertically; the bevel gears *l l'* give a motion of rotation to the spindle *d*.

ACTION.—When motion is communicated to the speed-cone *c* by a belt, it drives the bevel gear *l'*, and this drives *l*.

The hub of the bevel wheel *l* is hollow, and the drill-spindle *d* passes through it. By means of a 'spline' the spindle can slide freely through the hub of *l*, but is compelled to rotate with it no matter what its position vertically may be. The work rests on the table *b*, and the tool *e* is in contact with it. As the drill rotates, the tool is pressed down upon the work by the action of the feed-screw *f*, which rests upon the upper end of the drill-spindle and is connected with it by a collar, so that the spindle can turn without causing rotation of the feed-screw. As the work progresses, the tool is fed down or pressed down by turning the hand-wheel *h*, which causes the pinion *i* to rotate, and this in turn rotates the toothed wheel *k*. When the work is finished, a reverse motion of the hand-wheel *h* causes the feed-screw *f* to rise, carrying with it the drill-spindle and drill.

In all ordinary drills the feed is both automatic and by hand.

94. The Milling-machine—Parts—Action.

The milling-machine is a development of the principle of the lathe, and is used for forming any irregular surface whose elements in one direction are parallel to a plane director. In this machine the cutter rotates, while the work moves at right angles to the cutter and along a plane surface.

PARTS.—The machine, Fig. 62, consists of the bed *a*, which supports a frame, *b*, carrying a spindle and cone, *c*, with back gear, *d*, as in the lathe. To the frame is attached a horizontal arm, *e*, for the support of the outer extremity of the axis or arbor of the milling-tool *f*. This tool is fixed upon an axis or arbor, one end of which is supported by what may be called the live-centre, and the other end by the dead-centre at the extremity of the horizontal arm *e*. Below the cutter is a table, *g*, which moves at right angles

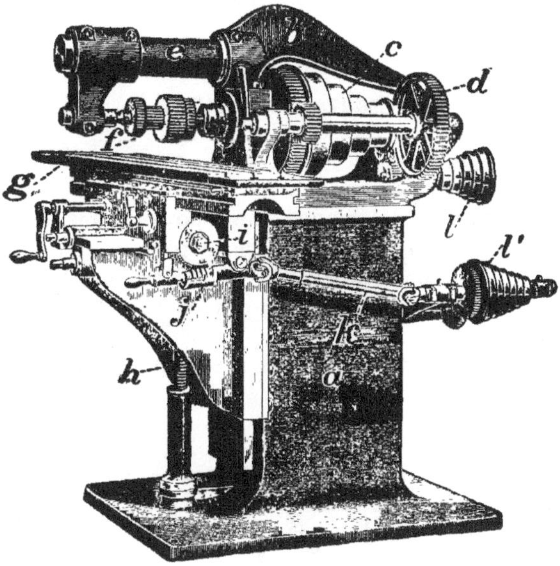

Fig. 62.

to the axis of the milling-cutter. This table is capable of adjustment vertically by the screw *h*, and is fed transversely by the feed-screw *i* driven by the worm-gear *j* through the shaft *k* and cone-pulleys *l l'*.

Action.—Motion is communicated to the cone-pulley *c* by a belt, and this causes the cutter *f* to rotate. Feed-motion is also communicated to the table *g* from the cone-pulley *c* through gear wheels to the cone *l*, and thence by a belt to *l'*. From *l'* it is communicated to the shaft *k* which drives the worm-wheel *j*, and this drives the feed-screw *i*.

With this machine it is not necessary to have a constant velocity ratio between the motion of the tool and that of the work, and hence belts instead of gearing are used for the feed. Also, since the cut is heavy owing to the large tool employed, a slow but powerful feed is required, and this is obtained with the worm-gear *j*. It is evident that the profile of the cutter may be of any figure within wide limits. Many varieties of these machines are used, and they are largely employed in the manufacture of the minor parts of small arms.

GUN-MANUFACTURE.

95. General Description of Modern Guns—Parts—The Forgings—Division of Operations.

DESCRIPTION.—All modern high-power guns are made of steel, and are composed of several parts united to form a whole, and the parts are so arranged as best to support the stresses upon them. The gun is therefore called a "built-up" gun.

PARTS.—The principal parts are (Fig. 63): the tube, *T*,

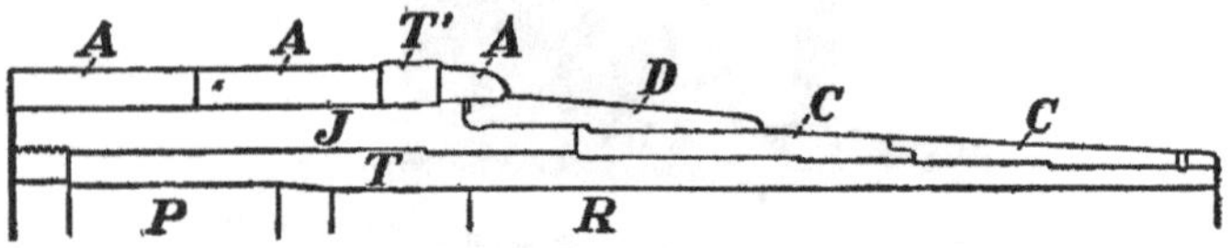

FIG. 63.

which iorms the interior of the gun and supports the other parts. This contains the powder-chamber, *P*, and the rifling, *R*. The jacket, *J*, is the next larger forging, and rests upon the exterior of the tube, carrying in rear the base-ring in which the threads of the breech-block engage. The *hoops* may be divided into two classes, the chase-hoops, *C* and *D*, and the reinforce hoops, *A*. The trunnion-hoop, *T'*, belongs to the latter class, and there may be one or more layers of each class according to the size of the gun. The hoops are arranged to break joints when two layers overlap, or to lock into each other when stiffness is required, as in the chase-hoops. The interior diameters of the jacket and hoops are less than the corresponding exterior diameters of the tube and the parts enveloped. This difference of diameters is called the shrinkage, and its amount, and the reason for using it, will be discussed later. These cylinders are put in place by heating them till they will pass over the part to be enveloped, and then cooling them in place.

FORGINGS.—The manufacture of the forgings and their treatment has been explained. At the gun-factory they are finish-bored, turned, and assembled to form the gun, and after assembling, certain operations are required upon the gun itself before it is ready for service.

DIVISION OF OPERATIONS.—The mechanical operations in gun-building are therefore naturally divided into:

1. Operations before assembling;
2. Operations after assembling.

96. Operations before Assembling—Tube—Warping—First Boring —Tool—Second Boring—Tool.

WARPING.—As received from the manufacturers, the tube is liable to be bent or warped, due to the oil-tempering.

The amount of this warping is ascertained by mounting the tube in a lathe, the ends being centred; and as the tube rotates, the deflection at the middle, or at the point where it is greatest, can be measured. If found to be considerable, it may require a readjustment of the axis of the tube in the lathe, or it may be so great as to cause rejection of the tube, though this latter seldom occurs.

FIRST BORING.—The tube is bored before being turned, in order that when turned there may be a uniform thickness of metal at every point in the same circumference. The first boring is done with a tool so arranged that it will run straight. This is necessary, because when received the bore of every tube is irregular to some extent, and the tube is generally warped or bent slightly. The bore must be exactly parallel to the axis of the lathe, as, in case of deviation from this line, the tool may run so far to one side as to spoil the tube. A deviation of about 0.25 inch in a length of 20 feet would reduce the thickness of metal on one side so much that the tube would be useless.

TOOL.—The tool used for this purpose is a semi-cylinder of cast iron, *A*, Fig. 64, carrying a cutting-tool of steel, *B*, in front. This semi-cylinder exactly fits a hole in the bore of the tube, which hole is previously bored very accurately with a small lathe-tool. The tool *A* is supported by a long

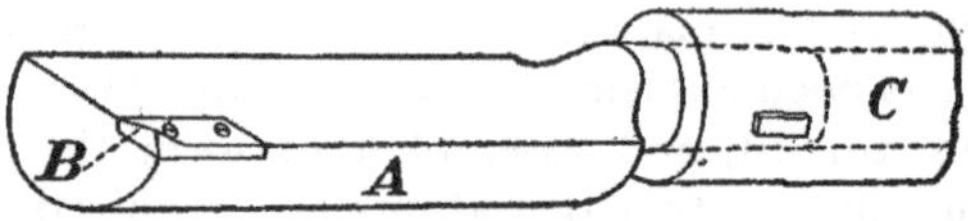

FIG. 64.

bar, *C*, called a boring-bar, which is pushed forward by a feed-screw, as in the ordinary lathe.

The tube is caused to rotate while the tool is pushed forward; and since the semi-cylinder *A* accurately fits the hole in the bore at starting, and is constantly forced down against it by the presssure of the cut on *B*, it produces a cylindrical surface along which *A* slides, without deviation. The length of *A*, being about three times the diameter of the bore, also corrects any tendency to deviation.

SECOND BORING.—The first boring gives a straight hole, but it is not smooth or regular. It is necessary now to use a tool which will remedy these defects.

TOOL.—The tool used for this purpose is called a wood reamer. It consists, Fig. 65, of a flat cast-iron head, *A*, car-

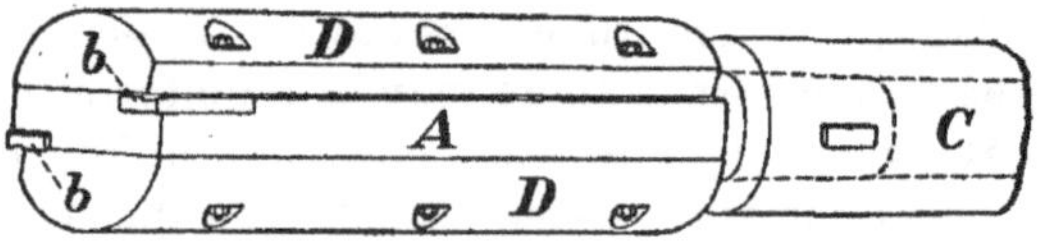

FIG. 65.

rying two cutters, *b b*, so that a cut is made at opposite extremities of a diameter. *D D* are two pieces of hard wood bolted to the cast-iron head, and turned to a diameter slightly greater than that of the hole to be made by the cutters *b b*. This packing *D*, guides the cutters, and keeps them steady, and being thoroughly oiled it polishes the bore.

The cutters are slightly wedge-shaped or conical, so that

they tend always to move towards the axis of the hole already bored. By having two cutters, each of them does one half the work of a single cutter, and hence the tool travels comparatively rapidly down the bore; and from this fact, and also because a light cut is taken, and the cutting-edge of the tool is long, so that the work is well distributed, it follows that the wear of the tool is slight, and the bore very smooth and uniform. Fig. 66 illustrates this, being exaggerated to show the principle.

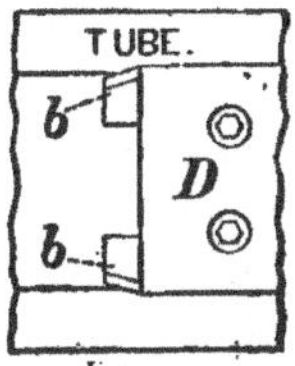

FIG. 66.

The tool is supported in the same bar, *C*, and fed forward as with the first tool, the tube rotating.

97. Boring and Turning Lathes—Back Rests—Bore of Tube—Turning.

LATHES.—In all these operations the tube is mounted in a boring and turning lathe. These lathes consist, Fig. 67, of the bed *B*, made very strong and much larger than is the ordinary lathe; the head-stock and cone-pulley *C*; the face-

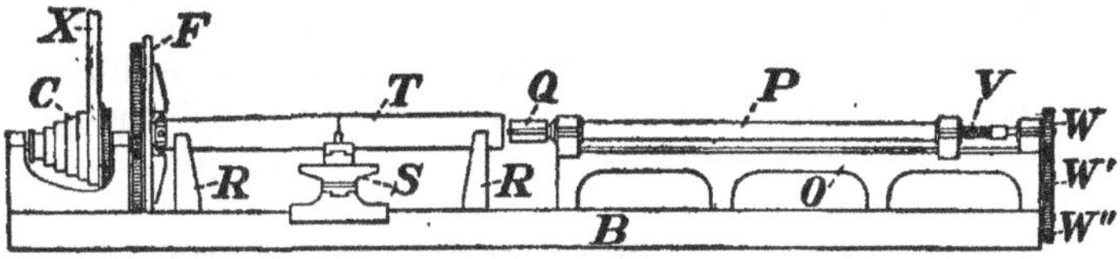

FIG. 67.

plate *F*; the slide-rest *S*, carrying a turning-tool; the back rests *R R*, forming intermediate supports for the tube *T*; the boring-bed *O*, supported on the bed proper, *B*, and carrying the boring-bar *P* with its tool *Q*; the feed-screw *V*, which lies inside the boring bar *P*; and the gears *W*, by which the feed-screw is driven.

Motion is communicated to all the parts by the belt *X*, acting on the cone-pulley. This causes the tube to rotate, and also communicates motion to a long shaft, not shown in figure, upon the end of which is the lower gear-wheel, *W''*. The motion is transmitted through *W'* to *W*, and thence to the feed-screw *V*, and by changing the gears any ratio between the velocity of rotation of the tube and that of translation of the tool Q can be obtained. The back rests RR can be adjusted to any diameter of forging, and the boring-bar moved forward or backward. It is necessary that there be only one source of motion, since if the feed-screw or slide-rest were driven independently of the cone-pulley, a change in speed of one would not cause a corresponding change in the others, and hence damage to tools, tube, or machine might result.

The slide-rest is driven by a second feed-screw not shown.

In this lathe, the work may be turned on the exterior while boring is in progress. It is best, however, not to make a heavy cut on the exterior during boring, as it may cause bending of the tube and consequent irregularity of bore. Each lathe is supplied with an oil-pump, by means of which a stream of oil is forced into the bore while the work is in progress. The chips or cuttings come out at the opposite end of the tube from that at which the tool enters. The same machines in general are used for boring and turning jackets and hoops, with some slight changes necessitated by the difference in size of the forgings.

BORE OF TUBE.—Before assembling, the tube is bored below its finished size, as the cooling of the jacket and hoops causes irregular contraction of the bore, and metal enough must be left to remove these irregularities and give a uniform bore.

TURNING.—After or during boring, the exterior of the tube is turned to the proper diameter. The exterior of the jacket and hoops is not turned before assembling, as changes in these diameters are caused by the shrinkage, and it is preferable to finish them after assembling.

98. Assembling—Furnace—Expansion—Cooling.

The parts having been turned and bored as explained are carefully measured to see that their dimensions are correct. A variation of 0.003 inch is allowed from prescribed diameters. If the dimensions are correct, the parts are ready for assembling.

FURNACE.—The jacket is first placed on the tube. To do this the jacket must be expanded sufficiently to allow the tube to pass readily through it. As a general rule, an expansion of 0.004 inch per inch of diameter is sufficient. That is, if the interior diameter of the jacket be 15.00 inches, it is to be expanded

$$15.00 \times .004 = 0.06 \text{ inch},$$

and the expanded diameter will be $15.00 + .06 = 15.06$ inches.

To obtain this expansion the jacket is heated in a furnace. This furnace consists essentially of a vertical cylinder of cast or wrought iron closed completely except at the top, where the forging is introduced and removed. This cylinder is surrounded by a fire-box so arranged that the heat shall be as uniform as possible at all points. This uniformity of heating is essential to prevent warping of the forging and consequent difficulty of assembling. The forging is protected from direct contact with the fuel, to insure uniformity of heating, and also to prevent dirt from collecting on it, as this would be difficult to remove.

EXPANSION.—The amount of expansion has been stated. The heat necessary to obtain this expansion varies slightly with different forgings, but ordinarily it does not exceed 600° F.

The requisite expansion is determined by noting the colors which form on the polished surface of the steel, as these colors pass through a regular gradation, from pale yellow to purple, blue, and black. The latter color is seldom exceeded.

Gauges are also made of the exact diameter to which the bore should expand. When the color indicates the proper expansion, the gauges are tried, and when they will enter the bore, the requisite expansion has been attained.

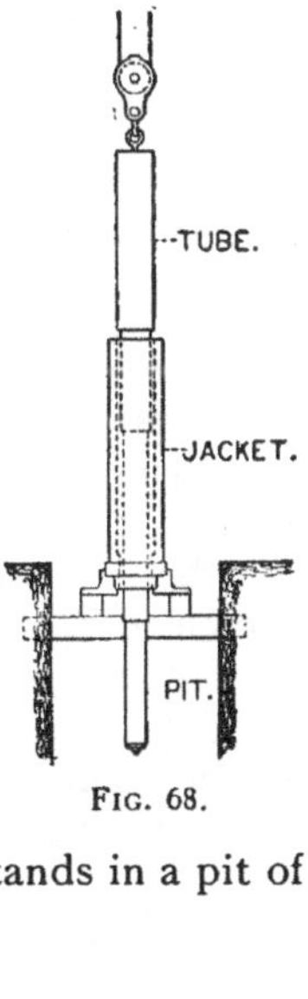

FIG. 68.

ASSEMBLING.—The furnace door is now opened, the jacket hoisted vertically out of the furnace by a crane, and placed on a casting, as shown in Fig. 68. This casting stands in a pit of sufficient depth to contain the tube. The tube is lowered slowly through the jacket till it is in place.

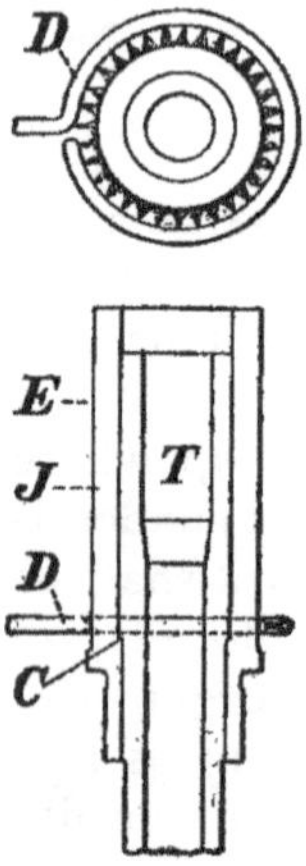

FIG. 69.

COOLING.—The heated forging is now cooled by the application of water as follows:

Fig. 69 shows a section of the tube of 8″ gun with hot jacket in place; *J* is the jacket, *T* the tube resting against a shoulder, *C*, in the jacket; *D* is a ring formed of pipe bent into a circle, the inside being perforated with small holes about ¼ inch apart.

This pipe is placed above the shoulder *C*, so that the jacket in cooling may contract or "draw" toward this shoulder, and hence insure a tight joint there. A current of water circulates through the pipe, and issues from the small holes on its interior against the hot jacket. By this means the cooling can be readily effected, the ring being gradually moved upward, toward the breech, as the cooling progresses. It is important that the parts below *E* be cooled first, as otherwise the jacket will grip at *E*, and on cooling and contracting longitudinally, it will compress the tube in this direction, and produce great longitudinal strains. The same process applies to hoops, the water being applied first at the joint between the cooled and the hot hoops, in order to cause contraction toward the joint, and keep the latter closed.

99. Operations after Assembling—Finish-boring—Rifling—Rifling-machine—Rifling-tool.

FINISH BORING.—The gun after assembling is placed in the boring-lathe, and finish-bored up to the true diameter. The wood reamer is used for the final boring. The powder-chamber, and the slope connecting this with the bore, are also finished.

RIFLING.—The next operation is rifling, or cutting the spiral grooves in the bore for giving rotation to the projectile. This operation requires a special machine and tool.

RIFLING-MACHINE.—This resembles to some extent the

boring and turning lathe already described, but differs in the following respects:

1. The gun does not rotate;

2. The cutting-tool has a motion both of rotation and of translation.

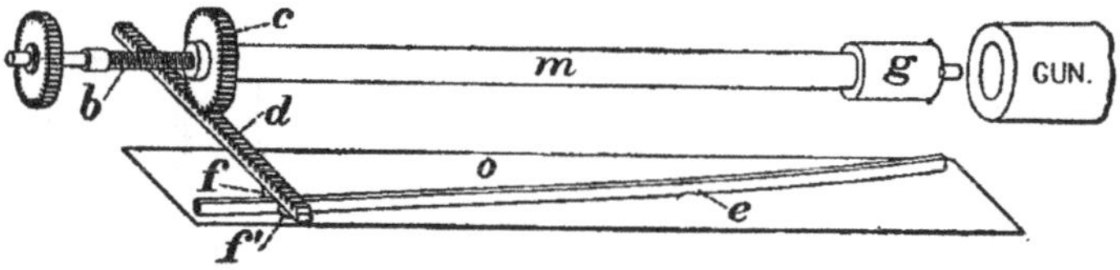

FIG. 70.

Fig. 70 shows the outlines of the rifling-machine. The gun is supported on a bed as for boring, Fig. 67, and the rifling-bar *m* is supported as in boring the tube.

The feed-screw *b* gives the motion of translation to the rifling-bar *m* and tool *g*.

To the side of the rifling-bed is bolted a table, *o*, which is horizontal, and on this table is bolted a "guide bar" *e*, made of flexible steel, and whose shape is that of the developed groove of the rifling. A toothed wheel or gear, *c*, is fixed to the rifling-bar, and a toothed rack, *d*, engages with this gear. At the outer end of the rack are two small rollers, *ff'*, embracing the steel guide-bar *e*. The rifling-bar *m* is free to turn about its axis while moving forward.

The action is as follows: When the rifling-bar is driven forward by its feed-screw *b*, it carries with it the toothed rack *d*.

The outer end of this rack travels along the guide-bar *e*, and as the roller *f* bears against this guide-bar, the rack is pushed inward or to the left in the figure. This causes the gear *c* to rotate to the right, carrying the rifling-bar with it, and thus the rifling-tool is caused to cut the proper groove in the gun.

RIFLING-TOOL (Fig. 71).—This is a cylindrical head, *c*, of metal accurately fitting the bore. Four radial arms, *d*, slide in grooves in the front face of the cylinder, and carry the cutters *k* on their outer ends.

The inner ends rest on a wedge, *e*, which has a sliding motion parallel to the axis of the cylinder. By sliding this wedge forward the radial arms and cutters are pushed out, and by sliding it backward they are pulled in. By this means the depth of the cut or feed is regulated.

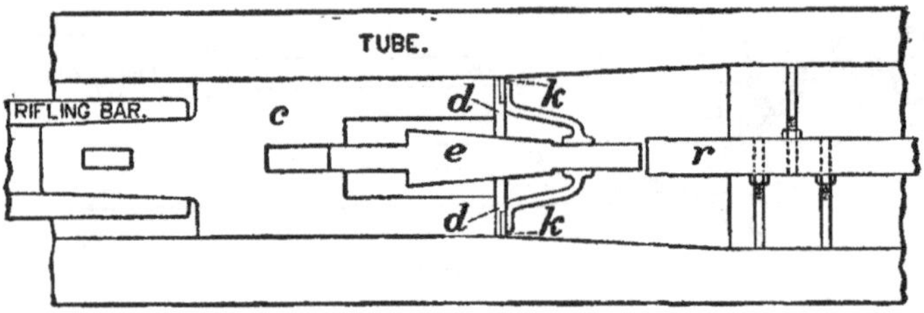

FIG. 71.

When the tool reaches the end of the groove in the gun, the projecting end of the sliding wedge strikes a rod, *r*, in the bore, and the cutters are thus drawn back, which prevents breaking them by dragging them over the cut; the motion of the machine is then reversed and the tool drawn out of the bore. As the rifling-bar and tool move forward, a stream of oil is forced on the cutters by a pump.

Arriving at the end of the cut, the cutters are automatically withdrawn as explained; and as the motion of the rifling-machine is reversed the bar and tool return, being guided in their return motion by the bearing of the roller *f'*, Fig. 70, upon the outside of the guide-bar. The sliding wedge is then adjusted for the next cut, and pushed out to the front, raising the cutters, and so on till the groove is finished. To cut the next groove, the rifling-bar is turned in its bearings a distance equal to the width of one land of the rifling, and the new groove cut as above described.

The remaining operations are *finish-turning*, *inserting the breech-screw*, *fitting the mechanism*, *marking*, and *weighing*, and are not different from the ordinary operations of a machine-shop.

ELASTIC STRENGTH OF GUNS.

100. Definitions—Case Considered—Radial Stress and Strain.

DEFINITIONS.—In the following discussion *stress* means the force acting in pounds or tons per square inch, and *strain* the effect of this force; this effect being either extension or compression, and expressed in inches per inch of length.

Elastic Strength.—The elastic strength of a cylinder or gun is the greatest stress to which it can be subjected without straining any part of the cylinder or gun beyond its elastic limit.

CASE CONSIDERED.—To show the stresses acting upon a cylinder, and the strains produced by them, let us consider the case of a single cylinder, closed at both ends, and acted upon by an interior pressure only, the exterior pressure being that of the atmosphere, and consequently so small that it may be neglected. This case corresponds to that of a gun composed of a single piece of metal, closed at one end by the breech, and at the other by the projectile, and acted on by the pressure of the powder-gas. It is evident that a normal stress is acting upon all parts of the interior of this cylinder, including the ends.

RADIAL STRESS AND STRAIN.—Take a ring of this cylinder, whose length is unity, Fig. 72, and consider a cube of this ring whose edges are unity.

Let p represent the normal stress upon the inner surface of this cube. Then the effect of this stress is to compress the cube in the direction of the radius, and to decrease the thickness of the wall of the cylinder. It also increases the length of the radius. Since the same is true for every unit-cube into which the ring may be divided, we conclude:

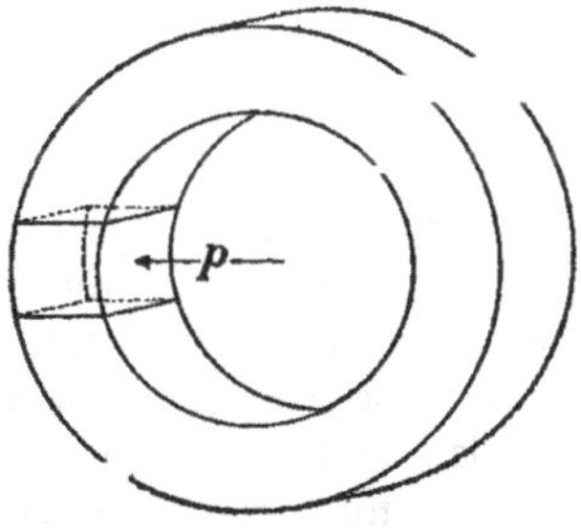

FIG. 72.

(1) That one effect of an interior stress upon a closed

tube is to strain the wall of the tube in the direction of the radius;

(2) That this stress decreases the thickness of the walls of the tube, and increases the interior radius,

This is called the radial stress, and its accompanying strain is the radial strain.

It must remembered that for the particular case considered the effect is as stated. But we may have both an interior and an exterior stress acting at the same time, or we may have an exterior stress acting alone, the interior stress being zero.

According to the relative values of the stresses acting we may have therefore a radial strain of extension or of compression, as will be shown later.

101. Tangential Stress and Strain—Longitudinal Stress and Strain —Conclusions.

TANGENTIAL STRESS AND STRAIN.—Consider again the same ring of metal as before, whose length is unity, and take any particular unit cube, as *a*, Fig. 73.

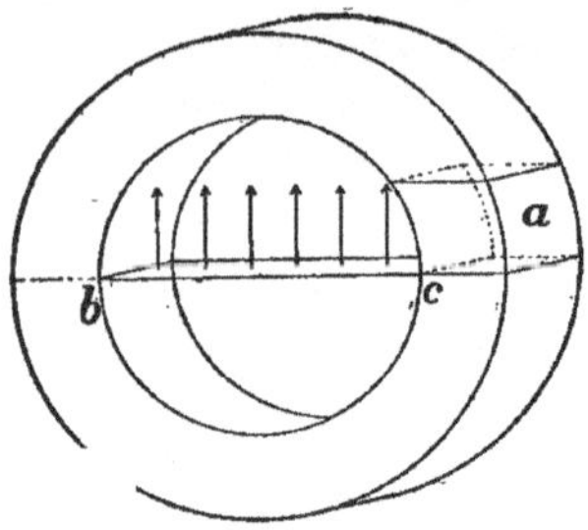

FIG. 73.

The stress p acts normal to the diametral plane bc, and its effect is to separate the cylinder into two halves along this plane. Hence the edges of the cube a parallel to the direction of the stress or normal to the plane bc are strained by this stress, and this is true for the whole cube; hence the effect is to elongate the cube in this direction. This is called the tangential or circumferential stress, or the hoop tension, and it produces a corresponding strain. Its amount is obtained by multiplying the intensity of the stress by the area over which it acts. The intensity is p, and for each side of the ring the area over which it acts is $r \times 1 = r$. Hence the resultant tangential force is pr. This force is resisted by the elasticity of the fibres, and it produces a corresponding stress in these fibres, which at any point is represented by t. Since the same may be shown for each of the unit cubes, the total effect of this stress is to strain or elongate the interior circumference of the cylinder in the direction of the tangent. This also increases the length of the radius. Hence we conclude:

(1) That another effect of the interior stress upon a closed tube is to strain the wall of the tube in the direction of the tangent.

(2) That the stress increases the interior radius of the tube.

As in the case of the radial stress, it must be remembered that this stress may decrease the circumference of the interior layer, or shorten the radius, depending upon the resultant of the forces acting.

It appears from the above discussion that the radius is changed by both the radial and the tangential stresses, and the two cases must not be confused.

LONGITUDINAL STRESS AND STRAIN.—In addition to the radial and tangential stresses acting on the unit cube, there is a third stress due to the pressures on the ends of the cylinder. This stress acts parallel to the axis of the cylinder, and its effect is to strain the elementary cube in the direction of this axis. Since this is true for each cube, the resultant strain is an elongation of the tube in this direction.

This is called the longitudinal stress, and it produces a corresponding strain.

CONCLUSIONS.—If we follow the same method of discussion for the case of an exterior and an interior stress acting at the same time, or for the case of an exterior stress acting alone, the interior stress being zero, similar results will be obtained.

Hence we conclude in general that when a single cylinder is acted on by exterior and interior stresses, their effect is to produce in the cylinder:

1. A radial stress, p, and its corresponding strain;
2. A tangential stress, t, and its corresponding strain;
3. A longitudinal stress, q, and its corresponding strain;

and that all these stresses exist at the same time and at every point of the cylinder.

102. Relations between Stresses and Strains when all the Forces are Tensions—Application to Cube in Gun-cylinder.

RELATIONS BETWEEN STESSES AND STRAINS.—Since all the stresses p, t, and q exist at the same time, and each produces its own strain, it is required to find the resultant strain due to these three stresses acting together.

For this purpose it is more simple to consider at first, three stresses of the same kind.

If a cubical elastic solid be acted on by a given stress in a direction normal to one of its faces, experiment shows that it produces a corresponding strain in that direction, and that it will also produce contrary strains in the two directions at right angles to the first, equal to one-third the first strain. Thus, Fig. 74, if the force p act on the cube in the direction shown, it will elongate the edges aa, bb, etc., and will contract the edges ac, ab, and this contraction will be one third the elongation of aa, bb.

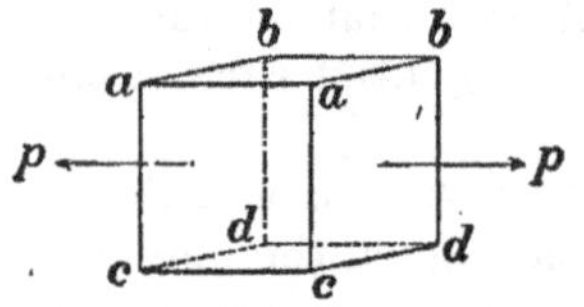

FIG. 74.

This law holds only within the elastic limit. Consider the general case of a cube acted on by the three stresses X, Y, and Z, at right angles to the faces of the cube, and suppose these stresses to be tensions.

Let λ be the resultant strain in the direction of the stress X;

μ, that in the direction of Y;

ν, that in the direction of Z;

E_0 the modulus of elasticity of the cube.

The stress X, by a preceding principle (see equation (169)), produces a strain in its own direction equal to

$$\frac{X}{E_0}.$$

The stress Y decreases this strain by the amount

$$\frac{1}{3}\frac{Y}{E_0},$$

and the stress Z by the amount

$$\frac{1}{3}\frac{Z}{E_0}.$$

Hence the total strain in the direction of X is

$$\lambda = \frac{1}{E_0}\left(X - \frac{Y}{3} - \frac{Z}{3}\right).$$

In the same way we have for the total strains in the direction of Y and Z

$$\mu = \frac{1}{E_0}\left(Y - \frac{X}{3} - \frac{Z}{3}\right);$$

$$\nu = \frac{1}{E_0}\left(Z - \frac{X}{3} - \frac{Y}{3}\right).$$

APPLICATION TO CUBE IN GUN-CYLINDER.—Referring now to the unit cube in the gun-cylinder, we have the same case, except that one of the stresses is a compression. Hence, substituting in the above equations t for X, $-p$ for Y, and q for Z, we have

$$\left.\begin{aligned} \lambda &= \frac{1}{E_0}\left(t + \frac{p}{3} - \frac{q}{3}\right) \\ \mu &= -\frac{1}{E_0}\left(p + \frac{t}{3} + \frac{q}{3}\right) \\ \nu &= \frac{1}{E_0}\left(q - \frac{t}{3} + \frac{p}{3}\right) \end{aligned}\right\} \quad . \quad . \quad . \quad . \quad (170)$$

In these equations λ is the strain in the direction of the circumference or tangent, μ the strain in the direction of the radius, and ν the strain in the direction of the axis of the cylinder, due to the action of the three forces p, t, and q at any point. These strains may be extension or compression according to the relative values and directions of the forces.

103. Lamé's First Law Connecting t, p, and q.

In equations (170) the values of λ, μ, and ν are unknown, and they are expressed in terms of t, p, and q, which are also unknown.

Hence in order to find the values of λ, μ, and ν in known terms it is necessary to establish certain equations of condition, by means of which t, p, and q may be replaced by known terms. This has been done by what are known as Lamé's formulas, from the name of the distinguished investigator of the subject of the elasticity of solid bodies, M. Lamé.

To deduce the first law it is assumed that the longitudinal stress q is constant throughout the cross-section of the cylinder.

This is not exactly true, but the results obtained upon this hypothesis are sufficiently exact in practice.

Assume the last of equations (170),

$$\nu = \frac{1}{E_0}\left(q - \frac{t}{3} + \frac{p}{3}\right).$$

In this equation q is constant by hypothesis, and the value of ν varies only with t and p. One third of the difference of these quantities is the only variable in the above equation; and since t and p vary together at different points, the variations in the value $\frac{t-p}{3}$ will be small, and may be neglected in comparison with q.

Hence we may assume without material error that ν is constant throughout the cross-section of the cylinder.

From this we have

$$t - p = 3(q - \nu E_0) = \text{constant}, \quad . \quad . \quad . \quad (171)$$

or

$$t - p = \text{constant}. \quad . \quad . \quad . \quad . \quad . \quad . \quad . \quad (172)$$

Therefore

$$\left.\begin{aligned} t - p = T_0 - P_0 \\ t - p = T_1 - P_1 \end{aligned}\right\} . \quad . \quad . \quad . \quad . \quad . \quad (173)$$

in which T_0, P_0, T_1, and P_1 are the values of t and p at the interior and exterior of the cylinder respectively.

From this we have Lamé's First Law:

The difference between the tension and the pressure is the same at all points.

104. Lamé's Second Law.

The second law, or second equation of condition for p, t, and q, is deduced as follows:

In Fig. 75

Let R_0, be the interior radius of the cylinder;

R_1, the exterior radius;

r, the radius of any circle of the section;

r', the radius of a circle of the section, exterior and near to that whose radius is r;

p, t, and q, the radial, tangential, and longitudinal stresses, respectively, at any point of the circle whose radius is r;

P_0 and T_0, the values of p and t for the interior of the cylinder;

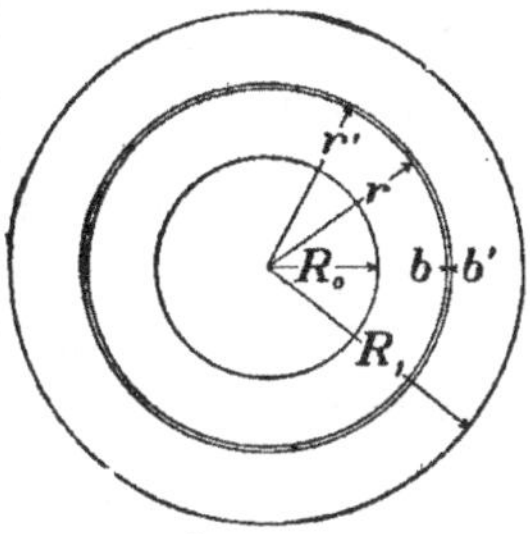

FIG. 75.

P_1 and T_1, the values of p and t for the exterior of the cylinder;

q, constant for all parts of the cross-section;

E_0, the modulus of elasticity.

Consider the cylindrical ring whose radii are r and r' and whose length is unity.

The interior pressure p, as previously shown in the discussion of tangential stress, produces a tangential stress on the interior of the ring equal to pr.

For the circle whose radius is r' the pressure p becomes p', and the force causing tangential stress on the exterior of the ring is $p'r'$.

There is therefore a difference in tension between the two parts of the ring equal to

$$p'r' - pr,$$

and this difference of tension is balanced by the product of the thickness of the ring $r' - r$, and the mean stress along bb', or along any other part of its thickness.

We have therefore the following equation for the whole ring, since the tension and pressure have opposite signs.

$$2p'r' - 2pr = -2\tau(r' - r), \quad . \quad . \quad . \quad . \quad (174)$$

τ being the mean stress throughout the thickness of the ring. From which

$$\frac{p'r' - pr}{r' - r} = -\tau. \quad . \quad . \quad . \quad . \quad . \quad (175)$$

Passing to the limit by making $r' = r$, in which case τ becomes t, we have

$$\text{limit of } \left(\frac{p'r' - pr}{r' - r}\right)_{r' = r} = \frac{d(pr)}{dr}, \quad . \quad . \quad . \quad (176)$$

$$\text{limit of } (-\tau)_{r' = r} = -t. \quad . \quad . \quad . \quad . \quad . \quad (177)$$

Hence

$$\frac{d(pr)}{dr} = -t. \quad . \quad . \quad . \quad . \quad . \quad . \quad (178)$$

From (171) we have

$$t = p + 3(q - \nu E_0). \quad . \quad . \quad . \quad . \quad . \quad (179)$$

Substituting the second member for t in (178), we have

$$\frac{d(pr)}{dr} = -p - 3(q - \nu E_0). \quad . \quad . \quad . \quad . \quad (180)$$

Differentiating; p and r being variables,

$$\frac{pdr + rdp}{dr} = -p - 3(q - \nu E_0). \quad . \quad . \quad . \quad . \quad (181)$$

Reducing,

$$-\frac{dr}{r} = \frac{dp}{2p + 3(q - \nu E_0)}. \quad . \quad . \quad . \quad . \qquad (182)$$

Integrating,

$$\log_e\left(\frac{1}{r}\right) = \frac{1}{2}\log_e[2p + 3(q - \nu E_0)] + \log_e C. \quad . \qquad (183)$$

Substituting the value of $p + 3(q - \nu E_0)$ from (179), we have

$$\frac{1}{r^2} = C^2(t + p), \quad . \quad . \quad . \quad . \quad . \quad . \qquad (184)$$

or

$$(t + p)r^2 = \frac{1}{C^2} = \text{constant}. \quad . \quad . \quad . \qquad (185)$$

From which we can write

$$\left.\begin{aligned}(t + p)r^2 &= (T_0 + P_0)R_0^2 \\ (t + p)r^2 &= (T_1 + P_1)R_1^2\end{aligned}\right\} \quad . \quad . \quad . \quad . \qquad (186)$$

and from these we have Lamé's Second Law:

The sum of the tension in the direction of the circumference, and of the pressure in the direction of the radius, varies inversely as the square of the radius.

Formulas (172) and (185) are known as Lamé's formulas.

105. Curve of Stresses in a Cylinder—Discussion.

Lamé's formulas enable us to determine the stresses existing at every point of the cross-section of a cylinder subjected to the action of exterior and interior forces. A curve showing the relation between the radii and the stresses for all points of the cross-section is called a curve of stress, and is thus determined.

Assume equations (173) and (186):

$$\begin{aligned}(t + p)r^2 &= (T_0 + P_0)R_0^2; \\ t - p &= T_0 - P_0; \\ (t + p)r^2 &= (T_1 + P_1)R_1^2; \\ t - p &= T_1 - P_1.\end{aligned}$$

Combining these equations, and eliminating T_1, T_0, and p, we obtain

$$t = \frac{P_0R_0^2 - P_1R_1^2}{R_1^2 - R_0^2} + \frac{R_1^2R_0^2(P_0 - P_1)}{R_1^2 - R_0^2}\frac{1}{r^2}. \quad . \quad . \quad (187)$$

Combining again and eliminating T_1, T_0, and t, we have

$$p = -\frac{P_0R_0^2 - P_1R_1^2}{R_1^2 - R_0^2} + \frac{R_1^2R_0^2(P_0 - P_1)}{R_1^2 - R_0^2}\frac{1}{r^2}. \quad . \quad (188)$$

Since q is assumed constant throughout the cross-section it is not considered in this discussion.

Equations (187) and (188) give the values of t and p at any circumference whose radius is r. From these equations we can construct the curves of stress. To illustrate, take equation (187).

Differentiating, we have

$$\frac{dt}{dr} = -\frac{2R_1^2R_0^2(P_0 - P_1)}{R_1^2 - R_0^2}\frac{1}{r^3}. \quad . \quad . \quad . \quad . \quad (189)$$

Differentiating (189), we have

$$\frac{d^2t}{dr^2} = \frac{6R_1^2R_0^2(P_0 - P_1)}{R_1^2 - R_0^2}\frac{1}{r^4}. \quad . \quad . \quad . \quad . \quad . \quad (190)$$

From (187), we have for the values of t for the interior and exterior of the cylinder, by making $r = R_0$ and $r = R_1$, respectively,

$$T_0 = P_0\frac{(R_0^2 + R_1^2)}{R_1^2 - R_0^2} - P_1\frac{2R_1^2}{R_1^2 - R_0^2}; \quad . \quad . \quad (191)$$

$$T_1 = P_0\frac{2R_0^2}{R_1^2 - R_0^2} - P_1\frac{(R_1^2 + R_0^2)}{R_1^2 - R_0^2}. \quad . \quad . \quad (192)$$

Now considering the two forces P_0 and P_1, we may have the following cases:

1. $P_0 > P_1$;
2. $P_0 = P_1$;
3. $P_0 < P_1$.

First. $P_0 > P_1$.

Equation (189) shows that $\frac{dt}{dr} < 0$; hence t decreases algebraically as r increases. Equation (190) shows that $\frac{d^2t}{dr^2} > 0$; hence the curve of stress is concave upwards.

If we lay off values of r along the axis OX, and of t along OY, the resulting curve will be the curve of stress for the particular stress considered, and may be any one of the curves b, d, c, a in Fig. 76. If T_0 and T_1 are both positive, we have curve b, which is the general case. If $T_0 = -T_1$, we have curve d. If $T_0 = 0$, we have curve c. If T_0 and T_1 are both negative, we have curve a.

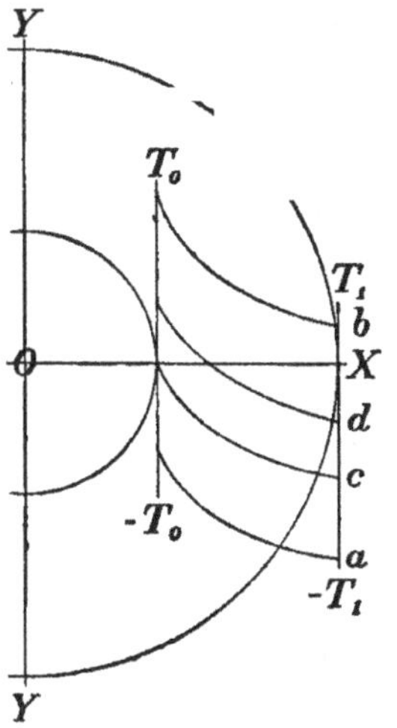

FIG. 76.

Second. $P_0 = P_1$.

In this case

$$\frac{dt}{dr} = 0, \quad \frac{d^2t}{dr^2} = 0,$$

and the curve of stress becomes a right line parallel to the axis of OX.

Equation (191) shows that $T_0 < 0$, and hence the right line will be below the axis of OX, as in Fig. 77.

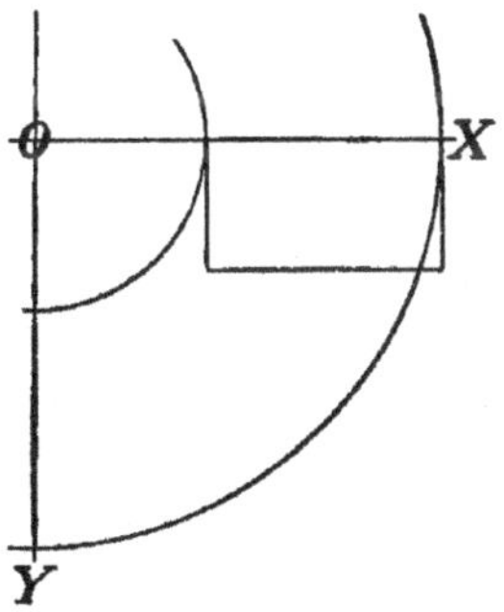

FIG. 77.

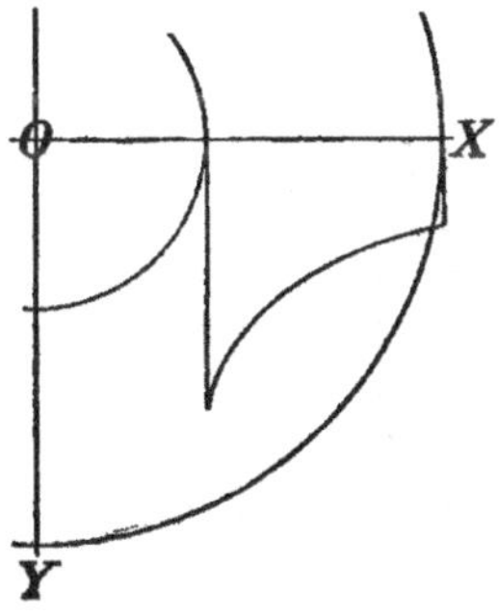

FIG. 78.

Third. $P_0 < P_1$.

In this case

$$\frac{dt}{dr} > 0, \quad \frac{d^2t}{dr^2} < 0,$$

hence t increases algebraically with r, and the curve is convex upwards. Equation (191) shows that $T_0 < 0$, and hence the curve will be as in Fig. 78.

106. Conclusion from Curves—Method of Strengthening Cylinder.

CONCLUSION.—Similar results will be obtained by discussing equation (188), and an examination of all the curves thus obtained will show that in general the greatest stresses are at the interior of the cylinder, and the object of modern gun-construction is to strengthen this interior layer.

METHOD OF STRENGTHENING CYLINDER.—Take a gun composed of one piece of metal, such as the old cast-iron guns. When fired, the interior pressure is P_0, and the exterior pressure is zero, since the pressure of the air may be neglected.

FIG. 79.

Then $P_0 > P_1$, and the curve of tensions is b, Figs. 76 and 79. If P_0 is great, the inner layer may be deformed or ruptured before the exterior layers are brought to their limit of endurance. Suppose, however, that before firing we cause a pressure, P_1, to act upon the exterior of the cylinder. Then we have the case where $P_0 < P_1$, and the state of stress in the cylinder before firing is as in Figs. 78 and 80. That is, all the layers are compressed, those at the interior more than those at the exterior.

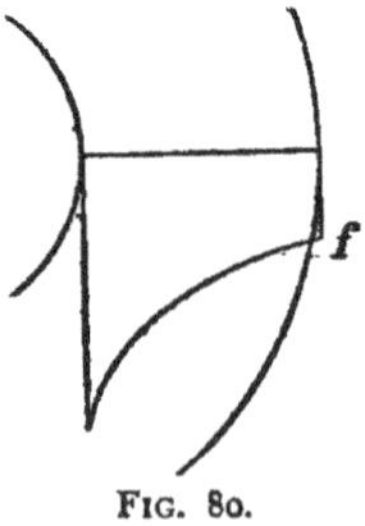

FIG. 80.

Now when the gun is fired, we have the condition $P_0 > P_1$, and the curve of tensions would be b, Fig. 79, were it not that we have already a curve of tensions, f, Fig. 80, in the cylinder.

Hence the new curve of tensions is the resultant of the two curves b and f, or AB, Fig. 81.

That is, before the inner layer can be put in tension, a force must be exerted sufficient to overcome the preliminary or initial compression, and after this any excess of the force will produce tension. The result is shown in Fig. 81. $A'B'$ is the curve of tensions, supposing no previous stress acting on the cylinder. CD is the curve of initial compression; AB is the resultant curve, which is obtained by taking differences of ordinates, and is the curve representing the actual tensions.

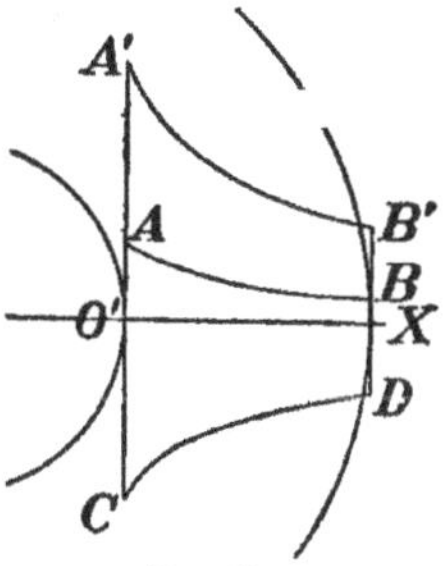

FIG. 81.

This method is called the method of initial compression, and is now universally used in gun-construction.

107. Values of the Strains at any Point in a Cylinder in Terms of the Radii and Pressures.

The general laws of the stresses in a cylinder have been determined, and the curves of stress constructed. The object of the present discussion is to find the values of the strains λ, μ, and ν at any point in a cylinder in terms of the exterior and interior pressures and radii, and of the modulus E_0, these strains being within the elastic limit of the metal.

For this purpose assume equations (170):

$$\lambda = \frac{1}{E_0}\left(t + \frac{p}{3} - \frac{q}{3}\right);$$

$$\mu = -\frac{1}{E_0}\left(p + \frac{t}{3} + \frac{q}{3}\right);$$

$$\nu = \frac{1}{E_0}\left(q - \frac{t}{3} + \frac{p}{3}\right).$$

It may be shown analytically, and the result has been proved by actual experience in gun-building, that in considering the resistance of the cylinder we can simplify the problem by neglecting the longitudinal force q, and considering at first only the forces t and p, and afterwards consider separately the force q, when the longitudinal strength of the gun is to be determined. This is equivalent to making $q = 0$ in the above equations. These equations then become

$$\left.\begin{aligned} \lambda &= \frac{1}{E_0}\left(t + \frac{p}{3}\right); \\ \mu &= -\frac{1}{E_0}\left(p + \frac{t}{3}\right); \\ \nu &= \frac{1}{E_0}\left(-\frac{t}{3} + \frac{p}{3}\right). \end{aligned}\right\} \quad . \quad . \quad . \quad . \quad . \quad (193)$$

Substitute the values of t and p from (187) and (188) in (193), and we have

$$\lambda = \frac{2(P_0R_0^2 - P_1R_1^2)}{3(R_1^2 - R_0^2)E_0} + \frac{4R_1^2R_0^2(P_0 - P_1)}{3(R_1^2 - R_0^2)E_0}\frac{1}{r^2}; \quad (194)$$

$$\mu = \frac{2(P_0R_0^2 - P_1R_1^2)}{3(R_1^2 - R_0^2)E_0} - \frac{4R_1^2R_0^2(P_0 - P_1)}{3(R_1^2 - R_0^2)E_0}\frac{1}{r^2}; \quad (195)$$

$$\nu = -\frac{2(P_0R_0^2 - P_1R_1^2)}{3(R_1^2 - R_0^2)E_0}. \quad . \quad . \quad . \quad . \quad . \quad . \quad . \quad . \quad (196)$$

These equations are general and give the values of the strains λ, μ, and ν at any point r in a cylinder.

To find the values of the strains at any particular point we substitute for r the value of the radius at that point.

Thus for the interior of the cylinder, substitute R_0 for r, and for the exterior R_1 for r.

108. Maximum Values of Strains in a Cylinder.

If any part of a gun-cylinder is subjected to a stress beyond its elastic limit, this part becomes deformed.

Hence other parts will be called upon to bear stresses different from those for which they were calculated, and the result will be that after a few rounds the whole structure may be deformed or destroyed. We then use the following principle, which is the foundation of the modern theory of gun-construction:

No fibre of any cylinder in the gun must be strained, under any circumstances, beyond the elastic limit of the metal of that cylinder.

From this principle we can determine the maximum stress to which a cylinder can be subjected.

It has been shown that the inner layer of a cylinder is subjected to the greatest stress. Hence if this layer does not pass its elastic limit, every other layer, and consequently the cylinder itself, will be safe.

The strains at the inner layer will be obtained from (194) and (195) by making $r = R_0$. Considering for the present only λ and μ, since ν is constant, we have, making $r = R_0$,

$$\lambda = \frac{(4R_1^2 + 2R_0^2)P_0 - 6R_1^2P_1}{3(R_1^2 - R_0^2)E_0}; \quad \ldots \quad (197)$$

$$\mu = -\frac{(4R_1^2 - 2R_0^2)P_0 - 2R_1^2P_1}{3(R_1^2 - R_0^2)E_0}. \quad \ldots \quad (198)$$

These are any strains within the elastic limit.

Let θ_0 be the elastic limit of the cylinder for tension;

ρ_0, the elastic limit for compression, in pounds or tons per square inch.

Then, by equation (169), the elongation at the elastic limit will be

$$\frac{\theta_0}{E_0},$$

and the compression at the elastic limit

$$\frac{\rho_0}{E_0},$$

and, by the principle previously stated, these are the maximum strains that can be allowed. Since, in general, λ is extension and μ compression, λ and μ must be equal to $\frac{\theta_0}{E_0}$ and $\frac{\rho_0}{E_0}$, respectively, at the limit. If, however, λ becomes compression and μ extension, λ must be placed equal to $\frac{\rho_0}{E_0}$, and μ to $\frac{\theta_0}{E_0}$, as will be seen later. We have then, equating these values,

$$\lambda = \frac{\theta_0}{E_0} = \frac{(4R_1^2 + 2R_0^2)P_0 - 6R_1^2P_1}{3(R_1^2 - R_0^2)E_0}; \quad \ldots \quad (199)$$

$$\mu = -\frac{\rho_0}{E_0} = -\frac{(4R_1^2 - 2R_0^2)P_0 - 2R_1^2P_1}{3(R_1^2 - R_0^2)E_0}. \quad . \quad . \quad (200)$$

109. Limiting Interior Pressures—Discussion.

LIMITING INTERIOR PRESSURES.—Solving (199) and (200) for P_0, we find the pressures which will produce the strains $\lambda = \frac{\theta_0}{E_0}$ and $\mu = \frac{\rho_0}{E_0}$, and these will be the maximum interior pressures which the cylinder will stand. These values are

$$P_{0\theta} = \frac{3(R_1^2 - R_0^2)\theta_0 + 6R_1^2P_1}{4R_1^2 + 2R_0^2}; \quad . \quad . \quad . \quad (201)$$

$$P_{0\rho} = \frac{3(R_1^2 - R_0^2)\rho_0 + 2R_1^2P_1}{4R_1^2 - 2R_0^2}. \quad . \quad . \quad . \quad (202)$$

$P_{0\theta}$ will cause the inner layer of the cylinder to elongate till it reaches the elastic limit θ_0, and $P_{0\rho}$ will cause the inner layer to be compressed radially till it reaches the elastic limit ρ_0.

These values of $P_{0\theta}$ and $P_{0\rho}$ will differ, and the smaller value marks the limit of stress to which the cylinder can be safely subjected.

For instance, if $P_{0\theta} < P_{0\rho}$, the limit θ_0 will be reached, while that for compression ρ_0 will not be. The cylinder has therefore more compressive strength than can be used, since if we increase $P_{0\theta}$ till it is equal to $P_{0\rho}$, we pass the limit θ_0, which is contrary to the principle stated.

DISCUSSION.—In a single cylinder the most common case is that in which P_1, the exterior pressure, is that of the atmosphere, and may be neglected. P_1 is therefore zero, and under this condition equations (201) and (202) become

$$P_{0\theta} = \frac{3(R_1^2 - R_0^2)\theta_0}{4R_1^2 + 2R_0^2}; \quad . \quad . \quad . \quad . \quad . \quad (203)$$

$$P_{0\rho} = \frac{3(R_1^2 - R_0^2)\rho_0}{4R_1^2 - 2R_0^2}. \quad . \quad . \quad . \quad . \quad . \quad (204)$$

Since for all metals used in gun-construction $\theta_0 =$ or $<$ ρ_0, $P_{0\theta}$ will always be less than $P_{0\rho}$, and hence it alone will be considered.

(1) Required the thickness of wall necessary to resist a given interior pressure $P_{0\theta}$. Solving (203) for $(R_1 - R_0)$, the thickness, we have

$$R_1 - R_0 = H = R_0\left(\sqrt{\frac{3\theta_0 + 2P_{0\theta}}{3\theta_0 - 4P_{0\theta}}} - 1\right). \quad . \quad . \quad (205)$$

(2) To show the relation between the thickness of the cylinder and its resistance, we have from (203)

$$P_{0\theta} = \frac{3\left(1 - \frac{R_0^2}{R_1^2}\right)\theta_0}{4 + \frac{2R_0^2}{R_1^2}} \quad . \quad . \quad . \quad . \quad . \quad . \quad . \quad (206)$$

Suppose the cylinder to be one calibre thick. Then

$$R_1 = 3R_0,$$

and

$$P_{0\theta} = .63\theta_0.$$

If the cylinder be of infinite thickness, $R_1 = \infty$, and

$$P_{0\theta} = .75\theta_0,$$

which shows that an increase in thickness from one calibre to infinity, increases the strength of the cylinder only from $.63\theta_0$ to $.75\theta_0$.

Hence we conclude that a single cylinder is not materially strengthened by increasing its thickness beyond one calibre, and also that the greatest possible value for the interior pressure in a single cylinder without initial compression is less than

$$.75\theta_0.$$

110. Limiting Exterior Pressures—Thickness of Cylinder—Exterior Strains.

LIMITING EXTERIOR PRESSURES.—It has been shown that, in order to strengthen the cylinder, we apply an exterior pressure P_1, and produce a compression of the cylinder, this compression being greatest at the interior.

What is the limiting value of this exterior pressure?

Its limiting value is that which will compress the inner layer up to its elastic limit; and it is determined as follows:

The interior pressure being zero, or $P_0 = 0$, the strains λ and μ at the interior become, from (197) and (198),

$$\lambda = \frac{-2R_1^2P_1}{(R_1^2 - R_0^2)E_0}; \quad . \quad . \quad . \quad . \quad . \quad (207)$$

$$\mu = \frac{2R_1^2P_1}{3(R_1^2 - R_0^2)E_0}. \quad . \quad . \quad . \quad . \quad . \quad (208)$$

The first, being negative, shows that the inner layer is compressed tangentially; the second, being positive, that the wall of the cylinder is extended in a radial direction or increased in thickness. As before, the limiting compressive strain is

$$\frac{\rho_0}{E_0},$$

and that for extension

$$\frac{\theta_0}{E_0};$$

and the values of λ and μ must not exceed these respectively. Hence

$$\lambda = -\frac{\rho_0}{E_0} = \frac{-2R_1^2P_1}{(R_1^2 - R_0^2)E_0};$$

$$\mu = \frac{\theta_0}{E_0} = \frac{2R_1^2P_1}{3(R_1^2 - R_0^2)E_0}.$$

Solving for P_1, we have

$$P_{1\rho} = \frac{(R_1^2 - R_0^2)\rho_0}{2R_1^2}; \quad . \quad . \quad . \quad . \quad . \quad (209)$$

$$P_{1\theta} = \frac{3(R_1^2 - R_0^2)\theta_0}{2R_1^2}. \quad . \quad . \quad . \quad . \quad (210)$$

The negative sign is omitted in (209), as it indicates compression simply. These equations are useful in determining the limiting value of the exterior pressure which the tube or inner cylinder will support, when we are considering what pressure P_1 can be applied to its exterior to strengthen it;

and since $P_{1\rho}$ is generally less than $P_{1\theta}$, equation (209) only is used.

THICKNESS OF CYLINDER TO RESIST EXTERNAL PRESSURE.—From equation (209) the thickness of wall of cylinder necessary to resist a given exterior pressure $P_{1\rho}$ may be obtained.

Solving it for $R_1 - R_0 = H'$, the thickness, we have

$$R_1 - R_0 = H' = R_0\left(\sqrt{\frac{\rho_0}{\rho_0 - 2P_{1\rho}}} - 1\right). \quad . \quad . \quad (211)$$

STRAIN AT EXTERIOR OF CYLINDER.—The only strain of importance at the exterior of the cylinder is that in the direction of the circumference, or λ. Referring to the general value for it at any point r, equation (194), and making $r = R_1$, we have for its value at the exterior

$$\lambda = \frac{6R_0^2P_0 - (4R_0^2 + 2R_1^2)P_1}{3(R_1^2 - R_0^2)E_0}. \quad . \quad . \quad . \quad (212)$$

When $P_0 = 0$,

$$\lambda = \frac{-(4R_0^2 + 2R_1^2)P_1}{3(R_1^2 - R_0^2)E_0}, \quad . \quad . \quad . \quad . \quad . \quad (213)$$

a contraction.

When $P_1 = 0$,

$$\lambda = \frac{2R_0^2P_0}{(R_1^2 - R_0^2)E_0}, \quad . \quad . \quad . \quad . \quad . \quad . \quad . \quad (214)$$

an extension.

111. Longitudinal Strength.

To determine this, suppose the cylinder closed at one end by the breech, and at the other end by the projectile, as in a gun. The total pressure acting on the bottom of the cylinder is then

$$\pi R_0^2P_0.$$

The area of cross-section of the cylinder is

$$\pi(R_1^2 - R_0^2),$$

and the pressure $\pi R_0^2 P_0$ is resisted by the elasticity of this cross-section. Supposing the pressure to be uniformly distributed, the stress per unit area of cross-section is

$$q = \frac{\pi R_0^2 P_0}{\pi(R_1^2 - R_0^2)} = \frac{P_0 R_0^2}{R_1^2 - R_0^2}. \quad . \quad . \quad . \quad (215)$$

Substitute in the third of equations (170) for t and p their values from (187) and (188), and for q its value from (215), and we have for the value of the strain ν in the direction of the axis of the cylinder,

$$\nu = \frac{P_0 R_0^2 - 2P_1 R_1^2}{3(R_1^2 - R_0^2)E_0}. \quad . \quad . \quad . \quad . \quad . \quad (216)$$

The maximum value of this strain must be equal to $\frac{\theta_0}{E_0}$ as before, hence

$$\nu = \frac{\theta_0}{E_0} = \frac{P_0 R_0^2 - 2P_1 R_1^2}{3(R_1^2 - R_0^2)E_0}.$$

Solving for P_0, we have

$$P_{0\theta} = \frac{3(R_1^2 - R_0^2)\theta_0 + 2R_1^2 P_1}{R_0^2}. \quad . \quad . \quad . \quad (217)$$

If $P_1 = 0$,

$$P_{0\theta} = \frac{3(R_1^2 - R_0^2)\theta_0}{R_0^2}. \quad . \quad . \quad . \quad . \quad . \quad . \quad (218)$$

For the thickness of wall necessary to resist the pressure P_0 acting parallel to the axis, we have, solving (218) for $R_1 - R_0$,

$$R_1 - R_0 = H'' = R_0\left(\sqrt{\frac{P_{0\theta} + 3\theta_0}{3\theta_0}} - 1\right). \quad . \quad (219)$$

This discussion applies to the older guns made of a single piece of metal. With modern built-up guns the breech-block is in an outer cylinder or jacket, and a new formula must be deduced for the longitudinal strength in such a gun.

112. General Principles of Built-up Guns—Method of Applying Exterior Pressure.

General Principles.—It has been stated, in discussing the resistance of a single cylinder, that it may be strengthened by applying a force P_1 to the exterior of the cylinder. This force, as shown, produces a compression of all the layers of the inner cylinder, the interior layer being compressed to the greatest extent, as it should be, since it is extended more than any other layer by the action of the powder-gas.

The layers of the cylinder being thus subjected to tangential compression, this compression must first be overcome before the inner layers can be brought to a state of tension. Hence part of the powder-pressure is exerted to bring the inner layers to a neutral state of strain, and any excess of pressure over that required for this purpose will cause tension in the inner layers. It is evident, however, that since the cylinder will safely support a certain interior pressure $P_{0\theta}$ or $P_{0\rho}$ without this preliminary compression, it will support a much greater interior pressure with the aid of this compression.

Method of Applying Exterior Pressures.—The method of applying this exterior pressure is by placing over the inner cylinder an exterior one, whose interior diameter is slightly less than the exterior diameter of the inner cylinder. The exterior cylinder is applied as has been explained in Gun-construction. Upon cooling it contracts upon the inner cylinder, and if the difference of diameters is properly regulated it will produce the required pressure.

It is evident, however, that in compressing and strengthening the inner cylinder, the outer cylinder is itself extended and weakened; but this extension or weakening of the outer cylinder, when properly regulated, can be supported without damage to the structure.

According to Lamé's second law the sum of the tension and pressure varies inversely as the square of the radius. Hence a value of $(T_0 + P_0)$ which would be large at the interior would be very much diminished at the radius R_1. This principle is applied to any number of cylinders placed one over the other. The differences of diameters of any two adjacent cylinders is called "the shrinkage," the resulting gun a built-up gun, and the cylinder a compound cylinder.

113. Calculations for Compound Cylinder—States Considered—Nomenclature.

CALCULATIONS.—Suppose the cylinders assembled with the proper shrinkage. It is required—

1. To calculate the maximum resistance of the compound cylinder;

2. To calculate the shrinkage, so that when assembled the pressure exerted by the exterior upon the interior cylinders shall be such as to give to the compound cylinder its maximum resistance.

STATES OF CYLINDER.—When the powder-pressure is acting, the cylinder is said to be "in action;" when the powder-pressure ceases to act, the cylinder is "at rest." It is evident, however, that the system is not free from stress when at rest, owing to the shrinkage; and it is necessary to consider the stresses both in action and at rest, as will be seen later.

NOMENCLATURE.—For simplicity of discussion, consider that the compound cylinder is made up of two cylinders only. The inner cylinder will be designated as the tube, and the exterior as the jacket.

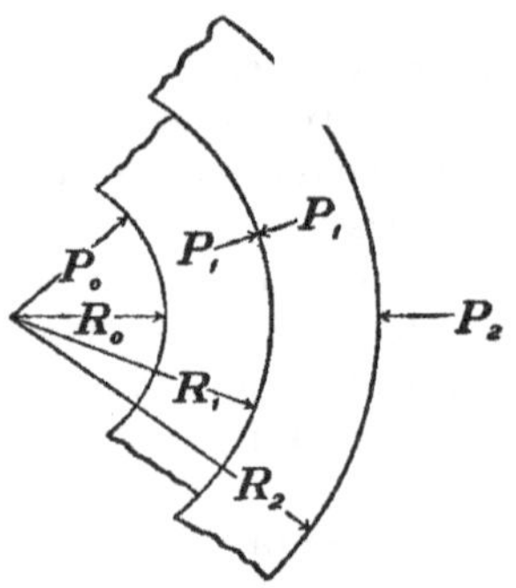

FIG. 82.

In Fig. 82 let

R_0, R_1, R_2 be the radii of the interior, middle, and exterior surfaces of the cylinders, R_1 being the radius of the surface of contact between the tube and jacket;

P_0, P_1, P_2, the normal pressures at the interior, middle, and exterior surfaces, respectively, when the system is in action;

p_0, p_1, p_2, variations in P_0, P_1, P_2, produced by any cause whatever, such as a change from a state of action to that of rest;

θ_0, θ_1, elastic limits of tube and jacket, respectively, for tension;

ρ_0, ρ_1, same for compression;

E_0, E_1, moduli of elasticity of tube and jacket respectively. These are generally assumed as equal, hence $E_0 = E_1$;

P_1', the normal pressure acting at the surface of contact of the two cylinders when the system is at rest.

For a single cylinder the radii have been denoted in the previous discussions by R_0 and R_1, and the pressures by P_0 and P_1.

As a general rule, if n denote the number of the cylinder, counting from the interior, its radii and pressures are $n - 1$ and n, for the interior and exterior respectively. Thus, for the fourth cylinder we have R_3, R_4, P_3, and P_4, and by applying this rule the equations deduced for two cylinders may be applied to any number.

114. Resistance of Compound Cylinder in Action.

In the case of a compound cylinder in action the tube is acted on by an interior pressure P_0 and by an exterior pressure P_1. The jacket is acted on by an interior pressure P_1 and by an exterior pressure P_2, which is that of the atmosphere, and therefore regarded as zero. The jacket is therefore a single cylinder acted on by an interior force, P_1, and its resistance is given by equation (203).

Making, in (203),

$$P_{0\theta} = P_{1\theta}, \quad R_0 = R_1,$$
$$R_1 = R_2, \quad \theta_0 = \theta_1,$$

we write

$$P_{1\theta} = \frac{3(R_2^2 - R_1^2)\theta_1}{4R_2^2 + 2R_1^2}, \quad . \quad . \quad . \quad . \quad . \quad (220)$$

$P_{1\theta}$ being alone considered, because, as previously shown, it is always less than $P_{1\rho}$. (See equations (203), (204).)

This pressure $P_{1\theta}$ will extend the inner layer of the jacket to its elastic limit, and hence it is the greatest pressure which can be safely applied to the interior of this cylinder.

The pressure $P_{1\theta}$ just found also acts upon the exterior of the tube, and P_0 acts upon the interior. Hence we have the case, already discussed, of a cylinder acted upon by two forces, and equations (201) and (202) apply, viz.,

$$P_{0\theta} = \frac{3(R_1^2 - R_0^2)\theta_0 + 6R_1^2 P_{1\theta}}{4R_1^2 + 2R_0^2}; \quad . \quad . \quad . \quad (221)$$

$$P_{0p} = \frac{3(R_1^2 - R_0^2)\rho_0 + 2R_1^2 P_{1\theta}}{4R_1^2 - 2R_0^2}. \quad . \quad . \quad . \quad (222)$$

The smaller value must be selected, and this value marks the limiting pressure which the tube, and consequently the compound cylinder, will safely support.

When this interior pressure acts, it raises the inner layer of the tube to its elastic limit for tension or compression, according as $P_{0\theta}$ or P_{0p} is the less. At the same time it produces the pressure $P_{1\theta}$ at the surface of contact. Hence when the maximum interior pressure is acting it raises the inner layers of both cylinders to their elastic limits.

Equation (220) is solved first, since it contains only known quantities in the second member. The resulting value, $P_{1\theta}$, is then substituted in both (221) and (222), and both of these equations can then be solved, the smaller value being taken as explained.

Collecting these equations for convenience, we have for calculating the maximum pressures which a compound cylinder composed of a tube and jacket will safely support in action

$$P_{1\theta} = \frac{3(R_2^2 - R_1^2)\theta_1}{4R_2^2 + 2R_1^2}; \quad . \quad . \quad . \quad . \quad . \quad . \quad . \quad . \quad (220)$$

$$P_{0\theta} = \frac{3(R_1^2 - R_0^2)\theta_0 + 6R_1^2 P_{1\theta}}{4R_1^2 + 2R_0^2}; \quad . \quad . \quad . \quad (221)$$

$$P_{0p} = \frac{3(R_1^2 - R_0^2)\rho_0 + 2R_1^2 P_{1\theta}}{4R_1^2 - 2R_0^2}. \quad . \quad . \quad . \quad (222)$$

115. Longitudinal Strength of Compound Cylinder.

In a gun composed of two cylinders the jacket carries the breech-block, in order to free the tube as much as possible from longitudinal stress.

The total pressure upon the breech-block is, as before,

$$\pi R_0^2 P_0.$$

This acts upon the area of cross-section of the jacket, which is $\pi(R_2^2 - R_1^2)$; hence the stress per unit area of this cross-section is

$$q = \frac{\pi R_0^2 P_0}{\pi(R_2^2 - R_1^2)} = \frac{P_0 R_0^2}{R_2^2 - R_1^2}. \quad . \quad . \quad . \quad (223)$$

The values of t and p from (187) and (188) become for the outer cylinder, by making the following changes in the nomenclature,

$$P_0 = P_{1\theta}, \quad R_0 = R_1;$$

$$P_1 = 0, \quad R_1 = R_2;$$

$$t = \frac{P_{1\theta} R_1^2}{R_2^2 - R_1^2} + \frac{R_2^2 R_1^2 P_{1\theta}}{R_2^2 - R_1^2} \frac{1}{r^2}; \quad . \quad . \quad . \quad . \quad (224)$$

$$p = -\frac{P_{1\theta} R_1^2}{R_2^2 - R_1^2} + \frac{R_2^2 R_1^2 P_{1\theta}}{R_2^2 - R_1^2} \frac{1}{r^2}. \quad . \quad . \quad . \quad (225)$$

Substituting these values of t, p, and q in the third of equations (170), we have

$$v = \frac{3P_0 R_0^2 - 2P_{1\theta} R_1^2}{3(R_2^2 - R_1^2)E_1}. \quad . \quad . \quad . \quad . \quad . \quad . \quad . \quad (226)$$

The maximum safe longitudinal strain upon the jacket is $\frac{\theta_1}{E_1}$. If the value of v calculated by (226) is less than this, the jacket will not be overstrained longitudinally by P_0. If greater than $\frac{\theta_1}{E_1}$, the pressure P_0 must be reduced.

The limiting value for the pressure is obtained, as before, by placing

$$v = \frac{\theta_1}{E_1} = \frac{3P_0 R_0^2 - 2P_{1\theta} R_1^2}{3(R_2^2 - R_1^2)E_1}.$$

Solving for P_0, we have

$$P_0 = \frac{3(R_2^2 - R_1^2)\theta_1 + 2P_{1\theta}R_1^2}{3R_0^2}. \quad . \quad . \quad . \quad (227)$$

116. The System at Rest—Reasons for Considering It—Variations of Pressure.

The formulas previously deduced give the maximum pressures which the compound cylinder will safely support in action; and in order that these pressures may exist, the jacket must be applied by shrinkage upon the tube. The pressure thus produced on the tube will strengthen it.

It may be, however, that this pressure which is applied to the exterior of the tube will be so great that when the powder-pressure ceases to act it will compress the inner layer of the latter to such an extent as to cause this layer to pass its elastic limit. Thus the tube may be injured by the exterior force which is applied to strengthen it.

It follows, therefore, that although the compound cylinder would support certain pressures in action if the requisite exterior pressure could be applied, it may be impossible to apply this pressure to the exterior of the tube at rest, and therefore, before we can determine whether equations (220), (221), and (222) represent allowable pressures in action, it is necessary to consider the effects of the pressure at rest upon the tube.

When the powder-pressure acts, we have the forces P_0 at the interior and P_1 at the surface of contact of the cylinders.

These can be calculated by (220), (221), and (222).

When the powder pressure ceases to act, the interior pressure becomes zero, and the variation of pressure at the interior is from $+ P_0$ to 0. This difference between the pressure in action and at rest gives the variation of pressure. Hence for the interior we have

$$p_0 = 0 - P_0, \quad \text{or} \quad p_0 = - P_0. \quad . \quad . \quad . \quad (228)$$

This is further evident by considering that the algebraic sum of the pressure in action and the variation of pressure must be the pressure at rest.

For the surface of contact of the two cylinders the pressure at rest is P_1' and in action it is P_1; hence the variation of pressure p_1 at that surface is, as before,

$$p_1 = P_1' - P_1. \quad . \quad . \quad . \quad . \quad . \quad . \quad (229)$$

117. Limiting Value of Exterior Pressure on Tube—System at Rest.

The limiting value of the exterior pressure upon the tube for the state of rest is that value which will compress the inner layer of the tube to its elastic limit, and it is given by equation (209).

$$P_{1\rho} = \frac{(R_1^2 - R_0^2)}{2R_1^2}\rho_0, \quad . \quad . \quad . \quad . \quad . \quad . \quad (230)$$

and no greater pressure than this can be allowed to exist at the exterior surface of the tube at rest.

Now the pressure actually existing at this surface "at rest" is, from (229),

$$P_1' = P_1 + p_1. \quad . \quad . \quad . \quad . \quad . \quad (231)$$

This value of P_1' depends on P_1, the pressure at the exterior of the tube "in action," and also upon p_1, the variation of the pressure at that surface in passing from the state of action to that of rest. It is necessary therefore to calculate P_1' by (231) and compare it with $P_{1\rho}$, the maximum admissible pressure calculated by (230). If $P_1' > P_{1\rho}$, it follows that it will strain the tube at rest beyond its elastic limit, and hence it cannot be allowed. The value $P_{1\rho}$ must then be adopted in place of P_1' and be substituted for it in (231). This substitution in (231) will produce a corresponding change in the value of P_1, and this change in P_1 will also change P_0. (See equations (220), (221), (222).) On the other hand, if P_1', from (231), be less than $P_{1\rho}$ from (230), P_1' must be used and not $P_{1\rho}$, because although the tube will support the pressure $P_{1\rho} > P_1'$ at rest, if this value of $P_{1\rho}$ be substituted for P_1' in (231) it will cause an increase in the value of P_1, the pressure on the exterior in action.

But it has been shown previously that the value of P_1 from equation (220) represents the *greatest* pressure which the jacket will endure in action without passing its elastic limit.

Hence this pressure must not be increased.

We are therefore limited on the one hand to the value P_{1p}, equation (230), which must not be exceeded at rest, and on the other to the values P_0 and P_1, equations (220), (221), (222), in action which must not be increased.

118. Calculation of p_1 in Equation (231).

In order to calculate P_1' from (231) we must know p_1, since P_1 is given by (220). To calculate p_1 we proceed as follows:

When variations of pressure occur at any surface they produce corresponding changes in the dimensions of the surface at which they act, and these changes depend directly upon the variations of pressure which cause them. The changes of dimensions in the direction of the circumference of the cylinder are the greatest (see equations (194) and (195)), and hence it is only necessary to consider these.

For the jacket, the exterior pressure is always zero, and the variation of pressure at the interior is p_1. Equation (197) gives the change of the inner layer of a cylinder in the direction of the circumference due to the forces P_0 and P_1.

In the present case $P_0 = p_1$, and $P_1 = 0$. Also $R_1 = R_2$, $R_0 = R_1$, $E_0 = E_1$, hence

$$\lambda = \frac{(4R_2^2 + 2R_1^2)p_1}{3(R_2^2 - R_1^2)E_1}. \qquad (232)$$

For the tube we have the variation of pressure p_1 acting upon the exterior surface, and p_0 upon its interior. Equation (212) gives the tangential change at the exterior of a cylinder due to the two forces P_0 and P_1. To adapt it to the present case make

$$P_0 = p_0; \qquad P_1 = p_1.$$

Making these substitutions, we have

$$\lambda = \frac{6R_0^2 p_0 - (4R_0^2 + 2R_1^2)p_1}{3(R_1^2 - R_0^2)E_0}. \qquad (233)$$

Since the exterior surface of the tube and the interior surface of the jacket are in contact, they form virtually but one surface, and whatever change occurs in one will occur also in the other. Hence the two values of λ in (232) and (233) are equal.

Equating these and solving for p_1, we have, since $E_0 = E_1$,

$$p_1 = \frac{R_0^2(R_2^2 - R_1^2)p_0}{R_1^2(R_2^2 - R_0^2)}. \quad \ldots \ldots (234)$$

And since, from (228),

$$p_0 = -P_0,$$

we can find p_1 in terms of P_0 from 234.

119. True Value of P_0.

The true value of P_0 is that which is safe for the system both in action and at rest.

It has been shown that if $P_1' < P_{1p}$, the values of P_0 given by (221) and (222) are safe.

If $P_1' > P_{1p}$, these values are not safe and the true value of P_0 is calculated as follows:

The equation expressing the limiting value for the exterior pressure system at rest is

$$P_{1p} = P_1 + p_1, \quad \ldots \ldots (235)$$

in which the value of P_{1p} is obtained from equation (230). Hence

$$P_1 = P_{1p} - p_1. \quad \ldots \ldots (236)$$

Substituting the value of p_1 from (234), in which $p_0 = -P_0$,

$$P_1 = P_{1p} + \frac{R_0^2(R_2^2 - R_1^2)P_0}{R_1^2(R_2^2 - R_0^2)}. \quad \ldots (237)$$

The equations expressing the limiting values of the interior pressure for the state of action are (221) and (222).

Substituting the value of P_1 from (237) in (221) and (222) and taking the smaller value, we have the true value of P_0, which will be safe both in action and at rest, since it has been obtained by combining two equations which contain the conditions of safety both for action and for rest.

The results of the substitution are

$$P_{0\theta} = \frac{3(R_1^2 - R_0^2)\theta_0 + 6R_1^2 P_{1p}}{(4R_1^2 + 2R_0^2) - 6R_0^2 \dfrac{(R_2^2 - R_1^2)}{R_2^2 - R_0^2}}; \quad . \; . \; . \quad (A)$$

$$P_{0p} = \frac{3(R_1^2 - R_0^2)p_0 + 2R_1^2 P_{1p}}{(4R_1^2 - 2R_0^2) - 2R_0^2 \dfrac{(R_2^2 - R_1^2)}{R_2^2 - R_0^2}}. \quad . \; . \; . \quad (B)$$

Having the true value of P_0 from (*A*) or (*B*), we find the true value of p_1 from (234), and this value of p_1 in (235), with the value of P_{1p} from (230), will give the true value of P_1.

120. Calculation of the Shrinkage.

Shrinkage is the difference of diameters of two adjacent cylinders. This is called the actual or absolute shrinkage. Dividing the absolute shrinkage by the diameter, we have the shrinkage per unit of diameter, or the relative shrinkage.

In Fig 83 let

OA be the interior radius } of tube,
OB the exterior radius }
OC the interior radius } of jacket,
OD the exterior radius }

before the cylinders are assembled. Then

$$2BC = 2(OB - OC)$$

is the absolute or actual shrinkage, and

$$\frac{2BC}{2OC} = \frac{2(OB - OC)}{2OC} = \frac{BC}{OC}$$

is the relative shrinkage.

When the cylinders are assembled, the surface of contact will take a position such as *E E E*. The jacket will compress the exterior of the tube by the amount *BE*, and the interior of the jacket will be extended by the amount *CE*, and we have

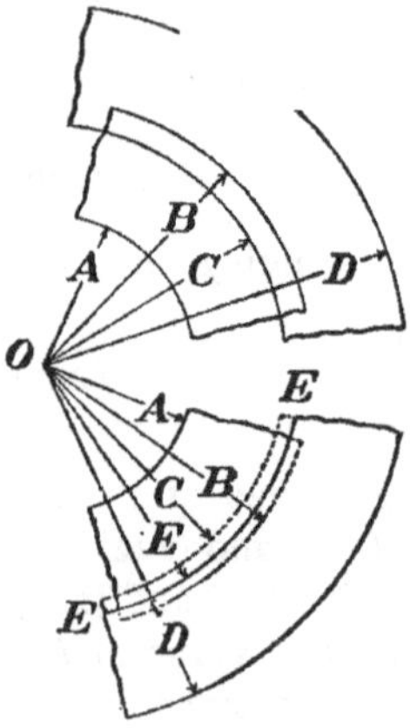

FIG. 83.

$$BE + CE = BC,$$

By this compression the force P_1' is exerted upon the exterior of the tube and the interior of the jacket. It is required to find BE and CE, and the shrinkage will then be known.

CALCULATION.—The value $\frac{BE}{OC}$ is the compression per unit of length of circumference, or of radius, of the exterior of the tube, produced by the force P_1'. This relative compression is strictly $\frac{BE}{OB}$, since OB is the original exterior radius, but the error is so slight that it may be neglected. This compression is given by equation (213), since $P_0 = 0$, hence

$$\frac{BE}{OC} = \lambda = \frac{(4R_0^2 + 2R_1^2)P_1'}{3(R_1^2 - R_0^2)E_0} \quad . \quad . \quad . \quad . \quad (238)$$

The value $\frac{CE}{OC}$ is the extension of the interior of the jacket per unit of length of circumference, or of radius, produced by the force P_1'.

It is given by equation (199) by making

$$R_1 = R_2; \quad P_1 = 0;$$
$$R_0 = R_1; \quad E_0 = E_1;$$
$$P_0 = P_1'.$$

Hence

$$\frac{CE}{OE} = \lambda = \frac{(4R_2^2 + 2R_1^2)P_1'}{3(R_2^2 - R_1^2)E_1} \quad . \quad . \quad . \quad . \quad (239)$$

The negative sign is omitted in (238), since it simply indicates compression.

Hence denoting the relative shrinkage by ϕ, we have

$$\phi = \frac{BE + CE}{OC} = \frac{2R_1^2(R_2^2 - R_0^2)P_1'}{E_0(R_1^2 - R_0^2)(R_2^2 - R_1^2)} \quad . \quad (240)$$

Steps.—The different steps in the calculation of the shrinkage may be thus summarized:

1. Calculate $P_{1\theta}$, $P_{0\theta}$ and P_{0p} by (220), (221), and (222). Use smaller value of P_0.

2. Calculate P_{1p} from (230) and p_1 from (234), making in the latter $p_0 = -P_0$, the value obtained from (221) and (222).

3. Find P_1' from (231), and compare this value with that of P_{1p} obtained from (230).

If P_1' from (231) is greater than P_{1p} from (230), steps 4, 5, and 6 will be as follows:

4. Calculate $P_{0\theta}$ and P_{0p} from A and B. Take smaller value.

5. Recalculate p_1 from (234), making $p_0 = -P_0$, the value found from A and B.

6. Find P_1 from (237), using P_{1p} from (230).

7. Calculate the relative shrinkage by (240). The value of P_1' to be used in (240) must correspond to the adopted value of P_0, being either $(P_1 + p_1)$ from (231) or P_{1p} from (230) according as P_0 is retained as originally found in step 1 or is changed as indicated in steps 3 and 4.

8. The absolute shrinkage is obtained by multiplying the relative shrinkage by the interior diameter of the jacket. Hence if S denote the absolute shrinkage,

$$S = \phi \times 2R_1.$$

9. The exterior diameter of the tube should then be

$$2R_1' = 2R_1 + S. \qquad (241)$$

121. Measurements in Gun-construction — Thickness of Wall — Length of Bore.

MEASUREMENTS.—The value of the shrinkage having been calculated by (240), the exterior diameter of the tube is given by (241). The exterior of the tube is then turned to this diameter, an error of 0.003 inch only being allowed.

After turning, the exterior diameters are measured at every inch of length of tube; if too large, they are reduced; if too small, they cannot be corrected, except by using a smaller jacket. Hence it is important not to turn below size.

The interior diameters of the tube are also measured at each inch of length.

The tube and jacket are now assembled, and, when cool, the interior diameters of the tube under the jacket are again measured.

The pressure of the jacket upon the tube will produce a contraction of the bore of the latter, and this contraction is given by equation (207), making $P_1 = P_1'$, since this latter is the pressure at rest. The measured contraction should agree with the calculated value; and if it does, we have a proof of the accuracy of the measurements, and of the correctness of the formulas.

The agreement is generally very close.

THICKNESS OF WALLS.—From previous calculations it has been shown that there is very little gain in tangential resistance by increasing the thickness of the cylinder beyond one calibre. This rule is generally followed in modern guns for the thickness of wall over the powder-chamber.

The thickness at other points along the chase is obtained by a consideration of the powder-pressures at the different points, and these are given by the pressure-curve, whose construction has been explained. It is also necessary to so adjust the thickness of the different parts, that the weight of the gun shall not exceed the limit generally allowed for the different calibres, and that the axis of the trunnions or the centre of gravity of the gun shall be at a distance from the breech, equal to about $\frac{2}{5}$ the total length of the gun.

The weights of guns are as follows, nearly:

8-inch	14 tons
10-inch	28 "
12-inch	52 "
12-inch mortar	13 "

In general, the shape of the chase conforms to that of the pressure curve, and the resistance at different sections along the gun is calculated so that at any section it shall always be greater than the powder-pressure by a certain coefficient or factor of safety. For the 12-inch gun the elastic resistance is about 24 tons per square inch, and the powder-pressure 16 tons, at the chamber, so that the factor of safety is $\frac{24}{16} = 1.5$.

LENGTH OF GUN.—For a given calibre, charge of powder, weight of projectile, etc., we can calculate by Sarrau's formulas the value of u for a required initial velocity V, and may so adjust the elements of loading that the maximum pressure shall be constant and equal to, say, 15 tons for this velocity. Generally modern guns are from 35 to 45 calibres long.

122. Wire Guns.

In the built-up gun it has been shown that when in action, the inner layers of the tube and jacket are strained to their elastic limits respectively. None of the other fibres are strained up to that limit, and hence the total strength of the metal is not utilized. If instead of two cylinders we have four, assembled with proper shrinkage, the total thickness of the gun being constant, it is evident that the inner layers of each of the four cylinders would be strained to their elastic limits and hence more of the total strength of the metal would be utilized. As the number of cylinders increases, the strength utilized will be greater, till we finally approach the limit where the cylinders are infinitely thin, and the whole thickness of metal in each is strained to its limit.

Practical reasons, however, prevent the carrying out of this method, because the longitudinal strength of the cylinders decreases with the thickness; the expense of boring and turning the cylinders is great, and it would be impossible to bore and turn very thin cylinders accurately.

For these reasons, it has been proposed to substitute wire, for the rings or hoops of the built-up gun. This wire is wrapped round an inner tube with a certain tension, so that the tube is compressed initially as in the case of the built-up gun, and the wire extended.

The advantages claimed for the wire gun are:

1. The tension of the layers of wire can be so regulated that each wire will be strained to its elastic limit when the system is in action, and we approach the condition of infinitely thin cylinders.

2. The wire being very small in section, any physical defects can be detected, and hence all the metal in the gun will be sound.

3. A high elastic limit can be given to the wire, and hence it will have a greater tangential strength than a forged steel hoop.

4. In order to utilize the high elastic limit of the wire, the tube may be compressed at rest beyond its elastic limit.

The objections are:

1. Compressing the inner layer of the tube beyond its elastic limit violates the fundamental principle of modern gun-construction; and if this is *not* done, the wire gun cannot in general be stronger tangentially than the built-up gun, since the strength of the tube marks the limit of the strength of the system.

2. The coils of wire have no longitudinal strength, and hence the longitudinal strain must be supported, as in the built-up gun, by a jacket, and the attachment of this jacket to the tube presents difficulties.

3. The wire gun is not as stiff longitudinally as the built-up gun, since the wire does not support the tube so firmly as the hoops. This is a question of importance with modern long guns.

123. Description of Wire Guns—Woodbridge—Crozier—Brown Segmental.

WOODBRIDGE (Fig. 84).—This gun consists of an inner tube, *t*, wrapped with wire as shown. Over the rear part of the tube is a jacket, *j*, made of longitudinal steel bars of wedge-shaped cross-section.

This jacket is wrapped with the wire *w*, under such tension as to strongly compress the inner tube at rest. The longitudinal thrust is transmitted to the jacket as follows: The jacket is screwed to the tube in rear; the trunnion-hoop *t'* bears against a thin hoop *h*, and this against a collar *c* screwed to the jacket in front. Hence the pull of the breech block in rear is transmitted to the rear of the jacket, and the thrust of the trunnions to the front of it.

The calibre of the gun is 10 inches.

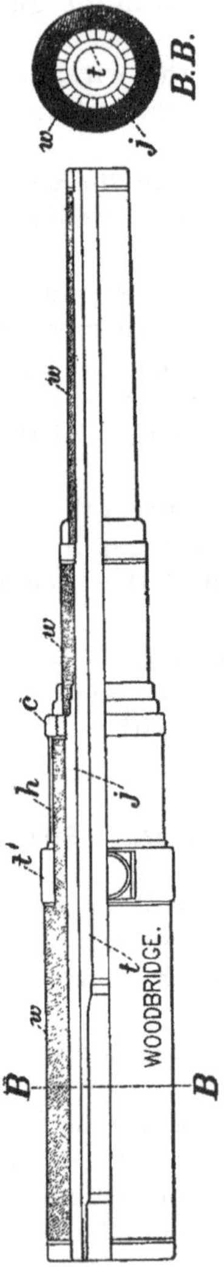

FIG. 84.

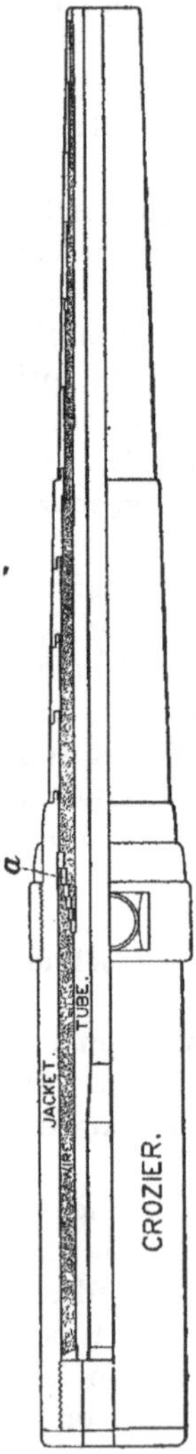

FIG. 85.

CROZIER (Fig. 85). — In this gun the tube is compressed initially beyond its elastic limit by the wire.

The principal features are:

1. The wire on the chase is covered on the exterior by thin hoops, put on with very slight shrinkage, so as to give stiffness, and protect the wire.

2. A jacket of cast iron or cast steel is used for cheapness, to carry the breech-block and support the longitudinal strain. It is put on with very little shrinkage, the tangential strength of the gun depending on the tube and wire alone.

3. The jacket and tube are connected, and the motion of either prevented, by a series of rings or steps, *a*, abutting against each other.

BROWN SEGMENTAL (Fig. 86).—In this gun there is first, a small lining tube, *a*, which extends beyond the trunnions. The metal of this tube has a high elastic limit, 112,000 lbs. The main tube, *b*, is made of wedge-shaped steel bars, of about the same elastic limit. This outer tube is wrapped with wire, and compressed to such an extent that its interior is under compression even in action. This prevents the joints between the bars from opening. The jacket is light, and is not in contact with the wire except at the trunnions and breech. The figure shows the method of attachment of

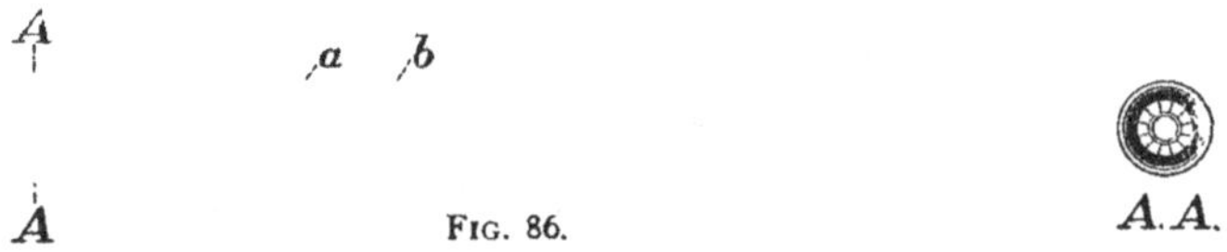

FIG. 86.

the breech-block. The pull of the block at the breech and the thrust of the trunnions in front are borne by the jacket.

Relative motion of tube and jacket is prevented by the connection in rear.

These guns have been made and tried in this country. In Europe the systems of Schultz, Longridge, Armstrong, and others have been tried.

MEASUREMENTS.

124. Necessity for—Measurements Required—Standard Comparator.

NECESSITY FOR.—In a modern gun it has been shown that the stresses and shrinkages are functions of each other. Hence, if the correct shrinkage be not given to the gun, it will not properly support the stress to which it is subjected, and may be either strained beyond its elastic limit, or not strained up to that limit according to the actual value of the shrinkage as given in construction. After these shrinkages are calculated theoretically, their application to a particular gun, depends on the accuracy of the measurements made by the inspector during construction. Hence the necessity for accurate measuring instruments, and skill in their use.

MEASUREMENTS REQUIRED.—In general the following measurements are required in gun-manufacture:

1. Interior diameters;
2. Exterior diameters;
3. Lengths;
4. Measurements by templets and gauges.

STANDARD COMPARATOR.—In this case, as in all others where accuracy is required, all measurements must be referred to a common standard. This standard is called a "comparator," and its general principles may be explained as follows:

A stiff bed or body, *a*, of cast iron, Fig. 87, rests upon three levelling-screws with rounded points of support.

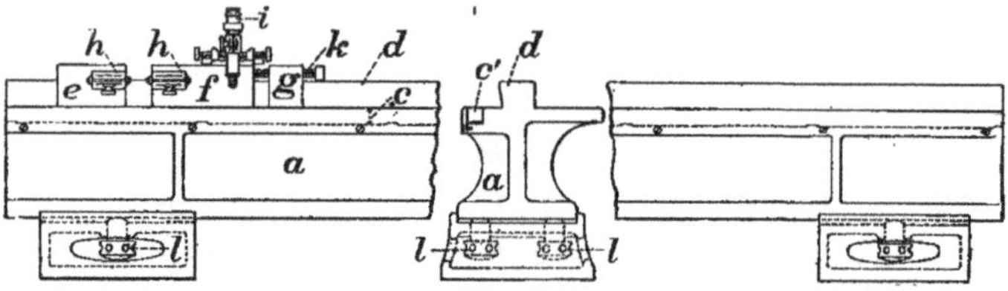

FIG. 87.

In this bed is a groove or recess, *c*, in which rests a standard bar or rod, *c'*, accurately graduated in inches and decimal parts of an inch. On top of the bed is the rib *d*, which forms the guide for the heads *e*, *f*, and *g*, which slide along it. These heads can be fixed by clamp-screws in any position along the bed. *e* is called the fixed head, *f* the sliding head, and *g* the auxiliary head. *h h* are two sockets which carry steel points, and these points can be adjusted lengthwise in the sockets, and clamped by the clamp-screws; *i* is a microscope reading 0.0001 inch; *k*, a tangent screw connecting *f* and *g*.

Use.—The primary object of this instrument is to lay off exact lengths. To do this, the graduated bar being in its recess *c*, bring the ends of the steel points *h h* in contact. Then adjust the graduated bar and microscope, till the zero-line of the eyepiece of the microscope coincides with the zero of the graduated bar. Clamp the fixed head, *e*, and slide the sliding and auxiliary heads, till the microscope is over the nearest division of the graduated bar corresponding to the length to be measured.

The auxiliary head *g* is then clamped, and the sliding head *f* moved by the tangent-screw *k* till the microscope reads the required part of the inch. The distance between the points *h h* will then be the length required.

125. Measurement of Interior Diameters of Short Hoops—Measuring-points—Use.

The interior diameters to be measured may be—

1. Those of a comparatively short hoop;
2. Those of a long hoop or tube.

In the first case, when the length of the hoop is such that all parts of it can be reached by hand, measuring points or rods are used.

MEASURING-POINTS.—For diameters from two to ten inches, the points are made of steel, and consist of a fixed point, *a* (Fig. 88), and a micrometer-point, *b*. The fixed point

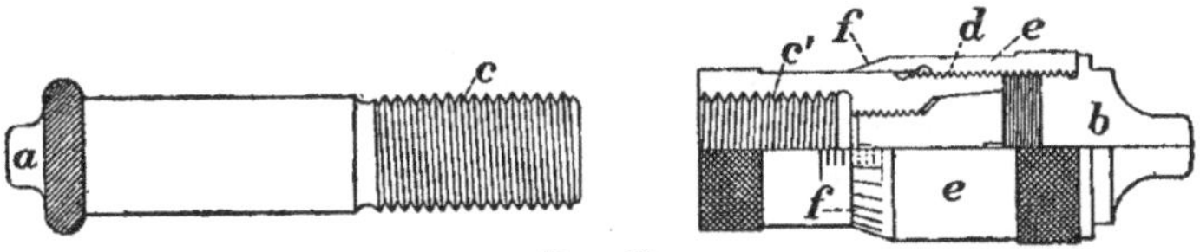

FIG. 88.

varies in length according to the diameter to be measured, there being a number of them. Each one is threaded at the end, *c*, and the micrometer-point screws on this thread by the corresponding thread, *c'*. The screw *d* of the micrometer is accurately cut, so that one turn of the head *e* corresponds to a certain decimal part of one inch, generally 0.025. The circle *f* is then graduated to read 0.001 inch.

For diameters beyond 10 inches, the heat of the hand is found to affect the measurements, as it causes considerable expansion in a long steel rod. The rod, also, if made sufficiently light to be readily handled, would lack stiffness. For these reasons the measuring-points for larger diameters are made as follows: *a* (Fig. 89) is a holder of wood, bored

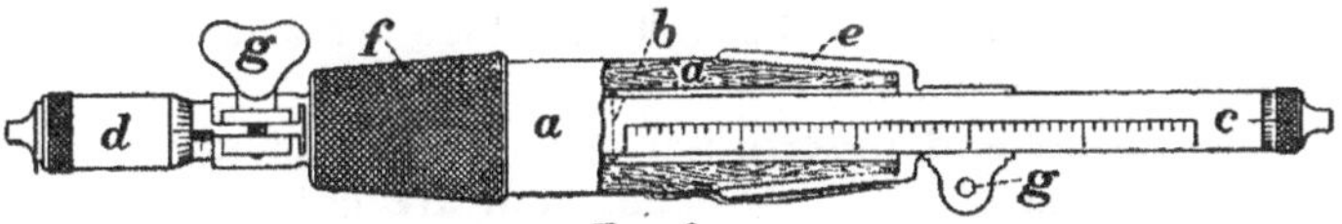

FIG. 89.

out in the middle, *b*, for the reception of the fixed and micrometer points *c* and *d*. Metal ferrules, *e* and *f*, of the shape shown, are fitted to the ends of the holders, and are provided with clamp-screws, *g*, to clamp the points *c* and *d*.

These points are essentially the same as before, the only difference being that the lengths of the holders vary, and the same points *c* and *d* are fitted to different holders.

USE.—Suppose a given diameter, say 12.50 inches, is to be measured. The standard comparator is first set to 12.50 inches, as just explained. A holder of proper length is then selected, and the points *c* and *d* fixed in it. The end of the fixed point *c* has an adjustment by which, having set the micrometer-point at zero, the length of the whole rod can be altered till it is exactly 12.50 inches. The interior diameter of the hoop can now be measured, and the differences, if any, in thousandths of an inch, read off on the micrometer-scale on the point *d*. There are also other methods of adjusting the rod.

126. Measurement of Interior Diameters of Long Tubes—The Star Gauge—Setting the Star Gauge.

THE STAR GAUGE.—In the case of long tubes, all parts of which are not readily accessible, some means must be adopted of making the measurements at a distance from the operator. The instrument used for this purpose is called a "star gauge." Its principal parts (Fig. 90) are a long hol-

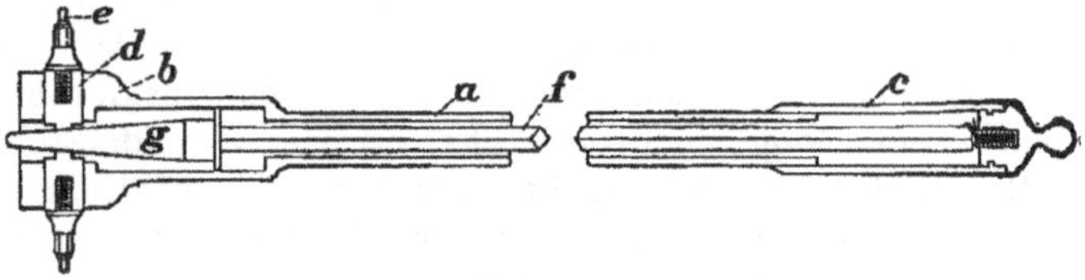

FIG. 90.

low brass rod, *a*, called the staff, to which are attached the head, *b*, and the handle, *c*.

The figure and description are intended to give only a general idea of the instrument, and are not accurate in details, as the instrument is too complicated to be fully described here.

The head *b* has three or more sockets, *d*, which are pressed inward upon the cone *g* by spiral springs, not shown in the figure. Into these sockets are screwed the star-gauge points *e*. There are generally three of these, 120° apart, varying in length, for the different calibres to be measured, so that by screwing the proper points into the sockets *d*, any diameter can be measured. The handle *c* is at the other extremity of the staff *a*. In the older forms of star gauge it had a sliding motion along the staff. With the new instruments motion is given by a micrometer screw. Extending through the staff is a square steel rod, *f*, united at one end to the handle *c*, and terminating at the other end in a cone, *g*. This cone has a known taper, and a forward movement of one inch corresponds to a certain definite increase in its diameter. This increase is marked on a scale upon the handle.

Use.—When the handle *c* is pushed forward, the cone *g* also moves the same amount, since it is connected with the handle by the steel rod *f*. When the cone moves forward, it pushes out the sockets *d*, resting upon its surface, and this forward motion of the handle and cone continues, till the points *e* come in contact with the walls of the bore to be measured. The amount of this outward movement of the points can then be read on the scale on the handle, and by comparing this with the original position of the points the size of the bore becomes known.

SETTING THE STAR GAUGE.—As with the measuring-points previously described, it is necessary to "set" the star gauge before use; that is, to establish an origin or datum to which all measurements are referred. Suppose the bore to be measured is 10.00 inches in diameter. Accompanying the instrument is a series of rings very accurately bored to the different sizes likely to be required in practice. The 10.00-inch ring is selected, and the standard comparator set to that length, a measuring-point adjusted to it, and the ring then tested by this point to see if it is exactly 10.00 inches. If not, the error is noted and corrected for. The 10.00-inch points having been screwed into the sockets of the star gauge, the ring is held so that when the handle is moved forward, the points will all touch the ring. While the points are in this position, the handle is adjusted so that the reading is zero.

The ring is then removed, the instrument inserted in the bore, and the readings of the scale taken for every inch of length of bore. These readings are added to, or subtracted from, the original diameter of the ring, according as they are greater or less than zero, and the results give the true diameter of the bore.

127. Exterior Diameters—Calipers—Arm—Support—Action. The instruments used for measuring exterior diameters are called calipers. One form consists, Fig. 91, of the arm *a*, the measuring-points *b*, *c*, and the support *d*.

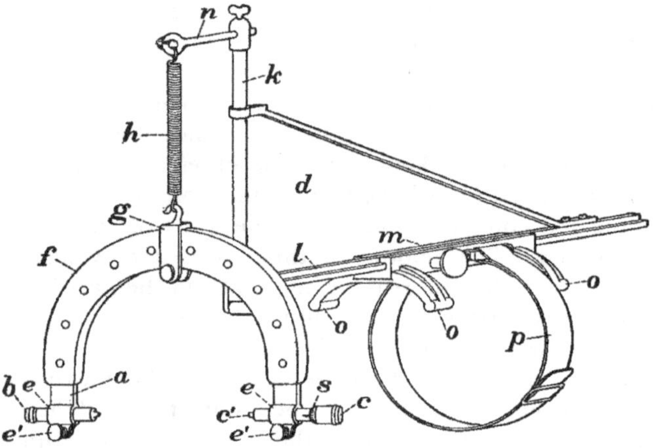

FIG. 91.

ARM.—The arm *a* is made of steel in a semicircular form, and as light as possible consistent with stiffness. The arm terminates in sockets, *e*, at each end, which are provided with clamp-screws *e'*.

To increase the stiffness of the arm and protect it from variations of temperature in use, it is covered with wood, *f*.

The measuring-points *b*, *c* pass through the sockets *e* in the arm, and are clamped in position by the clamp-screws *e'*. The point *b* is called the *fixed* point, as it does not change its position relatively to the arm when once clamped; the point *c* is the *measuring* or *micrometer* point, and having been clamped in the socket *e*, its extremity, *c*′, is capable of a small motion by means of a micrometer-screw, whose construction has been previously explained. This point carries a scale, *s*, reading to thousandths. The points when in position are always in a straight line.

The arm with its points is suspended from its support, *d*, by the hook *g* and spiral spring *h*.

SUPPORT.—The support of the calipers consists of a standard, *k*, fixed to a bar, *l*. This bar slides longitudinally upon a base, *m*. The standard *k* carries a rod, *n*, to which the spiral spring *h* is attached, and to this spring the hook *g*. The whole support rests on the exterior of the tube to be measured, being brought parallel to the axis by the feet *o o*, and held in this position by the leather strap *p*, which is buckled tightly around the tube.

ACTION.—Suppose a diameter of 15.00 inches is to be measured. Set the standard comparator to this length, and having determined from it the length of a measuring-point of exactly 15.00 inches, set the micrometer-point *c* at zero, and adjust the points *b* and *c* in the sockets till the distance between them is exactly 15.00 inches. Raise or lower the caliper-arm till the points *b* and *c* are slightly above a horizontal plane through the axis of the tube. The bar *l* may then be moved along the tube parallel to its axis, sliding on the bed *m*, and measurements made for every inch of length. The bar *l* will slide for a length of 12 inches. The leather strap *p* must then be loosened, the whole support moved forward this distance, and the strap again tightened, when measurements may be made as before, till the whole is completed.

128. Measurement of Lengths—Step Gauge—Surface Lengths.

The accurate measurement of lengths is very difficult to make, and as each particular case requires a special arrangement, only general ideas can be given.

STEP GAUGE.—One of the most frequent measurements required is the length of the recess or step, *ab*, in a hoop. If this be too short, the hoop will not come in contact with the preceding one when shrunk on; and if too long, an opening will be left at the shoulder, which leaves the tube unsupported. To measure this length, an instrument called a step gauge is used. This consists, Fig. 92, of a steel blade, *c*, sliding through a socket in a body, *d*. These blades are of different lengths, corresponding to the different hoops to be measured. On the end of the blade is fixed a steel templet, *f*, which exactly fits the shoulder in the hoop. The templet being held against the shoulder *b*, while the body is pressed against the face of the hoop at *e*, the length can be read off on the scale.

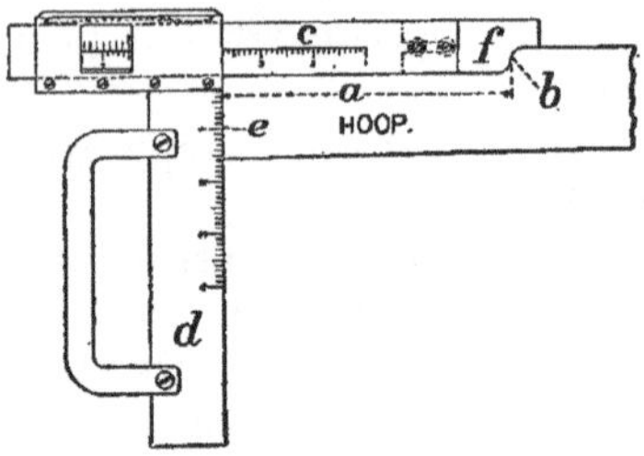

FIG. 92.

SURFACE LENGTHS.—In each shrinkage operation, the changes in diameter and length due to that operation are measured. The changes in diameter are measured with the points, star gauge, or calipers.

For measuring the changes in length, the following plan is adopted:

Two holes are made with a punch in the exterior surface of the tube or hoop, and their exact distance apart before shrinkage measured as follows:

An instrument, Fig. 93, consisting of a main body, *a*, carries a fixed head, *b*, and two movable points, *c* and *d*. The point *c* is attached to a sliding head, *c'*, which carries a micrometer-screw, *e*.

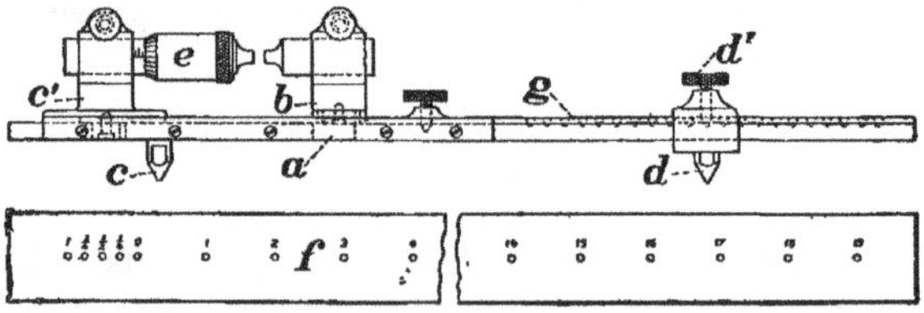

FIG. 93.

g is an extension-bar, having holes at intervals of 0.25 inch.

Accompanying the instrument are *reference-bars*, *f*, which have holes in them exactly one inch apart, and at the left end one inch is graduated into $\frac{1}{4}$, $\frac{1}{2}$, $\frac{3}{4}$ inch.

When the holes are punched in the surface of the hoop as before explained, their distance apart is measured approximately with a scale. Suppose this distance to be 18.40 inches.

Move the point *d* along the extension-bar *g* till it will enter the 18-inch hole, and clamp it, the screw *d′* passing into one of the holes in the bar *g*. Place the instrument on the reference-bar *f*, the point *c* entering the $\frac{1}{2}$-inch hole in it, and the point *d* resting in the 18-inch hole. The distance between the points *c* and *d* is now 18.50 inches.

Fix the micrometer *e* at zero, and move the points *b* and *e* till they are in contact. Now place the instrument on the hoop to be measured, the point *c* in one punch-mark and *d* in the other, make contact again with *e*, and subtract the reading of the micrometer-scale from 18.50 for the distance apart of the holes.

After shrinkage the same process gives the distance apart of the punch-marks; and the difference before and after, the change due to shrinkage.

129. General Principles of Measurements—Touch—Interior Diameters of Short Hoops.

In the above descriptions all the complicated details of the instruments have been omitted, and only the general method of their operation and use given. The templet measurements require no special notice. A few general principles relating to the method of using these instruments must be understood.

TOUCH.—The accuracy of all measurements with these instruments depends upon the skill of the operator, and hence practice is necessary to obtain satisfactory results. In most cases the sense of touch is relied upon to determine when proper contact of the measuring-points with the surface to be measured, is obtained.

Various mechanical devices, such as electrical indicators, etc., have been tried to determine when proper contact has been obtained, and can be used to advantage when a large amount of measuring is to be done.

INTERIOR DIAMETERS.—In measuring these (Figs. 94

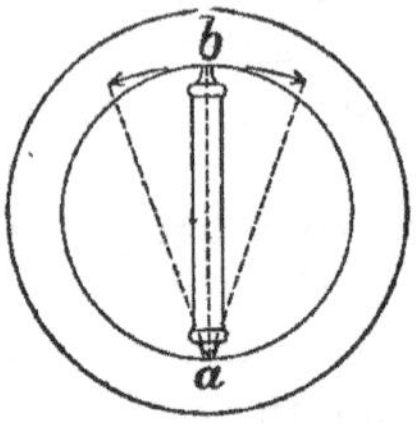

FIG. 94.

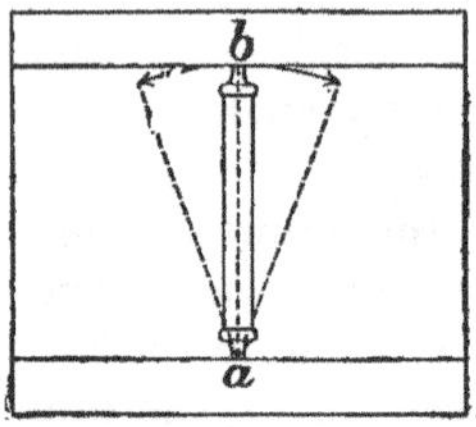

FIG. 95.

and 95), the hoop is placed horizontally, and the lower fixed point of the measuring-rod *a* is held by the operator at the point whose diameter is to be measured. It is evident from Fig. 95 that the diameter is the shortest line from *a* to *b*, and hence if the upper point of the measuring-rod be moved from *b* in the direction of the arrows, it will cease to touch the surface of the hoop.

From Fig. 94, the diameter *ab* is the longest line in the cross-section; and if the point be moved to either side of *b*, it will jam against the surface of the hoop. Hence in determining an interior diameter at any point with the measuring-rods, hold the fixed point of the rod firmly against the lower surface of the hoop at the point where the diameter is required. Move the micrometer-point in two directions at right angles to each other, one along the axis of the hoop, the other across the axis, till a point is found where contact occurs due to both these motions.

The reading of the rod will give the diameter.

130. Interior Diameters of Long Hoops—Exterior Diameters—Vernier Scale.

INTERIOR DIAMETERS OF LONG HOOPS.—These are measured with the star gauge, and in order that they may be correct the points must move at right angles to the axis of the bore.

By the construction of the instrument these points must move at right angles to the staff of the star gauge, and hence it becomes necessary that the staff be placed accurately in the axis of the bore. For this purpose the gun or tube is carefully levelled, and various supports are used in the bore, which insure centering of the star-gauge staff. Exterior rests are also provided to support that part of the staff outside the bore.

EXTERIOR DIAMETERS.—In the measurement of exterior diameters, the same principles apply as to interior diameters. In Fig. 96 the diameter *c* is the longest line in the cross-

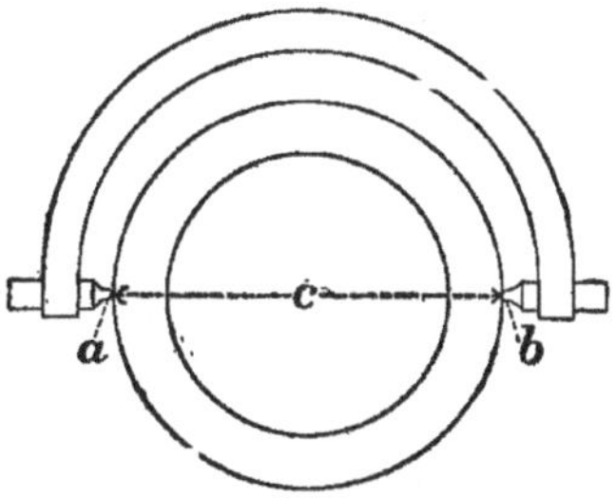

FIG. 96.

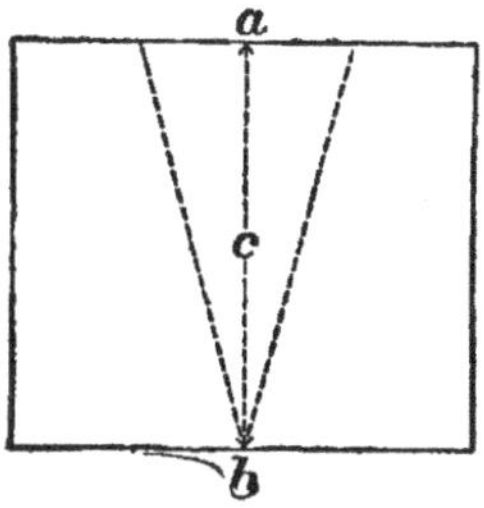

FIG. 97.

section, and the shortest line in the longitudinal section, Fig. 97. Hence the fixed point of the calipers is held at *a*, and the measuring-point *b* moved in two directions at right angles, as in case of the interior diameters, till proper contact is made.

VERNIER SCALE.—A very useful instrument in these

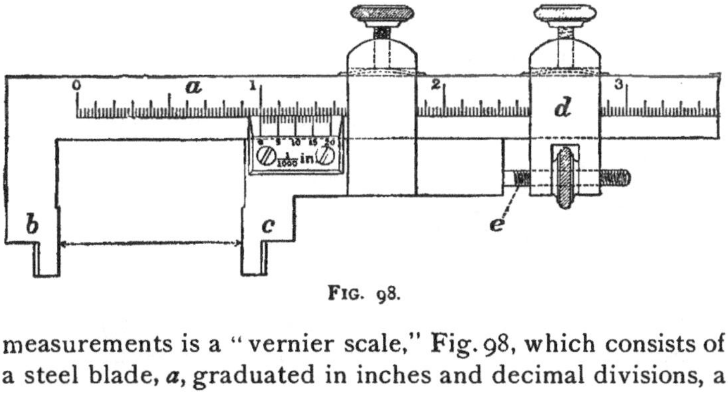

FIG. 98.

measurements is a "vernier scale," Fig. 98, which consists of a steel blade, *a*, graduated in inches and decimal divisions, a fixed jaw, *b*, a movable jaw, *c*, and an auxiliary jaw, *d*, with its tangent screw, *e*. The principle is the same as that of the standard comparator, lengths being measured between *b* and *c*. The advantage of this instrument is that it can be carried to any part of the shops, and when its error is determined by the standard comparator, it can be used in place of the latter with great convenience. Its disadvantage is that it is affected by changes of temperature when carried to different places in the shop, and when handled.

DESCRIPTION OF CANNON.

1. In U. S. Service.

131. Classification—Hotchkiss Mountain Rifle.

CLASSIFICATION.—Cannon may be classified according to the service for which they are intended, into mountain, field, siege, or sea-coast guns; according to the kind of fire they deliver, into guns, howitzers, and mortars; according to the kinds of projectiles used, into smooth-bore and rifled; and according to the methods of loading, into muzzle- and breech-loaders. As all modern guns are breech-loading rifles, it is most convenient for discussion to consider them according to the service for which they are intended.

Machine and rapid-fire guns will be considered later.

MOUNTAIN GUNS.

HOTCHKISS MOUNTAIN RIFLE.—This is the only gun of this class in service. It is made as light as practicable, so that it can be carried on the back of a mule, its weight being 116 lbs. Its carriage weighs 220 lbs., and two men can pack, unpack, and mount it.

The gun, Fig. 99, is made of steel in a single forging, the

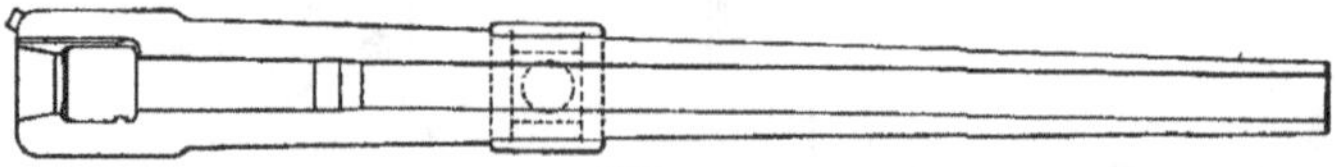

FIG. 99.

trunnion-hoop being screwed on. The calibre is 1.65 inches; weight of shell loaded 2 lbs. 10 ozs.; of powder-charge, $5\frac{1}{2}$ ozs. The initial velocity is 1275 ft.-secs.

Breech Mechanism.—The mechanism is a simple form of the Krupp. It consists of a rectangular steel block, *b*, Fig.

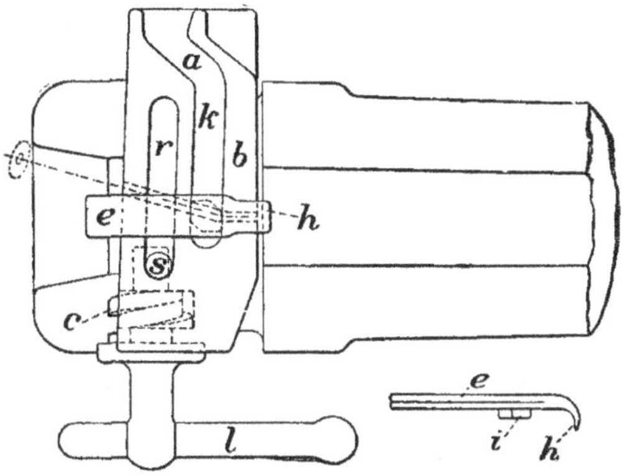

FIG. 100.

100, with rounded corners; its front face being at right angles to the axis of the bore, and its rear face slightly inclined to that axis.

This block slides transversely in a recess in the breech, and when withdrawn leaves the breech open for loading. It is locked in the firing position by a cam, *c*, entering a corresponding recess in the breech, and this cam is operated, and the block withdrawn and pushed home, by the lever-handle, *l*. *e* is the extractor for withdrawing the empty cartridge-case. It is a prismatic bolt, sliding in a groove in the upper part of the breech, parallel to the axis of the bore, and terminating in front in a hook, *h*. A tenon, *i*, on the under side of the extractor, fits in the extractor-groove, *k*, cut in the top of the breech-block. This groove is straight for some distance, and then curves quickly to the rear. When the block is withdrawn it moves in guides which are parallel to its rear face, and which consequently give it a motion such that the extractor is at first gradually withdrawn, thus removing the empty case from its seat in the chamber.

The tenon of the extractor then enters the inclined part, *a*, of the groove in the block, and the extractor, with the cartridge-case, is drawn quickly backwards, thus ejecting the case to the rear. The motion of the breech-block is arrested by a stop-bolt, *s*, which is screwed through the upper part of the breech, and enters the groove *r* on the top of the block.

Ammunition.—The ammunition is contained in a metallic cartridge-case, and as this forms a gas-check, no accurate fit of the parts of the mechanism is required, and the breech-block works freely in its slot.

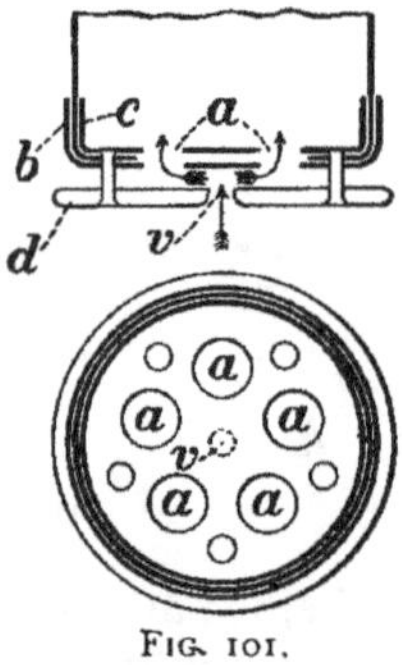

FIG. 101.

The charge is fired with an ordinary friction primer. The head of the case is formed by a cup, *c*, inside, Fig. 101, having five holes, *a*, in it. The exterior is strengthened by a cup, *b*, having five holes corresponding to *a* and a sheet-iron disk, *d*, riveted to the cups and case, and having a central hole, *v*. The flame from the primer passes through the hole *v*, and thence through *a* to the charge. The gas-pressure from the charge forces the cups *b* and *c* backwards, closing the hole *v* in *d*, and preventing the escape of gas. The projectiles are shell and canister. In order to use shrapnel, a heavier gun of 3-inch calibre has lately been adopted, weight 218 lbs.

FIELD GUNS.

132. U. S. Field Artillery—3.6-inch B. L. Mortar—3.2-inch B. L. Field Gun, Light—3.6-inch B. L. Field Gun, Heavy.

The field artillery in the U. S. service consists of the 3.6-inch mortar, 3.2-inch light field gun, and 3.6-inch heavy field gun.

Common Features.—See Figs. 102, 103, and 104. These pieces are all built of gun-steel; are breech-loaders, and have rifled bores. They have conical gas-check seats, *c*, and cylindrical powder-chambers, *d*, of larger diameter than the bore, and these chambers are connected with the bore by a conical slope, *e*, forming the seat for the rotating band of the projectile, and by which it is centered in the bore.

In front of this powder-chamber slope is a second conical slope, *f*, which is formed by cutting away the tops of the lands of the rifling to a certain depth at the origin or beginning of the rifling, and gradually decreasing the depth of this cut to zero, at a certain distance from the origin, this distance varying with the size of the piece. As a rule one half of the lands are cut away at the origin. Thus for the 3.6-inch gun the depth of the rifling groove is 0.04 inch, and the lands are cut away 0.02 inch at the origin. The object of this rifling slope is to allow the band of the projectile to enter gradually to its full depth into the groove, and thus diminish the strain due to forcing. It also facilitates the loading of the projectiles, and tends to prevent the escape of gas over the band, as the latter is forced readily and quickly to the bottom of the groove.

3.6-INCH MORTAR.—This is a short piece intended for vertical fire against troops protected by intrenchments or irregularities of ground from the direct fire of the field guns.

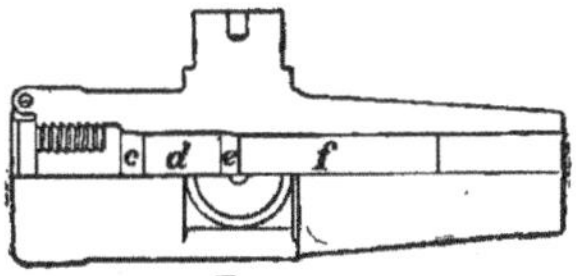

FIG. 102.

It is made of a single piece of steel (Fig. 102), and is designed to use the same kind of powder and the same projectile as the 3.6-inch field gun.

It is mounted upon a cast-steel carriage, and the weight of piece and carriage are so adjusted that they can be readily moved by hand.

3.2 AND 3.6-INCH FIELD GUNS.—The 3.2-inch gun (Fig. 103) is intended for use as a horse-artillery gun for rapid movements, and the 3.6-inch (Fig. 104) for the light or field battery.

Common Features.—The two guns are exactly similar in construction, and each consists of an interior tube, and a jacket, assembled by shrinkage.

The tube is inserted in the jacket from the front; a shoulder, *a*, on the tube resting against a corresponding one on the jacket.

Forward movement of the tube in the jacket is prevented by the shoulder *b*, as shown in Figs. 103 and 104. The threads for the breech-block are cut in the rear end of the jacket, which thus supports the longitudinal stress.

The principal dimensions, etc., are given in the table page 253, with those of the siege guns, and of these, the weight of piece, charge, and projectile should be remembered.

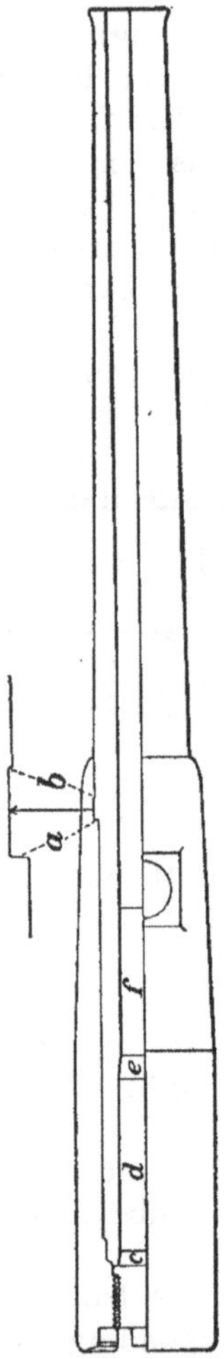

FIG. 103.

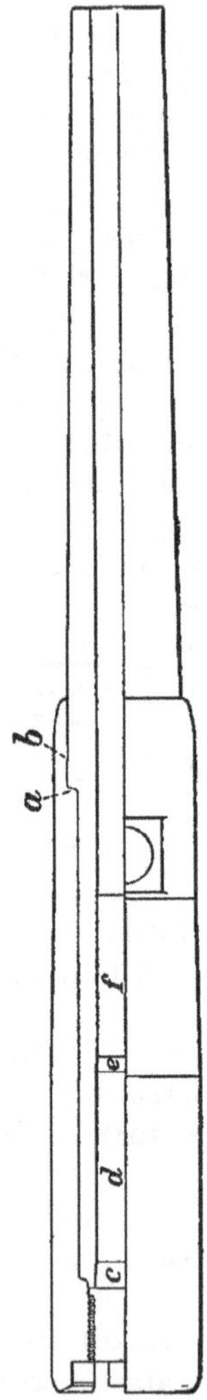

FIG. 104.

133. Breech Mechanism of Field Artillery — Principal Parts — The Breech-block.

PRINCIPAL PARTS. — The breech mechanism comprises those parts which are necessary to open, close, and lock the breech, to prevent the escape of gas in firing, and through which the charge is ignited.

The principal parts are:

1. The breech-block, which closes the breech and, by its bearing on the fixed parts of the gun, supports the gas-pressure when the charge is fired.

2. The obturator, which prevents the escape of gas.

3. The carrier-ring, which guides the block as it is withdrawn from the breech, supports its weight when withdrawn, and by which it is swung round, out of the way, for loading.

4. The lever-handle or other device by which the block is rotated after firing, and its threads or bearings disengaged from those in the breech of the gun.

5. The vent, by which fire is communicated to the charge; and the vent-closer, by which premature discharge is prevented.

THE BREECH-BLOCK.—In all guns in the U. S. service, except the Hotchkiss mountain-gun already described, the breech-block belongs to the French or interrupted-screw system. That is, screw-threads are cut around the exterior cylindrical surface of the block, and around the corresponding interior cylindrical surface of the breech-recess. To avoid the delay in unscrewing the block and screwing it home after and before each discharge, the circumferences of the block and breech-recess are divided in the field-guns into six equal sectors, and the screw-threads on every alternate

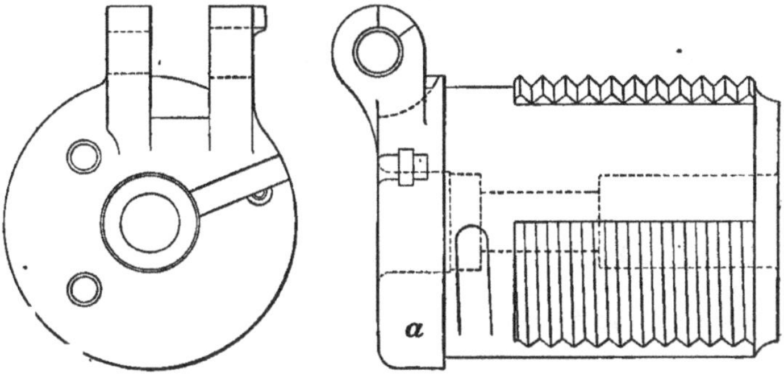

FIG. 105.

sector removed, thus leaving on the block, and in the breech-recess, three threaded and three slotted sectors of equal width. By this arrangement the block can be pushed in or pulled out of its recess, the threaded sectors on the block sliding in the slotted sectors of the breech-recess. After it is pushed home to within one sixth of a turn of the thread, or one sixth of the pitch, a rotation through an angle of 60° will cause the threads on the block to engage in those in the jacket, and the threads thus engaged are found to have ample strength to resist the pressure of the powder-gas. The threaded and slotted sectors of the block are partly shown in Fig. 105. The exterior diameter of the block at the threaded portion, is greater than that of the powder-chamber, in order to give as large a surface as possible for the screw-threads and thus increase their relative strength, and also to leave a large opening in the breech to facilitate the insertion of the projectile and charge. The length of the block is greater than its exterior diameter, to give a greater number of threads, and thus distribute the pressure of the powder-gas over a greater number of them, and reduce the stress on each, and consequently the tendency to strip. The front face of the block is plane, and the rear face has certain projections whose uses will be explained. The diameter of the unthreaded portion in front is less than that of the threaded portion, in order that it may enter for a short distance into the gas-check seat in the rear end of the tube. The rear end is not threaded, and has a shoulder, *a*, upon it, which fits tightly against the rear face of the carrier-ring when the breech is closed, and thus prevents the entrance of dust in transportation. The interior is bored out for the reception of the parts of the obturator, and certain grooves are made on the exterior whose object will be explained.

134. The De Bange Obturator—Action—Remarks.

The obturator prevents the escape of gas around the threads of the breech-block and through the mechanism. Two obturators are used in the field service: the De Bange, with the 3.2 and 3.6 rifles, and the Freyre, with the 3.6 mortar.

THE DE BANGE OBTURATOR.—This consists (Fig. 106) of a central spindle or stem, *a*, terminating in front in a large head, *b*, called, from its shape, the "mushroom-head"; the vent, *c*, with a copper bushing, *d*, in front, and the primer-seat, *e*, in rear; two steel cups, *f f'*, called gas-check cups, and between them a plastic pad, *g*, made of asbestos and tallow, strongly compressed by hydraulic pressure before its insertion, and covered with canvas, the outer edges of the pad being protected by two thin strips of copper, *m*; an obturator-nut, *h*, held in place by a spline-screw, *k*, which is halved into the nut and spindle; and a spiral spring, *j*, bearing against a shoulder in the breech-block, and against the front of the obturator-nut, *h*.

When in place in the gun, the spindle *a* passes through the axis of the breech-block, the outer surface of the pad *g*

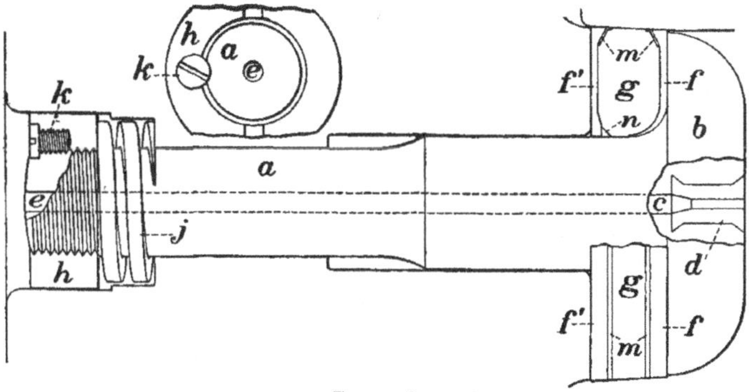

FIG. 106.

rests against the gas-check seat in the gun, and is held between the elastic gas-check cups *f f'*. The mushroom-head *b* is in the powder-chamber.

ACTION OF THE DE BANGE OBTURATOR.—When the charge is fired, the gas-pressure acts normal to the surface of the mushroom-head, and the latter, with its spindle *a*, is forced to the rear. The pressure is thus transmitted to the gas-check cups *ff'*, and the elastic pad *g*, being held by the front of the block and pressed between the cups *ff'*, is forced to expand radially and pressed firmly against the walls of the gas-check seat, preventing the escape of gas. An elastic packing-ring, *n*, also expands under the pressure, and, fitting tightly against the spindle *a*, prevents the flow of the tallow of the pad, and thus avoids the sticking of the pad to the spindle. When the pressure is removed, the action of the spring *j* keeps all the parts, cups and pad, in place.

The pad has the shape shown in order to have as small a surface as possible in contact with the spindle and the walls of the gas-check seat, to avoid sticking, and to furnish in the middle a reservoir of material that may be pressed outward and inward and thus secure perfect contact.

REMARKS.—The canvas cover of the pad prevents its breaking in use. The shape of the gas-check cups is such as to avoid all sharp angles which would cut the pad, and to retain the edges of the cups for preventing the flow of the tallow. The elastic packing-ring *n* also assists to prevent this flow. The spline-screw *k* prevents the unscrewing of the obturator-nut *h*, which is liable to occur when the breech-block is rotated for withdrawing, owing to the sticking of the pad to the walls of the gas-check seat.

135. The Freyre Obturator—Action—Remarks.

THE FREYRE OBTURATOR.—This obturator (Fig. 107) is used with the 3.6-inch mortar, and consists of a central spin-

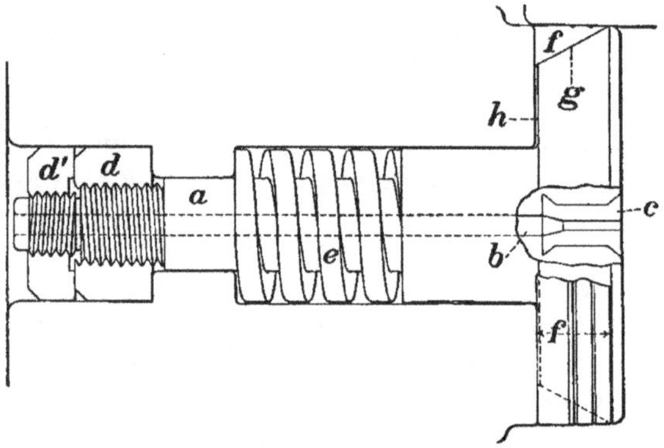

FIG. 107.

dle, *a*, terminating in front in a large flat disk; the vent *b* and its copper bushing, *c*, in front, and the primer-seat in rear; the threads and nuts, *dd'*, the rear nut, *d'*, being a locking-nut with a left-hand thread, while the obturator-nut, *d*, has a right-hand one, and hence *d'* prevents *d* from unscrewing during the rotation of the block; the spiral spring *e*, bearing on a shoulder on the spindle *a* and a corresponding shoulder in the block, and tending to push the spindle constantly forward; and the gas-check ring *f*. The exterior surface, *g*, of the head of the spindle is conical, and ground to an exact fit with the interior surface of the conical gas-check ring, *f*. This ring *f* is made of steel of high elastic limit, and when in place, rests against the front face of the breech-block as shown, and its length parallel to the axis of the bore is such that when in this position there is a space, *h*, between the head of the spindle and the front face of the breech-block. When in place, the spindle *a* passes through the axis of the breech-block, the outer surface of the elastic gas-check ring *f* rests against the walls of the gas-check seat, and the front surface of the spindle-head is in the powder-chamber.

ACTION OF THE FREYRE OBTURATOR.—When the charge is fired, the gas, acting normal to the spindle-head, presses it backward into the conical ring *f*, the space *h* allowing this movement. The ring, *f*, being held against the face of the breech-block, is thus forced to expand radially by the wedging action of the spindle-head, and is pressed firmly against the walls of the gas-check seat, preventing the escape of gas around the exterior of the ring. The tight fit of the two conical surfaces prevents any escape between the ring and head. When the pressure of the gas is removed, the elasticity of the ring *f* and the action of the spring *e* return the spindle-head and ring to their former positions.

REMARKS.—This obturator has the following advantages:

1. Being of metal, it is very slightly affected by changes of temperature, weather, etc.

2. It occupies very little space in the powder-chamber; and hence when space and consequently weight are important, as with this mortar, it is used.

Its disadvantage is that it is liable to get out of order. A blow struck on the thin edge of the ring *f* in loading, or closing the breech, would allow the gas to escape; and after a channel is once formed for the gas, the obturator is useless. This accident is liable to occur in field service, and to guard against it to some extent the spindle-head is made to project well beyond the front edge of the ring. This projecting portion would ordinarily receive any blow which might injure the edge of the ring *f*.

It will be observed that the spiral spring *j* in the De Bange obturator, Fig. 106, acts in the opposite direction to that of the Freyre, Fig. 107. The reason is as follows: The gas-check ring in the Freyre being elastic, when proper compression of this ring is once secured by the nuts *dd'*, it will be retained unchanged. With the De Bange the pad is not elastic, and hence if properly compressed before firing this compression will change after firing, due to the great pressure upon it. Hence a constant tension is always required with the De Bange to keep the cups and pad in place, while with the Freyre the spring is used to push the spindle forward, and assist in restoring the ring to its former position.

136. Carrier-ring—Object of Latch.

CARRIER-RING.—This ring guides the block, supports its weight when it is withdrawn from the breech, and enables it to be swung round to one side of the gun, out of the way of loading. It consists (Fig. 108) of a ring of steel, *h*,

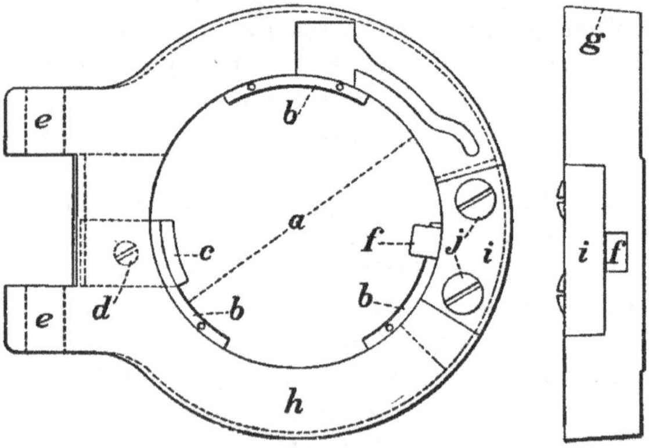

FIG. 108.

which surrounds the breech-block, and through which the breech-block slides parallel to the axis of the piece. The breech-block occupies the space *a*. On the interior there are three lugs, *b*, the exact width of the slotted sectors of the block. These lugs bear in the slotted sectors, and furnish guides for the block when it is drawn to the rear, so that it is compelled to move parallel to the axis of the gun.

On the left-hand side is a stop, *c*, which travels in a groove in the breech-block, and limits the motion of translation of the latter when it is withdrawn from the breech, and also its motion of rotation, when turned to lock into the threads of the breech of the gun, or unlocked for withdrawal. This stop passes through the carrier-ring and is secured by a screw, *d*. The stop may occupy any other convenient position, and may be a simple stud, as in the case of the mortar, where the stop is at the top of the carrier-ring.

Two lugs, *e e*, are for the purpose of attaching the carrier-ring to the jacket, by a pin which passes through holes in the lugs, and corresponding holes in the jacket. This pin forms the axis around which the carrier-ring, with the block, swings, when the breech is open for loading. The exterior surface, *g*, of the carrier-ring is conical, to secure a good fit in the breech.

OBJECT OF LATCH.—When the threads of the breech-block are disengaged from the corresponding threads in the breech, the block is pulled to the rear through the carrier-ring. It is evident, however, that this pull upon the block will cause the carrier-ring to swing around the pin passing through the lugs *e*, and this will tend to jam the block, and prevent its movement to the rear. To avoid this, the carrier-ring must be locked to the gun while the block is moving to the rear. When the travel of the block is finished, the carrier-ring must be unlocked from the gun, in order that it may be swung round with the block to the loading position. These objects are accomplished by the latch *f*, shown in Fig. 108 and in detail in Fig. 109.

137. Description of Latch and its Working.

The latch consists of a piece of metal shaped as shown in Fig. 109. The lower inner end, *a*, fits against one of the slotted sectors of the breech-block, and is constantly pressed down upon it by the action of the flat spiral spring *b*, acting on a shoulder, *c*, of the latch and a corresponding shoulder, *d*, of the latch-recess. All the working parts are covered by the latch-plate *i* (see Fig. 108), secured to the exterior of the carrier-ring by two screws, *j*, so that in case any part breaks it may be readily removed and repaired. The front of the carrier-ring next the breech has a hole, *g*, cut through it, and opposite this is a recess, *h*, in the corresponding face of the latch.

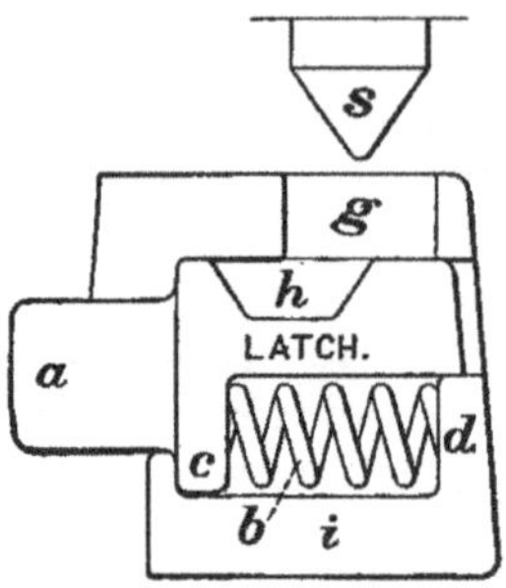

FIG. 109.

Action of the Latch.—Suppose the breech closed and the gun fired. At the rear end of the breech-block is a transverse groove, *a*, Fig. 110, which is on a level with the slotted sector of the block at *c*, and gradually increases in depth to

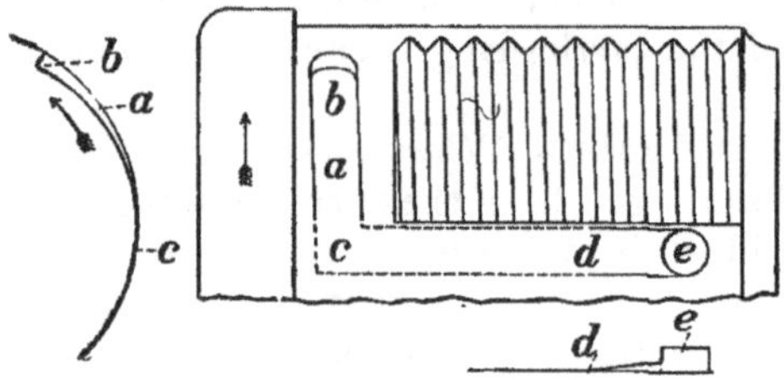

FIG. 110.

its end, *b*. The depth of this groove at *b* is such that when the inner end of the latch rests in it, the action of the flat spiral spring, *b*, Fig. 109, will force the latch down sufficiently far to release it from the jacket. The inner end of the latch rests at *b* during firing, and hence at this time the carrier-ring is unlocked from the jacket, and there is no strain on the latch. After firing, the first operation in opening the breech, is to rotate the block in the direction of the arrow. As the latch is in the carrier-ring, it does not rotate with the block, and hence the action of the groove *a* is to push up the latch into its recess in the breech, and thus lock the carrier-ring to the jacket. The inner end of the latch now stands at *c*. The block is now withdrawn, the inner end of the latch sliding along the slotted sector, and keeping the carrier-ring locked to the jacket. This continues up to the point *d*.

At this point, the path of the inner end of the latch begins to descend along a gradual slope from *d* to *e*. Hence by the action of the flat spiral spring, the latch begins to move out of the jacket, being forced inward into the groove *de*. At the end of the travel of the block, corresponding to the point *e*, the depth of the groove *de* becomes sufficient to allow the entire withdrawal of the latch from its recess in the jacket, and hence at this instant the carrier-ring becomes unlocked, and the block and ring can be swung round for loading.

After Loading.—After loading, when the block and carrier-ring are swung round to close the breech, the pressure of the hand is applied to the rear end of the block. This pressure would tend to move the block forward through the carrier-ring, and hence jam the gas-check against the breech. The block must therefore be locked positively to the carrier-ring in the loading position, and this is done as follows:

The extremity *e* of the groove *a* in the block, terminates in a cylindrical hole, into which the inner end of the latch drops at the end of its motion. The block therefore cannot move with reference to the carrier-ring, till the latch is lifted from this hole *e*. This is accomplished as follows:

There is a conical stud, *s* (see Fig. 109), projecting from the rear face of the base ring, which, as the carrier-ring is closed, passes through the hole *g*, Fig. 109, in the front face of the carrier-ring, and enters the recess *h* in the front of the latch. This stud, bearing against the inclined end of the recess *h*, owing to the shape of the two surfaces, raises the latch slightly till it clears the cylindrical hole *e*, Fig. 110, and stands at such a height that when the block is pushed forward, the lowest part of the inclined surface *de* will pass under the inner end of the latch, and thus cause it to move up the inclined surface and push the latch home.

138. The Lever-handle.

This is a device for rotating the breech-block. In the mortar there are two of these handles fixed to the block at opposite extremities of a diameter (Fig. 111).

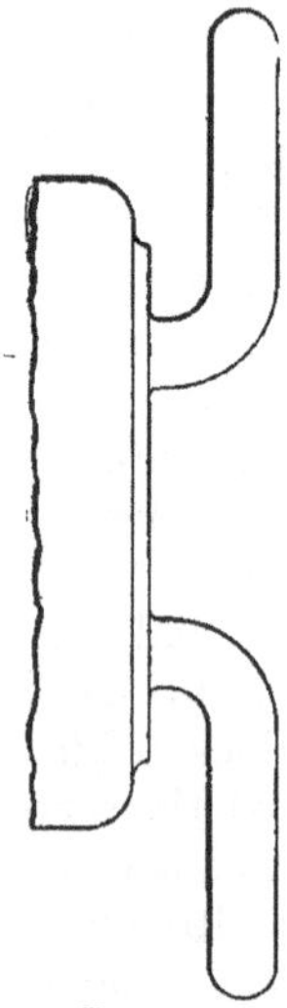

FIG. 111.

In the 3.2 and 3.6-inch guns there is one handle, *h*, pivoted to the upper part of the block by a pin, *e*, Fig. 112. This handle is raised vertically for rotation; and when lowered for firing, its lower end fits into a recess, *c*, in the end of the jacket, for additional security against accidental opening. To limit the vertical motion of the handle when it is raised, a stop, *a*, is placed upon the pin *e*, which abuts against a corresponding stop, *b*, on the lug *l*. The head *d* of the lever-handle, is eccentric, and forms a cam, with the following objects: When the block is in the firing position, this cam *d* enters a corresponding recess, *r*, in the rear face of the carrier-ring, and thus locks the block to the ring, and with the end of the lever-handle, as before explained, prevents any rotation of the block in firing. When, after firing, the lever-handle is raised and the block rotated, if an attempt be made

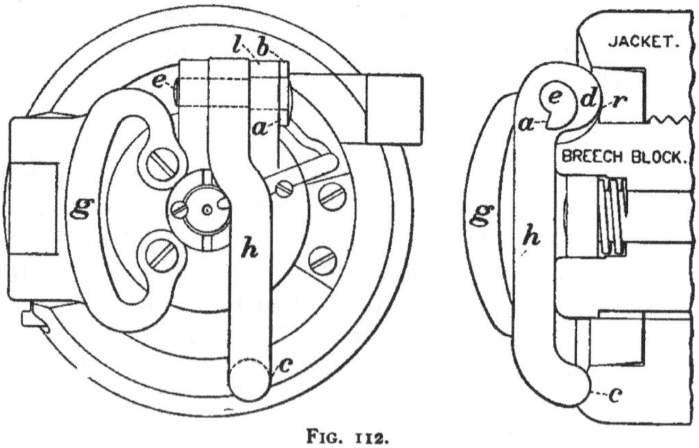

FIG. 112.

to withdraw the latter, it sometimes fails on account of the sticking of the pad in its seat. If the lever-handle be now lowered, the surface of the cam *d* bears against the rear face of the carrier-ring, since no recess is cut for it in this position, and thus exerts a powerful leverage, sufficient to start the pad from its seat.

The lever-handle is made to work tight between its lugs in the block, in order that it may not fly up from its recess in the carrier-ring, by the shock of discharge.

In the 3.2 and 3.6 guns there is a fixed bronze handle, *g*, attached to the breech-block for the purpose of withdrawing it.

139. The Vent-cover.

This is a device to prevent the insertion of a primer, and the premature discharge of the piece, before the breech-block is locked. It must be so arranged that the vent will be closed at all times, except when the threads of the block are engaged in those of the breech.

3.6 *Mortar.*—For the 3.6 mortar the device is as follows: A handle, *a*, Fig. 113, is attached to a shaft, *b*, which fits

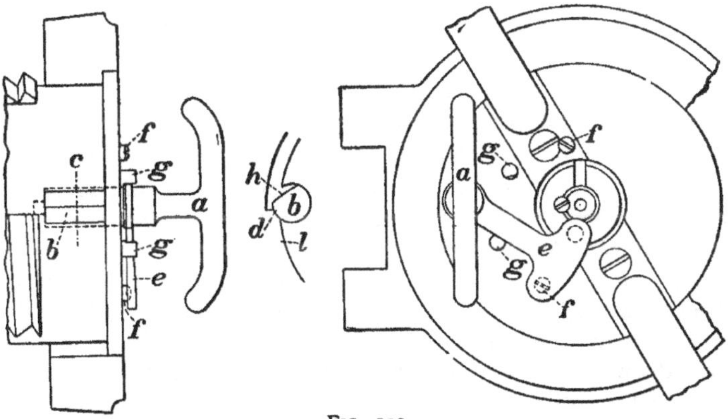

FIG. 113.

into a recess on the left side of the breech-block. The shaft is shown in cross-section at *c*. When turned into the position shown in the figure, so that the vent is open, the corner *d* projects through the block, and binds against the edge of one of the lugs *l* in the carrier-ring, so that when in this position, with the vent uncovered, the breech-block cannot be rotated to open the breech. The piece *e* is attached to the shaft *b*, and closes the vent. When open or closed, its ends rest on two pins, *f*, which retain it in position, and its motion in opening or closing is limited by two studs, *g*. In order to rotate the breech-block, the vent must first be closed by turning the shaft *b* upwards by the handle *a*. The corner *d* of the shaft then no longer bears on the edge of the lug in the carrier-ring, and the surface *h* forms part of the exterior curved surface of the block.

3.2 *and* 3.6 *Guns*.—For these guns a radial slot, *a*, Fig. 114,

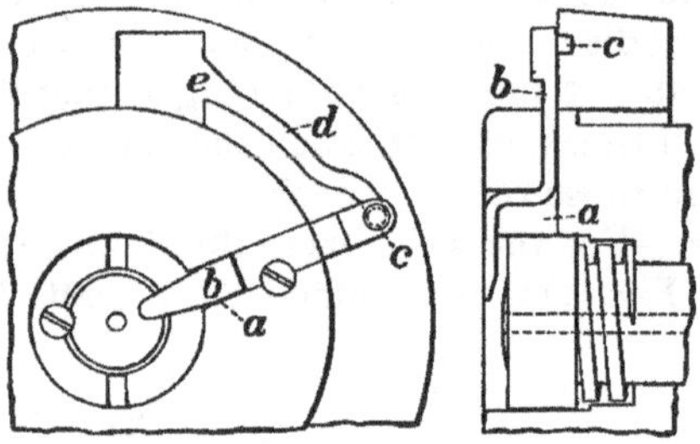

FIG. 114.

is made in the rear part of the breech-block which projects outside the carrier-ring. In this slot slides a piece of metal, *b*, having a pin, *c*, projecting from its forward face next the breech of the gun. A groove, *d*, is cut in the rear face of the carrier-ring, which is eccentric at its lower end, and the pin *c* bears in this groove.

When the block is pushed home, the pin enters the groove at *e*, and its weight keeps it over the vent, as it stands in a vertical position during the time the block is withdrawn. As the block is rotated to the right in closing, the vent is still covered, due to the bearing of the pin *c* in the concentric part of the groove *d*. At the last instant of rotation, however, the pin *c* enters the eccentric part of the groove *d*, the vent-closer is lifted, and the vent uncovered.

140. Action of Mechanism of 3.6-inch Mortar.

1. Suppose the breech closed and ready for firing. In this position, the threads of the block are engaged in those of the gun, the gas-check is in its seat, the vent-cover has been moved to the right, or downward, by hand, thus uncovering the vent. By this motion of the vent-closer, the corner of the shaft, as before explained, has been caused to project beyond the surface of the block, and to bind against the edge of one of the lugs of the carrier-ring, so that the block cannot be rotated while the vent is open. The inner end of

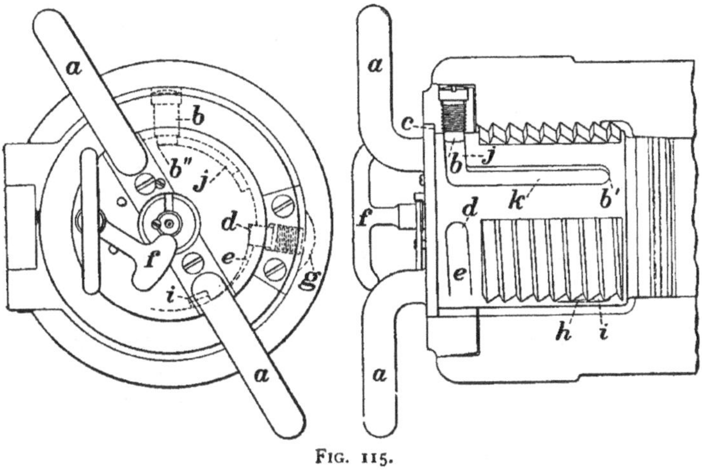

FIG. 115.

the carrier-ring latch is in the extremity *d* of the transverse groove *e*, Fig. 115, the outer end has left its recess, *g*, in the gun, and the carrier-ring is unlocked from the jacket.

2. After firing, the elasticity of the spiral spring and of the gas-check ring acts to move the ring from its seat in the gun, and restore it to its former position before firing.

The vent-closer *f* is now turned upward, closing the vent, and at the same time unlocking the block from the carrier-ring, so that it may be turned to the left by the handles *a a*. The block is then turned to the left 60°. By this operation, the threaded sectors on the block come into the slotted sectors in the breech, so that the block can be pulled to the rear through the carrier-ring. While the rotation of the block is taking place, the inner end of the carrier-ring latch is moving up the inclined groove *e*, and the outer end of the latch has been pushed into the recess *g* in the jacket, thus locking the carrier-ring to the jacket. The block is now pulled to the rear through the carrier-ring till its motion is stopped by the stop *b* in the carrier-ring striking against the front shoulder, *b'*, of the longitudinal groove *k*. During this motion of the block the inner end of the carrier-ring latch is bearing on the surface of one of the slotted sectors, and the carrier-ring remains locked to the gun. Near the end of the motion of the block, however, the inner end of the latch begins to descend along an inclined groove, *h*, in the surface of the block, and the outer end, due to the action of the flat spiral spring before described, is withdrawn from its recess, *g*, in the gun. At the end of the travel of the block, this withdrawal of the latch is complete, and the inner end of the latch drops into a cylindrical hole, *i*, at the end of the inclined groove, thus locking the block to the carrier-ring. The block and carrier-ring are now swung round by hand out of the way for loading.

3. To close the breech the block and carrier-ring are swung round into place. As the carrier-ring closes against the breech, a conical stud, *s*, Fig. 109, on the rear face of the latter enters a recess in the front of the latch, and lifts the inner end of the latter out of the cylindrical hole *i* in the block. The block is now pushed forward by hand, sliding through the carrier-ring, the latch is pushed up into its recess in the jacket by the action of the inclined surface *h* on the block, and the forward motion of the block is continued till the rear end of the groove *k* strikes against the stop *b*. The block is then rotated to the right 60°, engaging its threads in those of the breech. At the same time the inner end of the latch moves down the inclined transverse groove *e*, and the upper end of the latch is withdrawn from its recess *g* in the gun, thus unlocking the carrier-ring from the gun. The rotation of the block to the right is limited by the stop *b* striking against the shoulder *b''* at the end of the transverse groove *j*. The vent-closer is then turned down by hand,

thus opening the vent and locking the block to the carrier-ring, and the mechanism is in its firing position.

The action of the mechanism of the 3.2 and 3.6 field-guns is exactly similar, except that the vent-closer is automatic.

SIEGE-GUNS.

141. 5-inch Gun—7-inch Howitzer—7-inch Mortar.

Siege-guns are intended for attacking and defending permanent inland works, and the land fronts of sea-coast fortifications.

In the U. S. service the pieces are:

The 5-inch siege-gun;
The 7-inch howitzer;
The 7-inch mortar.

Common Features.—These guns, like the field-guns, are built of gun-steel, and are breech-loading with rifled bores. They have conical gas-check seats, and cylindrical powder-chambers of larger diameter than the bore, with which they are connected by a conical slope for centering the projectile. They have also the conical rifling slope or forcing-cone, formed as explained in the field-guns, and for the same purpose.

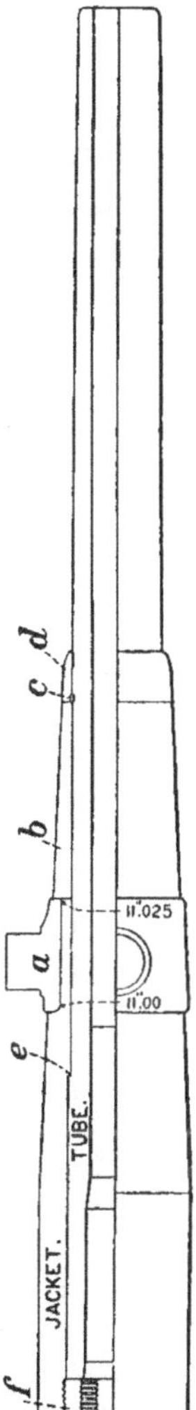

FIG. 116.

5-INCH SIEGE-GUN.—This gun is intended for direct fire in siege operations. It is built up (Fig. 116) of a tube, jacket, trunnion-hoop, *a*, sleeve, *b*, locking-ring, *c*, key-ring, *d*, and base-ring, *f*. The tube is inserted into the jacket from the rear. The peculiarity of this gun is the manner of assembling the trunnion-hoop. It would be preferable to have the jacket and trunnion-hoop in one piece, as in the field-guns, but difficulties in making such a forging of the required physical qualities prevent this, and hence the jacket is extended under the trunnion-hoop to give better support to the tube, and the trunnion-hoop assembled over the front of jacket as shown.

The surface of contact of jacket and trunnion-hoop is a cone, with the larger base to the front,and this tends to lock the trunnion-hoop in place, and prevent any forward motion.

Relative motion of tube and jacket is prevented by the shoulder *e* and base-ring *f*. The locking-ring *c* prevents forward movement of the sleeve *b*, which is important, as the trunnion-hoop abuts against *b*, and hence brings a thrust upon it when the piece is fired.

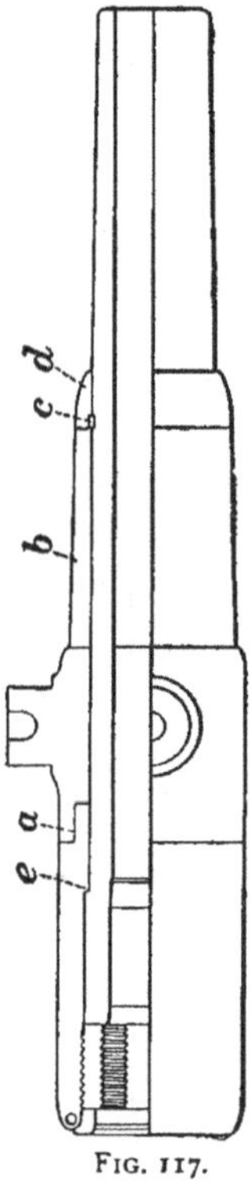

FIG. 117.

7-INCH HOWITZER.—This is a comparatively short, light piece, of large calibre, intended to carry a shell, with a large bursting charge, and to give a high-angle or curved fire, and reach troops sheltered by a parapet, and also to breach masonry protected by an earthen cover, to destroy earthworks, etc.

It is built up of a tube, a jacket, a trunnion-hoop, a sleeve, a locking-ring, a key-ring, and a base-ring, assembled by shrinkage (Fig. 117). The construction is shown in the figure. The tube is inserted into the jacket from the rear, and has a shoulder at *e* which prevents forward motion. The longitudinal stress is transmitted from the trunnions to the jacket through the locking-lip *a*, and forward motion of the sleeve *b* is prevented by the locking-ring *c*. The key-ring *d* is shrunk over the locking-ring.

7-INCH MORTAR.—This is a short rifled piece intended to carry the same shell as the 7-inch howitzer, and give a vertical fire. It is built of a single piece of forged gun-steel (Fig. 118), and resembles the 3.6-inch field-mortar.

Breech Mechanism.—The breech mechanism of the 5-inch siege-gun and 7-inch howitzer, are similar to that of the 3·2-inch gun already described. The breech mechanism of the 7-inch mortar differs from that of the 3.6 mortar in the following particulars:

PRINCIPAL DIMENSIONS, WEIGHTS, ETC., OF FIELD AND SIEGE ARTILLERY.

TABLE I.—BREECH-LOADING ORDNANCE, U. S. LAND SERVICE

[Guns, howitzers, and mortars of 3 inches calibre and upwards.]

	Hotchkiss Mountain-gun, Steel.	Field-artillery, Steel.			Siege-artillery, Steel.		
		Light Gun, model 1890.	Heavy Gun, model 1891.	Mortar, model 1890.	Gun, model 1800.	Howitzer, model 1890.	Mortar, model 1892.
Calibre, inches	3	3.2	3.6	3.6	5	7	7
Weight: Pounds	218	805	1,181	244	3,660	3,710	1,715
Tons	......	......	......	......	......	......	......
Total length, feet	3.76	7.31	7.79	2.05	12.15	8.475	4.86
Length of bore, calibres	13.48	25.20	23.50	5.19	27	12.67	7
Diameter over powder chamber, inches	6.7	9.00	9.80	7.80	15.0	16.70	13.80
Diameter of powder-chamber, inches	3.15 to 3.21	3.60	3.90	3.80	5.7	7.20	7.25
Thickness over powder-chamber, calibres	0.55	0.75	0.76	0.48	0.82	0.66	0.45
Number of cylinders comprised in the thickness	1	2	2	1	2	2	1
Maximum tangential resistance, pounds per sq. in.	23,850	38,080	39,160	22,050	37,520	35,120	22,000
Rifling: Number of grooves	24	24	26	20	30	42	28
Width of grooves, inches	0.276	0.3	0.3162	0.4454	0.3736	0.3736	0.6354
Depth of grooves, inches	0.026	0.04	0.04	0.045	0.05	0.05	0.055
Width of lands, inches	0.117	0.1188	0.1188	0.12	0.15	0.15	0.15
Twist, calibres	1 in 25.59 uniform	1 in 50 to 1 in 25	1 in 50 to 1 in 25	1 in 40 to 1 in 25	1 in 50 to 1 in 25	1 in 50 to 1 in 25	1 in 40 to 1 in 15
Total capacity of bore, cubic inches	292	689.3	894.5	200.2	2,824.5	3,520.1	1,950.9
Capacity of powder-chamber, cubic inches	27.12	110.3	149.2	33.2	395.6	315.7	182.8
Length of powder-chamber, inches	3.72	10.7	12.275	2.835	15.15	7.465	4.28
Travel of projectile in bore, calibres	12.243	21.81	20.08	4.47	23.96	11.61	6.40
Powder-charge: Kind	(a)	(b)	(b)	(b)	(b)	(a)	(c)
Weight, pounds	0.875	3.5	4.1875	15 ozs.	12.5	10	5.5
Density of loading	0 9	0.8780	0.777	0.833	0.875	0.877	0.833
Projectile: Weight, pounds	12	13.5	20	20	45	105	125
Ratio to weight of piece	1 to 19	1 to 60	1 to 59	1 to 12	1 to 81	1 to 35	1 to 13
Pressure in powder-chamber, pounds per square inch	14,560	35,000	35,000	16,000	35,840	28,224	18,000
Muzzle velocity, feet-seconds	870	1,685	1,550	650	1,830	1,085	690
Muzzle energy, foot-tons	63	266	333	59	1,045	857	412
Penetration of steel: Muzzle, inches	1.4	3.8	3.9	1.1	6.2	3.8	2.2
3500 yards, inches	......	......	......	......	2.5	2.4	......

a I. K. granular. b U. F. sphero-hexagonal. c I. B. sphero-hexagonal.

1. It has the De Bange gas-check.

2. The vent-closer is a sliding piece which is moved over the vent by turning a handle similar to that in the 3.60 mortar. This turning of the handle to open the vent locks the breech-block to the carrier-ring.

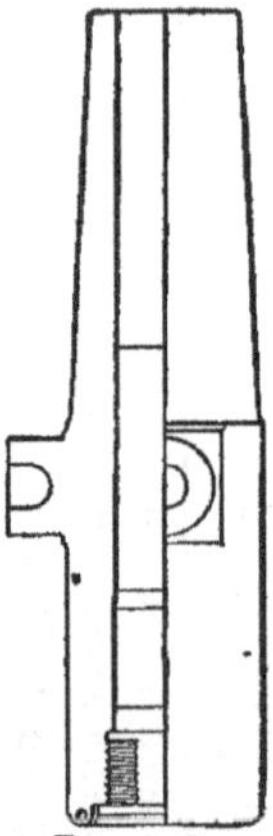
FIG. 118.

The principal data relating to the field and siege artillery are given in the table on page 253.

SEACOAST GUNS.

142. Calibres—Common Features—8-inch Rifle.

CALIBRES.—The guns at present adopted for the U. S. seacoast service are:

8-inch, 10-inch, 12-inch } steel B. L. rifles;
12-inch steel B. L. mortar;
12-inch rifled mortar, with cast-iron body and steel hoops.

COMMON FEATURES.—The guns are intended for direct fire against armored ships; the mortars, for vertical fire against the decks of war-ships. The seacoast guns, with the exception of the 12-inch mortar with cast-iron body, are built up of gun-steel, and are breech-loading with rifled bores. They have conical gas-check seats, cylindrical powder-chambers, which are connected with the bore proper by a conical slope, and they have also the rifling slope, called the forcing-cone, already described in the field and siege services.

THE 8-INCH GUN, Fig. 119, is composed of a tube, *T*, inserted into the jacket from the rear, a jacket, *J*, two *C* or chase hoops, one *D* hoop, four reinforce or *A* hoops (A_2 being the trunnion-hoop), and a base-ring, *R*.

Relative motion of tube and jacket is prevented by the shoulder *a* and the base-ring. The other shoulders on the tube reduce its thickness by successive steps from rear to muzzle. C_1 and C_2 hoops are locked together by a locking-lip, *g*, as shown, Fig. 120, the smaller diameter of the lip, C_2 being expanded sufficiently by heat to pass over the larger diameter of C_1. This prevents relative motion of C_1 and C_2 hoops. The C hoops in all guns have a tendency to move forward, probably due to the vibration of the chase and other causes, and to prevent this, four pins, f, Fig. 121, pass through the C_2 hoop radially into the tube. The D hoop overlaps the joint between jacket and C_1 and by means of the shoulders at c and c', locks the C_1 hoop to the jacket, and hence the jacket and C_1 are not locked together by a lip. The small ring e is called a filling-ring. It is necessary, because in assembling the D hoop it is desirable to make a tight contact at the shoulders c and c'. The rear end of D is therefore made of such length as when hot to fill the space from c' to jacket-shoulder. Hence when cold there will be an opening at b, which is filled

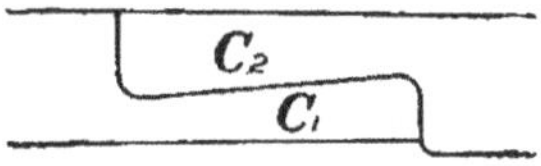

FIG. 120.

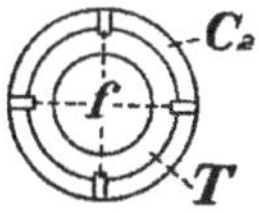

FIG. 121.

by turning out a groove, and driving into it a split ring of metal, e. This gives stiffness to the chase. The trunnion or A_2 hoop abuts against a shoulder on the jacket at d. The longitudinal strain due to firing is then distributed along the jacket from the base-ring to the shoulder d.

The reinforce or A hoops are not locked, because there is no tendency to slide in these hoops, and the reinforce does not require stiffening.

As a rule, it may be observed that the locking of hoops together is for two purposes:

1. To obtain longitudinal stiffness.
2. To prevent sliding.

Hence reinforce-hoops need no locking, and chase-hoops are not locked when the joint between them is overlapped by an exterior hoop.

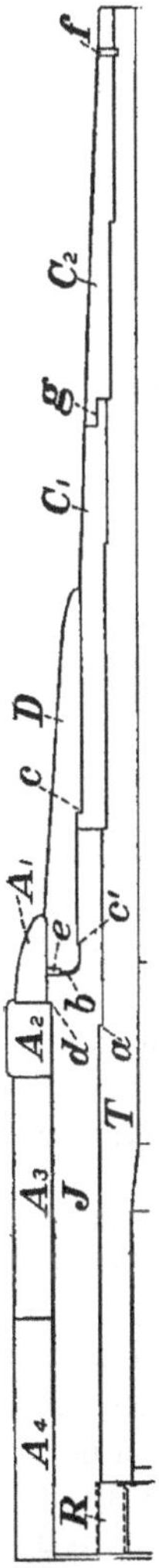

FIG. 119.

143. The 10 and 12-inch Rifles.

In these guns (Fig. 122) the tube is inserted into the jacket from the front, as in the field-guns, and the breech-screw threads are cut in the jacket.

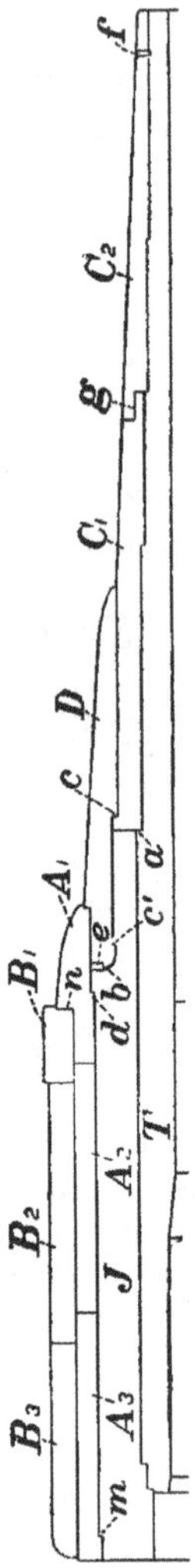

FIG. 122.

The parts of the 10 and 12-inch guns are:

One tube;
One jacket;
Two *C* hoops;
One *D* hoop;
Three *A* hoops;
Three *B* hoops;
Four securing-pins, *f*;
One filling-ring, *e*.

The following features of construction may be noted: There are relatively few pieces, and consequently the hoops are very long. This gives great longitudinal stiffness. The chase-hoops are locked together by a locking-lip, *g*, and the sliding of these hoops prevented by the four pins *f*.

In the 10 and 12-inch guns there is no shoulder in the jacket to prevent forward motion of the tube. This motion is therefore prevented by the bearing of the C_1 hoop against the shoulder *a* on the tube, and the C_1 hoop is held in place by the *D* hoop locking over two shoulders, one on the jacket at *c'*, and the other on the C_1 hoop at *c*.

The *D* hoop is shrunk on over these shoulders as shown in the figure, locking the jacket and C_1 hoop together; and as tight joints must be made at *c* and *c'*, the length of the *D* hoop must be such as to exactly fill the space between *b* and the shoulder on the jacket at *c'* when hot. Hence when cold it will leave an open joint at *b*, and this is filled by turning out a groove and putting in the filling-ring *e*. It will be observed that this same construction is used in the 8-inch

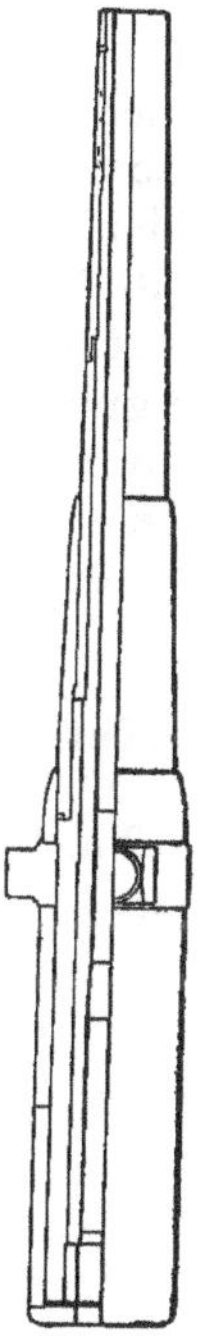
FIG. 123.

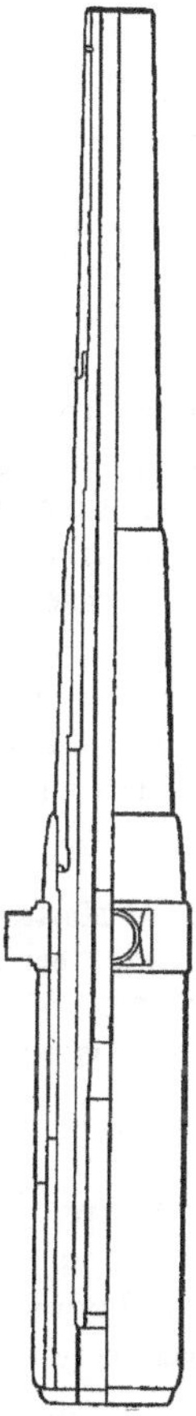
FIG. 124.

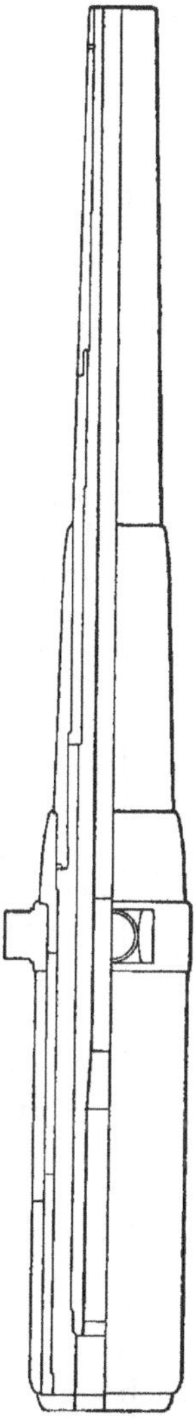
FIG. 125.

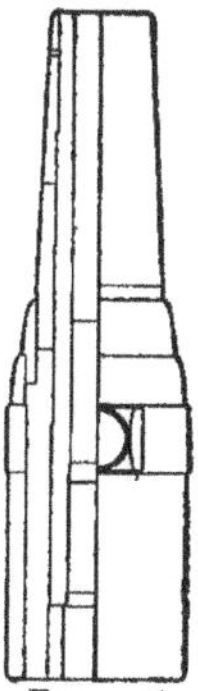
FIG. 126.

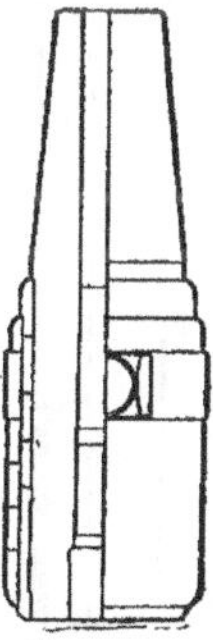
FIG. 127.

gun; but it is of greater importance here, as the C_1 hoop is depended on to hold the tube in place, while in the 8-inch gun a shoulder on the jacket does this. The A_2 hoop has a shoulder, *m*, near its rear end which fits over a corresponding shoulder on the jacket. By the shrinkage of the A_2 hoop on the jacket it has a firm hold on the latter, and hence this shoulder on the jacket, bearing against the shoulder on A_2, strengthens the jacket longitudinally, since it distributes a portion of the pull of the breech-block to the A_2 hoop. The forward thrust of the trunnion-hoop B_1 is transmitted to the shoulder *n* on the A_1 hoop, and from the A_1 hoop to the jacket by the shoulder *d*. Thus the jacket takes the longitudinal strain in all cases. Figs. 123 to 127 show the 8, 10, and 12-inch guns drawn to the same scale and giving their relative sizes, and also the 12-inch mortars, which are to be described.

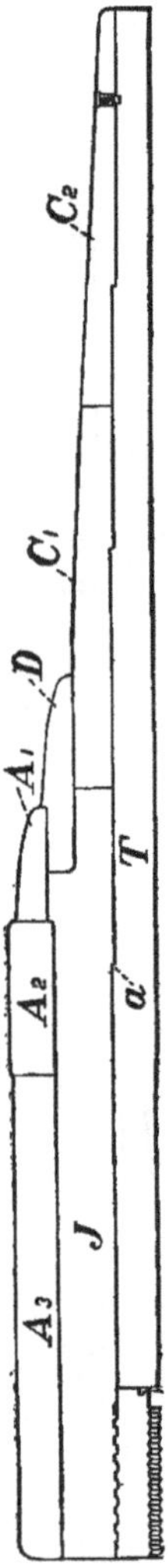

FIG. 128.

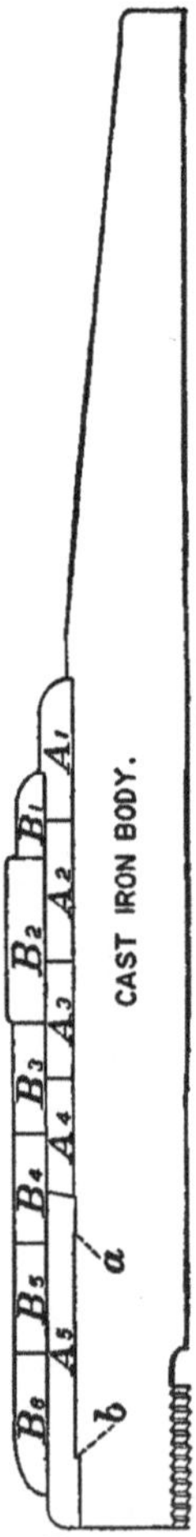

FIG. 129.

144. 12-inch Steel Mortar—12-inch Cast-iron Mortar, Steel-hooped.

12-INCH STEEL MORTAR.—This mortar, Fig. 128, is composed of—

One tube;
One jacket;
Two *C* hoops;
One *D* hoop;
Three *A* hoops;
One base-ring.

The tube is inserted into the jacket from the rear, as in the 8-inch gun, and the shoulder *a* prevents motion of tube. As the piece is short, and therefore stiff, the *C* hoops are not locked, but four radial securing-pins are inserted in the muzzle-hoop to prevent sliding.

12-INCH MORTAR WITH CAST-IRON BODY, STEEL-HOOPED.—This mortar, Fig. 129, was designed before the 12-inch steel mortar, with the object of procuring a high-power B. L. mortar which would be cheap and could be made in large quantities. The value of mortar-fire depends on grouping a large number of mortars in one place and under the control of one person, who can thus drop a number of projectiles in a given area, and compensate by the number of shots for the lack of accuracy. Hence the necessity for cheapness. It is found, however, that the steel mortar has greater power and endurance, and it is possible that the manufacture of cast-iron mortars will be abandoned.

The mortar consists of—

The cast-iron body;
Five *A* hoops;
Six *B* hoops.

The only point in the construction that requires notice is that the A_6 hoop is shrunk on the cast-iron body over two shoulders, *ab*. This is for the purpose of strengthening the cast iron longitudinally against the pull of the breech-block, and is similar to the method adopted in the 10 and 12-inch guns.

145. Breech Mechanism—Block—Obturator—Anti-friction Washers and Spring.

The breech mechanisms of the 8, 10, and 12-inch guns are essentially the same. That for the mortar differs in some respects from the guns.

The essential parts of the mechanism are:

The breech-block;
The obturator;
The console or tray;
The device for rotating and withdrawing the breech-block;
The vent and vent-closer.

BREECH-BLOCK.—This resembles the blocks already described for the field and siege artillery. In the 8-inch gun there are three threaded and three slotted sectors; in the 10 and 12-inch, four.

The rear end of the block is left unthreaded for some

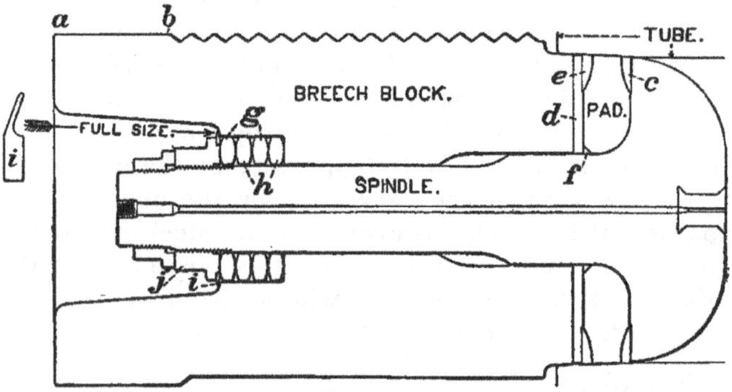

FIG. 130.

distance, *ab*, Fig. 130. The cylinder of metal thus formed, fits accurately into the breech-recess, when the block is home, and prevents the entrance of sand or dust, which might cause jamming of the threads.

OBTURATOR.—This is the De Bange system modified, and differs from that used in the field and siege services as follows (Fig. 130):

The front cup is replaced by a split steel ring, *c*, shown in detail in Fig. 131, fitted against the outer portion of the mushroom-head. The rear gas-check cup is replaced by a flat disk of steel, *d*, fitting tightly against the front of the block. A split steel ring, *e*, similar to *c*, takes the place of part of the outer edge of the gas-check cup formerly used, and another split ring, *f*, fits against the spindle.

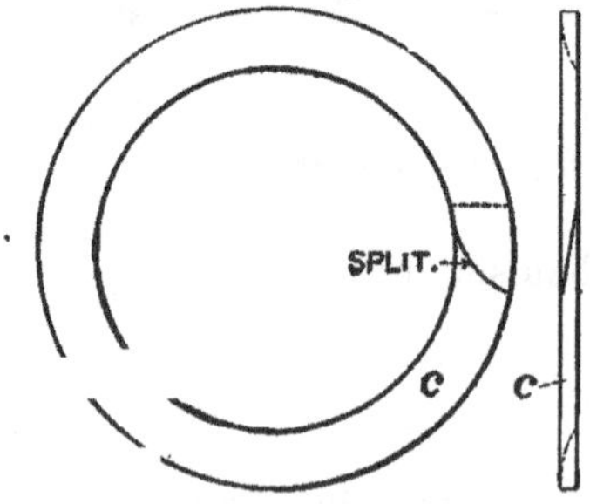

FIG. 131.

When the gas-pressure acts, these rings expand, and prevent the flame and gas from reaching the covering of the gas-check pad, and also from penetrating in the direction of the spindle. When the pressure is relieved the rings resume their normal size, and tend to cause the pad to leave its seat in the gun, and thus prevent sticking.

ANTI-FRICTION WASHERS AND SPRING.—In all guns using the De Bange fermeture, the pad is liable to stick in its seat after firing, and render the breech difficult to open. This is provided for in the field and siege guns by the cam action of the head of the lever-handle, as has been explained.

In the sea-coast service, the lever-handle is not used for rotating the block, as it is not sufficiently powerful, and the following arrangement is adopted to overcome the sticking of the pad:

By referring to the field-gun mechanism, it will be seen that there is a spring, *j*, between the obturator-nut *h*, Fig. 106, and the shoulder in the block. Hence when the block is rotated, it moves back $\frac{1}{6}$ of the pitch of its thread, compressing this spring and allowing the pad to remain fast in its seat. At the end of the rotation the cam action of the lever-handle draws out the pad from its seat.

In the large guns, the spring is replaced by two anti-friction washers of steel, *g*, and two of brass, *h*, of the shape shown in Fig. 130. Each alternate washer is of brass, so that two metals of the same kind shall not rub against each other. A cup-shaped spring *i* rests against a shoulder on the front end of the obturator-nut *j*, and bears on the shoulder of the block. This spring acts as a cover to keep out dust. The action is as follows:

When the block is rotated, it moves back $\frac{1}{6}$ or $\frac{1}{8}$ of the pitch of its thread. This brings a pressure upon the washers, which is thence transmitted to the obturator-nut, and by this pressure the pad is loosened in its seat. The object of the anti-friction washers is to allow this rotation of the block to occur independently of the spindle, and this is done by diminishing the lever-arm of the friction.

The value of this arm is, from mechanics, $\frac{r^3 - r'^3}{r^2 - r'^2}$, in which r and r' are the exterior and interior radii of a ring. By giving a double-convex section to the washers, r is decreased and r' increased; and hence the moment of the friction with reference to the axis of the spindle is sufficiently decreased to allow the spindle to stand fast while the block rotates.

146. Apparatus for Rotating and Withdrawing the Breech-block —The Tray—The Translating Screw.

ROTATING DEVICE.—The lever-handle cannot be used for rotating the breech-block in sea-coast guns owing to its

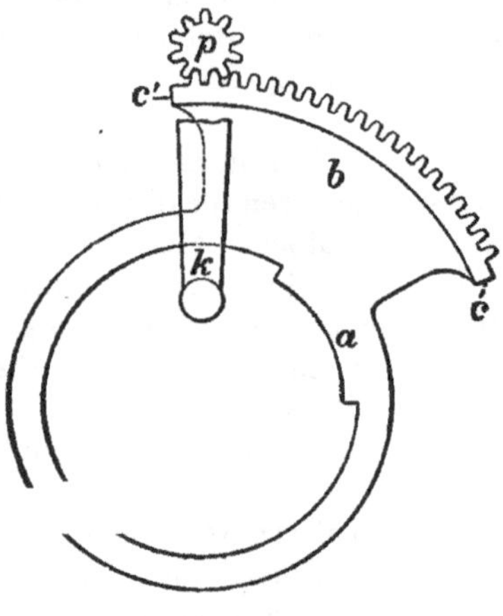

FIG. 132.

lack of power. The device adopted is called the "rotating-ring," and is shown in Fig. 132.

It is a ring of steel encircling the breech-block, and having a lug, *a*, the exact width of one of the slotted sectors, projecting on the interior. This lug enters one of the slotted sectors of the block, and the remainder of the interior circle is of such diameter that the breech block will slide through it. On the exterior there is a projecting toothed sector, *b*, which gears into a pinion, *p*. When the pinion is rotated by a crank, *k*, motion is communicated to the rotating-ring, and through the lug *a* to the block. The rotating-ring is held in place against the rear face of the breech, by a steel plate, called the "breech-plate," which allows rotation, but prevents any other motion. The rotation of the ring and block is limited to an angle of 60° or 45°, according to the gun, by the surfaces *c c'* striking against corresponding surfaces in the breech-plate. The rotation of the block having been completed, it can be

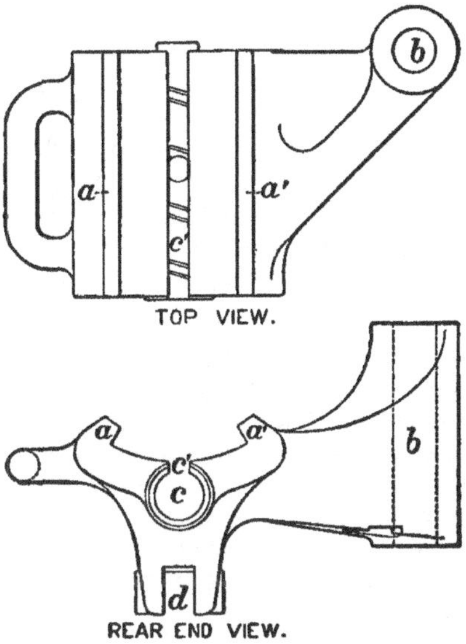

FIG. 133.

withdrawn from the breech, sliding through the rotating-ring.

THE TRAY OR CONSOLE.—In the field and siege guns, the block, when withdrawn, is supported by a carrier-ring. In the sea-coast guns, this method will not answer, as the carrier-ring does not furnish sufficient bearing-surface to support the block. A tray is therefore used for this purpose, and is shown in Fig. 133.

It is made of brass, and is hinged to the rear face of the breech, by a steel pin passing through the hub *b*.

The tray swings around this pin, and with the block may be rotated to the right, out of the way for loading.

Fig. 134.

TRANSLATING-SCREW.—Near the middle of the tray is a hole, *c*, which is threaded, and a slot *c'* is cut in its top, parallel to the axis of the hole. In this hole *c*, works a double-threaded screw called the translating-screw, Fig. 134, the threads being right and left handed, and one of them narrower and more shallow than the other. The shallow thread engages in the corresponding thread in the hole *c*, Fig. 133, so that the screw when turned, will move in or out of the tray. On the rear end of the breech - block there is a projecting stud, *b*, Fig. 135, called the translating-stud. When the breech - block has rotated one sixth of a turn to the left, this stud moves down, and engages in the larger thread of the translating-screw. Hence when this screw is rotated, it withdraws the block from its recess with a motion equal to the sum of the pitches of the two threads, for each revolution of the screw.

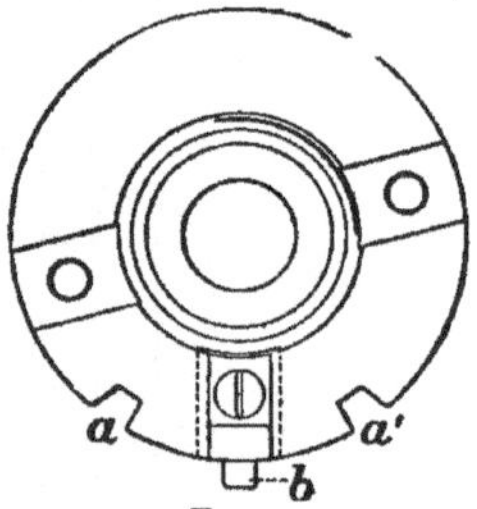

Fig. 135.

147. Remaining Parts of Breech Mechanism—Guide-rails—Guide-grooves—Side-latch—Tray Latch.

On each side of the tray (Fig. 133) are two projections, *aa'*, equidistant from the screw. They are called the "guide-rails."

On the under side of the breech-block, at equal distances from the translating-stud, are two grooves, *aa'*, Fig. 135, called "guide-grooves." These grooves do not extend the whole length of the block, but end abruptly at shoulders. When the block is withdrawn by the translating-screw, the guide-grooves slide on the guide-rails, which thus furnish the bearing for the block, and the block continues its movement to the rear till it is suddenly stopped by the striking of the shoulders of the grooves *aa'* on the front ends of the guide-rails of the tray.

During this motion of the block the translating-stud *b*, Fig. 135, travels in the slot *c'*, Fig. 133, being always engaged in the larger thread of the translating-screw.

SIDE LATCH.—When the block has reached the end of its travel, the tray and block are swung by hand to the right for loading.

To hold the tray and block in this position, and prevent the accidental closing of the breech, by the swinging in of the tray, a catch is provided on the under side of the tray, and a side latch on the breech of the gun. This latch catches the tray as it swings around, and retains tray and block till the latch is lifted by hand.

TRAY-LATCH.—When the tray and block are swung around after loading, into the position for the insertion of the block in the breech, the block is moved forward along the tray by the translating-screw.

But in order that the block may enter its recess in the breech, the tray must fit accurately against the rear face of the breech, so that the guide-rails shall be parallel to the axis of the bore.

Again, as soon as the block enters its recess and begins to bear on it, the thrust of the translating-screw will tend to move the tray back from the breech. There must be some arrangement, therefore, to latch the tray against the breech, and hold it in that position till the block is home, and this is the object of the tray-latch. This latch fits into a recess, *d*, Fig. 133, in the lower part of the tray, and engages in a corresponding recess in the breech of the gun. It is shown in Fig. 136. It is constantly acted on by the spring-lock *c*, which keeps it engaged in the recess in the breech of the gun. The upper end of this lock bears against the translating-screw in the tray, and hence the lock can rise and the tray be unlatched only when the end of the translating-screw is beyond the lock *c*. This happens when the block is withdrawn. The sudden shock of the block striking against the guide-rails in its outward motion, is

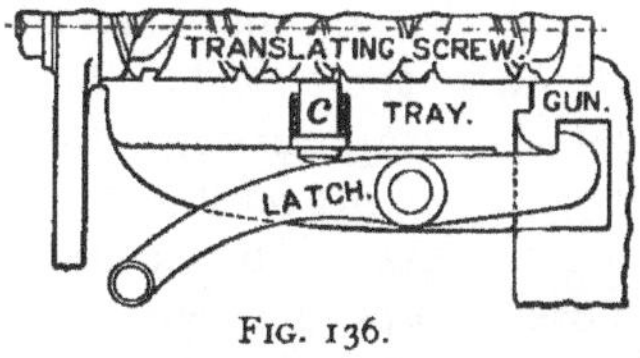

FIG. 136.

communicated to the latch, and acting obliquely, a force perpendicular to the axis of the latch is developed, of sufficient intensity to disengage it automatically from the breech.

148. Vent-cover—Action of Breech Mechanism.

VENT-COVER.—This consists of a flat piece of steel, *a*, Fig. 137, pivoted loosely in the breech-block.

The head of this piece bears against the inner surface of the breech-recess as the block rotates, and hence it cannot move around the pivot, but remains in the same radial position covering the vent. At the end of the rotation of the block, when the threads are engaged in those of the breech-recess, the head drops into a groove cut for it in the breech recess, and the vent-closer assumes the position *a*′, uncovering the vent. When the block is rotated to the left for unlocking, the first motion brings the head to its bearing against the breech-recess, and thus closes the vent.

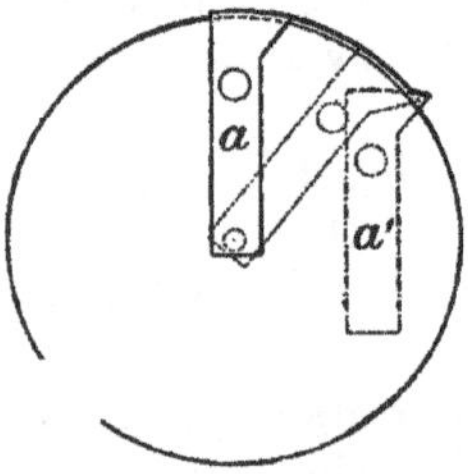

FIG. 137.

ACTION OF BREECH MECHANISM—*Breech Closed.*—In this position of the block its threads are engaged in those of the breech-recess, the gas-check is in its seat in the tube, the vent-cover has moved to the right, uncovering the vent, the tray is latched to the breech by the tray-latch, and the translating-screw is in its recess as far as it will enter.

To Open the Breech.—Turn the crank attached to the pinion of the rotating-ring, in the direction indicated by the arrow on the breech. This motion is communicated to the rotating-ring through the toothed sector, and from this ring to the breech-block, by the lug which enters its slotted sector. As the block begins to rotate to the left, the vent-closer closes the vent as explained. When the block has rotated one sixth of a turn, the translating-stud enters the thread of the translating-screw in the tray. This screw is then rotated by its crank, and the block withdrawn from its recess in the breech. At the end of its travel the shoulders on the block strike against the ends of the guide-rails on the tray, and the shock disengages the tray-latch from the

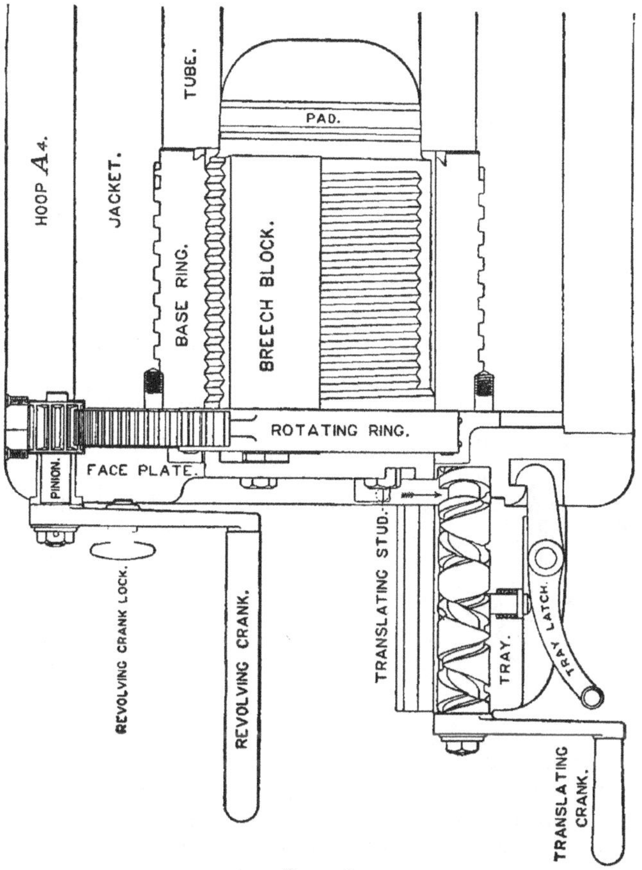

FIG. 138.

breech. The tray and block can now be swung around for loading.

To Close the Breech.—Lift the side latch; swing the tray and block around to the left till the tray-latch is engaged in its recess in the breech. The block is now driven home by the translating-screw.

Rotate the block to the right by the rotating-crank, and when the rotation is finished, the block will be in the position first described and the gun ready for firing.

The mechanism for the 10 and 12 inch guns is so similar to this that no special description is necessary.

149. Breech Mechanism of 12-inch Mortar.

In this piece the mechanism differs from that of the guns as follows:

Rotating Device.—On the rear face of the breech-block is fixed a steel plate, *k*, Fig. 139, called a face-plate. The upper

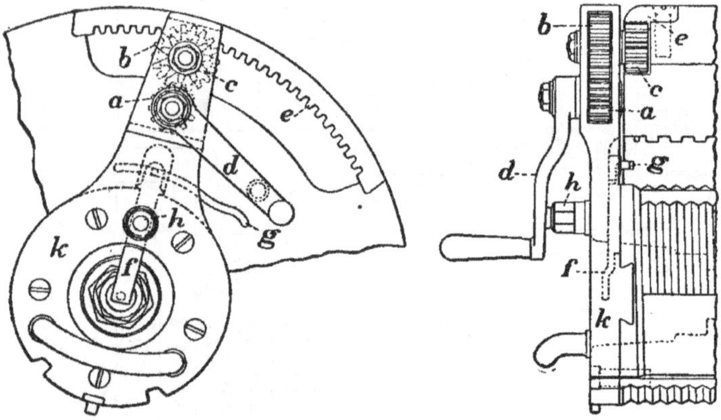

FIG. 139.

end or stem of this face-plate is cut out, and carries two gears, *a*, *b*, and on the exterior a third gear, *c*, on the same shaft with *b*. Motion is communicated to these gears by the crank *d*. On the rear face of the breech is a circular rack *e*, with which the gear *c* engages, when the block is pushed home. It is evident that a rotation of the crank *d* will cause the face-plate and block to rotate to the right, or a reverse motion of the crank, when the breech is closed, will cause a rotation of the block to the left. The block is withdrawn by a translating-screw as before. The tray-latch is the same in principle as in the guns, the only change in construction being that the spring-lock acts in front of the pivot, and the latch rises when it is disengaged.

Vent-closer.—This resembles that used in the 3.2-inch gun, and consists of a piece of metal, *f*, sliding in a slot in the face-plate. A pin projecting from the front of this sliding-piece, bears in a groove *g* in the rear face of the breech, which is concentric for some distance with the axis of the breech-block, and at its lower extremity becomes eccentric. Its action in uncovering the vent is the same as in the case of the 3.2-inch guns.

Action of Mechanism.—When the block is closed, the head of the face-plate is at the right-hand end of the rack *e*, the crank *d* is parallel to the axis of the face-plate, and is held in place by a spring-lock *h*. After firing, the crank *d* is turned, and the gear *c* engaging in the rack *e*, rotates the block to the left.

The rotation of the block is limited, by the striking of the sides of the face-plate against the ends of the circular recess in which the rack is placed. The block is now withdrawn by the translating-screw, and block and tray swung round for loading.

To close the breech the operations are reversed.

150. Improved Mechanism—Continuous Rotation.

Objections to Ordinary Mechanism. — In the mechanism already described, one crank is necessary to rotate the block, and at the end of this movement, the power must be transferred to another crank for withdrawing the block. When the block is withdrawn, the power must be applied to the handle of the tray to swing the block and tray around. We have thus three separate and distinct motions, involving loss of time, and complication of mechanism.

IMPROVEMENTS.—In the latest improved mechanism, the object is to effect by the application of power to one crank, and by its continuous movement, the rotation, translation, and swing, of the block and tray.

This mechanism, as applied in our service, is called the Farcot, from its inventor, and consists (Fig. 140) of the following parts:

On the right side of the rear end of the breech-block is a circular-toothed sector *a*. A cut is made in one of the threaded sectors of the block parallel to its axis, and this circular rack is extended along the block as shown in the figure (plan), the width of this longitudinal rack *b* being that of the thickness of the wheel *c*, while the width of

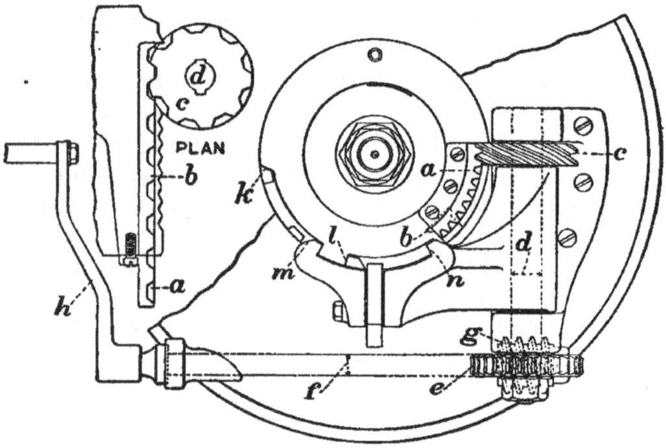

FIG. 140.

the circular sector *a* is one sixth of the circumference of the block. On the top of the hinge-pin *d*, is mounted a worm-gear *c*, whose teeth fit the corresponding teeth of the sector *a*.

At the bottom of the hinge-pin *d* is a second worm-gear, *e*. A horizontal crank-shaft, *f*, has at the right end a worm, *g*, gearing into the worm-gear *e*, and at the left end a crank, *h*.

Action of Mechanism. — When the crank *h* is turned, motion is communicated to the worm-gear *c* through *g* and *e*, and the action of *c* on the sector *a* rotates the block one sixth of a turn, till the shoulders *kl* on the block strike against the guide-rails *mn*.

The block then being no longer able to turn, the teeth of the wheel *c*, engaging in those of the rack *b*, along the block, will force the block to the rear out of the breech-recess. The cut in the threaded sector of the block, for the reception of the rack *b*, is made so deep that the worm-wheel *c* binds against the edges of this cut in travelling along the rack,

TABLE II.—BREECH-LOADING ORDNANCE, U. S. LAND SERVICE.

	Seacoast Guns, Steel.					Seacoast Mortars.	
	Model 1888, M.	Model 1888, M.	Model 1888, M.	Model 1892.	Proposed.	Cast iron Steel-hooped.	Steel.
Calibre, inches	8	10	12	12	16	12	12
Weight:							
Pounds	32,372 / 32,480	67,200	116,480	128,719	280,000	31,920	29,120
Tons	14.5	30	52	57.5	125	14.25	13
Total length, feet	23.21	30.60	36.66	40.0	49.67	10.75	11.76
Length of bore, calibres	32	34	34	37.83	35	9	10
Diameter over powder-chamber, inches	30	38.5	46.2	46.4	62	41.75	38
Diameter of powder-chamber, inches	9.5	11.8	14.2	14.5	18.8	12.4	12.5
Thickness over powder-chamber, calibres	1.08	1.13	1,125	1.10	1.148	1.18	1.02
Number of cylinders comprised in the thickness	3	4	4	4	4	3	3
Maximum tangential resistance, pounds per sq. in.	51,980	53,090	53,000	52,640	53,000	29,490	50,200
Rifling:							
Number of grooves	48	60	72	72.0	96	68	72
Width of grooves, inches	0.3736	0.3736	0.3736	0.3736	0.3736	0.379	0.3736
Depth of grooves, inches	0.06	0.06	0.06	0.06	0.07	0.07	0.06
Width of lands, inches	0.15	0.15	0.15	0.15	0.15	0.175	0.15
Twist, calibres	1 in 50 to 1 in 25	1 in 50 to 1 in 25	1 in 50 to 1 in 25	1 in 50 to 1 in 25	1 in 50 to 1 in 25	1 in 40 to 1 in 25	1 in 40 to 1 in 20
Total capacity of bore, cubic inches	14,125	28,977	50,049	55,829	121,487	12,554	13,947
Capacity of powder-chamber, cubic inches	3,597	7,064	12,092	12,798	29,341	1,990	2,636
Length of powder-chamber, inches	50.75	65.09	77.33	78.58	106.06	15.75	20.8
Travel of projectile in bore, calibres	25.66	27.51	27.58	31.29	28.37	7.66	8.225
Powder charge:							
Kind	(*a*)	(*b*)	(*c*)	(*d*)	(*d*)	(*e*)	(*e*)
Weight, pounds	125	250	450	487	1,060	80	105
Density of loading	0.9619	0.9797	1.0285	1.0535	1	1.1128	1.1026
Projectile:							
Weight, pounds	300	575	1,000	1,000	2,370	800 / 1,000	800 / 1,000
Ratio to weight of piece	1 to 108	1 to 117	1 to 116	1 to 129	1 to 118	1 to 40	1 to 36
Pressure in powder-chamber, lbs. per square inch	37,000	37,000	37,000	38,000	37,000	27,500	30,000
Muzzle velocity, ft.-secs.	1,950	1,975	1,975	2,100	1,975	1,020	1,140
Muzzle energy, foot-tons	7,907	15,548	27,040	30,570	64,084	5,770	7,207
Penetration in steel:							
Muzzle, inches	16.0	20.4	24.9	27.1	33.8	8.2	9.7
3500 yards, inches	10.6	14.6	18.7	20.6	27.5	....	....

a U.R. brown prismatic. *b* W.H. brown prismatic. *c* V.P. brown prismatic. *d* brown prismatic; *e* V.M. brown prismatic.

and hence any tendency of the block to rotate is overcome. When the block reaches the end of its travel, it strikes against the ends of the guide-rails and releases the tray-latch from its recess in the breech by the shock, as before explained. As the block is not able to move further on the tray, but is free to swing with the tray around the pin *d*, the pressure of the teeth of *c* against those of the rack will cause this swinging to take place, thus opening the breech for loading.

A reversal of these motions closes the breech for firing.

The table on page 271 gives the details with reference to the seacoast guns and mortars in the U. S. service.

151. Old Guns in U. S. Service—3-inch Wrought-iron Rifle—4.5-inch Siege-gun—4.2-inch Parrott Siege-gun—8-inch Converted Rifle—15-inch Rodman Smooth-bore.

3-INCH WROUGHT-IRON RIFLE (Fig. 141).—This gun was used during the war of 1861–65, and is still found in service. It is made by wrapping boiler-iron around a wrought-iron mandrel, heating the resulting cylinder to a welding-heat, and passing it through the rolls. The gun is then bored, turned, and rifled. The object of the construction is to have the fibres of the wrought iron in the direction of the tangential

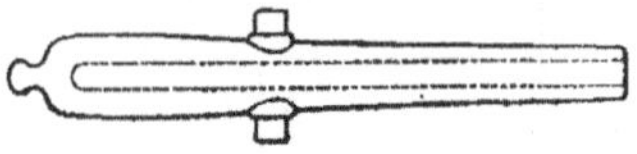

FIG. 141.

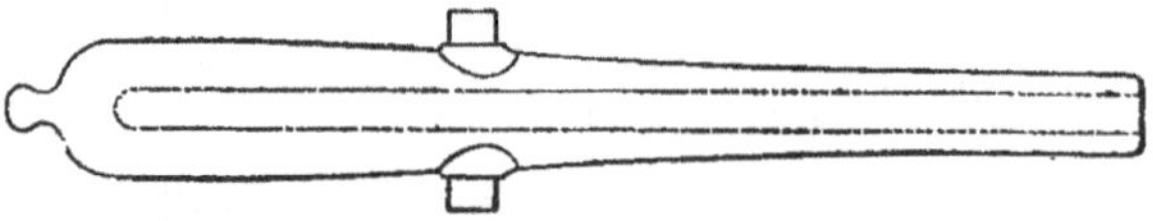

FIG. 142.

stress. The objection to the construction is the liability to false welds.

4.5-INCH SIEGE-GUN (Fig. 142).—This gun is made of cast iron, cast solid, and bored, turned, and rifled. It has given very good results, but is uncertain in strength, like all guns cast on this plan, and several accidents have occurred which have caused it to be abandoned.

4.2-INCH PARROTT SIEGE-GUN (Fig. 143).—This gun is also made of cast iron, but is reinforced at the breech by a

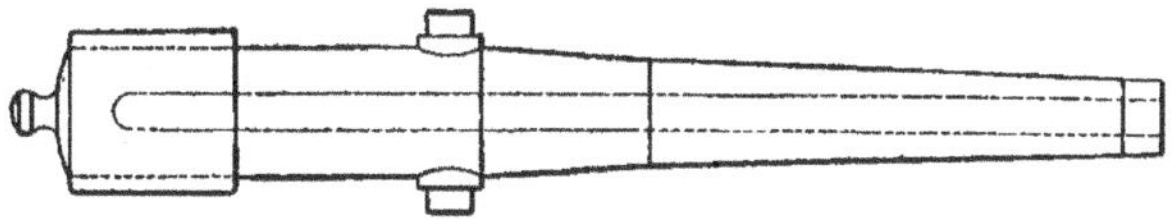

FIG. 143.

heavy jacket of wrought iron. This jacket was made by coiling a hot bar of wrought iron around a mandrel into a spiral, and welding the coils into a cylinder by blows from a hammer parallel to the axis of the cylinder. The cylinder was next bored, and then shrunk upon the exterior of the breech of the gun. At the time these guns were made, nothing was known about the theory of shrinkage as at present applied to guns, and hence the shrinkage was not properly regulated, and was very often a source of weakness, especially at the junction of the front end of the cylinder with the gun. In spite of this, however, these guns have proved very serviceable.

8-INCH CONVERTED RIFLE (Fig. 144).—These guns were made for the purpose of utilizing a large number of old 10-

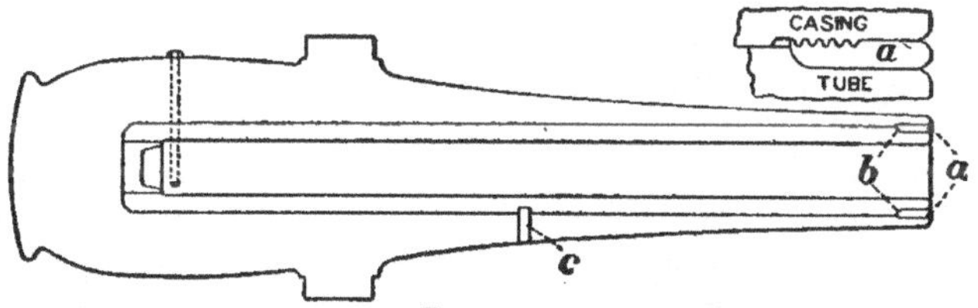

FIG. 144.

inch Rodman cast-iron guns which were on hand, the idea being to render them more accurate and powerful, by converting them into rifled guns. This was done by boring out the 10-inch gun to a larger diameter, and inserting a tube into this bore. This tube was held in place by a collar, *a*, screwed into the cast-iron body and resting against a shoulder, *b*, on the muzzle end of the tube. The tubes were made at first by coiling bars of wrought iron around a mandrel, and welding them by axial blows. This method was abandoned on account of the false welds in the tube, which sometimes cracked and separated from this cause. The tubes were finally made of steel, and numbers of these guns are still in service. The rotation of the tube, due to the action of the projectile, is prevented by a pin, *c*, which passes through the cast iron body and enters the tube.

15-INCH RODMAN SMOOTH-BORE.—This gun is still retained in service, and is intended to be used with large charges of mammoth powder, as a secondary gun, for comparatively short ranges, and against light armor. It is cast hollow on the Rodman plan; its projectile weighs 450 pounds, and the gun about 22 tons.

152. Foreign Guns—Krupp Mechanism—Locking-screw.

All heavy guns are built upon the same principles as those already explained, and hence a description of the guns of different countries is unnecessary. The only departure from the system above described, is in the case of the breech mechanism.

KRUPP MECHANISM.—While the French or interrupted-screw system has been adopted by most of the foreign nations, Germany, and some others, use the Krupp system.

It has stood the test of service and has been well and favorably known for many years, and hence will be described here.

The jacket, *a*, Fig. 145, extends to the rear of the tube, and carries the fermeture. A slot is cut transversely in the jacket just in rear of the tube. This slot, in front, is perpendicular to the axis of the bore, and is a plane surface, with corners rounded to avoid sharp angles. In rear, the surface of the slot is cylindrical, and the axis of the cylinder is inclined to that of the bore. Two guides, *b b'*, are parallel to the axis of the cylinder. In this slot slides a breech-block, *k*, whose shape corresponds to that of the slot. It has two recesses for the guides *b b'*, and in the upper face, a third recess, in which rests a long screw *c*, called the translating-screw. This screw is held in two collars in the breech-block, and works in a half-nut, *d*, on the gun. When the screw *c* is turned by a wrench, such as *e*, the block is drawn out of its recess or pushed home.

LOCKING-SCREW.—In order to obtain a rapid motion in opening and closing the breech, the screw *c* is cut with a quick pitch. Consequently, there is very little power to press the gas-check firmly home, or, in opening the breech, to overcome any sticking that may occur. It is also necessary to have some method of locking the breech-block to the jacket in firing, to prevent accident.

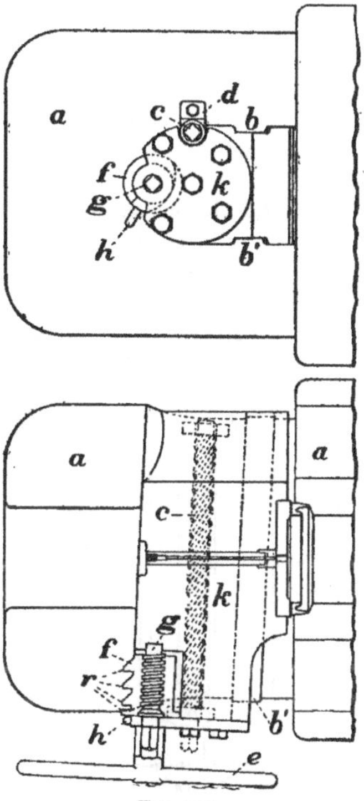

FIG. 145.

All these objects are accomplished by the locking mechanism. This consists of a nut, *f*, and a screw, *g*. The nut has a series of rings, *r*, formed on its exterior surface. The outer ring is complete, the others are partially cut away. When the nut is turned so that the cut-away portions of the rings are in rear, the surface of the nut coincides with that of the rear of the block. When turned 120°, the parts of the rings not cut away project beyond the block and enter corresponding cuts in the breech. The nut has a small amount of travel along its screw *g*.

Action.—The translating-screw *c* leaves the block not quite forced home. The nut *f* is at the bottom of its recess in the block, nearest the axis of the gun, and the cut rings of the nut are turned to the rear. The wrench *e* is now applied to the screw *g* and the screw turned. This will cause the nut *f* to move along the screw, it being unable to turn because of the cut parts of the rings bearing on the back of the transverse slot in the breech. As soon, however, as the rings come opposite the cuts in the breech, the nut will turn, its rings entering the corresponding cuts in the breech, and after turning 120°, the pin *h* bears against a shoulder on the block, and stops the rotation of the nut. As the screw still turns, the effect will be to cause the rings to bear against the cuts in the breech, and thus force the block home. At the same time the rings bearing in the cuts lock the block. A reversal of these operations opens the breech.

153. The Gas-check — General Features of the Mechanism—Advantages and Disadvantages.

GAS-CHECK.—With the Krupp mechanism, it is evident that neither the De Bange nor the Freyre gas-checks can be used, since both of them must be drawn back from their seats in the gun, being attached to the breech-block. The Krupp block slides across the breech, and hence it is necessary to use a gas-check which can be left in the gun. The Broadwell ring is used. It consists (Fig. 146) of two parts: the obturating-ring, *a*, and the obturator-plate, *b*. The exterior surface, *cc'*, of the ring is spherical, so that it can be readily seated in the gun, and returned to its place if it should become unseated. The surface *c'd* is plane, with a series of grooves to act as an air-packing, as before explained, and also to collect any dirt that may be on the surface of the obturator-plate. The obturator-plate *b* is of hardened steel, and is fitted into the face of the breech-block. The hollow *e* collects fouling, which, if the whole

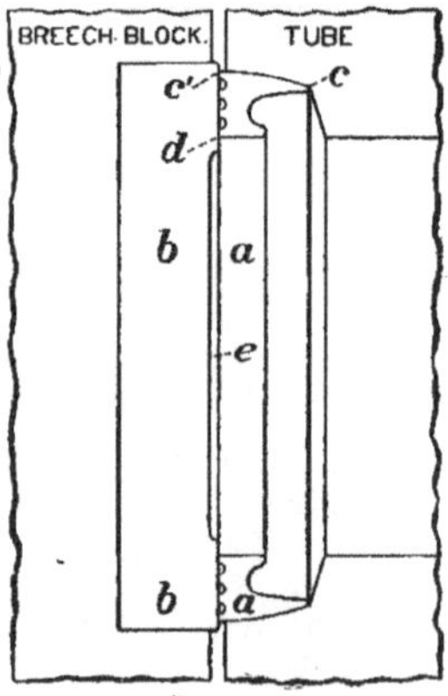

FIG. 146.

front surface were plane, would be drawn against the edge of the obturating-ring when the block is withdrawn, and thereby increase the liability to fouling of the surface *c'd*. The surfaces *c'c* and *c'd* are those which must be kept sealed against the escape of gas. The surface *c'd* is especially difficult to seal, and hence the necessity for the heavy pressure given by the locking-screw, to set the obturator-plate firmly against the ring.

Action.—The gas acts upon this ring to force the thin edge *c* against the walls of the bore, and also to press the ring backwards against the obturator-plate, forcing down especially the edge *d*.

GENERAL FEATURES OF MECHANISM.—The locking-screw just described is supported at its outer extremity by a plate, *k*, Fig. 145, called the locking-plate. The travel of the block is limited by a chain, or by a stop-bolt which passes through the upper part of the breech, and projects into a groove in the block. The jacket is bored out in prolongation of the bore, for the insertion of the projectile and charge in loading, and this hole is also made through the breech-block, so that when the block is withdrawn the hole through it is also in prolongation of the bore. The rounded shape of the rear of the block, and of the slot, gives strength by avoiding sharp corners, and the inclination of the axis of the cylinder to that of the bore, with that of the guides, gives a component motion of the block parallel to the axis and gradually seats the block firmly, while by a slight motion outward, all the parts become free and the block is easily withdrawn.

ADVANTAGES.—The Krupp mechanism is very simple and not liable to get out of order. It has been thoroughly tested, and found to be reliable. If it becomes stuck or wedged in the gun, it may be more easily removed than the screw, as it is more accessible.

DISADVANTAGES.—It requires a heavier forging for its jacket than the screw system, and consequently increases the weight of the gun for the same length of bore. The

Broadwell ring is not as good a gas-check as the De Bange or Freyre.

The longitudinal stress is not uniformly distributed over the cross-section of the jacket, and this is seen by a tendency of the gas-check seat to become oval, the longer axis being parallel to that of the slot.

It is more exposed to a front fire when open.

It tends to guillotine the cartridge.

CHAPTER IV.

PROJECTILES AND ARMOR.

PROJECTILES.

154. Classification—Solid Shot—Chilled Shot—Steel Shot.

CLASSIFICATION.—Projectiles may be classified according to their structure, as solid shot, shell, and case-shot; according to their use, into field, siege, and sea-coast projectiles; and according to their shape, as spherical and oblong.

Spherical projectiles are now obsolete.

SOLID SHOT.—Solid shot were formerly used for armor-piercing, and are still used in small arms against animate objects. The advantages of solid shot are, that they have greater weight for the same volume, and hence greater energy for a given velocity; and where it is necessary to concentrate energy upon a given area, as in attacking an armor-plate, they were generally employed.

The disadvantages are that for attacking armor, the projectile must possess great hardness to penetrate, and great toughness to resist breaking up on impact, and if the shot be made solid, it is subjected to initial strains due to casting or forging, which cannot be removed; the metal in the interior is not sound, and hence we obtain weaker projectiles when solid than when they have an interior cavity or core. This cavity removes the unsound metal, if the projectile is of cast iron, or if of steel, allows it to be treated, so that the strains can be removed and toughness attained. Such shot are generally called "cored shot."

The only solid projectiles at present in general use are for small arms.

CHILLED SHOT.—With the introduction and improvement of wrought-iron armor, cast-iron projectiles became useless, as they were broken on impact. This led to the introduction of chilled cast-iron projectiles. The Palliser projectile, so called from its inventor, Major Palliser of the British Army, was for a long time quite celebrated, and very effective against wrought-iron armor.

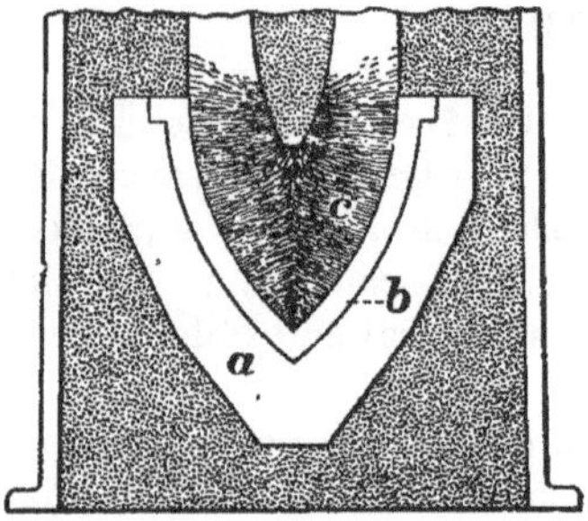

FIG. 147.

It was made by casting the ordinary cored shot in a chill, Fig. 147.

The body of the projectile is cast in sand to give toughness, and the head in a cast-iron mould or chill *a*, so called because it carries off the heat of the parts in contact with it so rapidly as to cause chilling, and produce great hardness.

The exterior of this chill conforms generally to the shape of the head, to insure uniform cooling, and it is lined with a movable lining, *b*.

The latter soon becomes worn from contact with the heated metal, and is removed and replaced by a new lining, thus preserving the body of the chill *a*. The head of the chilled shot is shown at *c*.

STEEL SHOT.—As armor improved in its resisting qualities, the chilled cast-iron projectile was broken on impact, and steel shot were substituted. The best of these are made of

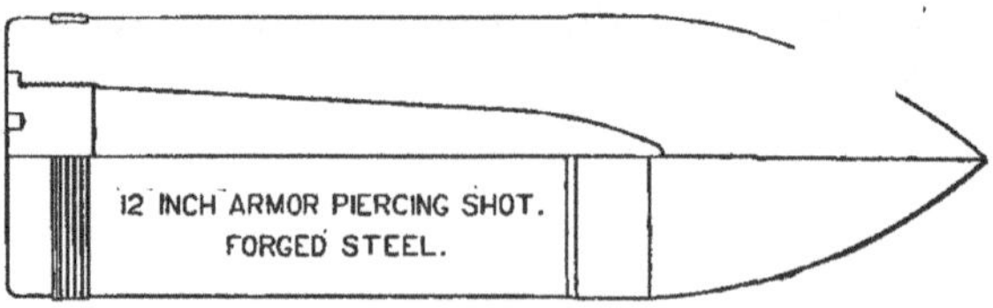

FIG. 148.

chrome steel, forged and tempered. Two processes, known as the Holtzer and Firminy, are so far the best, but they are secret, and nothing is known of them. The projectiles made by these processes give the best results when used against modern steel armor, but they are very expensive, and hence attempts are now being made, with some appearance of success, to replace them by cast-steel projectiles, which are tempered by a secret process.

Fig. 148 shows a forged-steel Holtzer armor-piercing cored shot.

155. Shell—Definition—Shell for Sea-coast Service—Deck-piercing Shell.

DEFINITION.—A shell is a hollow projectile, containing a bursting charge of gunpowder, or some high explosive, and a fuze to ignite this charge at some point of its flight, or upon impact.

Shells are used in the sea-coast, siege, and field services, and their construction depends on the purpose for which they are intended.

SHELLS FOR SEA-COAST SERVICE.—In the sea-coast service, shells are used in high-powered guns for attacking armor, or in mortars with high angle-fire for piercing the decks of vessels.

Against Armor.—If the shells can be made strong enough to penetrate armor, they are preferred to shot, because they burst after penetration, and acting in a confined space on a ship, cause great destruction. For this purpose the walls of the shell must be strong, and hence the cavity small. The cavity being small, will not contain a large bursting charge of powder, and the walls of the shell being strong, the gases from this charge may not develop sufficient pressure to rupture them.

On this account, and because of its greater destructive effect, a high explosive as a bursting charge is necessary.

Gun-cotton has been tried as a bursting charge for these shells. While it has given good results in some cases, it is liable to premature explosion from shock and friction, and if desensitized by moisture or by paraffine, it requires a strong primer of dry gun-cotton to detonate it, and this primer is liable to detonation by shock. The same principle applies to nearly all the high explosives which have been tried, and hence the problem of a suitable high explosive for armor-piercing shell is not yet solved. This has led to the introduction of various methods of firing high explosives, as the pneumatic dynamite gun, etc.

With armor-piercing shell, it is sought by various means to delay the action of the bursting charge till penetration is complete, as by wrapping the charge in flannel, using delayed-action fuzes, etc.

AGAINST THE DECKS OF VESSELS.—The problem of penetrating the sides of armored vessels being so difficult, attempts are made to perforate their decks. When the weight of guns, machinery, and armor carried by ships of the pres-

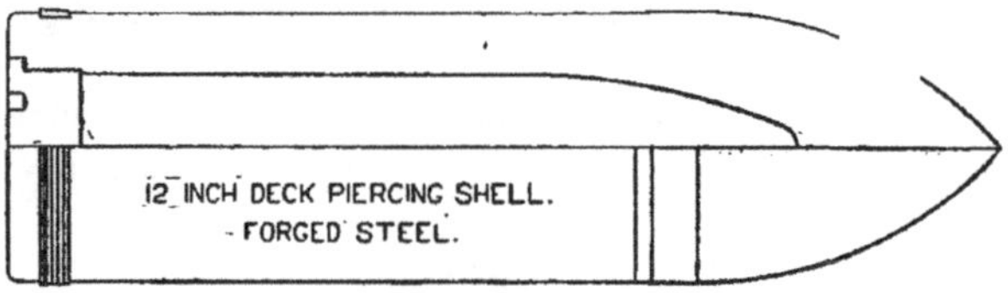

FIG. 149.

ent day is considered, the available weight left for deck protection is comparatively small, and hence a thickness of about 4½ inches of protective deck is about all that can be carried. Against these decks, the vertical fire of shell from heavy

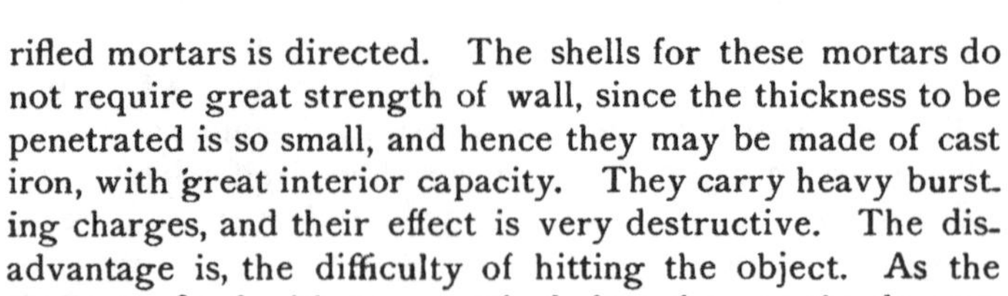

FIG. 150.

rifled mortars is directed. The shells for these mortars do not require great strength of wall, since the thickness to be penetrated is so small, and hence they may be made of cast iron, with great interior capacity. They carry heavy bursting charges, and their effect is very destructive. The disadvantage is, the difficulty of hitting the object. As the shells are fired with comparatively low charges, the dangers of premature explosion from shock are lessened, and recent experiments at Sandy Hook have shown that high explosives can be fired from these mortars with safety. At present these shells are made of forged steel.

Figs. 149 and 150 show a steel deck-piercing shell and a cast-iron shell for the 12-inch mortar.

156. Siege Shell—Field Shell.

SIEGE SHELL.—The shells for siege purposes are somewhat similar to those for deck piercing. They are used in direct and curved fire, and against earth or masonry. Their object, therefore, is to displace the earth and masonry, and as no great strength is required against these obstacles, the siege-shell are made of cast iron. In firing against masonry, it is necessary not only to penetrate, but also to remove the broken fragments, so that the next shot may fall upon a fresh surface. For this purpose large bursting charges are required, and hence large cavities, and comparatively thin walls. Very long shell are sometimes supplied for this purpose, and are called "torpedo shell."

Against earth, the maximum displacement is required, and some experiments made with gun-cotton as a bursting

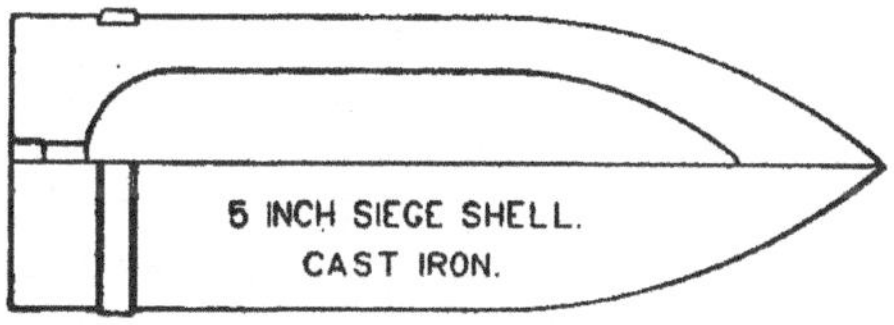

FIG. 151.

charge show that it is very effective. The 5-inch siege shell is shown in Fig. 151.

FIELD SHELL.—In field artillery the objects to be attacked have little resistance, as they are generally light field entrenchments, buildings, or troops, and hence the effect depends on the number of fragments into which the shell bursts.

The number of these fragments will depend on the brittleness of the material, and the pressure of the gases from the bursting charge, at the time when rupture occurs.

The natural tendency of a shell is to burst in a meridian plane, or a plane through its longer axis, since the total pressure of the gases normal to this plane is greater than that normal to the transverse plane. If the pressure of the gases is developed slowly, as from a bursting charge of large-grained powder, rupture will occur as just indicated, and we will have a few large fragments, and the effect will be limited. To avoid this it is necessary—

1. To use as a bursting charge, fine-grained powder of high gravimetric density. By this means the pressure is rapidly developed, and the largest possible weight of charge is contained in a given volume.

2. To prevent rupture in a longitudinal plane, the interior of the shell is sometimes grooved spirally, to weaken it, and give more fragments. A remaining velocity of 500 ft.-secs. is generally considered sufficient to disable or kill a man, and a fragment weighing about 1 ounce with this velocity is effective. With the 13.5-lb. shell this would give 216 effective fragments, and with the 20 lb. shell 320. In practice these results cannot be obtained.

Owing to the irregularity of their action, field-shells are seldom used against animate objects, except at very long distances, or when under cover. They are used, however, with percussion fuzes, to obtain the range quickly. By firing a percussion shell with a certain elevation, so as to strike in front of the target, and again with an increased elevation, so as to strike beyond it, and observing the points of burst, the target is thus enclosed in a fork, and by working between these limits, the true elevation is soon obtained. The 3.2-inch shell is is shown in Fig. 152.

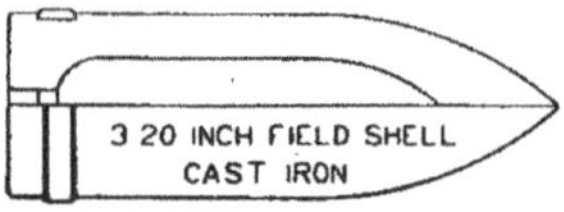

FIG. 152.

157. Case-shot—Grape—Canister.

DEFINITION.—Case-shot may be defined to be a collection of particles enclosed in a case or envelope, the latter being intended to rupture in the gun, or at some point in flight, and liberate the enclosed particles.

According to the place of rupture of the envelope, case-shot may be divided into—

1. Grape;
2. Canister,

whose envelope is broken in the gun, by the shock of discharge; and,

3. Shrapnel,

whose envelope is broken at some point of the flight of the projectile.

GRAPE.—This projectile is no longer used, but is interesting historically. It consists of three layers of balls, each layer containing generally three balls (see Fig. 153) held in place by top and bottom plates, *a* and *b*, of iron; a central bolt and nut, *c*; and two intermediate rings, *dd*. It was used in the sea-coast service with smooth-bore guns, against the masts and rigging of ships, and against men; also in the siege and field services against animate objects in mass, at distances too great for smaller projectiles.

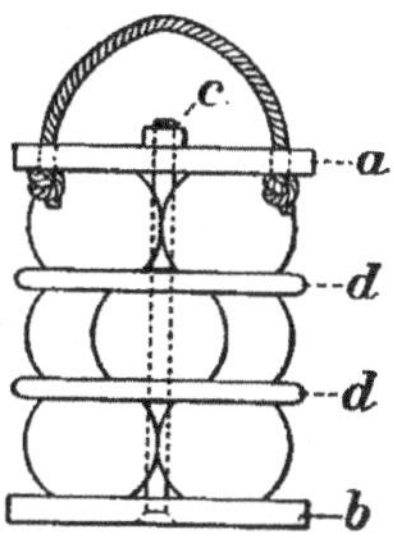

FIG. 153.

CANISTER.—This consists (Fig. 154) of a number of spherical bullets of lead hardened with antimony, or of cast iron, contained in a can; hence the name. The envelope is closed by a top and bottom plate of iron, and is intended only for convenience in transportation, and in loading.

It was used principally in the field service with the old smooth-bore guns, against animate objects at close range.

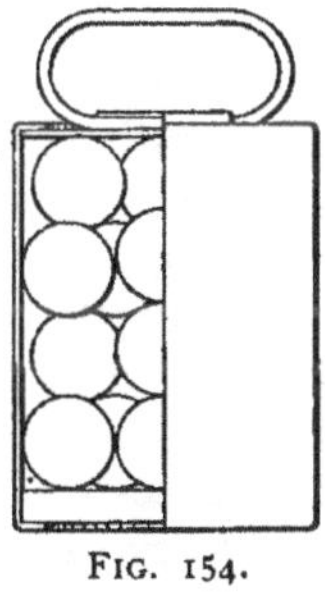
FIG. 154.

In both these projectiles, the case ruptured in the bore, and the projectiles scattered at the muzzle, forming a cone of dispersion, with its apex at the latter point.

For rifled guns it was necessary to prevent the case taking the grooves, and thus giving it the rifled motion and increasing the lateral dispersion of the projectiles, which was already great.

FIG. 155.

For this purpose the case was made stronger, since with the tin case, the projectiles were forced sidewise by the shock of discharge, expanding the case, and forcing it into the grooves. In our service this was done by adopting the Sawyer canister, the case of which is made of malleable cast iron, weakened by spiral cuts, as shown in Fig. 155, so as to insure its breaking up in the gun.

All these projectiles were used at short range, and since the fighting range has been increased, owing to the longer range and higher ballistic power of small arms, they are little used at the present day. A few rounds of canister are sometimes carried with the field gun for emergencies.

158. Shrapnel—Cone of Dispersion—Causes which Affect it.

SHRAPNEL.—This projectile is now the most important in field artillery, and is employed to the exclusion of all others. It consists essentially of a case or envelope containing small round projectiles, and a bursting charge, and fuze. The charge is sufficient to rupture the envelope at a given point of flight of the projectile, the fuze being arranged to ignite the charge at that point. After the rupture of the envelope, the contained projectiles move on with a velocity which is the resultant of that due to discharge, and to the bursting charge, and act from the point of burst to the target, as canister. The object of the envelope then is to convey the small projectiles to within striking distance of the target, where they are liberated, and each particle acts. The projectile is used entirely against animate objects, and its advantages over the shell are that the division of the particles is made beforehand, and each one is of the proper size to exert a disabling or killing effect.

CONE OF DISPERSION.—When rupture of the case occurs, each contained particle describes its own path, and the paths thus described, taken together, form the elements of the "cone of dispersion." The intersection of this cone with the ground is an irregular oval, and its area will vary with—

CAUSES WHICH AFFECT IT.—1. The angle of elevation of the piece;

2. The velocity of translation of the shrapnel before bursting;

3. The velocity of rotation of the shrapnel before bursting;

4. The position of the bursting charge;

5. The height above the ground at which the shrapnel bursts.

Angle of Elevation.—If this be large, other things being equal, the angle of fall will be large, and the plane of intersection being more nearly normal to the mean axis of the cone of dispersion, the area of the oval will decrease; the converse is true for small angles of elevation (see Fig. 156).

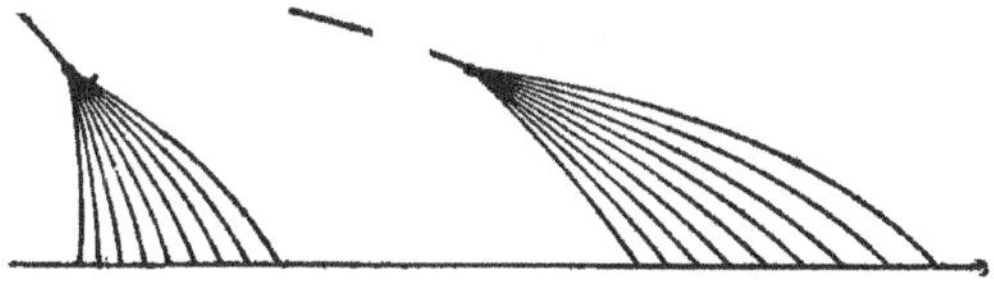

FIG. 156.

Velocity of Translation.—The greater this velocity, the greater will be the velocity of the particles in the plane of fire, and consequently the longer the oval in this direction.

Velocity of Rotation.—This causes the particles to move at right angles to the plane of fire, and hence increases the lateral dispersion of the particles, and the width of the oval.

Position of Bursting Charge.—This may be in front, or in rear of the particles. If in front, it decreases the velocity of translation of the particles, and hence decreases the length of the oval, and for this reason its effect is injurious. For other reasons, however, the position is a good one, as will be seen.

When in rear, it increases the velocity of the particles in the plane of fire; but there are objections to this position.

Height of Burst.—It is evident that, for a given inclination, and for given velocities in the plane of fire and laterally, the higher the point of burst, the greater the area of the oval.

The constant tendency with shrapnel is to increase the velocity of the particles in the plane of fire, and to decrease that at right angles to this plane; and the best possible condition for the efficiency of this projectile is when the area of the oval is such that each bullet will hit a man. The best position for the point of burst is about 6 yards above and 50 yards in front of the target.

159. Construction of Spherical Shrapnel—Early Shrapnel—U. S. Spherical Case—Boxer Spherical Shrapnel.

Spherical shrapnel is now obsolete, but the history of its development shows clearly the directions in which improvements have been made.

EARLY SHRAPNEL.—The projectile was invented by Colonel Shrapnel of the British Army, about 1803. In its early form it was simply a spherical shell filled with bullets, and the bursting charge was contained in the interstices between them. The objections to this arrangement were:

1. When the projectile was fired the balls spread sidewise, and tended to deform or burst the shell. Hence the latter was made with thick walls to resist this force, and this decreased the interior capacity, and consequently, the number of bullets which it would contain.

2. The powder, being loose among the bullets, was subjected to trituration and friction in handling, and hence was liable to accident. Since the space between the bullets was large, the density of loading of the bursting charge was low, and hence a large bursting charge was required to rupture the envelope. This scattered the fragments too much, and rendered the action of the fuze more irregular.

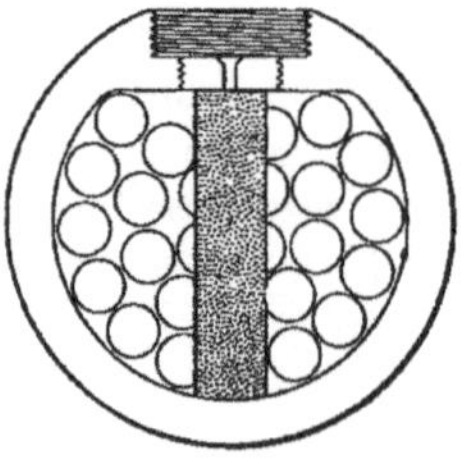

FIG. 157.

U. S. SPHERICAL CASE. — These defects suggested the improvements which were made in the spherical shrapnel used during the Civil War (Fig. 157):

1. To prevent the spreading of the bullets, the shell was first filled with them, and melted sulphur was then poured in, filling the interstices between the bullets. They were thus converted into a solid mass, and as their tendency to spread sidewise was thus destroyed, the case was made thinner, and consequently held more bullets.

2. To diminish the bursting charge, a cylindrical hole was bored through the bullets and sulphur, and in this the bursting charge was placed. Thus the density of loading was increased, and a small bursting charge could be used with less uncertainty in the action of the fuze.

The objections to this arrangement were that the sulphur caused the bullets to stick together, and prevented their separation after the bursting of the case, and that the effect of the bursting charge was to increase the lateral dispersion of the particles.

Boxer Spherical Shrapnel.—Most of the defects in the above shrapnel were remedied in the Boxer Spherical Shrapnel (Fig. 158), invented by Colonel Boxer of the English Army.

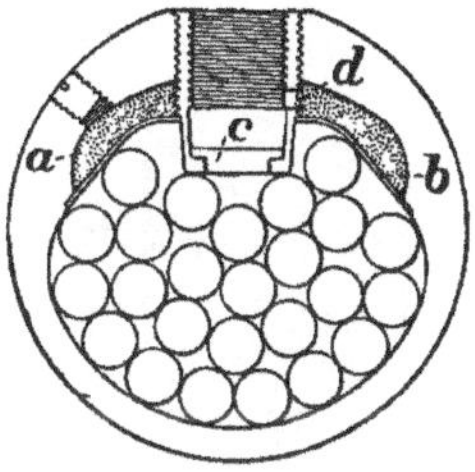

Fig. 158.

In this projectile, the bursting charge was placed in a chamber, *a*, formed by introducing a wrought-iron diaphragm, *b*, into the mold before casting, and allowing the cast iron to cool around it. The bullets were introduced through the opening *c*, the upper end of which carried the fuze, the flame from which reached the charge through a hole, *d*. The sulphur used as packing in the U. S. shrapnel, was replaced by coal dust.

This shrapnel possessed the following advantages:

1. The bullets did not adhere to the matrix after bursting.

2. A small bursting charge could be used.

3. The diaphragm *b* weakened the case, so that it would burst readily.

4. As soon as the shrapnel left the piece (since its forward portion, which contained the fuze, was lighter than

the rear portion), the lighter portion would turn to the rear, leaving the centre of gravity in advance of the centre of figure. This brought the bursting charge in rear, and hence, on explosion, it acted to increase the forward velocity of the bullets, and its tendency to scatter was very small.

160. Oblong Shrapnel—Boxer—Modern Shrapnel—Position of Bursting Charge.

BOXER OBLONG SHRAPNEL.—At this point the development of spherical shrapnel ceased, owing to the introduction of rifled guns and oblong projectiles. The first oblong shrapnel of any importance was that of Colonel Boxer. This consists (Fig. 159) of a cast-iron body, *a*; and a wooden head, *b*, covered with sheet-iron, *c*, riveted to the cast-iron body. The bursting charge is contained in the chamber *d* in rear, and over this chamber, separating the charge from the bullets, is a cast-iron disk, *e*. The central tube *f* is filled with powder, and conveys the flame from the fuze *g*, to the bursting charge *d*. The balls are held together by melted resin, and a paper lining prevents the adhesion of the matrix to the walls of the envelope.

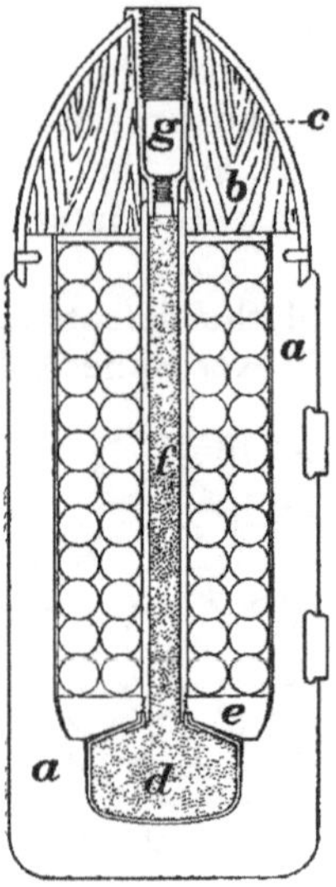

FIG. 159.

This shrapnel has the following advantages:

1. Those common to all oblong projectiles—of greater range and accuracy, and, for a given cross-section, containing a larger number of projectiles than the corresponding spherical shrapnel.

2. The charge, being in rear, acts, as with the Boxer spherical shrapnel, to increase the forward velocity of the bullets after rupture.

3. The head and its attachments, being relatively weak, give way easily, and the bullets are swept out to the front by the rear disk *e*; the action in this respect being like the discharge of canister.

The disadvantages are:

1. The body being of cast iron, the walls are made comparatively thick to withstand the shock of discharge, and this reduces the interior capacity, and consequently the number of bullets which the shrapnel contains.

2. The wooden head takes up room which can be better utilized.

3. The delay caused by the communication of fire from the fuze to the bursting charge may interfere with the action of the shrapnel, and cause it to pass beyond its proper point of burst before exploding.

4. The effect of the pressure of the gases from the central tube, is to cause an increase in the lateral spread of the bullets, which is objectionable.

MODERN SHRAPNEL.—These disadvantages suggest the improvements which have been made in modern shrapnel. Several of these are now under trial in this country. The following changes have been made in them, in comparison with the Boxer oblong shrapnel:

1. To give sufficient strength of wall to withstand the shock of discharge, the body is made of drawn-steel tubing, and the head and base are welded on by electricity. As this is an expensive construction, wrought-iron tubing has been substituted for the steel, and the head and base are made of cast iron screwed into the wrought-iron body.

2. In case the bursting charge is in front, as in one of the shrapnel undergoing trial, a cast-iron chamber takes the place of the wooden head of the Boxer, and contains the bursting charge and fuze.

3. To lessen the delay caused by communication of fire through the central tube to the rear bursting charge, this tube is enlarged on the interior, and made of brass tubing so as to give a larger channel for the passage of the flame, and it does not occupy more space in the shrapnel, as the exterior diameter of the tube is not changed.

4. To prevent adherence of the balls after rupture, the matrix is made of cast iron, indented so as to hold the balls in place and form a solid mass with the projectile, and yet

so arranged as to break up into fragments when the shrapnel bursts, which add their effect to that of the balls.

POSITION OF BURSTING CHARGE.—When the bursting charge is in front, we have the following advantages:

1. It occupies less space in the shrapnel, since no central tube is required;

2. It acts promptly to burst the case, and hence the point of burst can be more accurately fixed;

3. It occupies space in the shrapnel which it is difficult to fill with bullets.

Its disadvantage is:

1. It decreases the velocity of the fragments in the plane of fire, instead of increasing it.

When the bursting charge is in rear, it has the advantage:

1. It increases the velocity of the fragments in the plane of fire.

Its disadvantages are:

1. It occupies increased space in the shrapnel;

2. It causes delay in bursting;

3. It increases the lateral spread of the fragments;

4. It is more expensive in construction.

For these reasons it is probable that the front charge will be adopted, but it is not yet settled.

161. Description of Modern Shrapnel—Steel-welded—Frankford Arsenal.

STEEL-WELDED.—The steel-welded shrapnel (Fig. 160) consists of a steel tube, *a*, to which the base *b* and head *c* are welded by electricity. The charge is in rear, and is separated from the bullets by a disk, *d*. The central tube communicates fire to the charge from the fuze. The bullets are held in place by a matrix of resin, melted and poured in after the former are in place.

THE FRANKFORD ARSENAL.—This shrapnel (Fig. 161) consists of a wrought-iron tube, *a*, to which the base *b* of cast iron is screwed. The rotating band *c* fits in a groove cut on the rear end of the tube *a*. The head *d* is also of cast iron, carries the bursting charge and fuze, and is screwed to the body. The bullets are held in place by a skeleton matrix of cast iron, consisting of a top and bottom plate, and a series of intermediate plates.

These intermediate plates are made in segments, so that the whole can be built up layer by layer, and inserted in the body of the case. The head and base are then screwed on, the band being inserted in its groove before the latter is screwed home. This shrapnel is much cheaper than the steel one, and has given good results at the Proving Ground.

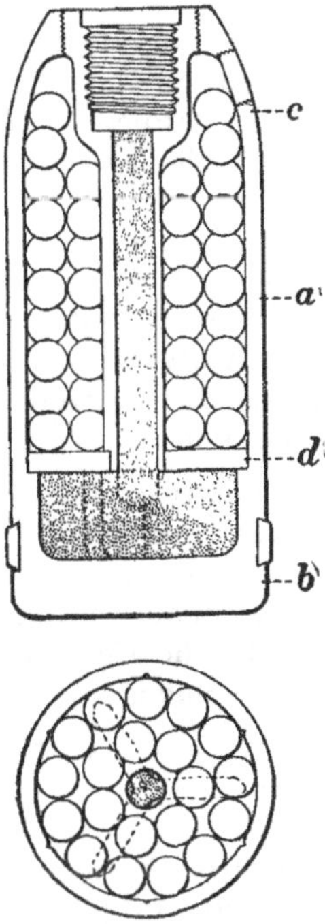

FIG. 160.

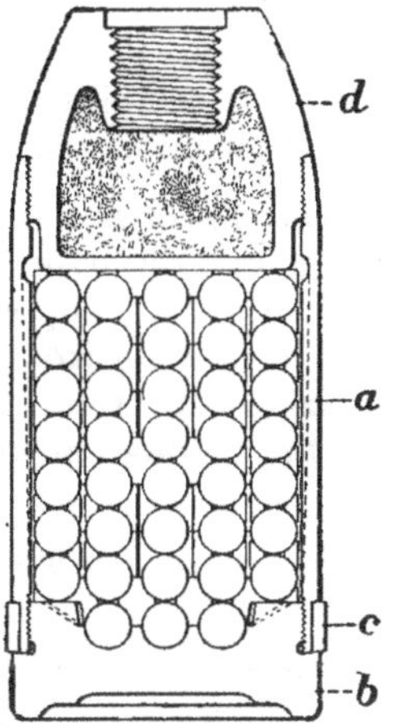

FIG. 161.

The Hotchkiss shrapnel is similar to this.

162. Necessity for Rotation of an Oblong Projectile—Energy of Rotation Required.

NECESSITY.—It has been shown that an oblong projectile when rotating about its longer axis, will move through the air in the general direction of that axis.

Without this motion of rotation about the longer axis, the resultant resistance of the air, acting with a certain lever arm, will cause the projectile to rotate about a short axis through the centre of mass.

The effect of this would be to cause great irregularity of motion, owing to the varying surface presented to the air by the projectile, during this rotation.

The rotary motion about the longer axis is imparted to the projectile by cutting spiral grooves in the surface of the bore, and by placing a device upon the projectile which will fit in these grooves, and thus cause the projectile to take up the rifled motion.

ENERGY OF ROTATION.—The question as to the amount of energy of rotation about the longer axis which the projectile must have, to enable it to maintain its proper position during flight, requires for its determination analytical methods which are too complex to be given here. A general discussion will show upon what principles it depends.

163. General Discussion of the Rotation of a Projectile—Value of *R*.

The general discussion of the motion of rotation of an oblong projectile, based upon Euler's equations, shows that for a projectile rotating from left to right, as in our service, the longer axis in the time t will deviate to the right of the plane of fire through an angle, ϕ, whose value is

$$\phi = \frac{Rl}{I\omega}t, \quad \dots \dots \dots \quad (242)$$

in which (see Fig. 162) R is the resistance of the air acting at the centre of pressure, l its lever-arm with reference to a horizontal axis through the centre of gravity, I the moment of inertia about the longer axis of the projectile, and ω the angular velocity about the same axis.

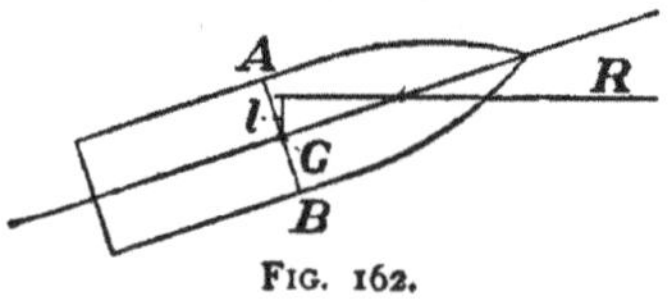

FIG. 162.

In order that the projectile may be stable, Rl must be small and $I\omega$ large. The methods of decreasing R will be explained. In order that l may be small, the centre of mass and centre of pressure must coincide as nearly as possible. The best position for the centre of mass is determined by experimental firing, the projectile being so weighted that this centre can be changed.

To increase I, the diameter of the projectile must be increased, its weight remaining constant; or its mass or weight may be increased, if its dimensions are constant, by increasing the density of the material of which it is made.

ω may be increased by giving a more rapid twist to the grooves in the gun, but this is limited by the increased strain brought upon the gun, and upon the projectile.

Equation (242) shows, generally:

1. As l increases, ϕ increases; and since l depends on L, the total length of the projectile, if we increase the length of the projectile, we must increase I or ω, or both.

Therefore generally a long projectile must have greater angular velocity about its longer axis, than a short one of the same calibre.

2. If two projectiles have the same length but different diameters, the value of I will be greater for the larger projectile, and hence ω may be less. That is, the projectile of greater diameter will require less angular velocity about its longer axis, than the projectile of smaller diameter and the same length.

3. If we have two similar projectiles of different densities, the dense projectile will require less angular velocity about its longer axis, since its mass, and hence its moment of inertia, is greater. Also, a shell will be more stable and require less angular velocity than a similar shot of the same weight, since its radius of gyration is greater.

4. Since $\phi = \frac{Rl}{I\omega}t$ measures the deviation of the longer axis in the time t, the reciprocal, $\frac{I\omega}{Rl}$, may be taken as the measure of the capacity of this axis to resist deviation; and for a given value of R at any time t it is evident that by increasing the value of the ratio $\frac{I\omega}{l}$ we increase the stability of the projectile.

VALUE OF R.—The resistance of the air, R, varies with the form, cross-section, and velocity of the projectile, and with the density of the air. The resistance being R, the retardation produced by this resistance will be

$$\frac{R}{M},$$

M being the mass of the projectile.

This value of the retardation has been determined by experiment, as will be explained in Exterior Ballistics.

In these experiments, the expression for $\frac{R}{M}$ has been assumed to be (equations (263) and (265), Exterior Ballistics)

$$\frac{R}{M} = \frac{A}{C}f(v) = A\frac{\delta}{\delta_1}\frac{cd^2}{W}f(v), \quad . \quad . \quad . \quad . \quad (243)$$

in which A, $\frac{\delta}{\delta_1}$, and c are constants, whose values are explained in Exterior Ballistics, and $f(v)$ is some function of the velocity of the projectile. Hence we may write

$$\frac{R}{M} = K\frac{d^2}{W}f(v). \quad . \quad . \quad . \quad . \quad . \quad . \quad (244)$$

For a given value of v, the retardation increases with the factor $\frac{d^2}{W}$; and hence this factor must be made as small as possible, by increasing W and decreasing d. The reciprocal of this factor, $\frac{W}{d^2}$, may then be taken as a measure of the capacity of the projectile to overcome the resistance of the air, just as $\frac{I\omega}{l}$ measures its stability.

164. Sectional Density—How it May be Increased—Effect of its Increase on the Gun.

SECTIONAL DENSITY.—The factor $\frac{W}{d^2}$ is called the "sectional density" of the projectile. The area of base being $\frac{1}{4}\pi d^2$, $\frac{W}{\frac{1}{4}\pi d^2}$ will be the weight of the projectile per unit area of base, and hence $\frac{W}{d^2}$ is taken as a measure of this weight, the constant factor $\frac{1}{4}\pi$ being omitted. The sectional density is very important in considering the motion of a projectile in air, and also in the gun.

If two projectiles have the same initial velocity, but different sectional densities, that having the greater sectional density will be less retarded by the air, equation (244), and consequently will lose less velocity. Hence for a given range, its time of flight will be less, and being exposed to the action of the air, and other deviating causes, for a less time, its accuracy will be greater.

If the two projectiles be fired with the same angle of elevation and the same initial velocity, that having the greater sectional density will have the greater range, since it retains more velocity at the end of each successive interval of time.

For the same initial velocity, the trajectory or path of the projectile having the greater sectional density, will be flatter or less curved than that of the other, because since its velocity is greater at any point of its path, its time of passage over a given distance is less, and consequently the time during which the force of gravity acts upon it to produce curvature is less. This gives greater accuracy of fire.

An increase of sectional density therefore increases—

1. The accuracy;
2. The range;
3. The flatness of the trajectory.

How it May be Increased.—The sectional density may be increased by increasing W or by decreasing d. W may be increased by keeping the calibre constant, and increasing the length of the projectile. This has been done with modern projectiles, for large guns, till the length is $3\frac{1}{2}$ to 4 calibres.

It may also be increased by increasing the density of the metal of which the projectile is made. This is done by using lead for small-arm projectiles, but this material does not possess sufficient hardness for projectiles for larger guns.

The sectional density may also be increased, by fixing the weight W, and decreasing the calibre, or d. This method has been adopted for small arms, the calibre and weight of projectile having both been reduced in such proportions as to increase the sectional density.

EFFECT OF INCREASE OF SECTIONAL DENSITY ON THE GUN.—Let P represent the maximum pressure per square inch on the base of the projectile;
M, the mass of the projectile.

Then we have

$$M\frac{dv}{dt} = \pi r^2 P, \quad . \quad . \quad . \quad . \quad . \quad . \quad . \quad (245)$$

from which

$$\frac{dv}{dt} = \frac{\pi r^2 P}{M} = \frac{\pi r^2 P}{\frac{W}{g}} = \frac{Pg}{\frac{W}{\pi r^2}}. \quad . \quad . \quad . \quad (246)$$

Replacing r by its value $\frac{1}{2}d$,

$$\frac{dv}{dt} = \frac{Pg}{\left(\frac{W}{\frac{1}{4}\pi d^2}\right)}. \quad . \quad . \quad . \quad . \quad . \quad . \quad . \quad (247)$$

As the sectional density $\frac{W}{d^2}$ increases, $\frac{dv}{dt}$ decreases, and hence to obtain an increase of acceleration, the value of P, or the pressure on the projectile, and consequently that upon the gun, must increase. Since the maximum pressure is fixed by the strength of the gun, equation (247) limits the value of the sectional density, for a given acceleration. The initial velocity is

$$v = \int \frac{dv}{dt} dt; \quad . \quad . \quad . \quad . \quad . \quad . \quad . \quad (248)$$

and for a given value of P, this velocity will decrease, from equation (247), as the sectional density increases. Hence when rifled guns were first introduced, using the old quick powders, the pressures could not be increased, and consequently the initial velocities of the projectiles decreased. When slow-burning powders were adopted, with longer bores, the sectional density of projectiles was increased, and also the initial velocities, with less maximum strain on the gun. The reason for this has been explained in Interior Ballistics.

165. Rifling—Kinds—Uniform—Increasing.

RIFLING.—In order to give to the projectile the angular velocity ω required in equation (242), it is necessary to cut spiral grooves in the bore of the gun, and to attach a device to the projectile which will fit these grooves. The spiral groove in the gun is called the rifling.

Let v denote the velocity of the projectile at any point of the bore;
ϕ, the angle made by the tangent to one of the grooves, with an element of the bore;
r, the radius of the bore.

The velocity of the projectile along the groove, is the resultant of two components, v, and $v \tan \phi$, at right angles to each other.

The actual velocity of rotation of a point on the surface of the projectile is ωr, and this is equal to the component $v \tan \phi$. Hence

$$\omega r = v \tan \phi; \quad \therefore \ \omega = \frac{v}{r} \tan \phi. \quad . \quad . \quad . \quad (249)$$

UNIFORM RIFLING OR TWIST.—If the value of ϕ be constant for the whole length of the bore, the rifling or twist is said to be uniform.

In this case the angular velocity varies directly with v and inversely with r. The objection to uniform rifling is as follows:

When the projectile starts from its seat, and during the first part of its path in the bore, the pressure of the powder-gas rises to its maximum, and the gun is subjected to the greatest stress at this time.

With the uniform rifling, the angular velocity ω is impressed upon the projectile at this time also. Hence, while the gun is subjected to its greatest stress, due to the starting of the projectile, it is also subjected to its greatest stress in giving rotation to the projectile.

After rotation is once acquired, the stress due to this cause falls off very rapidly. Therefore, with the uniform twist, both these stresses act together.

INCREASING TWIST.—If, however, the angular velocity ω be imparted gradually to the projectile, it moves from its seat more readily, and the strain on the gun at first is thus diminished. When the powder pressures fall off along the bore, the twist, or value of ϕ, gradually increases, till it reaches its final value necessary to impart the angular velocity ω to the projectile. In this case the stresses are more uniformly distributed along the bore, and the gun strained less at the origin of motion, while the final velocity of rotation is the same.

The twist in this case is called an increasing twist, as the value of ϕ increases gradually from the breech. In modern guns the curve of the rifling, when developed on a plane surface, is a semi-cubic parabola, whose equation is

$$y^{\frac{3}{2}} = 2px.$$

To give steadiness of rotation to the projectile, the twist increases from the breech to a point about two calibres from the muzzle, and from this point to the muzzle it is uniform.

166. Twist in Terms of Calibre—Kinds of Grooves.

TWIST IN CALIBRES.—The twist is generally expressed in terms of the calibre, as one turn in ten calibres, etc.; meaning that the projectile makes one complete turn in passing over a length of bore equal to ten calibres, etc. Suppose the groove to be developed (Fig. 163), and let a be the development of one turn of the uniform groove, n the number of calibres in which the projectile makes one complete turn, and r the radius of the projectile, then the distance $AB = 2nr$ and $BC = 2\pi r$, and

$$\tan \phi = \frac{2\pi r}{2nr} = \frac{\pi}{n},$$

for the value of the tangent of the angle of the rifling. For the increasing groove, ϕ is variable, but for any point, its value is $\frac{\pi}{n}$. In our service, the rifling of sea-coast guns is increasing, from one turn in 50 calibres at the breech, to one turn in 25 calibres at a distance of about two calibres

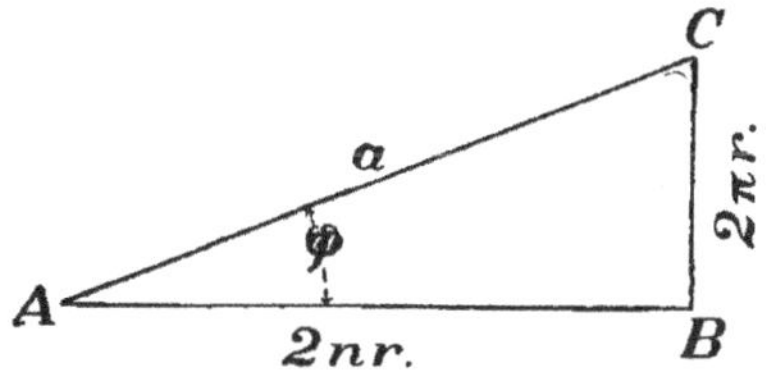

FIG. 163.

from the muzzle. For the field-guns, the rifling was formerly uniform, but in the later models an increasing twist has been adopted.

KINDS OF GROOVES.—The number, depth, and width of grooves depend on the rotating device. By a *groove* is understood the spiral cut made in the bore, and by a *land*, the space between two adjacent grooves.

When rifling was first introduced, the grooves were few in number, and as the points of contact of the projectile with the bore were also few, these points required considerable strength. The grooves were therefore made correspondingly deep and wide. This decreased the strength of the gun, as it increased the diameter of the bore subjected to the action of the powder-pressure. With a change in the rotating device, the grooves increased in number, and decreased in depth and width. This is called polygroove rifling, and adds greatly to the strength of the gun. In spite of this, the grooves are sources of weakness, as the action of the powder-gas tends to erode them at the junction of lands and grooves, and all sharp corners must be avoided.

In small arms the presence of grooves adds to the difficulty of cleaning the bore, and the grooves in these guns are made as shallow as possible.

The shape of the groove in the sea-coast guns in our service is shown in Fig. 164, which gives the grooves of the 8″ rifle.

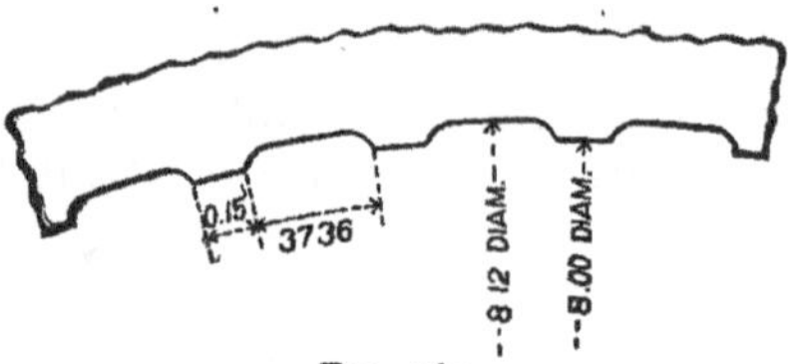

FIG. 164.

The number of grooves in sea-coast guns, is six times the calibre of the gun in inches. Thus the 8-inch rifle has 48 grooves and lands; the 10-inch, 60, etc.

167. Rotating Devices—Studded System—Flanged System.

ROTATING DEVICES.—The spiral grooves having been cut in the bore of the gun, it is necessary to attach some device to the projectile which will fit into these grooves, and communicate the required motion of rotation to it. Although muzzle-loading projectiles are practically obsolete, a few such guns still remain in our service, and a description of the means employed to give rotation to their projectiles will show the development of such devices.

Since muzzle-loading projectiles are of less diameter than the bore, the rotating device must be made either to fit the grooves before firing, or to do so after firing, by the action of the powder-pressure. Accordingly, the rotating devices for muzzle-loading projectiles are divided into:

1. The studded or flanged system.
2. The expanding system.

STUDDED PROJECTILES.—This system was generally used for muzzle-loading projectiles in Europe, and especially in England. The projectile (Fig. 165) was provided with studs made of a soft metal, such as zinc or copper, to avoid wearing the lands of the rifling. These studs were arranged in two or three rows, depending on the length of the projectile, and at an inclination equal to the angle ϕ of the grooves. They were inserted into undercut holes in the projectile, and subjected to pressure, by which the soft metal was forced to fill the holes. (See Fig. 165.)

The advantage of this system is that the projectiles are certain to take up the rifled motion.

The disadvantages are:

1. The projectiles must be adjusted to each particular twist; and if two guns have the same calibre, but a different twist of rifling, different projectiles must be used for them.

2. They cannot be used with an increasing twist.

3. Owing to the relatively small number of studs, the pressure upon each is great, and they are liable to shear. To avoid this they must be made strong, and this necessitates increased depth and width of rifling grooves, and a corresponding weakening of the gun.

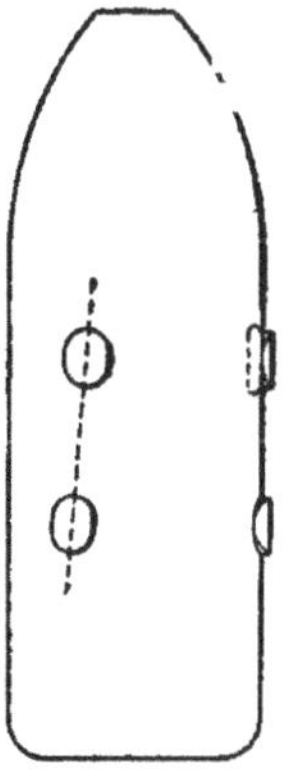

FIG. 165.

4. The stud-holes in the projectile weaken the latter, and their irregular surface increases the resistance of the air to its motion.

5. Unless a gas-check is provided on the base of the projectile, the escape of the gas between the projectile and the bore erodes the latter.

THE FLANGED SYSTEM.—In this system, flanges or ribs, fitting the grooves, were used instead of studs, the flanges being made generally of soft metal, except in case of the Whitworth. The principal example of this class is the Whitworth projectile, whose cross-section is a hexagon, and whose plane faces are inclined at an angle equal to that of the rifling (Fig. 166). The bore of the gun is rifled to correspond (Fig. 166*a*).

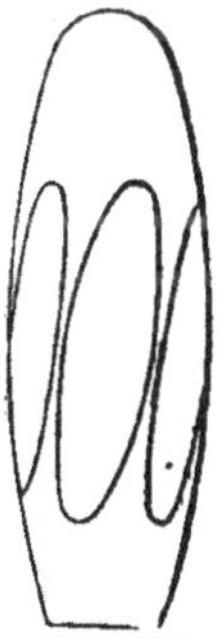

FIG. 166.

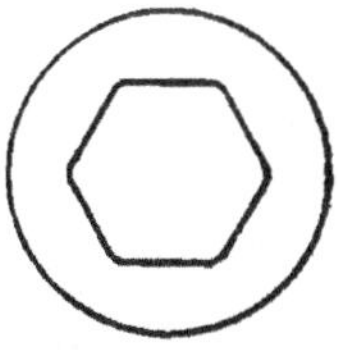

FIG. 166*a*.

In this case, the fit of the projectile in the bore was very accurate, and any slight fouling interfered with the loading. As the flanges were of hard metal, they could not yield, and hence any obstruction was liable to burst the gun. These projectiles have given remarkable results as regards accuracy and penetration, but they are no longer used.

168. Expanding System—Hotchkiss.

EXPANDING SYSTEM.—This system has been largely used in the United States for muzzle-loading projectiles.

It consists in placing upon the rear end of the projectile, or upon its cylindrical body, a band of soft metal, such as lead or brass, which is expanded by the action of the powder-gas, and forced into the grooves when the gun is fired.

The advantages of this system are:

1. It may generally be used with either a uniform or an increasing twist.

2. Projectiles of the same calibre, having different rotating devices of this class, will fit any gun of that calibre, and are easily loaded.

3. By the expansion of the rotating device the escape of gas between the projectile and bore is prevented, since the band acts as a gas-check.

The disadvantages are:

1. In some of the devices the gas was uncertain in its action, occasionally failing to produce expansion, and also tearing off the rotating device, or causing it to "strip;" thus failing to give the rotary motion to the projectile, and when fired over the heads of friendly troops, causing accident to them from the fragments of the band.

2. It was expensive, and required careful handling to prevent damage to the rotating device, and consequent inability to load.

3. It failed to centre the projectile in the bore.

The principal examples of this class are:

The Hotchkiss;
The Parrott;
The Eureka;
The Butler.

THE HOTCHKISS.—This projectile consists (Fig. 167) of a body, *a*; a base, *b*; and a jacket of lead, *c*, of the same diameter as the body of the projectile.

When fired, the pressure of the gas forces the base *b* up on *a*, and thus the lead jacket *c* is expanded into the grooves.

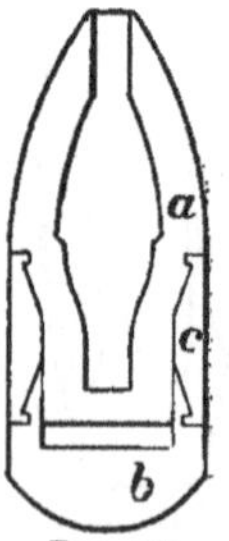

FIG. 167.

169. Expanding System—Parrott—Eureka—Butler.

PARROTT.—In the Parrott system a brass ring or band, *a*, is cast upon the base of the projectile (Fig. 168), leaving a circular channel or groove, *b*, between the ring and the base. The gas acting in the channel, forces the ring outward into the grooves. Frequently, however, the ring was torn off the base of the projectile, as its hold was not sufficient.

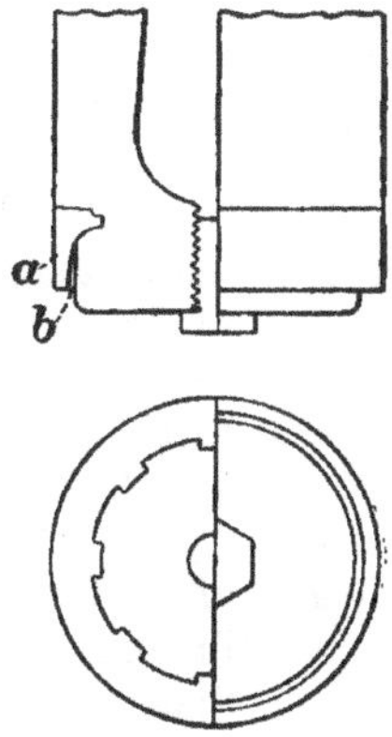

FIG. 168.

EUREKA.—In this system a brass cup, *a* (Fig. 169), is placed on the base of the projectile. This base is made in the form of a frustum of a cone, with the smaller base to the rear. It has several longitudinal grooves, *b*, cut in it, into which corresponding projections on the interior of the cup fit, and these prevent the rotation of the cup around the axis of the projectile, so that the rotary motion communicated to the cup by the rifling, is imparted to the projectile. The cup is curved where it rests against the rear end of the projectile, and to prevent stripping, it is held in place by the screw-bolt *c*. When the piece is fired, the gas-pressure forces the cup forward on the frustum of the base, till its curved surface rests against the rear of the projectile.

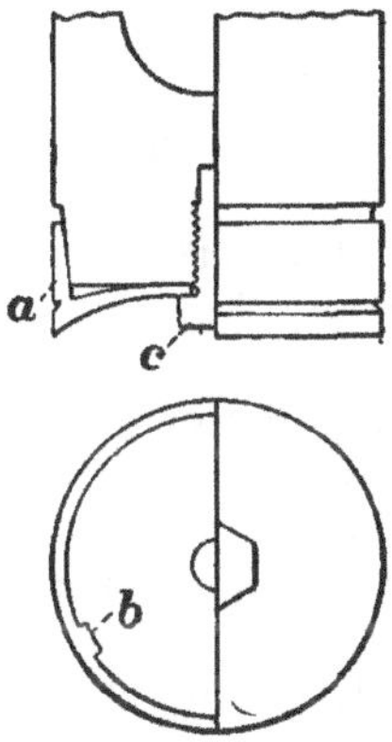

FIG. 169.

This causes the sides of the cup to expand and forces them into the grooves. It is a very satisfactory muzzle-loading rotating device, and is still in use in our service.

BUTLER.—This system was invented by Major Butler of the Ordnance Department, and consists (Fig. 170) of a brass ring *a*, having a lip or groove *b* in it. It is screwed to the base of the projectile to prevent stripping. When the piece is fired, the gas-pressure acts in the groove *b*, and forces the outer portion *c* of the ring outward into the grooves, and the inner portion *d* against the base of the projectile, thus insuring the adherence of the ring to the projectile. It is one of the best of these devices, and is still in service.

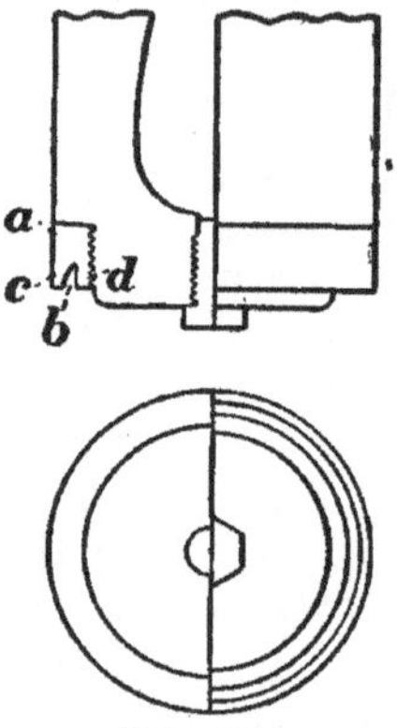

FIG. 170.

170. Breech-loading Projectiles—First Rotating Device—Hotchkiss Projectile—Copper Bands.

ROTATING DEVICES FOR BREECH-LOADING PROJECTILES.—With breech-loading guns the powder-chamber is larger than the bore, and hence the projectile may have a rotating device larger than the bore, and this may be compressed into the grooves by the action of the powder-gas. This is called the "compression system," to distinguish it from the studded or flanged, and the expansion systems. Its advantages are:

1. The projectile is certain to take the grooves, since its rotating device is compressed into them.

2. The rotating device being larger than the bore before firing, it acts as a gas-check, and prevents any flow of gas between the projectile and the bore.

3. It may be so shaped as to fit accurately the chamber of the gun before firing, and thus perfectly centre the projectile, or make its axis coincide with that of the bore. This gives increased accuracy of fire, and it is impossible to accomplish it with any muzzle-loading system.

Its only disadvantage is perhaps a slightly greater strain on the gun due to the increased pressure necessary to force the band into the grooves, but with the slow-burning powder used in modern guns this may be neglected. In fact it has been shown in Interior Ballistics that this increased resistance increases the muzzle velocity.

FIRST ROTATING DEVICE.—When breech-loading guns were first introduced, the rotating device was a jacket of lead (Fig. 171) cast on the body of the projectile.

The objections to this are that it is difficult to make the lead adhere to the projectile, and it becomes detached in flight. This is dangerous in firing over friendly troops, and the energy communicated to this lead jacket is lost; also the contact of the hot metal with the body of the projectile in the process of casting, is apt to injure the structure of the latter.

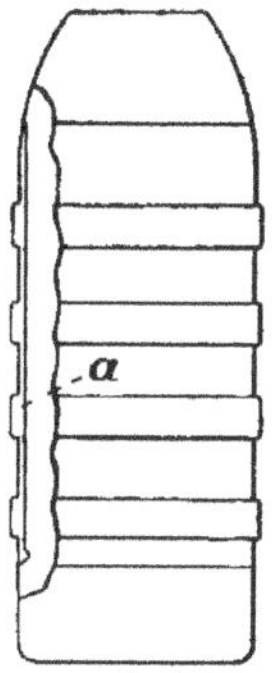

FIG. 171.

One rotating device of this class, the Hotchkiss, is still in use for small-calibre projectiles.

HOTCHKISS B. L. PROJECTILE.—The body of the projectile *a*, Fig. 172, is grooved circumferentially for about one calibre in length, as shown, and over these grooves *c*, is placed a band of brass, *b*.

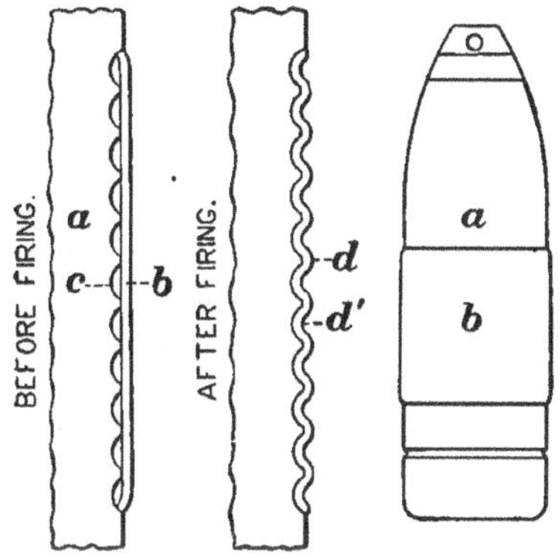

FIG. 172.

Longitudinal notches are also made in the grooves *c*, to prevent slipping of the band around the axis of the projectile.

When the gun is fired, the gas compresses the band into the grooves on the body of the projectile, and it takes a corrugated shape. The lands cut through the projections *d* of the band, and the metal thus displaced is forced into the grooves *d'*.

COPPER BANDS.—The lead jacket was next replaced by two bands of copper, Fig. 173, placed at equal distances from the centre of gravity of the projectile. The front band *a*, was used to support the forward portion of the projectile, and its diameter was slightly less than that of the bore between lands. The rear band *b*, was of a larger diameter, and was forced into the grooves, giving the rotation.

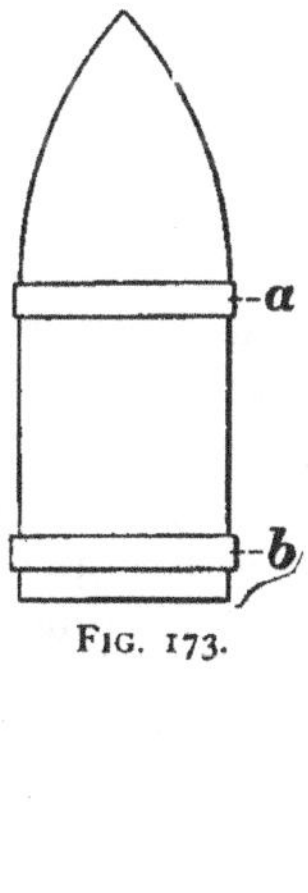

FIG. 173.

171. Rotating Device at Present—Profile of Band—Placing of Band —Position of Band.

RECENT DEVICE.—It was found that the front bearing or supporting band was not necessary, and in modern projectiles it is replaced by a slight swell, *a*, Fig. 174, at the base of the ogival head. The surface *a* is turned to a diameter slightly less than that of the lands, and the body of the projectile left as it comes from the casting or forging process.

This is less expensive, and leaves the metal on the exterior, which is stronger than any other part of the projectile.

The rotating band is made of copper, fitted in an undercut groove, as shown, Fig. 174.

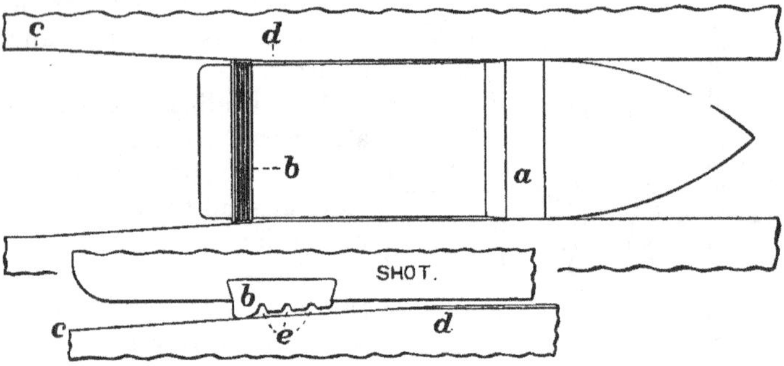

FIG. 174.

DETAILS OF ROTATING BAND.—In modern guns, the powder-chamber is joined to the bore by a long conical slope, *cd*, Fig. 174. The exterior of the rotating band has a slope slightly greater than this, so that when the projectile is in place, it will be accurately centred by the rear portion of the band, and this part of the band will also act as a gas-check, completely closing the interior of the bore. A number of grooves or cannelures, *e*, are turned on the exterior of the rotating band, to diminish the amount of metal to be cut through by the lands, allow space into which the portion cut out may be forced, and at the same time give the necessary length of bearing surface on the lands, by retaining the width of the band unchanged. The exterior diameter of the band at the rear, is slightly greater than that of the bore, measured from the bottom of the grooves.

For the field projectiles, the band is more simple, being a plain ring of copper with the front and rear faces bevelled, Fig. 175.

FIG. 175.

PLACING THE ROTATING BANDS IN POSITION.—This is generally done in our service, by hammering the band into place. The band may be made in two semicircles, or in a single piece of copper, whose length is just sufficient to encircle the projectile. In either case the cross-section before insertion is as in Fig. 176 at *a*.

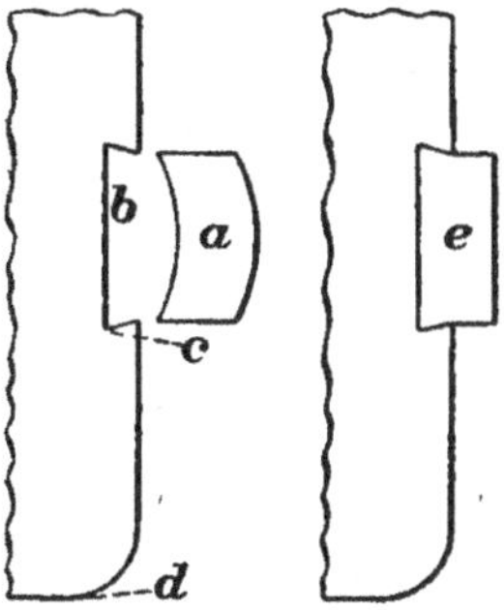

FIG. 176.

When inserted in the undercut groove *b* in the projectile, it is hammered, or subjected to pressure, till it takes the position shown at *e*, completely filling the groove.

It is then turned in the lathe to the proper dimensions.

POSITION OF REAR BAND.—The position of the rotating band has great influence upon the range and accuracy of the projectile, as has been shown by numerous experiments. It must be so placed that the distance *cd*, Fig. 176, will be sufficient to resist the shearing effect of the rifling, which would tend to strip the band off to the rear. This having been provided for, the best position of the band is determined by experiment.

172. Form of Projectiles—Head—Spherical Density—Weight.

FORM.—Numerous experiments have been made to determine the form of projectile that will best overcome the resistance of the air.

The result of these experiments shows that the resistance is affected by the shape of that portion of the head where it joins the cylindrical body, and also by the rear of the projectile, since the shape of these surfaces affects the flow of the air along its sides. (See "Exterior Ballistics.")

HEAD.—The heads of all modern projectiles are ogival, the radius of the ogive being from 2 to 3 calibres. The more pointed form gives less resistance, but introduces elements of weakness, and hence the above radii mark the limits thus far. The head is described with radii as shown in Fig. 177.

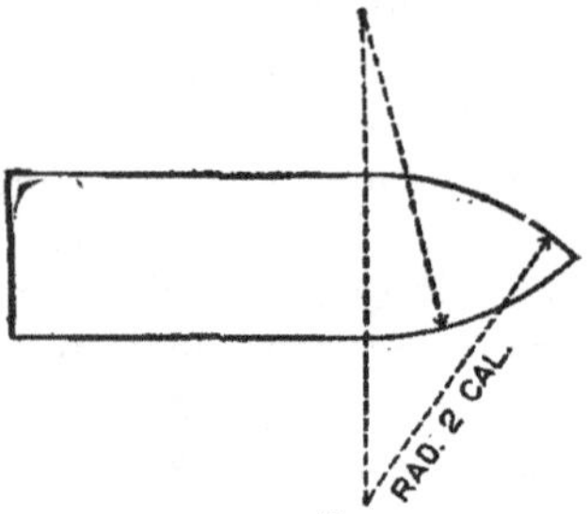

FIG. 177.

SPHERICAL DENSITY.—Let W be the weight of a solid spherical projectile, whose radius is r, and W' the weight of the oblong projectile of the same radius.

The ratio

$$S = \frac{W'}{W} \quad . \quad . \quad . \quad . \quad . \quad . \quad . \quad . \quad (250)$$

is called the spherical density of the oblong projectile, and it measures the number of times the weight of the sphere is contained in that of the oblong projectile. We have

$$W = \tfrac{4}{3}\pi r^3 \delta,$$

in which δ is the weight of a cubic inch of the metal of the projectile, and r is in inches. Making $\delta = \frac{1}{4}$ lb., which is its approximate value, and taking $\pi = 3$, we have

$$W = r^3;$$

and substituting in (250), we have

$$S = \frac{W'}{r^3}, \quad . \quad . \quad . \quad . \quad . \quad . \quad . \quad . \quad (251)$$

which is generally taken as the measure of spherical density.

This value has increased from 2.0, when oblong projectiles were first introduced, to 3.0 in 1880, and 4.7 in 1894.

WEIGHT.—For the weight of a spherical projectile we have

$$W = r^3,$$

or the weight of a spherical projectile in pounds is equal to the cube of its radius in inches.

For an oblong projectile we have, equation (251),

$$W' = S \times r^3,$$

or the weight of an oblong projectile in pounds is equal to its spherical density, multiplied by the cube of the radius in inches. Hence having the date of manufacture of a projectile, its spherical density is known, and from this its weight. This rule gives very close approximations, and avoids the necessity of remembering anything except the spherical density. Since the weight is proportional to the cube of the calibre, the quotient $\frac{W}{r^3}$ will be constant for all similar projectiles.

173. Manufacture of Projectiles—Pattern—Flask—Molding—Gate and Riser.

PATTERN.—Cast-iron projectiles are made as follows: A pattern is first made of the shape of the projectile to be cast. Its diameter is slightly greater than that of the projectile, to allow for the shrinkage or contraction of the metal in cooling. It is also slightly conical, instead of cylindrical, on the exterior, to permit its ready withdrawal from the mold. In Fig. 178 the pattern is made in two parts, and each part is conical from *a* to *b*. The spindle *c* is used to support the pattern in molding, and to mark the position of the core in the mold.

It terminates in a conical bearing, *d*.

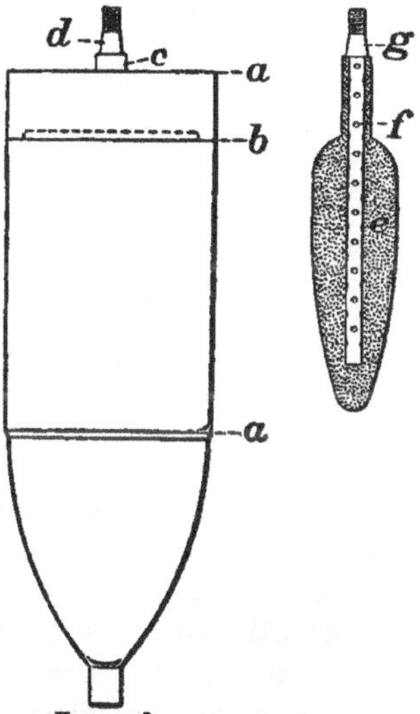

FIG. 178.

Core.—To form the interior cavity in a corea shot or shell, a second pattern or core *e*, Fig. 178, is required. This core is made of a mixture of sand and other substances which render it adhesive, and is formed upon a hollow spindle, *f*, which terminates in a conical bearing, *g*, of the same size as *d* on the shot-pattern spindle. The spindle *f* being hollow, and having holes along it, allows the gases formed by the contact of the melted metal with the core, to escape.

FLASK.—The pattern for the shot or shell is placed in a box called a flask, Fig. 179. This flask is made in parts

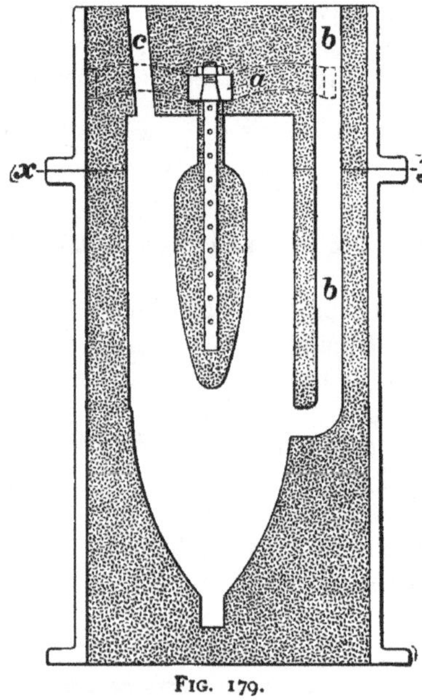

FIG. 179.

corresponding to the pattern of the projectile, and these parts are bolted together before casting. It contains a cross-bar, *a*, at the top, with a conical hole in it, into which the conical bearing of the spindle fits as shown.

MOLDING.—To form the mold, the part of the pattern containing the conical spindle, is seated in the cross-bar *a*, and this part of the flask, with pattern, placed on a board, the plane *xy* down. Cylindrical sticks *b* and *c* are placed as shown, and molding composition, composed of sand, clay, and carbonaceous material, rammed in. This part of the flask is now inverted, the remainder of the pattern, the flask, and the cylindrical stick *b* put in place, and molding composition rammed in from the opposite direction, filling the flask. The parts of the flask are then separated along *xy*, and the pattern and sticks withdrawn.

The core, having been separately molded, is then put in place, its spindle occupying the position shown, and being centred by the conical bearing in the cross-bar *a*.

The parts of the flask are next bolted together.

GATE AND RISER.—The channel *b* is called the gate. The metal is poured through it into the mold. In large projectiles, it generally enters low down, and in a tangential direction, to give a rotary motion to the metal as it enters the mold, and thus sweep the scoria and impurities to the centre and top.

The channel *c* is called the *riser*. It allows the escape of gas from the mold, and the collection of the scoria, and also allows fresh metal to be poured in to fill up cavities and make up for shrinkage due to cooling.

174. Operation of Casting—Kind of Iron Used—Position of Head and Base of Projectile in Mold—Steel Projectiles.

OPERATION OF CASTING.—The cast iron is melted either in cupola or reverberatory furnaces, and run into ladles, from which it is poured through the gate *b*, into the mold. The pouring is continued till the metal fills the riser *c*, and fresh portions of melted metal are added to the riser as the latter sinks.

As soon as the iron has cooled sufficiently, or set, the flask is removed, the core broken up, the spindle drawn out, and the projectile covered with the molding composition, to allow it to cool slowly. If a chilled projectile is to be cast, the chill is inserted in the mold before casting, as previously explained.

KIND OF IRON.—For small projectiles, since they cool rapidly, and become very hard, soft iron is employed. For large projectiles, a harder and tougher iron is used. The selection of the proper kinds of iron can only be determined by experience, and it is usual to prescribe a certain tensile strength for each kind of projectile, and to test specimens from a number of them, to see that they come up to the required standard.

POSITION OF HEAD AND BASE IN MOLD.—In Fig. 179 the projectile is shown cast head down, with the core-spindle projecting through the base. This position gives great density to the head, and less to the base. For armor-piercing shot, and in all cases where great strength of head is required, the casting is made in this position. But for shell, especially those with a base fuze, it is important to have a strong and sound base, as if it is weak or spongy, the gas from the powder-charge would penetrate the base, and burst the shell in the gun. In this case, therefore, the shell would be cast base down.

STEEL PROJECTILES.—Forged steel projectiles are cast as ingots, and are forged, and bored and turned to proper dimensions.

They are then tempered by a secret process, and are now made of such hardness and toughness, that they will penetrate the best armor-plates whose thickness does not exceed $1\frac{1}{8}$ times their diameter, without cracking or deforming the projectile.

Cast-steel projectiles are also tempered by a secret process.

175. Inspection and Proof of Projectiles—Quality of Metal—Shape and Dimensions—Eccentricity—Ballistic Test.

The objects of inspection are:

1. To test the quality of the metal;

2. To see that the shape and dimensions agree with those specified;

3. To see that the centre of gravity is on or near the longer axis of the projectile.

QUALITY OF METAL.—This is determined by testing specimens taken from different lots. For cast-iron projectiles, the soundness is tested by striking the projectile with a hammer, and a punch is used to determine the depth of any holes that may be discovered. Finally the shot or shell is subjected to water or steam pressure, applied to its interior. Any cracks or cavities will be detected by the escape of the water or steam. Chilled shot are struck with a hammer, at the junction of head and body.

For Steel Shot and Shell chemical analyses are made, to determine the composition. After the final treatment, the shot are cooled to about 40° F., and then suddenly heated by plunging them into a water-bath, at a temperature of 212° F. When they become uniformly heated to this temperature, they are suddenly plunged, with their axes horizontal, half way into a bath of water, at a temperature of 40° F. Any great initial strain to which they may have been subjected in tempering, will be detected by this treatment. The shell are subjected to an interior hydraulic pressure of 1000 lbs. per square inch.

SHAPE AND DIMENSIONS.—These are determined by templets and gauges. For example, the profile of the projectile is determined by using a templet of sheet iron or steel, correctly made, and applying it to the exterior of the projectile as in Fig. 180, *a* being the templet.

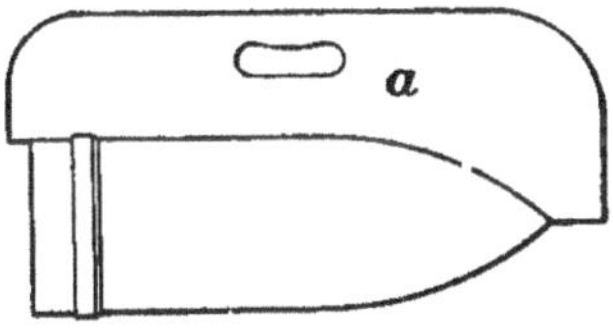

FIG. 180.

FIG. 181.

The diameters are determined by two rings, called ring gauges, Fig. 181. One of these has a diameter equal to the maximum that is allowed, and the other the minimum.

The maximum ring must pass over all the projectiles, and the minimum over none.

ECCENTRICITY.—When the centre of gravity of the projectile does not lie on the longer axis, the projectile is eccentric.

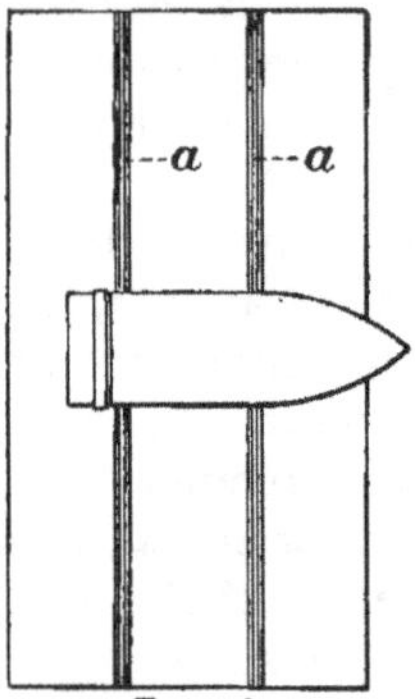

FIG. 182.

This eccentricity, if large, affects the flight of the projectile, causing irregularity, and hence its limits must not exceed a certain amount. To detect it, the projectiles are first placed on a rolling-table, which is an iron table having two parallel ribs, *a a*, Figs. 182 and 183, at a distance apart slightly less than the length of the cylindrical body of the projectile.

This table being leveled, if the centre of gravity of the projectile, does not coincide with the longer axis, the projectile when rolled on the ribs, as shown, will come to rest with its centre of gravity below that axis. If no eccentricity exists, the projectile will remain indifferently

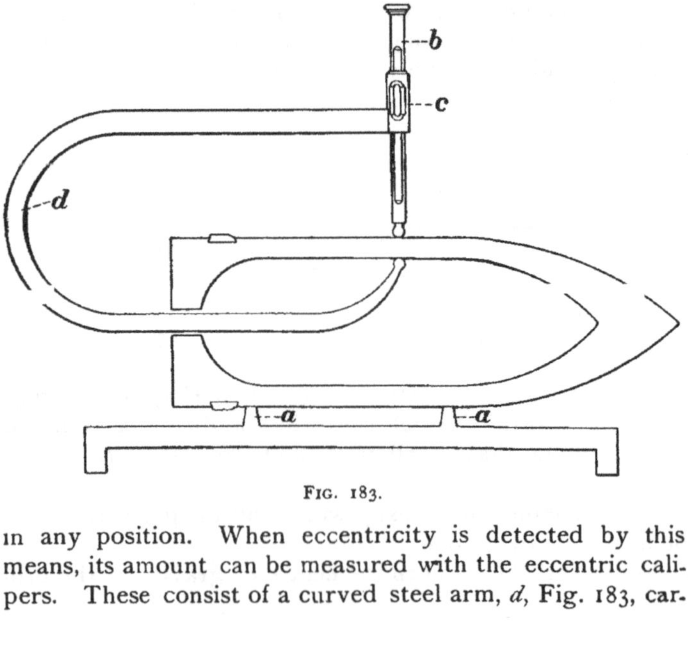

FIG. 183.

in any position. When eccentricity is detected by this means, its amount can be measured with the eccentric calipers. These consist of a curved steel arm, *d*, Fig. 183, carrying a sliding point, *b*, and a scale, *c*. The point is graduated in inches, and *c* is a vernier scale. The thickness of wall is read off on the scale, and that of the opposite wall also. One half the difference of the readings, gives the eccentricity.

BALLISTIC TEST.—Steel shot and shell are subjected to actual firing tests against armor-plates. For this purpose a certain number are taken from each lot manufactured. The shot are fired with a striking velocity of 1625 feet-seconds against a steel plate 1⅛ times the calibre of the gun in thickness, and the 12″ mortar shell against a 4½-inch steel plate, at an angle of 60°. The shot and shell must penetrate completely in each case, without breaking up.

ARMOR.

176. Kinds of Armor.

Armor may be divided into—

1. Chilled cast iron ;
2. Compound ;
3. Steel.

Wrought-iron armor is now obsolete.

Chilled Cast-iron Armor.—This armor, on account of its great weight, is used only on land, in the form of turrets.

It is manufactured by Gruson, of Germany, and is cast in large blocks of the proper shape, the outer face of the blocks being chilled, and thus acquiring great hardness. These blocks are built into turrets, whose form is shown in the text-book of the Engineering course. It depends for its great resistance upon the following:

1. Its intense surface hardness prevents the entrance of the projectile ;
2. Its great mass distributes the effect of the blow ;
3. Its curved form deflects the projectile.

Its resistance to penetration is greater than that of any other armor.

Compound Armor.—Wrought-iron armor was the first kind adopted, and it had sufficient resistance to keep out the ordinary cast-iron projectiles, but was readily penetrated by those of chilled cast iron.

It became necessary, therefore, to harden the face of the armor in order to break up these projectiles. This led to the introduction of the compound armor, which has been extensively used in England, and is still employed there.

It is formed by welding a hard steel face to a wrought-iron back, and the armor is distinguished into two makes, according to the method of welding adopted.

The two methods of manufacture are:

1. *Cammell & Co.—Wilson's Patent.*—The firm of Cammell & Co. manufacture compound armor by the Wilson patent, which consists in forming a back of wrought iron, by forging or rolling. This back is placed in a furnace, raised to a welding heat, and while at this temperature a layer of melted steel is run on one of its faces. After partially cooling, the compound plate is removed from the furnace and passed between heavy rolls, to reduce it to the proper thickness, and improve its quality. The steel face is then treated to remove strains.

2. *Brown & Co.—Ellis Patent.*—This method consists in forming the wrought-iron back and hard steel face separately. These are then placed in a furnace parallel to each other, and a short distance apart, and raised to a welding heat. Melted steel is then run between the plates, welding them together.

177. Steel Armor—History—Improvements—Latest Steel Plates—Harvey and Tresidder Processes of Surface Hardening.

HISTORY.—Steel armor, when first introduced, was hard and brittle, and broke up under the action of projectiles. The percentage of carbon was then reduced, and the plates no longer broke up, but allowed the projectiles to penetrate. It was shown, however, by the Italian experiments at Spezzia in 1876, that the low steel plate was superior to those made of wrought iron. To prevent the penetration of projectiles, the face of the plate was tempered in oil, or oil-hardened, and the plate then annealed, to remove internal strains. As projectiles improved, however, this hardness was not great enough to resist penetration, and hence new improvements were made.

IMPROVEMENTS.—These consist generally—

1. In having better facilities for the mechanical treatment of steel in large masses, such as heavy hammers, forging-presses, etc.

2. Combination of the steel with other ingredients, such as nickel, which increases its toughness and tenacity, and consequently decreases its tendency to break up.

3. Special treatment of the steel, as the Harvey process, by which the face is made hard, while the back retains its toughness.

LATEST STEEL PLATES.—From the time of the Italian experiments in 1876, up to those at Annapolis in 1890, competitive trials have been going on between the steel and the compound armor.

The tests at the latter place, showed the superiority of the all-steel plate, and it has been adopted in our Navy.

The steel plates which have given the best results up to the present time, are known as high-carbon nickeled steel and low-carbon nickeled steel, referring to the relative quantity of carbon in the alloy, although it is small in each case. The high-carbon nickeled-steel plate has so far given the best results. All the plates are hardened on the surface by the Harvey process.

HARVEY PROCESS.—This consists in carbonizing to a higher degree the outer surface of the plate, for a certain depth, depending on the dimensions of the plate, and then hardening this surface.

The process is as follows: The plate is embedded in sand and clay in a furnace, leaving a certain thickness exposed. The furnace is then filled with carbonaceous material, well packed over the exposed portion of the plate, and the whole raised to a high temperature, which is maintained for some time. The material is then removed, and when the plate has cooled sufficiently, its surface is hardened by the application of cold water.

TRESIDDER PROCESS.—This process is used in England, and consists in heating the plate to a certain temperature, and applying cold water to the surface under heavy pressure, and through a number of small holes, thus producing numerous small streams of water acting together. The idea is that the force with which the water is applied, brings it directly into contact with the hot metal, thus cooling it rapidly, and preventing the formation of the envelope of steam (spheroidal state) around the particles of water, and the consequent slow rate of cooling, which would be the case if the water were applied without pressure.

178. Effect of Projectiles on Armor — Early Armor — Compound Armor—Steel Armor.

Armor yields in two ways:

1. By racking, or breaking up;
2. By punching.

EARLY ARMOR—*Punching.*—Wrought-iron armor was at first generally attacked by heavy projectiles, moving with low velocities, as with the old 15-inch smooth-bore. Although the armor itself was soft, and yielded naturally by punching, the effect of these projectiles was to break its fastenings, and cause it to rack. As guns and projectiles increased in power, however, this armor yielded entirely by punching. That is, the effect of the blow was to punch a hole through the armor, and if the bolts held, no part of the plate beyond that struck, was affected.

Racking.—The object of armor, however, being to keep out projectiles, this defect led to the introduction of the harder kinds, as shown. This hard armor, instead of yielding locally to the blow of the projectile, distributed the energy of that blow over a greater mass of the plate, and when the energy was sufficient, the plate broke up, or was racked.

COMPOUND ARMOR.—The effect of projectiles upon this armor, is to break up the hard steel face, and to punch the wrought-iron back. The punching effect is, however, very much diminished by the energy lost in breaking up the steel face. The difficulty with this armor seems to be that the welding of the steel face to the wrought-iron back is uncertain, and hence after a few blows, the steel face breaks up, and separates from the wrought-iron back. Also the face being elastic, and the back having no elasticity, the former, after being struck, tends to recover its first position, while the latter does not. This increases the tendency of the front and back to separate. It follows that a compound plate must have a rigid support in rear for the best results.

STEEL ARMOR.—This armor at first yielded by racking, owing to its hardness and brittleness. Oil-hardening, however, decreased its brittleness and increased its toughness, so that the effect of the improved projectile was to punch it.

The modern processes of surface-hardening, however, have combined hardness with toughness, so that, at the present day, the armor resists both racking and punching to a remarkable degree. The history of the improvement is obvious: decrease of hardness and brittleness to decrease racking, and afterwards a combination of hardness and toughness to prevent both punching and racking.

179. Backing—Fastenings for Old Armor.

BACKING.—That portion of an armored structure directly in rear of the plate, is called the backing, and the character of this backing depends upon that of the armor.

Chilled Cast Iron.—This armor has no backing, or rather the cast iron itself may be regarded as forming the backing for the hard chilled face, since the thickness and mass are great.

Compound Armor.—For this armor, a rigid backing is required, to give the best results. For land structures, a rigid backing may be used, since weight is no objection, but for ships, such a backing is impossible, and it would therefore seem that this armor cannot give the best results when used under such circumstances. It is, however, the standard armor of the British Navy. An objection to the rigid backing is that it tends to cause racking of the armor-bolts. When an elastic backing is used with this armor, it allows the plate as a whole to yield to the blow of the projectile. But, as before stated, the steel face, being elastic, returns to its former position after the blow, while the wrought-iron back does not, and hence there is a tendency for the face and back to separate. The rigid backing, on the other hand, has the opposite effect.

Steel Armor.—This armor, being elastic, requires an elastic backing, and not a rigid one, since the elastic backing allows the plate to yield to the force of the blow, and the elasticity of the plate causes it to return to its former position.

Fig. 184 shows the arrangement of backing as generally used, and may consist of one or two thicknesses of timber, *b*, placed against the steel sides, *a a*, of the vessel.

FASTENINGS FOR OLD ARMOR.—The original armor-bolt for wrought-iron armor, was shaped as shown in Fig. 185.

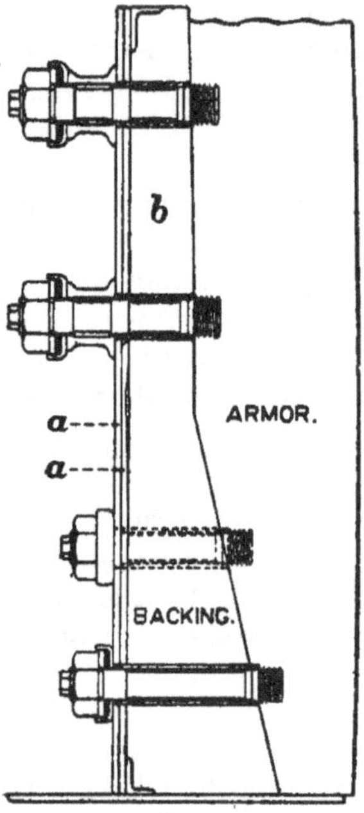

FIG. 184.

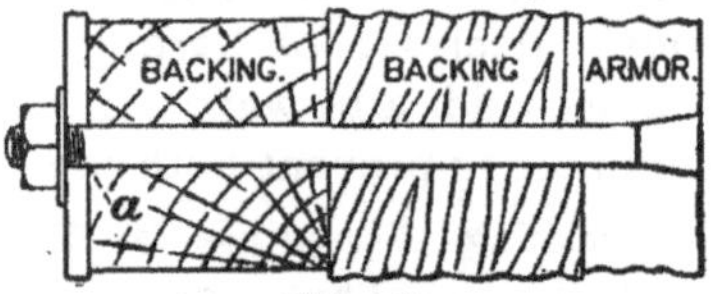

FIG. 185.

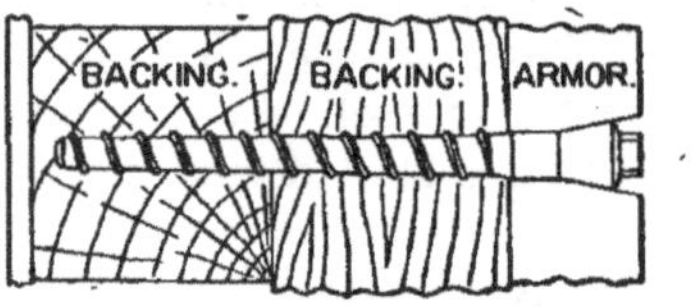

FIG. 186.

The objections to this bolt were:

1. If under water, it leaked.

2. When the plate was struck, the bolt would snap at the bottom screw-thread *a*, and the nut would fly off, acting as a projectile.

The French made the first attempt to remedy this, by using an ordinary wood-screw, Fig. 186, which screwed into the backing, but did not pass completely through it. This prevented leakage, and the flying of the bolt-heads about the deck.

The English changed the arrangement of the armor-bolt

by placing a rubber washer, *k*, under the nut, and cutting a plus thread on the bolt, as in Fig. 187. In this case the bolt

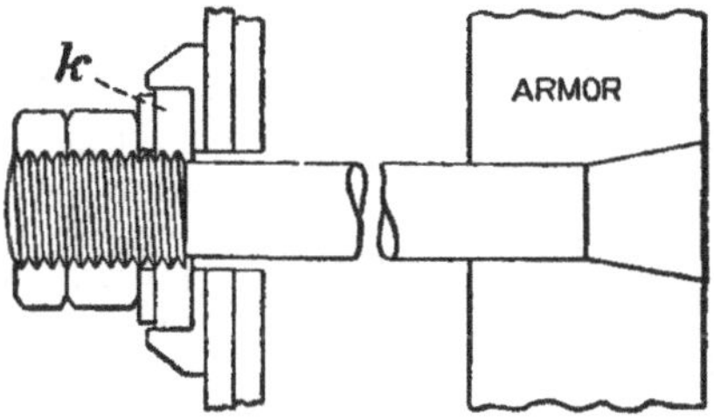

FIG. 187.

has the same strength throughout, the thread is not a source of weakness, and the rubber washer allows a certain play in the direction of the length, and thus prevents snapping from the sudden strain, and it also prevents leaking.

180. Improved Fastenings for Iron Armor — For Steel Armor— Tests of Armor-plates.

IRON ARMOR.—In addition to the strains brought to bear upon the bolt in the direction of its length, causing elongation and snapping in the older forms, there is a cross-strain due to the displacement of the armor sideways. To obviate this, the English spherical-headed bolt was devised, as shown in Fig. 188. This shape of head and nut, causes the strain to be always along the axis of the bolt; and to allow the bolt to take its normal position when the plate is displaced laterally, a clearance is made around it.

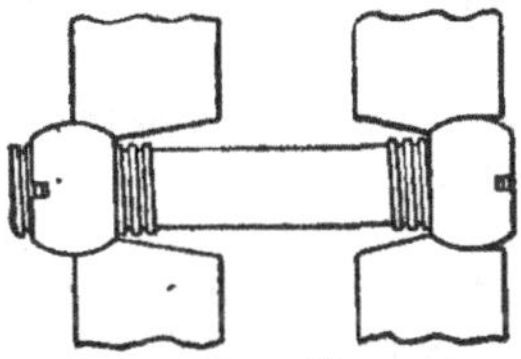

FIG. 188.

STEEL ARMOR.—With this armor, or with the compound armor, where it is necessary to preserve the steel or hardened face intact, the bolt must not pass through the plate. It is therefore screwed for a short distance into the back of the plate. In naval vessels, the backing is comparatively thin, and hence the bolt must be lengthened by some means, so that the stretch per unit of length may not be sufficient to break the bolt. The following arrangement is adopted in the Navy (Fig. 189).

a is the armor-plate; *b*, the backing; *c*, the armor-bolt, reduced in diameter as shown, to give lateral play in the iron pipe *d*, which passes through the backing and furnishes a seat for the bolt; *e* and *f*, rubber washers which are set up by the pressure of the nut *g*, and prevent leaking; *h*, a sleeve, whose object is to increase the length of the bolt, so that the elongation per unit length

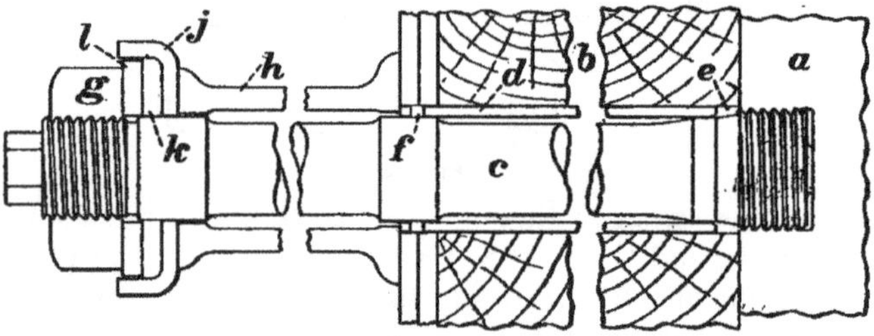

FIG. 189.

when the plate is struck, may prevent fracture; *j*, a cup-shaped washer, containing the rubber washer *k*, and the iron washer *l*. The washer *k* gives elasticity to the whole system. The sleeve *h* distributes the pressure of the nut *g* over a large area of the sides of the ship, and also, as stated, increases the length of the bolt. The number of bolts is greater for the steel and compound armor, than for the old wrought-iron armor.

TESTS OF ARMOR-PLATES.—The test of armor-plate prescribed by the Ordnance Department is as follows: One plate is selected from a lot, and is bolted to an oak backing 36 inches thick, properly supported, with rubber washers placed between the steel washers of the bolts in rear. The calibre of the gun is to that of the plate as 1 : $1\frac{1}{8}$; that is, for a 9-inch plate an 8-inch gun, etc. One armor-piercing shot is fired from this gun, so that the centre of the shot-hole shall not be nearer any edge of the plate than $2\frac{1}{2}$ calibres.

The projectile must have the following striking-energy:

For an 8-inch gun	3000 foot-tons
For a 10-inch gun..........	5000 "
For a 12-inch gun..........	7643 "

Under these conditions, the whole of the projectile must not get through the plate, nor must the plate break up, or pieces be detached, or cracks produced which expose the backing to view.

181. Penetration of Armor—Wrought Iron—Steel.

WROUGHT IRON.—Most of the formulas for penetration have been deduced for wrought-iron armor, on account of the length of time it has been in use, and the numerous experiments that have been made upon it.

In deducing these formulas two different hypotheses have been adopted:

1. That the projectile acted as a punch, separating a disk of metal from the plate. In this case, the resistance to be overcome, was the resistance of the metal to shearing along the circumference of this disk, and hence the energy of the projectile to overcome this resistance was estimated per inch of shot's circumference, and was obtained by dividing the total energy by the circumference, or $\frac{E}{\pi d}$.

2. That the projectile acted as a wedge, forcing the particles of the metal apart. In this case the penetration is proportional to the energy per unit of area of cross-section, or $\frac{E}{\pi r^2}$.

The principal formulas deduced under the first hypothesis are:

$$t^2 = \frac{E}{\pi d k}; \qquad \text{(Fairbairn's)} \quad . \quad . \quad . \quad (252)$$

$$t^{2.035} = \frac{E}{0.86\pi d}; \qquad \text{(English Admiralty)} \quad (253)$$

$$t^{1.868} = \frac{E}{k\pi d}; \qquad \text{(Muggiano)} \quad . \quad . \quad . \quad (254)$$

And under the second hypothesis:

$$t^{1.3} = \frac{E}{5.8168 d^{1.5}}; \qquad \text{(de Marre's)} \quad . \quad . \quad . \quad (255)$$

$$t = \frac{v}{608.3}\sqrt{\frac{p}{d}} - 0.14d; \quad \text{(Maitland's)} \quad . \quad . \quad . \quad (256)$$

$$t^{\frac{4}{3}} = \frac{E}{4.156d^{\frac{5}{3}}}; \quad \text{(Krupp's)} \quad . \quad . \quad . \quad . \quad (257)$$

$$t^{1.4} = \frac{pv^2}{2265464d}. \quad \text{(Gâvre)} \quad . \quad . \quad . \quad . \quad . \quad (258)$$

In these formulas

t is the thickness of wrought iron, in inches, which the projectile will penetrate;

E, the energy of the projectile in foot-tons;

d, its diameter in inches;

p, its weight in pounds;

v, its striking velocity in feet-seconds;

k, a constant.

STEEL ARMOR.—For steel armor, it was customary to calculate the penetration in a wrought-iron plate of the same thickness, and add a certain percentage of increase of resistance, varying from 10 to 30 per cent. This method is not satisfactory, as steel armor varies greatly in resistance according to treatment; and for the modern Harveyized plates, penetration seldom occurs, owing to the hard face, unless the gun greatly overmatches the plate.

The formulas generally used are those of de Marre, as follows:

For soft plates of Creusot steel, backed,

$$t^{0.7} = 0.0009787\frac{p^{\frac{1}{2}}v}{d^{\frac{3}{4}}}. \quad . \quad . \quad . \quad . \quad . \quad (259)$$

For the steel plates used as protection against steel shell from rapid-fire guns, unbacked,

$$t^{0.7} = 0.000734\frac{p^{\frac{1}{2}}v}{d^{\frac{3}{4}}}. \quad . \quad . \quad . \quad . \quad . \quad . \quad (260)$$

For the wood backing, with plate in front of it,

$$t^{0.6} = 0.006168\frac{p^{\frac{1}{2}}v}{d^{0.9}}. \quad . \quad . \quad . \quad . \quad . \quad (261)$$

Captain Orde Browne gives a rule which enables some idea to be formed of the relative powers of guns against armor, as follows: *The penetration of a projectile in wrought-iron armor is one calibre for every thousand feet striking velocity.*

For example, a 10-inch projectile, striking with a velocity of 1200 feet-seconds, will penetrate 1.2 calibres, or 12 inches.

CHAPTER V.

FUZES AND PRIMERS.

FUZES.

182. Definition—Classification—Time Fuzes—Requisites—Difficulties.

DEFINITION.—Fuzes are the means used to ignite the bursting charge of a projectile at any point of its flight, or upon impact.

CLASSIFICATION.—They are classified according to their mode of action, into—

1. Time;
2. Percussion;
3. Combination;
4. Delayed action fuzes.

TIME FUZES.—A time fuze is one which ignites the bursting charge at some fixed time after the projectile has left the muzzle, and it consists generally of a column of composition, whose rate of burning is known, which is set on fire by the discharge of the piece, and whose time of burning is regulated by the length of the column.

REQUISITES.—The requisites of a good time fuze are:

1. Its rate of burning should be uniform, and not affected by storage or changes of climate.

2. It must be safe in handling, and certain in its operation.

These conditions are very difficult to fulfil, and hence a good time fuze is perhaps the most difficult one to obtain.

Recent improvements have, however, done much to obviate the difficulties formerly experienced.

DIFFICULTIES.—The difficulties in making a reliable time fuze are:

1. To obtain a column of composition whose rate of burning is uniform.

The rate of burning of a composition depends upon its density, trituration, composition, and degree of moisture, as explained in Interior Ballistics. For the same composition, it is very difficult to obtain a uniform density throughout a long column. The old method of preparing a time-fuze was to place a small quantity of the composition in the fuze-case, and strike it a certain number of blows with a mallet and drift, and repeat this operation till the fuze was completed. This did not give uniform density.

A second method was to subject it to hydraulic-pressure, but this also failed. The method now in use gives better results, and will be explained.

2. For long times of burning, a long column of composition is required, and the results obtained by the old method were so unsatisfactory, that the ingredients of the composition were varied, in order to give a decreased rate of burning and thus shorten the column. This changed irregularly the rate of burning, increased the difficulty of preservation, and also increased the residue.

3. It was found impossible to guard against the changes, due to storage, climate, etc., as they affected both the composition, and the fuze-case.

4. After a uniform rate of burning is obtained, the pressure of the air on the composition, during flight, changes this rate.

5. With modern breech-loading guns, and high velocities, a small error in burning, increases the error in bursting of the projectile.

6. The flame from the powder-charge will no longer ignite the fuze.

183. Time Fuzes—Difficulties, How Overcome.

To overcome these difficulties, the following changes in manufacture have been made:

Uniform Rate of Burning.—To obtain a uniform rate of burning, and a column of composition of sufficient length to answer for the greatly increased ranges, a lead tube 0.62

inch in diameter is filled with mealed powder. The ends are then closed, and the tube, with the enclosed powder, drawn out by a process similar to that of wire-drawing, till the diameter is decreased to 0.152 inch. The rate of burning is tested by burning inch lengths of this compressed powder, and its rate is found to be very uniform. In this case, the pressure is applied in the direction of the shortest dimension of the column, so that it is more uniform in its effect; and if there be any difference in density, this difference, is neutralized by the burning of the column at right angles to the direction of the pressure, so that the same variation of density exists throughout each cross-section.

Changes of Climate, etc.—These have less effect upon the composition, because it is contained in a metallic case which is less subject to change.

Variation due to Pressure of Air in Flight.—This is corrected for, as far as possible, by graduating the fuze temporarily with the rate of burning as determined at rest, and then correcting this graduation, by actual test, on the firing-ground.

Ignition of Fuze.—With the old fuzes, and muzzle-loading projectiles, there was always a space between the surface of the bore and that of the projectile, through which the flame from the powder-charge could pass to ignite the fuze. This was called the "windage." With modern breech-loading projectiles, this space is closed by the band of the projectile, and hence the flame cannot pass through to ignite the fuze. The arrangement for accomplishing this with the modern time fuze, will be explained.

184. Older Forms of Time Fuze in Use—Mortar Fuze—Sea-coast Fuze—Bormann Fuze.

The older forms of time fuze still in use in the U. S. service are:

The mortar-fuze;
The sea-coast fuze;
The Bormann fuze.

MORTAR FUZE.—This is used in the old siege and sea-coast smooth-bore mortars. It consists (Fig. 190) of a conical case or plug, *a*, of wood graduated on the exterior into inches and tenths. On the interior there is a cylindrical cavity, *b*, bored out nearly to the bottom, and filled with the fuze-composition, driven as explained. The top of this cavity is enlarged at *c*, and filled with mealed powder moistened with alcohol, to insure ignition from the flame of the powder-charge. This is covered with a paper cap, *c'*, on which is marked the rate of burning in seconds per inch. The graduations begin at *d*, and stop at *e*. To prepare this fuze for use, the paper cap is removed, the fuze cut at the proper division, counting from the top, by sawing off the lower end, or better by boring a hole into the composition with a gimlet, as at *f*, as this prevents the composition from being dislodged by the shock of discharge, and finally by driving with a drift the fuze into the fuze-hole of the shell.

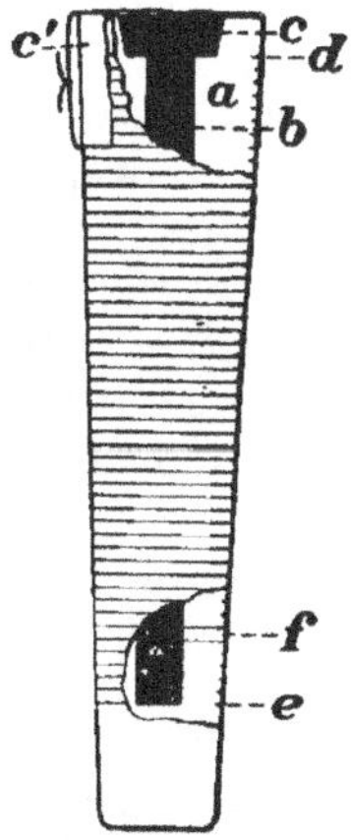

FIG. 190.

THE SEA-COAST FUZE.—This was used in the old smooth-bore guns, and at present in the 15-inch Rodman gun.

For ordinary firing, this fuze (Fig. 191) is composed of a

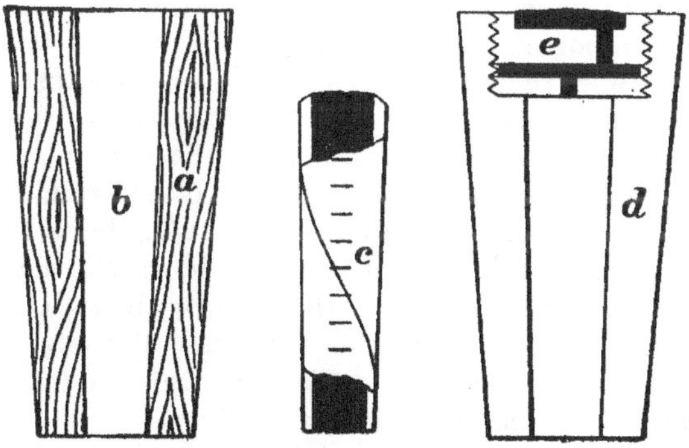

FIG. 191.

conical wood plug, *a*, with a conical hole, *b*. In this hole is placed the fuze, *c*, which is contained in a conical paper case.

The fuzes are of variable composition, and are marked on the exterior according to their time of burning. For ricochet firing over water, and for heavy charges, a brass fuze-plug, *d*, of the same shape is used, the distinctive feature being the water-cap, *e*, which is a brass cap, having a zigzag channel, filled with mealed powder. The shape of this channel renders the access of water difficult, and hence prevents the extinction of the composition.

THE BORMANN FUZE.—This was used with spherical shell and shrapnel, in the field service.

It consists (Fig. 192) of a pewter fuze-case, *a*, containing a ring of composition, *b*. Over this ring, *b*, lies an arc, *c*, graduated in seconds. At the zero end of the arc, this ring,

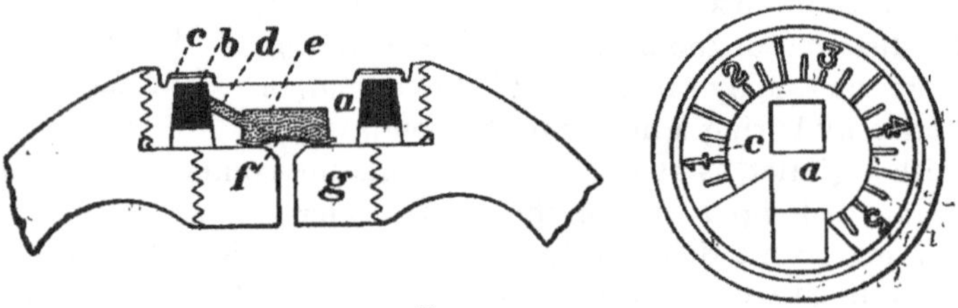

FIG. 192.

b, communicates with a channel, *d*, filled with fine powder, leading into a chamber, *e*, filled with the same powder, which is supported by a tin disk, *f*. The composition, *b*, is pressed into its recess in the direction of its shortest dimension, and burns around the ring at right angles to this direction. Hence the fuze possesses one of the good qualities of the modern time-fuze. The other end of the ring of composition, has no communication with the chamber *e*. Owing to its shape, and the material of which the case is made, this fuze is liable to be driven into the shell by the shock of discharge; and to prevent this, and increase the effect of the bursting-charge in the projectile, a wrought-iron disk, *g*, is screwed into the fuze-hole, below the fuze.

The action of the fuze is as follows: If required to burn any given time, say four seconds, the case is cut at the mark 4, exposing the ring of composition.

This composition then burns in both directions; but having no communication with the chamber *e*, in the direction toward 5, it will burn for four seconds, and then fire the charge.

The objections to the fuze are:

1. Its time of burning is too short for modern ranges;
2. It is difficult to ignite by the flame from the charge;
3. If once cut, it cannot be used for a greater time of flight.

The modern time fuze is not used, except in combination with the percussion fuze.

185. Percussion Fuzes—Requisites—Essential Parts.

A percussion fuze is one which is prepared for action by the shock of discharge, and which acts by the impact of the projectile.

REQUISITES.—The requisites of a good percussion fuze are, that it shall be safe in handling, and certain in its operation.

Safety in Handling.—This requires:

1. A safety device which will prevent accidental discharge in store, transportation and handling, or from accidental shock, such as dropping the projectile.

2. A safety device which will prevent accidental discharge in loading, and which will be released only on firing the piece.

These may be combined in one, as will be seen later.

Certainty of Action.—This requires:

1. That all the parts of the fuze be protected from clogging, by the action of the bursting-charge in transportation, and by other causes, such as dust, etc.

2. A safety device which will prevent relative motion of the parts of the fuze during flight.

ESSENTIAL PARTS.—The essential parts of every percussion fuze are:

A case, to contain all the moving parts and protect them;

A plunger, which is moved backward or forward on impact, and which fires the fulminating composition;

A fulminating composition, which is fired by the impact of the plunger;

The priming, which is a charge of powder ignited by the fulminate, and which ignites the bursting-charge;

A safety device in transportation;

A safety device in loading;

A safety device in flight.

186. Percussion Fuzes in U. S. Service—Classification—Hotchkiss Front Percussion Fuze—Action of Fuze—Safety Devices.

CLASSIFICATION.—Percussion fuzes are classed according to the position they occupy in the projectile, as:

1. Front fuzes, which are inserted in the head of the projectile, at the point of the ogive;

2. Base fuzes, which are inserted in the centre of the base.

The front-fuze is sometimes used with field projectiles, and generally when penetration is not required.

The base fuzes are used with armor-piercing projectiles, and generally where penetration is required, and where the head of the projectile must have great resistance.

The front fuze has the advantage of having the bursting charge of the projectile thrown toward it on impact, and there is no danger of the flame from the charge in the gun entering the cavity of the projectile, and causing premature explosion.

The percussion fuzes used in the U. S. service are the Hotchkiss Front and Hotchkiss Base Fuzes, or modifications of them.

HOTCHKISS FRONT PERCUSSION FUZE.—This fuze consists of a brass case, *a*, Fig. 193, threaded on the exterior for screwing into the projectile. The upper end is closed by a screw cap, *b*, carrying a projecting point, *c*. In the case *a*, is a plunger, *d*, composed of a brass case, *e*, and a lead body, *f*, in which is a brass wire, *g*. The central part of the lead body carries the priming charge of powder, *h*, and on the top of this priming is the fulminate, *i*. The brass case *e*, encloses the lead body *f*, to prevent the upsetting laterally of the plunger, by the shock of discharge, and its consequent wedging in the fuze-case.

When the plunger is inserted in the case *a*, the brass wires *g*, occupy the position shown, and the rear end of *a* is closed by a conical lead plug, *j*, bearing against the wires.

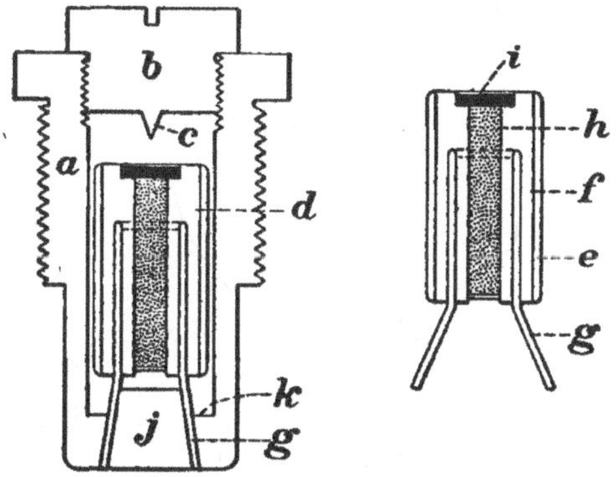

FIG. 193.

ACTION OF FUZE.—When the piece is fired, the shock of discharge causes the lead plug *j* to be dislodged from its seat in the fuze, and to fall into the cavity of the projectile. The plunger *d* then moves to the rear, and rests, during flight, upon the shoulder *k*, at the bottom of the fuze cavity. Upon impact, the plunger *d* is thrown forward, the fulminate *i*, striking the point *c*, and thus firing the priming charge *h*, which fires the bursting charge.

SAFETY DEVICES IN TRANSPORTATION AND LOADING.—These are combined in this fuze, and are formed by the conical lead plug, bearing on the brass wire *g*.

SAFETY DEVICE IN FLIGHT.—In all percussion-fuzes, the plunger has a tendency to move forward in the fuze cavity during flight. This is due to the fact that the projectile is retarded by the resistance of the air, while the plunger, not being subjected to this action, is not affected by it, and its only retarding force is friction, which is very small. If the plunger moves forward, it will either explode the fulminate during flight, or else if the sensitiveness of the fulminate be diminished to prevent this, it will be so close to the point *c*, on impact, that it may not acquire energy sufficient to cause the explosion at that time. In this fuze, the safety device in flight, is provided by the wires *g*, which spread outward when the lead plug *j* is dislodged, and thus prevent the forward motion of the plunger till impact.

187. Hotchkiss Base Percussion Fuze — Action of Fuze — Safety Devices.

HOTCHKISS BASE PERCUSSION FUSE.—The Hotchkiss base percussion fuze consists of a brass case, *a*, Fig. 194,

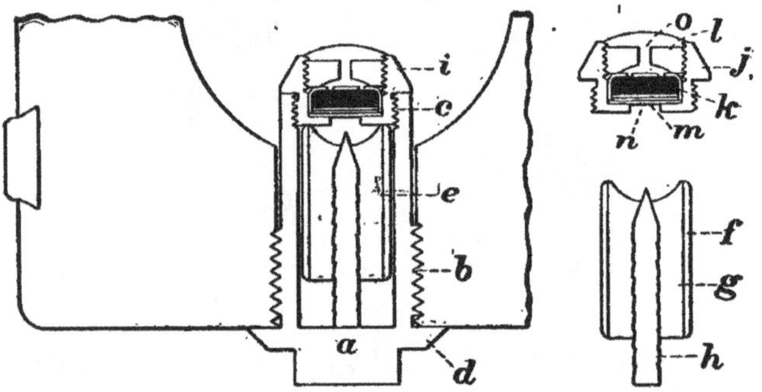

FIG. 194.

of the shape shown. It is threaded on the exterior at *b*, for screwing into the projectile, and on the interior at *c*, for the cap, which carries the fulminate. The flanges at *d* are made thin, to act as a gas-check, and prevent the entrance of the gas from the charge of the gun into the fuze-cavity, thus prematurely exploding the projectile.

In the case *a* is a plunger, *e*, composed of a brass jacket, *f*, a lead body, *g*, and a firing-pin, *h*. The combination of lead and brass is for the purpose previously described. The firing-pin, *h*, is made of steel, roughened on the outside, and the lead body is cast around it, so that before firing its point is slightly below the upper surface of the plunger.

The upper part of the fuze-case carries the screw cap *i*, which is composed of the part *j*, which screws into the fuze-case, the fulminate *k*, the screw cap *l* closing *j*, and the safety disk of copper *m*.

ACTION OF THE FUZE.—When the piece is fired, the shock of discharge causes the heavy plunger *e* to slide to the rear along the pin *h*, taking the position shown in Fig. 195.

The action of the brass casing of the plunger is to prevent the spreading of the lead body, and consequently cause the latter to take a firm hold on the firing-pin.

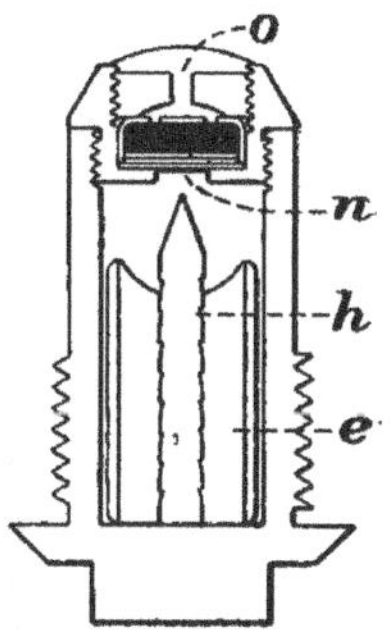

FIG. 195.

When the projectile strikes, the plunger is thrown forward, the point of the firing-pin passing through the hole *n* in the screw cap, strikes and explodes the fulminate, the flame from which passes through the hole *o* into the interior of the projectile, and ignites the bursting-charge.

SAFETY DEVICES IN TRANSPORTATION AND LOADING.—These are combined in this fuze and consist of:

1. The projecting pin *h* being held firmly in the lead body of the plunger, with its point below the upper surface of the latter, so that it requires the shock of discharge to force the plunger down over the pin, and allow its point to project.

2. Making the lengths of the plunger and projecting portion of the pin *h* such, that before firing they are held tightly in position in the case, and hence the plunger cannot acquire any motion which might force it along the pin.

SAFETY DEVICE IN FLIGHT.—This is provided by the copper disk *m* on the bottom of the fulminate cap, so that if the point of the pin should touch the cap in flight, it will not cause an explosion.

188. Combination Fuzes — Requisites — Frankford Arsenal Combination Fuze—Action of Fuze.

A combination fuze is one which contains both a time and a percussion fuze in the same case, and is intended to increase the chances of bursting the projectile, and of readily and quickly varying the kind of fire.

REQUISITES.—A good combination fuze must combine the requisites of both time and percussion fuzes, without being too bulky or too expensive.

THE FRANKFORD ARSENAL COMBINATION FUZE.—This fuze is used in the U. S. service for field shrapnel, and consists (Fig. 196) of a case, *a*, of bronze, the front portion of which carries the time fuze, and the rear portion the percussion fuze.

The time fuze is composed of the plunger *b*, the firing-

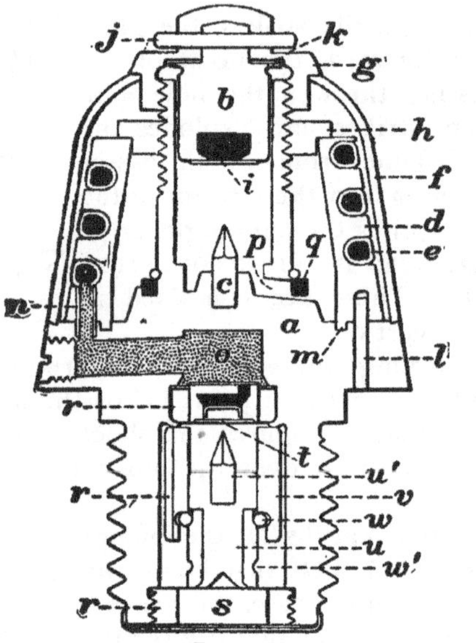

FIG. 196.

pin *c*, the cone *d*, the time-train *e*, the cover *f*, cap *g*, and clamping-nut *h*.

The plunger *b* is cylindrical in shape, and contains the fulminate *i*, in a recess at its base. Its upper extremity is pierced to receive a safety-pin, *j*, and there are five radial lugs, *k*, Fig. 197, which support the plunger on the top of the fuze body, and prevent it from falling against the firing-pin *c*, when the safety-pin at *j* is removed, before loading.

The firing-pin *c* is of steel, inserted into the body of the fuze at the bottom of the plunger channel.

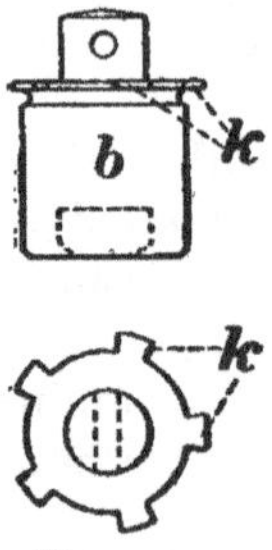

FIG. 197.

The cone *d* is an alloy of soft metal, held in place on the fuze-body by the clamping-nut *h*, and a groove *m* at the bottom, and is prevented from turning by a steel pin, *l*.

The lip *m* on the bottom of the cone, entering the groove in the body, acts as a gas-check to prevent ignition of the powder in the tube *n*. On the exterior of the cone *d*, is a left-handed groove which carries the time-train *e*, and this time-train communicates at its lower end with the priming-charge in the tube *n*, and thence with the chamber *o*.

The time-train *e* is formed, as previously described, of a lead tube, filled with mealed powder, and wire-drawn.

The cover *f* is of brass, and is held in place by the cap *g*, and prevented from turning by a small pin projecting from the body *a*, and fitting in a slot in its lower edge. On the exterior of the cover is a left-handed groove, corresponding to that on the time-cone *d*, and this groove is pierced with holes numbered from 1 to 15, corresponding to the number of seconds, the spaces between the holes being divided into five equal parts.

The percussion fuze is a modification of the Hotchkiss base fuze previously described, and consists of the primer in front; a plunger-spindle, *u*, carrying a firing-pin, *u'*; a plunger-sleeve, *v*; a safety-ring of brass, *w*; and a safety-disk of copper, *t*; the fuze being closed in rear by the screw-plug *s*, and this screw-plug, and the exterior of the plunger sleeve *v*, being grooved longitudinally, for the passage of the flame from the chamber *o* to the bursting-charge.

ACTION OF FUSE.—1. *As Time Fuze.*—Suppose the fuze is to burn 12 seconds. A hole is punched through the cover, time-train, and cone, into the interior of the fuze, at the 12-seconds mark. Just before loading, the safety pin is removed from the hole *j*. This allows the time plunger *b* to rest on the top of the fuze-body, where it is held by the five radial lugs, *k*. The projectile is now inserted in the gun. By the shock of discharge these five lugs are broken, and the time-plunger *b* is thrown to the rear, its primer striking the firing-pin *c*, which explodes the fulminate. The flame from the fulminate passes through the four radial holes *p*, at the base of the fuze body, and ignites the ring of compressed powder *q*. The flame from this powder, ignites the fuze composition at the hole marked 12, which has been punched through the time-cone *d*, and after burning for twelve seconds, this ignites the priming charge in the tube *n* and chamber *o*. The flame from this charge passes down along grooves *r* in the percussion fuze body and screw *s*, and ignites the charge.

2. *As a Percussion Fuze.*—When the piece is fired, the plunger sleeve *v* slides relatively to the rear, against the resistance of the safety ring *w*, and this ring is pushed from its groove, and along the spindle *u*. When the plunger sleeve *v* reaches its extreme rear position, the safety ring *w* slips into the groove *w'*, and, as its diameter has been increased by passing over the plunger spindle, it now fits into the groove in the plunger sleeve, and locks the spindle and sleeve together. The point of the firing-pin *u'* now projects beyond the plunger sleeve, and on impact, the sleeve and spindle are thrown forward, exploding the primer.

Safety Device in Transportation.—The safety pin *j*, for the time-fuze, and the plunger sleeve *v*, and safety ring *w*, for the percussion fuze.

Safety Device in Loading.—The radial lugs *k* for the time fuze, and the sleeve and ring, as before, for the percussion fuze.

Safety Device in Flight: Percussion Fuze.—The copper disk *t*.

189. Delayed-action Fuzes—The Merriam Delayed-action Fuze—Action of Fuze.

A delayed-action fuze is one which is prepared for action by the shock of discharge, and whose final action is retarded till the projectile has passed through the object or reached a certain position where its explosion will be most effective.

THE MERRIAM DELAYED-ACTION FUZE.—The principles of this class of fuzes will be best explained by the description of one of them which has been tried—the Merriam Fuze.

This fuze consists of a case or body, *a*, Fig. 198, threaded on the exterior for screwing into the base of the projectile. In the interior of the case are a hammer, *b*, in the form of a sphere, held in place by clips, *c*, which abut against a shoulder in the case and a circular recess in the ball *b*; two pistons, *d*, which are forced forward by the pressure of the gas of the powder-charge in the gun; a flat spring, *e*, which keeps the ball in place during flight; three small balls, *f*, which are held firmly in their seats below three percussion-caps, *g*; a valve, *h*, in front, which moves parallel to the axis of the fuze, and carries on its forward face a ring, *i*, of compressed powder; four radial chambers, *j*, carrying priming-charges of powder, and a screw, *k*, whose use will be explained.

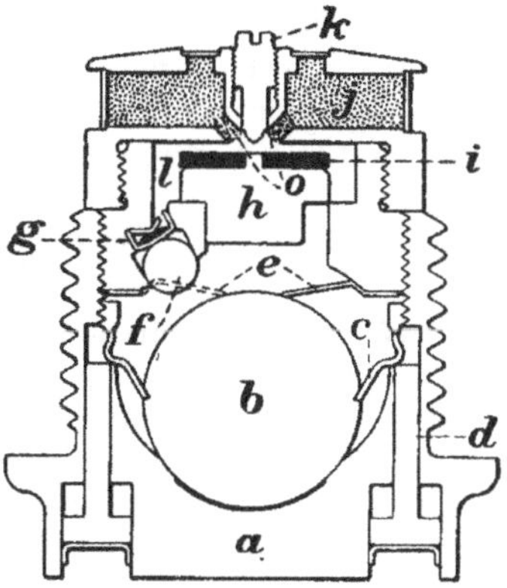

FIG. 198.

ACTION OF FUZE.—When the piece is fired, the pressure of the gas pushes forward the two pistons *d*, and these, striking the clips *c*, push them off the shoulders in the case. The ball *b*, is thus left free to move forward, but is prevented from doing so in flight, by the flat spring *e*. When the projectile strikes the object, such as an armor-plate, the ball *b* is thrown forward, and, striking one of the small balls *f*, drives it against its percussion-cap, exploding it. The flame from this cap passes into the chamber *l*.

When the ball *b* is thrown forward by the striking of the projectile, the valve *h* also moves forward at the same time, and bears against the front of the fuze-case, thus closing the openings *o o* which communicate with the priming-charges in the chambers *j*. The valve *h* reaches its seat against the front of the case before the ball *b* explodes the caps *g*, because it has a shorter distance to travel. The flame from the percussion-caps, entering the chamber *l*, ignites the compressed powder ring *i*, but, as this ring of powder is held between two closely-fitting surfaces, it can burn only on the edge. As long as the projectile is passing through the plate, or until it stops in the plate, it is being retarded, and the acquired energy of the valve will keep it in contact with the front face of the fuse. As soon, however, as the projectile passes through the plate, or stops in it, the valve *h* will move back and open the holes *o o*, and the charge will be fired. The screw *k*, holds back the valve *h*, when screwed down, and there is no delayed action in this case.

PRIMERS.

190. Definition — Classification — Requisites — Common Friction-Primer—Action.

DEFINITION.—Primers are the means employed to ignite the powder-charge in a gun.

CLASSIFICATION.—Primers are classified according to the method by which they are fired, into—

1. Friction;
2. Electric;

and each of these may be either common or obturating.

A common primer is one which ignites the charge, and is blown out of the vent, allowing the gas of the charge to escape through the latter.

An obturating primer is one which remains seated in the vent at discharge, and prevents the escape of gas through the vent.

The primers used with small-arm ammunition will be explained later. Those used with cannon are described here.

REQUISITES.—Primers should be safe in handling, not liable to damage or accident in store, and certain in action.

The requisite of safety in handling prevents the use of mercuric fulminate, except in the small-arm primers. As a general rule mercuric fulminate cannot be used where it is exposed to friction of any kind. Hence in fuzes it may be safely used, since all the parts are relatively fixed, and well protected; but this is not the case with primers.

COMMON FRICTION PRIMER.—This primer, (Fig. 199), is composed of two copper tubes, *a* and *b*, at right angles to each other; a copper wire, *c*, flattened and roughened at one end; a charge of powder filling the tube *a*; and a friction composition of antimony sulphide and potassium chlorate, filling the tube *b*.

The tubes *a* and *b* are each made from copper disks, *a′* and *b′*, by the successive action of punches and dies, by which the diameter and thickness of the tubes are decreased, and their length increased, as shown in the figure. After the proper length and diameter of each have been obtained, a hole is drilled in the side of the tube *a*, near its head, the

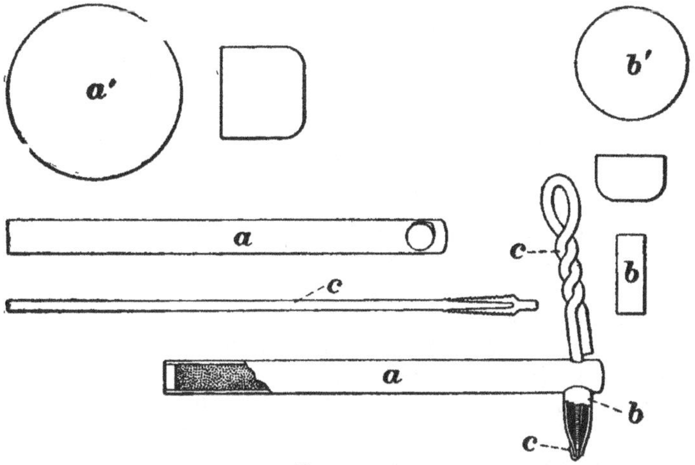

FIG. 199.

tube *b* soldered to *a*, the wire *c* inserted through the hole in *a*, its rough end resting in the tube *b*, which is then filled with the friction composition in a moist state, and the end of *b* closed on the wire, to hold the latter in place. The tube *a* is filled with small-arms powder, and its lower end closed with a wad of wax. The outer end of the wire *c* is formed into a loop, for the attachment of the hook of the lanyard.

ACTION.—When the wire is pulled by the lanyard, the roughened edges fire the friction composition in the tube *b*, and this ignites the powder in *a*.

191. Common Electric Primer—Action.

It is often necessary to fire at a distance from the gun, as in experiments; or from a central station where the object can be plainly seen; or where all the guns of a battery are to be fired simultaneously.

For this purpose the electric primer is used.

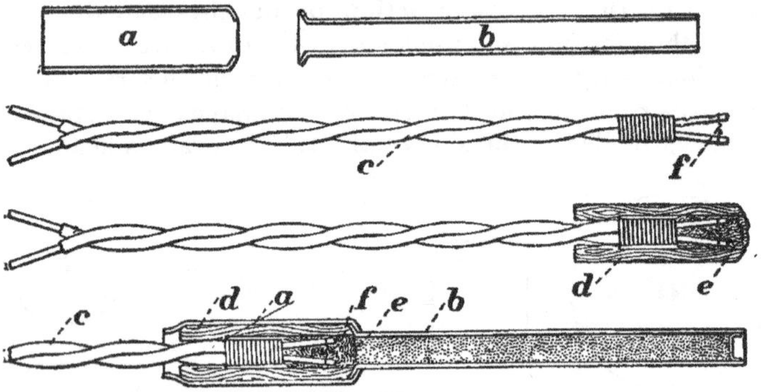

FIG. 200.

COMMON ELECTRIC PRIMER.—The common electric primer (Fig. 200), consists of two copper tubes, *a* and *b*, and two insulated copper wires, *c*, joined at one end by a small platinum wire, *f*.

These wires are inserted in a plug of wood, *d*, and are surrounded by a small quantity of dry gun-cotton, *e*. This plug of wood, with its wires and gun-cotton, is inserted into the tube *a*, and the outer end is closed down to hold it in place, and the opening filled with wax. The tube *b* is inserted in *a* beforehand, and soldered to it, as shown in the figure, and is then filled with small-arms powder, and the open end closed with wax.

ACTION.—When the circuit is closed, the current heats the fine platinum wire, *f*, and this fires the gun-cotton, which fires the powder in the tube *b*.

192. Obturating Friction Primer—Action.

With large guns, the long-continued action of the gases under high pressure, erodes the vent rapidly, if allowed to issue freely through it, and hence an obturating primer is necessary for these guns.

OBTURATING FRICTION PRIMER.—The obturating friction primer (Fig. 201), consists of a case, *a*, threaded on the exterior at *b*, to screw into the vent. A shoulder, *c*, limits the extent to which the case can be inserted. At the rear end, *d*, the case is square, to give a purchase for screwing it in and removing it. On the interior, the case is pierced with the hole *e* for the passage of the wire *f*. This passage is enlarged in front and has a cone-shaped surface at *g*. The front of the case is made thin at *h*, for a reason to be given later. The other parts of the primer are a brass wire,

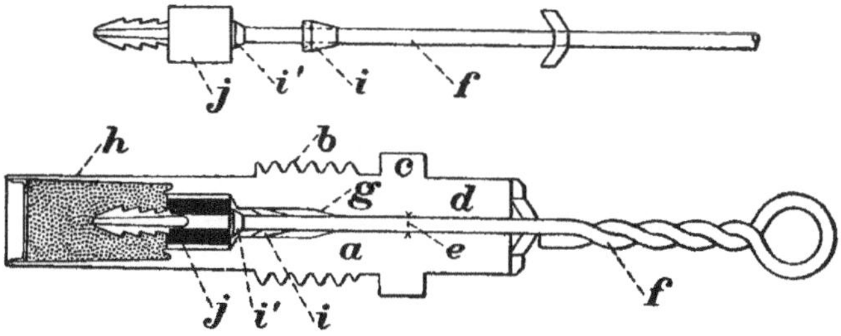

FIG. 201.

f, roughened at its forward end, and having a conical sleeve, *i*, loose upon it, and a conical enlargement, *i'*.

Upon this wire is secured pellet of friction-composition, *j*, and these parts, when inserted into the primer-case, occupy the positions shown; the rear end of the wire being twisted into a loop, for the attachment of the hook of the lanyard.

The front part of the case is filled with small-arms powder.

ACTION.—The primer is inserted into the vent and screwed home, till stopped by the shoulder *c*. When the wire is pulled to the rear by the lanyard, the roughened end fires the pellet of friction-composition *j*, and this fires the priming charge.

The gas may escape in two ways: first along the outside of the case and around the screw-thread; second, through the hole *e* in which the wire rests.

To prevent escape along the outside, the thin part of the case *h*, in front, expands under pressure, and fits tightly against the walls of the vent, thus forming a perfect gas-check. To prevent escape through the hole *e*, the drawing back of the wire *f*, in firing, brings the conical enlargement *i'* firmly against the front of the sleeve *i*, and the latter against its conical seat, *g*, in the case, and the gas-pressure keeps it in place, thus closing the hole *e*.

193. Obturating Electric Primer—Action.

OBTURATING ELECTRIC PRIMER.—This primer is used for the same reasons as the common electric primer, and con-

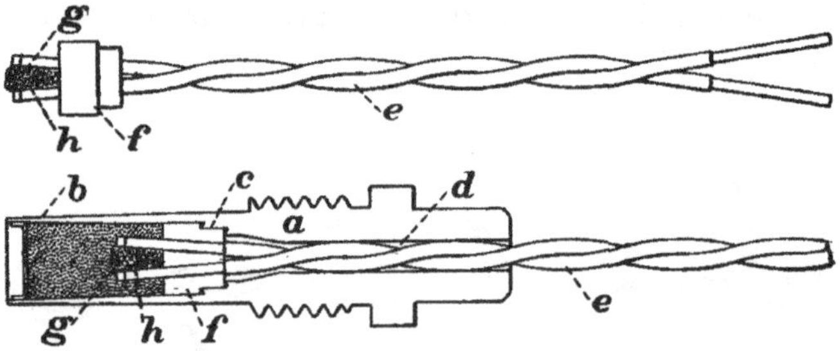

FIG. 202.

sists, (Fig. 202), of a case, *a*, exactly similar on the exterior to that of the obturating friction primer just described. On the interior, the forward part of the case *b* is made thin to serve as a gas-check, as before explained. A seat, of the shape shown at *c*, is made near the middle of the case, and a hole, *d*, allows the wires *e* to pass through. The other parts of the primer are, the two insulated wires, *e*, passing through a hard rubber plug, *f*, and connected at their forward ends by a piece of platinum wire, *g*. A small piece of gun-cotton, *h*, is wound round the platinum wire, and the whole inserted in the primer case, occupying the position shown.

The front of the primer case is filled with small-arms powder.

ACTION.—When the circuit is closed, the current heats the platinum wire *g*, and this fires the gun-cotton *h*, which in turn ignites the powder in the front of the primer case.

The escape of gas around the outside of the primer is prevented by the expansion of the thin portion of the case in front, as before. The escape through the hole *d* is prevented by the hard rubber plug *f*, which is forced into its seat by the pressure.

CHAPTER VI.

EXTERIOR BALLISTICS.

194. Definitions.

Exterior Ballistics treats of the motion of a projectile in air, after it has left the piece.

The Trajectory, a, Fig. 203, is the curve described by the centre of gravity of the projectile during its passage through the air.

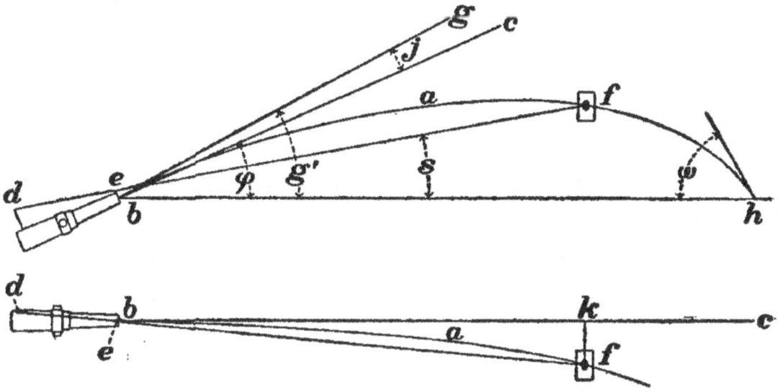

FIG. 203.

The Line of Fire, bc, is the prolongation of the axis of the piece.

The Plane of Fire is the vertical plane containing the line of fire.

The Line of Sight, def, is the straight line passing through the sights and the point aimed at.

The Plane of Sight is the vertical plane containing the line of sight.

The Angle of Sight, s, is the angle made by the line of sight with the horizontal.

The Angle of Departure, g′, is the angle made by the line of departure with the horizontal.

The Angle of Elevation, φ, is the angle made by the axis of the piece with the horizontal.

The angle of elevation generally differs slightly from the angle of departure, owing to the movement of the gun at discharge. This movement is due to the elasticity of the parts of the carriage, and the lack of accurate fitting of the trunnions in their beds, the play of the elevating device, etc.

The Jump, j, is the difference between the angle of departure and of elevation, and must be determined by experiment.

The Angle of Fall, ω, is the angle made by the tangent to the trajectory with the horizontal at the end of the range.

The Range, bh, is the horizontal distance from the muzzle to the point where the projectile strikes.

Initial Velocity is the velocity of the projectile at the muzzle.

Remaining Velocity is the velocity of the projectile at any point of the trajectory.

Final Velocity is the velocity of the projectile at the end of the range.

Drift, kf, is the departure of the projectile from the plane of fire, due to the resistance of the air, and the rotation of the projectile.

Direct Fire is from guns, with service charges, at all angles of elevation not exceeding 15°.

Indirect or Curved Fire is from guns, with less than service charges, and from howitzers and mortars, at all angles of elevation not exceeding 15°.

High-angle Fire is from guns, howitzers, and mortars at all angles of elevation exceeding 15°.

195. Forces Acting on a Projectile — Circumstances of Motion — Drift.

FORCES ACTING.—In the case of an oblong projectile, which is the only one considered, a motion of rotation about its longer axis is given to it, by the rifling of the gun, as it passes through the bore. When it leaves the bore, it is subjected to the action of gravity, and the resistance of the air. It is therefore a free body, having a motion of translation and of rotation impressed upon it, and acted on by the two forces above mentioned.

CIRCUMSTANCES OF MOTION.—The exact motion of the projectile under these circumstances is very complex, and is discussed in mechanics under the subject of "Rotation."

The general result of the action of the forces may be stated as follows:

When the projectile first issues from the piece, its longer axis is tangent to the trajectory. The resistance of the air acts along this tangent, and is at first directly opposed to the motion of translation of the projectile, and hence its resultant coincides with the longer axis, and it exerts no effort to overturn the projectile about its shorter axis.

The longer axis of the projectile, being a stable axis of rotation, tends to remain parallel to itself during the passage of the projectile through the air, but the tangent to the trajectory changes its inclination, owing to the action of gravity. The resistance of the air acting always in the direction of the tangent, thus becomes inclined to the longer axis of the projectile, and for projectiles in our service, and modern projectiles generally, its resultant intersects the longer axis, at a point in front of the centre of mass.

In Fig. 204, *G* being the centre of mass, and *R* the re-

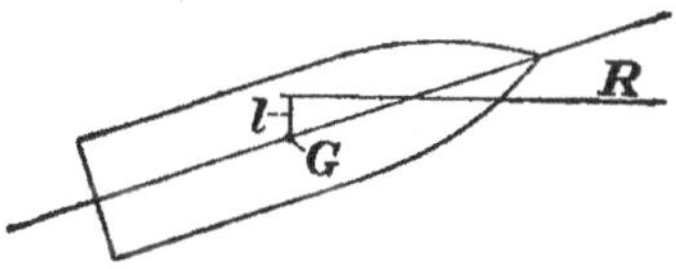

FIG. 204.

sultant resistance of the air, this resultant acts with a lever-arm *l*, to rotate the projectile about a shorter axis through *G*, perpendicular to the plane of fire.

If the projectile possesses sufficient energy of rotation about its longer axis under these circumstances, the rotation about the shorter axis will not occur, but the practical result will be, that for projectiles rotating from left to right, as in our service, the point of the projectile will move slowly to the right of the plane of fire. As soon as this motion of the point to the right occurs, it causes a relative change in the direction of the resistance of the air, and an oblique pressure is produced on the left side of the projectile, by which it is forced sidewise to the right, out of the plane of fire. At the same time, the resultant of this new oblique pressure, and of the rotation, causes the point of the projectile to move downward.

The result of the continued action of these forces is practically—

1. To cause the axis of the projectile to describe a cone about the tangent to the trajectory.

2. To force the projectile bodily to the right, and out of the plane of fire.

DRIFT.—This departure of the projectile from the plane of fire, due to the causes above mentioned, is called drift, and may be computed by Mayevski's formula, which will be given later. The actual motion of the projectile is more complex than that above given, and its full investigation requires analytical methods.

196. Form of Trajectory—Causes Affecting Resistance—Form—Cross-section—Density of the Air.

FORM OF TRAJECTORY.—From the above it appears, that the trajectory is not a plane curve, but one of double curvature. It is also shown by analytical methods, that the drift increases more rapidly than the range, and hence the projection of the trajectory on the horizontal plane, is convex to the horizontal projection of the line of fire, Fig. 203.

The trajectory ordinarily considered, is the projection of the actual curve upon the vertical plane of fire. This projection so nearly agrees with the actual curve that the results thus obtained are practically correct, and the advantage of considering it, instead of the actual curve, is, that we need consider only that component of the resistance of the air which acts directly along the longer axis of the projectile, and which is directly opposed to the motion of translation.

CAUSES AFFECTING RESISTANCE.—The resistance of the air to the motion of a projectile varies with—

1. Its form;
2. Its cross-section;
3. The density of the air;
4. The velocity of the projectile.

FORM.—Experiment shows that the ogival form of head offers less resistance than any other, and the radius of the ogive has been increased up to 2 and 3 calibres. Beyond this latter radius other considerations, such as strength to resist deformation, etc., enter. The resistance depends principally upon the form of the head near its junction with the cylindrical body of the projectile, as this affects the flow of the air over the projectile. The shape of the rear portion of the body also affects the resistance, and a projectile which is barrel-shaped in rear, such as the Whitworth, offers less resistance than one cylindrical in form, for the same reason as above. Practical considerations of ease of manufacture, facility of packing, etc., have, however, prevented the adoption of the Whitworth shape.

CROSS-SECTION.—Numerous experiments show that the resistance of the air varies directly with the area of cross-section of the projectile.

DENSITY OF THE AIR.—Experiment also shows that the resistance varies directly with the density of the air, and as this density varies with the temperature and pressure, readings of the thermometer and barometer must be taken, when accurate results are to be obtained. These readings are used to calculate the densities, as will be explained.

197. Relation between Velocity and Resistance—Experiments.

EXPERIMENTS.—The relation between the velocity of a projectile, and the resistance opposed to its motion by the air, has been the subject of experiment from the earliest times to the present day. The most notable experiments upon this subject are:

1. Robins in 1742 made the first experiments by means of the ballistic pendulum which he invented. His conclusions were, that up to 1100 ft.-secs. the resistance is propor-

tional to the square of the velocity; at 1100 ft.-secs. the law of the resistance changes; beyond 1100 ft.-secs. the resistance is nearly three times as great as if calculated by the law of the lower velocities.

2. Hutton in 1790 improved the ballistic pendulum, and made numerous experiments with large projectiles. His conclusions were, that the resistance increases more rapidly than the square of the velocity for low velocities, and for higher velocities that it varies nearly as the square.

3. General Didion made a series of experiments at Metz in 1839 and 1840 with the ballistic pendulum, and spherical projectiles of varying weights. His conclusions were, that the law of resistance is expressed by a formula of the general form

$$R \propto a(v^2 + bv^3),$$

a and b being constants. This formula held for short ranges, but not for heavy charges and high angles of elevation.

4. Experiments were therefore made again at Metz in 1857, and with electro-ballistic instruments. The conclusions from these experiments were, that the resistance varies as the cube of the velocity. Experiments by Prof. Hélie at Gâvre, in 1860 and 1861, gave practically the same result.

5. The most accurate experiments upon this subject were made by the Rev. Francis Bashforth in England, in 1865, and again in 1880. The advantage of these experiments is that they were made with a very accurate instrument, and with comparatively modern projectiles. The conclusions in general were, that the resistance varies with some power of the velocity, and that this power varies with the velocity, being generally as follows:

For velocities	between	900 and 1100 ft.-secs......	v^6
"	" between	1100 and 1350 ft.-secs......	v^3
"	" above	1350 ft.-secs....	v^2

6. The most recent experiments on the subject, and those now adopted for use, were made by Krupp in 1881 with modern guns, projectiles, and velocities. General Mayevski discussed the results of these experiments, and deduced expressions for the resistance as follows:

198. Method of Determining Resistance.

The resistance of the air is a force expressed in pounds per square inch, and it opposes the motion of the projectile in its passage.

The effect of this force is to retard the projectile. There are therefore two quantities to be determined:

1. The resistance, or pressure of the air, in pounds per square inch;

2. The retardation, or loss of velocity in feet per second, produced by this resistance.

METHOD EMPLOYED TO DETERMINE RESISTANCE.—The method generally employed to determine the resistance of the air, consists in measuring the velocities v_1 and v_2 of the projectile, at two points M_1 and M_2, situated at such a distance apart, that the path of the projectile, over this distance, may be regarded as a right line; and also so that the resistance may be considered constant over this distance. The energy of the projectile at the point M_1 is $\frac{1}{2}mv_1^2$, and at M_2, $\frac{1}{2}mv_2^2$. Their difference, $\frac{1}{2}m(v_1^2 - v_2^2)$, is the loss of energy over this distance due to the resistance of the air; and supposing this resistance constant, and calling the resistance ρ, and the path l, we have

$$\rho l = \frac{1}{2}m(v_1^2 - v_2^2). \quad . \quad . \quad . \quad . \quad . \quad (262)$$

This, being the mean resistance, corresponds to the mean velocity, or $\frac{v_1 + v_2}{2}$.

By properly selecting the points, and varying the velocity so as to include all service velocities, we obtain a series of values for the velocity and resistance, from which a curve can be constructed, giving the law of resistance for different velocities.

The distance between M_1 and M_2 must be chosen according to the velocities and projectiles used. Thus for low velocities, and large projectiles, the distance between the points must be greater, since the loss of velocity over a given path is less in this case, than for small projectiles moving with high velocities.

199. Modifications of General Method—Results—Resistance.

MODIFICATIONS.—When the curve of resistance obtained by the above general method is plotted, it is found that sudden changes occur in it for different velocities.

Also the above expression does not take account of variation in the form and cross-section of the projectile, or in the density of the air.

To have a general expression into which all these quantities enter, General Mayevski proceeded as follows:

Denoting the resistance as before by ρ, the retardation is

$$\frac{\rho}{M} = \frac{g}{W}\rho,$$

in which M is the mass of the projectile;
W, its weight in pounds;
g, the acceleration of gravity, 32.2 ft.-secs.

This expression was placed equal to $\frac{A}{C}f(v)$, in which A is a constant to be determined by experiment, C a factor called the "ballistic coefficient," and $f(v)$ some function of the velocity. Hence we have

$$\frac{g}{W}\rho = \frac{A}{C}f(v). \quad . \quad . \quad . \quad . \quad . \quad . \quad (263)$$

The Ballistic Coefficient C.—The value of this coefficient is

$$C = \frac{\delta_1}{\delta}\frac{W}{cd^2}, \quad . \quad . \quad . \quad . \quad . \quad . \quad (264)$$

in which

δ_1 is the standard density of the air;
δ, the density at the time of the experiment;
c, the coefficient of reduction;
d, the diameter of the projectile in inches;
W, its weight in pounds as before.

Substituting the value of C from (264) in (263), we have

$$\frac{g}{W}\rho = A\frac{\delta}{\delta_1}\frac{cd^2}{W}f(v). \quad . \quad . \quad . \quad . \quad . \quad (265)$$

For a given projectile, all the quantities which enter the ballistic coefficient C are known, and they take into account the cross-section and weight of the projectile, and the density of the air.

The form of the projectile enters in the coefficient of reduction c as follows: For projectiles of a standard form, or for those with which the experiments are made, the coefficient of reduction is taken as unity. For those differing from the standard, the retardation will be greater or less, as the form is less or more suited to overcome the resistance. Hence this coefficient will have values greater than unity for projectiles whose resistance is greater than the standard, and values less than unity for those whose resistance is less than the standard. For the older forms of projectiles in our service $c = 1$, for the new form $c = 0.9$ nearly.

The values of $\frac{\delta}{\delta}$ for all pressures and temperatures in practice are calculated and tabulated for use in Table III (Ballistic Tables).

The only remaining quantities in formula (265) are A and $f(v)$, and the object of the experiments is to determine the values of A and the exponent of v.

RESULTS.—As a result of the experiments, the general value (265) for the retardation assumes the following forms for different velocities:

For all velocities greater than 1330 ft.-seconds,

$$\frac{g}{W}\rho = \frac{A}{C}v^2; \quad \log A = \bar{4},1525284;$$

$$1330 \text{ ft.-secs.} > v > 1120 \text{ ft.-secs.}$$

$$\frac{g}{W}\rho = \frac{A}{C}v^3; \quad \log A = \bar{7}.0364351;$$

$$1120 \text{ ft.-secs.} > v > 990 \text{ ft.-secs.}$$

$$\frac{g}{W}\rho = \frac{A}{C}v^6; \quad \log A = \overline{17}.8865079;$$

$$990 \text{ ft.-secs.} > v > 790 \text{ ft.-secs.}$$

$$\frac{g}{W}\rho = \frac{A}{C}v^3; \quad \log A = \bar{8}.8754872;$$

$$790 \text{ ft.-secs.} > v > 100 \text{ ft.-secs.}$$

$$\frac{g}{W}\rho = \frac{A}{C}v^2; \quad \log A = \bar{5}.7703827.$$

RESISTANCE.—The corresponding resistance in pounds is obtained for each velocity by multiplying by $\frac{W}{g}$ since

$$\rho = \frac{W}{g}\frac{A}{C}f(v) = \frac{\delta}{\delta_1}\frac{cd^2}{g}Af(v). \quad . \quad . \quad . \quad (266)$$

200. Trajectory in Air—Nomenclature—Equations of Motion.

NOMENCLATURE.—Considering the motion of translation only, and that the resistance of the air is directly opposed to this motion, let (Fig. 205).

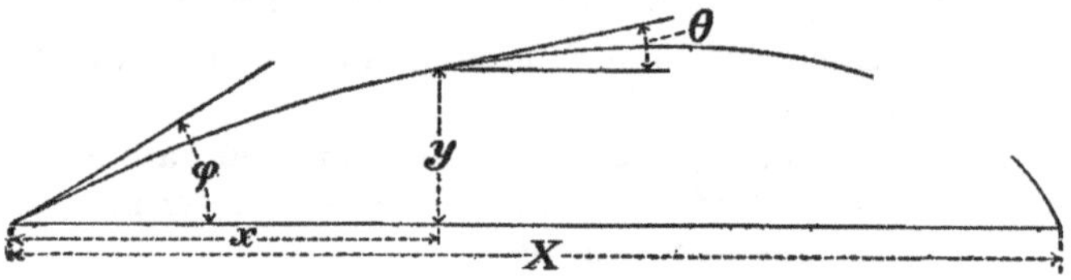

FIG. 205.

R be the retardation due to the resistance of the air, its value being given by equation (265);

V, the initial velocity;

v, the velocity of any point of the trajectory whose co-ordinates are x and y;

v_1, the velocity in the direction of x;

ϕ, the angle made by the tangent to the trajectory with the horizontal, at the origin; or the angle of elevation;

θ, the value of ϕ for any other point of the trajectory;

x and y, the co-ordinates of any point of the trajectory, in feet;

X, the whole range in feet.

EQUATIONS OF MOTION.—The only forces acting on the projectile after it leaves the piece, are the resistance of the air and gravity.

The resistance of the air is directly opposed to the motion of the projectile, and continually retards it. Gravity is supposed to act vertically, and retards the projectile in the ascending portion of the trajectory, while it accelerates it in the descending portion.

Considering the ascending portion, we have for the acceleration along x, since gravity has no component in that direction,

$$\frac{dv_1}{dt} = - R \cos \theta ; \quad \ldots \ldots \quad (267)$$

from this

$$dt = - \frac{dv_1}{R \cos \theta}. \quad \ldots \ldots \quad (268)$$

The velocity along x is

$$v_1 = \frac{dx}{dt} = v \cos \theta ; \quad \ldots \ldots \quad (269)$$

and along y,

$$\frac{dy}{dt} = v \sin \theta = v_1 \tan \theta. \quad \ldots \ldots \quad (270)$$

Substituting the value of dt from (268) in (269) and (270), we have

$$dx = - \frac{v_1 dv_1}{R \cos \theta}; \quad \ldots \ldots \quad (271)$$

$$dy = - \frac{v_1 \tan \theta \, dv_1}{R \cos \theta}. \quad \ldots \ldots \quad (272)$$

The acceleration along the radius of curvature is

$$\frac{v^2}{r} = g \cos \theta. \quad \ldots \ldots \quad (273)$$

Substitute in this for r its value $-\frac{ds}{d\theta}$ from calculus, and we have

$$\frac{v^2}{r} = v \frac{ds}{dt} \frac{1}{r} = - v \frac{ds}{dt} \frac{d\theta}{ds} = - v \frac{d\theta}{dt} = g \cos \theta, \quad . \quad (274)$$

hence

$$d\theta = -\frac{g \cos\theta \, dt}{v}. \quad . \quad . \quad . \quad . \quad . \quad (275)$$

Substitute in (275) for $\cos\theta$ its value from (267) and for v its value from (269), and we have

$$d\theta = \frac{g \cos\theta \, dv_1}{Rv_1}. \quad . \quad . \quad . \quad . \quad . \quad . \quad (276)$$

Collecting these equations, we have

$$\left.\begin{aligned} dt &= -\frac{dv_1}{R\cos\theta}; \\ dx &= -\frac{v_1 dv_1}{R\cos\theta}; \\ dy &= -\frac{v_1 \tan\theta \, dv_1}{R\cos\theta}; \\ d\theta &= \frac{g\cos\theta \, dv_1}{Rv_1}. \end{aligned}\right\} \quad . \quad . \quad . \quad . \quad . \quad (A)$$

If these equations could be integrated directly, they would give the values of x, y, t, and θ for any point of the trajectory. But as they are expressed in terms of R, v_1, and θ, three independent variables, the direct integration is impossible.

201. Method of Integrating Equations A.—1st Step.

1ST STEP.—The first step in the process of integration is to replace R by its value from equation (263),

$$R = \frac{g}{W}\rho = \frac{A}{C}f(v),$$

and to make

$$f(v) = v^n, \quad . \quad . \quad . \quad . \quad . \quad . \quad . \quad . \quad (277)$$

in which n represents the exponent of the power of v which is proportional to the retardation, for any particular velocity, and, according to Mayevski's experiments as shown, varies from 2 to 6.

From equation (269) we have

$$v_1 = v\cos\theta;$$

hence

$$v = \frac{v_1}{\cos\theta}; \quad \therefore \quad v^n = \frac{v_1^n}{\cos^n\theta}. \quad . \quad . \quad . \quad (278)$$

Making these substitutions in the value of R above, we have

$$R = \frac{g}{W}\rho = \frac{A}{C}\,\frac{v_1^n}{\cos^n\theta}. \quad . \quad . \quad . \quad . \quad (279)$$

Substituting this value of R in the first, second, and fourth of equations (A), we have

$$dt = -\frac{C}{A}\cos^{n-1}\theta\,\frac{dv_1}{v_1^n}; \quad . \quad . \quad . \quad . \quad (280)$$

$$dx = -\frac{C}{A}\cos^{n-1}\theta\,\frac{dv_1}{v_1^{n-1}}; \quad . \quad . \quad . \quad . \quad (281)$$

$$d\theta = \frac{gC}{A}\cos^{n+1}\theta\,\frac{dv_1}{v_1^{n+1}}. \quad . \quad . \quad . \quad . \quad (282)$$

Dividing both terms of (282) by $\cos^2\theta$, we have

$$\frac{d\theta}{\cos^2\theta} = \frac{gC}{A}\cos^{n-1}\theta\,\frac{dv_1}{v_1^{n+1}}; \quad . \quad . \quad . \quad (283)$$

or, since $\cos\theta = \frac{1}{\sec\theta}$,

$$\frac{d\theta}{\cos^2\theta} = \frac{gC}{A\sec^{n-1}\theta}\,\frac{dv_1}{v_1^{n+1}}. \quad . \quad . \quad . \quad . \quad (284)$$

Collecting these equations, we have

$$\left.\begin{aligned} \frac{d\theta}{\cos^2\theta} &= \frac{gC}{A\sec^{n-1}\theta}\,\frac{dv_1}{v_1^{n+1}}; \\ dt &= -\frac{C}{A\sec^{n-1}\theta}\,\frac{dv_1}{v_1^n}; \\ dx &= -\frac{C}{A\sec^{n-1}\theta}\,\frac{dv_1}{v_1^{n-1}}. \end{aligned}\right\} \quad . \quad . \quad . \quad (285)$$

202. Method of Integration of Equations (A)—2d Step.

2D STEP.—In equations (285), deduced by the 1st step, the first members are exact integrals. The second members are not, however, because they contain the two independent variables $\sec^{n-1}\theta$ and v_1. For all cases of direct fire the value of $\sec\theta$ differs but little from unity, since for angles of 15°, $\sec\theta = 1.035$, and for angles less than this, its value is still more nearly unity. Hence $\sec\theta$ can be replaced by unity without great error.

Siacci shows, however, by analysis, that a more correct value for direct fire is

$$\sec^{n-1}\theta = \sec^{n-2}\phi, \quad . \quad . \quad . \quad . \quad . \quad (286)$$

and this value has been universally adopted.

Substituting this value of $\sec^{n-1}\theta$ in equations (285), we have

$$\left.\begin{aligned} \frac{d\theta}{\cos^2\theta} &= \frac{gC}{A\sec^{n-2}\phi}\frac{dv_1}{v_1^{n+1}}; \\ dt &= -\frac{C}{A\sec^{n-2}\phi}\frac{dv_1}{v_1^{n}}; \\ dx &= -\frac{C}{A\sec^{n-2}\phi}\frac{dv_1}{v_1^{n-1}}. \end{aligned}\right\} \quad . \quad . \quad . \quad (287)$$

Taking the first of equations (287), multiply the numerator and denominator of the second member by $\sec^3\phi = \sec^2\phi\sec\phi$, and we have

$$\frac{d\theta}{\cos^2\theta} = \frac{gC\sec^2\phi\sec\phi\,dv_1}{A\sec^{n+1}\phi\, v_1^{n+1}}.$$

But $\sec^2\phi = \frac{1}{\cos^2\phi}$, and since ϕ is constant, $\sec\phi\, dv_1 = d(v_1\sec\phi)$; hence

$$\frac{d\theta}{\cos^2\theta} = \frac{gC\,d(v_1\sec\phi)}{A\cos^2\phi\,(v_1\sec\phi)^{n+1}};$$

and by the [illegible] process the other two equations may be placed i[illegible]n which the second members can readily be i[illegible]nce we have

$$\left.\begin{aligned}
\frac{d\theta}{\cos^2\theta} &= \frac{gC}{A\cos^2\phi}\,\frac{d(v_1\sec\phi)}{(v_1\sec\phi)^{n+1}}; \\
dt &= -\frac{C}{A\cos\phi}\,\frac{d(v_1\sec\phi)}{(v_1\sec\phi)^{n}}; \\
dx &= -\frac{C}{A}\,\frac{d(v_1\sec\phi)}{(v_1\sec\phi)^{n-1}}.
\end{aligned}\right\} \quad . \; . \; . \; . \quad (288)$$

Making

$$v_1 \sec\phi = \frac{v\cos\theta}{\cos\phi} = u, \quad . \; . \; . \; . \quad (289)$$

$$V_1 \sec\phi = \frac{V\cos\phi}{\cos\phi} = V, \quad . \; . \; . \; . \quad (290)$$

from equation (269), and integrating equations (288) between the limits ϕ and θ to which correspond V and u, we have

$$\tan\phi - \tan\theta = \frac{gC}{nA\cos^2\phi}\left[\frac{1}{u^n} - \frac{1}{V^n}\right]; \quad . \quad (291)$$

$$t = \frac{C}{(n-1)A\cos\phi}\left[\frac{1}{u^{n-1}} - \frac{1}{V^{n-1}}\right]; \quad . \; . \quad (292)$$

$$x = \frac{C}{(n-2)A}\left[\frac{1}{u^{n-2}} - \frac{1}{V^{n-2}}\right]. \quad . \; . \; . \; . \quad (293)$$

203. Simplification of Equations (291), (292), and (293)—Method of Calculating the Functions which Enter them.

SIMPLIFICATION.—To simplify equations (291), (292), and (293), make

$$I(u) = \frac{2g}{nAu^n} + Q; \quad . \; . \; . \; . \; . \; . \; . \quad (294)$$

$$I(V) = \frac{2g}{nAV^n} + Q; \quad . \; . \; . \; . \; . \; . \; . \quad (295)$$

$$T(u) = \frac{1}{(n-1)Au^{n-1}} + Q''; \quad . \; . \; . \quad (296)$$

$$S(u) = \frac{1}{(n-2)Au^{n-2}} + Q'''. \quad . \; . \; . \quad (297)$$

Making these substitutions, equations (291), (292), and (293) can be written,

$$\tan \phi - \tan \theta = \frac{C}{2 \cos^2 \phi} \left\{ I(u) - I(V) \right\}; \quad . \quad \text{(B)}$$

$$t = \frac{C}{\cos \phi} \left\{ T(u) - T(V) \right\}; \quad . \quad . \quad . \quad . \quad . \quad . \quad \text{(C)}$$

$$x = C \left\{ S(u) - S(V) \right\}; \quad . \quad . \quad . \quad . \quad . \quad . \quad . \quad . \quad \text{(D)}$$

$$u = v \frac{\cos \theta}{\cos \phi}, \text{ see (289).} \quad . \quad . \quad . \quad . \quad . \quad . \quad . \quad . \quad \text{(F)}$$

In equations (294), (295), etc. the expression $I(u)$, is called the inclination function, $T(u)$ the time function, and $S(u)$ the space function; Q, Q', Q'', etc., are arbitrary constants. The values of these functions may be calculated and tabulated for convenience, and the resulting tables are called "Ballistic Tables." Those used in the present course were calculated by Capt. James M. Ingalls, 1st Artillery, U. S. Army. By their use the calculation of these functions is avoided for any particular case, and the use of the formulas facilitated.

CALCULATION OF FUNCTIONS.—As an illustration of the method of calculation, take the $T(u)$ function, equation (296):

$$T(u) = \frac{1}{(n-1) A u^{n-1}} + Q''.$$

In this equation, n is the exponent of the power of v to which the resistance of the air is proportional; A, a constant determined by Mayevski's experiments, as explained; and Q'' an arbitrary constant.

For values of v greater than 1330 ft.-secs., we have $n = 2$ and $\log A = \bar{4}.1525284$. Hence for these velocities we have

$$T(u) = \frac{1}{[\bar{4}.1525284]\, u} + Q''.$$

The ballistic tables are so constructed that all the functions $S(u)$, $T(u)$, etc., reduce to zero for $u = 2800$ ft.-secs. Hence we have

$$T(u) = \frac{1}{[\bar{4}.1525284] \times 2800} + Q'' = 0,$$

and solving

$$Q'' = -2.5137.$$

When the velocity is 1330 ft.-secs., $n = 3$; $\log A = \bar{7}.0364351$; hence

$$T(u) = \frac{1}{2 \times [\bar{7}.0364351] \times \overline{1330}^2} + Q_1''.$$

But to avoid abrupt changes in the table, the value of $T(u)$ for 1330 ft.-secs. must be placed equal to that which would be obtained if $n = 2$, or from the first equation in which Q'' enters. This may be done, since Q_1'' is arbitrary, by placing

$$\frac{1}{[\bar{4}.1525284] \times 1330} - 2.5137 = \frac{1}{2 \times [\bar{7}.0364351] \times \overline{1330}^2} + Q_1''.$$

Solving with reference to Q_1'', we have

$$Q_1'' = +0.1791.$$

Therefore, for all values of u or v greater than 1330 ft.-secs. the value of $T(u)$ is calculated by the equation

$$T(u) = \frac{1}{[\bar{4}.1525284] \times u} - 2.5137,$$

and for all values of u or v between 1330 and 1120 ft.-secs. by the equation

$$T(u) = \frac{1}{2 \times [\bar{7}.0364351] \times u^2} + 0.1791.$$

Below 1120 ft.-secs. we make a similar change, equating the known value for 1120 with the new values for A and n, and determine the new arbitrary constant as before, and so on for all the functions.

204. Relation between x and y.

We have from equation (B), since $\tan\theta = \frac{dy}{dx}$,

$$\frac{dy}{dx} = \tan\phi - \frac{C}{2\cos^2\phi}\left\{I(u) - I(V)\right\},$$

or

$$\frac{2\cos^2\phi}{C}\left\{\frac{dy}{dx} - \tan\phi\right\} - I(V) = -I(u), \quad . \quad (298)$$

also,

$$v_1 = v\cos\theta$$

$$v_1\sec\phi = \frac{v\cos\theta}{\cos\phi} = u \text{ (see 289)},$$

$$d(v_1\sec\phi) = du.$$

Whence, substituting in the third of equations (288), we have

$$dx = -\frac{C}{A}\frac{du}{u^{n-1}},$$

or

$$\frac{dx}{C} = -\frac{du}{Au^{n-1}}. \quad . \quad . \quad . \quad . \quad . \quad . \quad (299)$$

Multiplying (299) and (298) together member by member, we have

$$\frac{2\cos^2\phi}{C^2}\left\{dy - \tan\phi\,dx\right\} - \frac{I(V)}{C}dx = \frac{I(u)\,du}{Au^{n-1}}.$$

Integrating and making x and y both zero at the origin, where $u = V$, we have

$$\frac{2\cos^2\phi}{C^2}\left\{y - x\tan\phi\right\} - \frac{I(V)}{C}x = -\frac{1}{A}\int_u^V\frac{I(u)\,du}{u^{n-1}}.$$

Making

$$A(u) = -\frac{1}{A}\int\frac{I(u)\,du}{u^{n-1}},$$

we have

$$\frac{2\cos^2\phi}{C^2}\left\{y - x\tan\phi\right\} - \frac{I(V)}{C}x = -\left\{A(u) - A(V)\right\}. (300)$$

From equation (D) we have

$$\frac{x}{C} = S(u) - S(V). \quad . \quad . \quad . \quad . \quad . \quad (301)$$

Dividing (300) by (301), member by member, we have

$$\frac{2\cos^2\phi}{C}\left\{\frac{y}{x} - \tan\phi\right\} - I(V) = -\frac{A(u) - A(V)}{S(u) - S(V)}, \quad (302)$$

or finally,

$$\frac{y}{x} = \tan\phi - \frac{C}{2\cos^2\phi}\left\{\frac{A(u) - A(V)}{S(u) - S(V)} - I(V)\right\}. \; . \quad (E)$$

In this equation $A(u)$ is called the altitude function. Collecting these equations, we write

$$\tan\theta = \tan\phi - \frac{C}{2\cos^2\phi}\left\{I(u) - I(V)\right\}; \quad . \quad . \quad . \quad (B)$$

$$t = \frac{C}{\cos\phi}\left\{T(u) - T(V)\right\}; \quad . \quad . \quad . \quad . \quad . \quad . \quad . \quad . \quad (C)$$

$$x = C\left\{S(u) - S(V)\right\}; \quad . \quad . \quad . \quad . \quad . \quad . \quad . \quad . \quad . \quad (D)$$

$$\frac{y}{x} = \tan\phi - \frac{C}{2\cos^2\phi}\left\{\frac{A(u) - A(V)}{S(u) - S(V)} - I(V)\right\}; \; (E)$$

$$u = v\frac{\cos\theta}{\cos\phi}. \quad . \quad . \quad . \quad . \quad . \quad . \quad . \quad . \quad . \quad . \quad . \quad . \quad . \quad (F)$$

These are the fundamental equations of Exterior Ballistics, and the object now is to explain their modifications and methods of use.

205. Modifications of General Formulas for the Whole Range X—For the Summit of the Trajectory.

MODIFICATIONS FOR RANGE X.—The range being the distance from the muzzle, to the point where the projectile in the descending branch of the trajectory, pierces the horizontal plane through the muzzle, we have for this point

$$x = X;$$
$$-\theta = \omega;$$
$$y = 0;$$
$$t = T:$$
$$u = u_\omega$$

and making these changes in (B), (C), (D), (E), and (F), we have

$$\tan \omega = \frac{C}{2 \cos^2 \phi} \Big\{ I(u_\omega) - I(V) \Big\} - \tan \phi ; \quad . \quad . \quad (B')$$

$$T = \frac{C}{\cos \phi} \Big\{ T(u_\omega) - T(V) \Big\} ; \quad . \quad . \quad . \quad . \quad . \quad (C')$$

$$X = C \Big\{ S(u_\omega) - S(V) \Big\} ; \quad . \quad . \quad . \quad . \quad . \quad . \quad . \quad (D')$$

$$\sin 2\phi = C \left\{ \frac{A(u_\omega) - A(V)}{S(u_\omega) - S(V)} - I(V) \right\} ; \quad . \quad . \quad . \quad (E')$$

$$u_\omega = v_\omega \frac{\cos \omega}{\cos \phi} . \quad . \quad . \quad . \quad . \quad . \quad . \quad . \quad . \quad . \quad . \quad . \quad (F')$$

Combining (B′) and (E′) and eliminating $I(V)$, we have

$$\tan \omega = \frac{C}{2 \cos^2 \phi} \left\{ I(u_\omega) - \frac{A(u_\omega) - A(V)}{S(u_\omega) - S(V)} \right\} . \quad . \quad . \quad (G)$$

SUMMIT OF TRAJECTORY.—For this point we have $\theta = 0$, and since

$$\sin 2\phi = 2 \sin \phi \cos \phi,$$

we have from equation (B)

$$I(u_0) = \frac{\sin 2\phi}{C} + I(V), \quad . \quad . \quad . \quad . \quad (303)$$

and from (F)

$$u_0 = \frac{v_0}{\cos \phi}, \quad . \quad . \quad . \quad . \quad . \quad . \quad . \quad . \quad (304)$$

in which u_0 and v_0 are the values of u and v for the summit of the trajectory.

Substituting in (E′) for

$$\frac{\sin 2\phi}{C} + I(V),$$

its value from (303), we have

$$I(u_0) = \frac{A(u_\omega) - A(V)}{S(u_\omega) - S(V)}, \quad . \quad . \quad . \quad . \quad (305)$$

and this value in (E′) and (G) gives

$$\sin 2\phi = C\left\{ I(u_0) - I(V) \right\}; \quad . \quad . \quad . \quad . \quad (306)$$

$$\tan \omega = \frac{C}{2 \cos^2 \phi}\left\{ I(u_\omega) - I(u_0) \right\}. \quad . \quad (307)$$

When ϕ and ω are both small, as in direct fire, we may without material error suppose

$$\phi = \omega,$$

hence

$$2 \cos^2 \phi \tan \omega = 2 \cos^2 \omega \tan \omega = \sin 2\omega;$$

and substituting in (307), we have

$$\sin 2\omega = C\left\{ I(u_\omega) - I(u_0) \right\}. \quad . \quad . \quad . \quad (308)$$

206. Auxiliary Formulas.

Equations (E), (E′), and (G) can be more readily used for calculation, if the quantities

$$\frac{A(u) - A(V)}{S(u) - S(V)} - I(V), \quad . \quad . \quad . \quad . \quad . \quad . \quad (a)$$

and

$$I(u) - \frac{A(u) - A(V)}{S(u) - S(V)}, \quad . \quad . \quad . \quad . \quad . \quad . \quad (b)$$

are calculated and tabulated for use.

These quantities are functions of $\frac{X}{C}$ and V, as may be shown in the following manner:

Suppose it is required to compute the height of trajectory y, by (E), angle ϕ by (E′), and angle ω by (G), having given the ballistic coefficient C, the initial velocity V, and the whole range X, or part of the range x.

In equation (E), ϕ and u are unknown. ϕ can be computed from (E′) when (u_ω) is known, and ω in (G) can be computed also when (u_ω) is known.

Hence (u_ω) is the only unknown quantity required to complete the solution. u_ω can be found from (D′), since

$$\frac{X}{C} + S(V) = S(u_\omega),$$

and u can be found from (D) since

$$\frac{x}{C} + S(V) = S(u).$$

It follows from this that the quantities (a) and (b) are functions of $\frac{X}{C}$, or $\frac{x}{C}$ and V, and therefore the values of these expressions when tabulated should have $\frac{X}{C}$ and V as arguments.

We therefore place

$$\frac{A(u_\omega) - A(V)}{S(u_\omega) - S(V)} - I(V) = A; \quad . \quad . \quad . \quad (309)$$

$$I(u_\omega) - \frac{A(u_\omega) - A(V)}{S(u_\omega) - S(V)} = B; \quad . \quad . \quad . \quad (310)$$

$$\frac{A(u) - A(V)}{S(u) - S(V)} - I(V) = a; \quad . \quad . \quad . \quad (311)$$

$$I(u) - \frac{A(u) - A(V)}{S(u) - S(V)} = b; \quad . \quad . \quad . \quad (312)$$

$$I(u) - I(V) = a + b = m, \quad . \quad . \quad . \quad (313)$$

and making the corresponding changes in equation (B), we have

$$\tan\theta = \tan\phi - \frac{mC}{2\cos^2\phi},$$

reducing

$$\tan\theta = \tan\phi\left\{1 - \frac{mC}{\sin 2\phi}\right\}. \quad . \quad . \quad . \quad (314)$$

In (E),

$$\frac{y}{x} = \tan\phi - \frac{aC}{2\cos^2\phi}.$$

reducing

$$y = x\tan\phi\left\{1 - \frac{aC}{\sin 2\phi}\right\}, \quad . \quad . \quad . \quad . \quad (315)$$

in (E′),

$$\sin 2\phi = AC, \quad . \quad . \quad . \quad . \quad . \quad . \quad . \quad . \quad . \quad (316)$$

in (G),

$$\tan \omega = \frac{BC}{2 \cos^2 \phi}, \quad . \quad . \quad . \quad . \quad . \quad . \qquad (317)$$

and for small angles of elevation, since

$$\phi = \omega,$$

we have, from (317),

$$\sin 2\omega = BC. \quad . \quad . \quad . \quad . \quad . \quad . \quad . \qquad (318)$$

Substituting in (314) and (315) for $\sin 2\phi$ its value from (316), we have

$$\tan \theta = \tan \phi \left\{ 1 - \frac{m}{A} \right\}; \quad . \quad . \quad . \quad . \qquad (319)$$

$$y = x \tan \phi \left\{ 1 - \frac{a}{A} \right\}. \quad . \quad . \quad . \qquad (320)$$

The auxiliary quantities a, b, A, B, m, are generally written $a = f(zV)$, $b = f(zV)$, $m = f(zV)$, etc.

207. Explanation of Ballistic Tables.

The values of the quantities $A(u)$, $S(u)$, $T(u)$, etc., have been calculated and tabulated as before explained, and their values are found in Ballistic Table I, for all velocities from 2800 to 400 ft.-seconds, for ogival projectiles. Table II gives the value of the corresponding quantities for spherical projectiles.

In these tables u is a general expression for velocity, so that if v or V be given, its value will be found in the column headed u in the tables.

To illustrate their use, find the values of the different functions from Table I, for a velocity of 1137.6 ft.-secs.

We have from the table

$$S(u) = S(1137.6);\ S(1137) = 6413.2;$$
$$S(0.6) = \underline{\quad 4.26} = 7.1 \times .6 = 4.26.$$
$$S(1137.6) = 6408.9$$

$$A(u) = A(1137.6);\ A(1137) = 341.73;$$
$$A(0.6) = \underline{\quad .636} = 1.06 \times .6 = .636.$$
$$A(1137.6) = 341.09$$

$$I(u) = I(1137.6);\quad I(1137) = 0.14942$$
$$\underline{I(0.6) \quad = \quad .000216} = 00036 \times .6 = .000216$$
$$I(1137.6) = 0.14920$$

$$T(u) = T(1137.6);\quad T(1137) = 3.736;$$
$$\underline{T(0.6) \quad = \quad .0036} = .006 \times .6 = .0036$$
$$T(1137.6) = 3.732$$

Conversely, having the values of the quantities $S(u)$, $A(u)$, etc., to find the corresponding values of u, we proceed as follows:

Find the value of u for

$$S(u) = 6430.7;$$
$$A(u) = 360.9;$$
$$I(u) = 0.1580;$$
$$T(u) = 3.720.$$

From Table I we have

$$S(u) = 6427.4, \quad u = 1135,$$
$$6430.7 - 6427.4 = 3.3.$$

Tabular difference for 1 ft.-sec. = 7.2.

$$7.2 : 3.3 :: 1 : x$$
$$x = 0.46 \text{ ft.-secs.},$$

hence

$$u \text{ for } S(u) = 6430.7 = 1135 - 0.46 = 1134.54 \text{ ft.-secs.};$$
$$A(u) = 360.45, \quad u = 1120 \text{ ft.-secs.};$$
$$360.9 - 360.45 = 0.45.$$

Tabular difference for 1 ft.-sec. = 1.15.

$$1.15 : 0.45 :: 1 : x;$$
$$x = 0.39 \text{ ft.-secs.};$$

hence

$$u \text{ for } A(u) = 360.9 = 1120 - 0.39 = 1119.61 \text{ ft.-secs.}$$

The same method applies to all other cases, and it is evident that the table is used like a table of logarithms.

208. Auxiliary Tables—Values of $f(zV)$, z, and V.

These tables are found in Ballistic Table I, and give the values of the quantities A, B, a, b, and m; and the tables are headed "auxiliary A," "auxiliary B," and "auxiliary m." The values of a and b are taken from the table for A and B, since a and b are general cases of A and B. The expressions for these quantities are given by equations (309) to (313), and their values were calculated and tabulated by Capt. Ingalls.

By referring to the tables it will be seen that the arguments are z and v. z is used for brevity in place of $\frac{x}{C}$ or $\frac{X}{C}$, and hence the value of z is

$$z = \frac{x}{C} \quad \text{or} \quad \frac{X}{C} \qquad (321)$$

In this table there are two columns of differences, Δ_z and Δ_v.

Δ_z corresponds to differences in the argument z, and Δ_v to those of the argument V.

USE OF TABLES.—We may have the following cases:

1. A given value of z and one of V, neither of which is found in the table; to find the corresponding value of A, B, or m.

2. A given value of A, B, or m and one of V, neither of which is found in the table; to find the corresponding value of z.

3. A given value of A, B, or m, and one of z, neither of which is found in the table; to find the corresponding value of V.

Suppose we have a value of z and one of V given, and the corresponding value of A, B, or m is required.

Let $f(zV)$ denote the value sought;

z_0 and V_0 the next smaller values of z and V found in the tables;

$f(z_0V_0)$ the value from table corresponding to z_0 and V_0.

For an increase of 100 in z, we find that $f(z_0 V_0)$ increases by Δ_z; hence, for an increase of $z - z_0$, the increase in $f(z_0 V_0)$ will be

$$100 : z - z_0 :: \Delta_z : x;$$

$$x = \frac{z - z_0}{100} \Delta_z.$$

Again, for an increase of 50 ft.-secs. in V, $f(z_0 V_0)$ decreases by Δ_v; hence, following the same rule, the decrease for $V - V_0$ will be

$$x' = \frac{V - V_0}{50} \Delta_v.$$

The true value of $f(zV)$ then will be

$$A = f(zV) = f(z_0 V_0) + \frac{z - z_0}{100} \Delta_z - \frac{V - V_0}{50} \Delta_v. \quad (322)$$

Suppose now we have given $f(zV)$ and V, and wish to find z.

Solving equation (322) for z, we have

$$z = z_0 + \frac{100}{\Delta_z} \left\{ \frac{V - V_0}{50} \Delta_v + f(zV) - f(z_0 V_0) \right\}. \quad (323)$$

Or, having $f(zV)$ and z, required V. Solving equation (322) for V, we have

$$V = V_0 + \frac{50}{\Delta_v} \left\{ \frac{z - z_0}{100} \Delta_z + f(z_0 V_0) - f(zV) \right\}. . \quad (324)$$

209. Examples of Use of Auxiliary Tables—Ballistic Coefficient—$\frac{\delta_1}{\delta}$.

EXAMPLE 1. Find the value of $A = f(zV)$ for $z = 1446.7$ and $V = 1224.4$.

In formula (322) we have

$$z_0 = 1400;$$
$$V_0 = 1200;$$
$$f(z_0 V_0) = .0352;$$
$$\Delta_z = .0028;$$
$$\Delta_v = .0025;$$
$$z - z_0 = 46.7;$$
$$V - V_0 = 24.4.$$

Hence

$$A = f(zV) = .0352 + \frac{46.7}{100} \times .0028 - \frac{24.4}{50} \times .0025,$$

$$f(zV) = .0352 + .0000876 = .0353.$$

In a similar manner $B = f(zV)$, and $m = f(zV)$ may be found by using the proper tables.

EXAMPLE 2. Find z for $B = f(zV) = 0.1430$, and $V = 1740$.

In formula (323) we have

$$z_0 \text{ for } V\,(1700) \text{ and } B\,(.1409) = 5100.$$

$$\begin{aligned} \Delta_z &= .0046\,; \\ \Delta_v &= .0054\,; \\ V - V_0 &= 40\,; \\ f(zV) &= 0.1430\,; \\ f(z_0V_0) &= 0.1409\,; \\ V_0 &= 1700. \end{aligned}$$

Hence

$$z = 5100 + \frac{100}{.0046}\left\{\frac{40}{50} \times .0054 + 0.1430 - 0.1409\right\},$$

$$z = 5239.5.$$

In a similar manner, having $f(zV) = A$ or m, z may be found, using the proper tables.

EXAMPLE 3. Find V for $m = f(zV) = 0.2400$ and $z = 5250$.

In formula (324) we have

$$V_0 \text{ for } m\,(.2331) \text{ and } z_0\,(5200) = 1750,$$

$$\begin{aligned} \Delta_v &= .0101\,; \\ \Delta_z &= .0072\,; \\ z - z_0 &= 50\,; \\ f(z_0V_0) &= .2331\,; \\ z_0 &= 5200\,; \\ f(zV) &= .2400. \end{aligned}$$

Hence

$$V = 1750 + \frac{50}{.0101}\left\{\frac{50}{100} \times .0072 + .2331 - .2400\right\}.$$

$$V = 1733.67.$$

In a similar manner, having $f(zV) = A$ or B, V may be found, using the proper tables.

Ballistic Coefficient.—The value of this coefficient is given by equation (264), and its calculation involves that of $\frac{\delta_1}{\delta}$.

Calculation of $\frac{\delta_1}{\delta}$.—The values of $\frac{\delta_1}{\delta}$ are given in Table III, for degrees Fahrenheit from 0° to 100°, and for heights of barometer from 28 to 31 inches. To find the value of $\frac{\delta_1}{\delta}$ for any intermediate values of F and H not in the tables, we proceed exactly as in the case of the auxiliary tables.

Example.—Find the value of $\frac{\delta_1}{\delta}$ for $F = 49°.6$ and $H =$ 29.30 inches. From Table III we have

For $F = 49°$ and $H = 29$ inches; $\frac{\delta_1}{\delta} = 1.012$;

$$\text{Difference } \frac{\delta_1}{\delta} \text{ for } 1° \ F = +.002;$$

$$\text{Difference for } 0°.6 = +.0012;$$

$$\text{Difference } \frac{\delta_1}{\delta} \text{ for 1 inch } H = -.034;$$

$$\text{Difference for 0.30 inch} = -.0102.$$

Hence

$$\frac{\delta_1}{\delta} \text{ for } F = 49°.6 \text{ and } H = 29.30 \text{ inches} = 1.012 + .0012 - .0102$$

$$\frac{\delta_1}{\delta} = 1.003.$$

PRACTICAL PROBLEMS.

210. Kind of Fire to which Formulas Apply—Problem I—Use of Equation D.

Kind of Fire.—The formulas above deduced apply strictly to direct fire only, where the values of ϕ and θ are so small that Siacci's value of sec θ may be used without appreciable error.

The formulas give, however, sufficiently accurate results for indirect or curved fire, and hence they are used for both direct and curved fire; but for mortar fire they must be modified, as will be explained.

PROBLEM I — USE OF EQUATION (D). — Assume equation (D),

$$x = C\,[S(u) - S(V)].$$

Since C is generally known, we have in this equation three quantities, x, u, and V, any two of which being given, the third can be found. Solving equation (D) for each of the three quantities, we can write

$$x = C\,[S(u) - S(V)];$$

$$S(u) = \frac{x}{C} + S(V);$$

$$S(V) = S(u) - \frac{x}{C};$$

or, since

$$z = \frac{x}{C},$$

the two latter can be written

$$S(u) = z + S(V),$$
$$S(V) = S(u) - z.$$

For the whole range X we have similar equations, changing u into u_ω and x into X. Collecting these equations, we have

$$\left.\begin{aligned} x &= C\,[S(u) - S(V)]; \\ S(u) &= z + S(V); \\ S(V) &= S(u) - z; \\ X &= C\,[S(u_\omega) - S(V)]; \\ S(u_\omega) &= z + S(V); \\ S(V) &= S(u_\omega) - z; \\ z &= \frac{x}{C} = \frac{X}{C}. \end{aligned}\right\} \quad . \quad . \quad . \quad (325)$$

These equations enable us to solve the following problems, which may be grouped under Problem 1.

Given.	Required.
C, u, V	x
C, V, x	u
C, x, u	V
C, u_ω, V	X
C, V, X	u_ω
C, X, u_ω	V

In this problem, if the angle of elevation does not exceed 10°, the values of u and v will be practically the same, but for angles greater than 10° the value of v must be calculated from that of u by equation (F),

$$v = u \frac{\cos \phi}{\cos \theta},$$

and for this purpose the value of θ must be known.

Its calculation will be explained later.

211. Problem 2.—Use of Equations (316) and (321).

Assuming the above equations, we have

$$\sin 2\phi = A\,C; \quad \ldots \ldots \quad (316)$$

$$z = \frac{x}{C} = \frac{X}{C}. \quad \ldots \ldots \quad (321)$$

1. Having C, ϕ, and V, find the whole range X.

From (321) we have

$$X = Cz.$$

In this equation X and z are unknown.

But $A = f(zV)$, and from (316)

$$A = \frac{\sin 2\phi}{C}.$$

Hence in the equation $A = f(zV)$, we have $f(zV)$ and V given to find z, which is obtained from equation (323), using auxiliary table A.

This value in equation (321) will give X.

2. Having C, ϕ, and X, find the initial velocity V. We have

$$A = f(zV),$$

in which A, z, and V are unknown. But from (316)

$$A = \frac{\sin 2\phi}{C},$$

and from (321)

$$z = \frac{X}{C}.$$

Hence we have $f(zV)$ and z given to find V, which is obtained from formula (324), using auxiliary table A.

3. Having C, V, and X, find the angle of elevation ϕ.

From (316) we have

$$\sin 2\phi = AC,$$

in which A and ϕ are unknown. But

$$A = f(zV),$$

in which

$$z = \frac{X}{C},$$

and V is given. Hence we can find A by formula (322), using auxiliary table A. This value of A in (316) gives ϕ.

We have, therefore, for Problem 2,

Given.	Required.
C, ϕ, V	X
C, ϕ, X	V
C, V, X	ϕ

212. Problem 3—Time of Flight.

1. Having C, ϕ, V, and x, find the time of flight for the range x.

From equation (C) we have

$$t = \frac{C}{\cos\phi}[T(u - (T(V)],$$

in which u, the velocity at the point x, and t, the time to that point, are unknown.

But we have, equation (D),

$$S(u) = z + S(V),$$

in which $z = \frac{x}{C}$ and V are known. Hence u can be determined, and this value of u in equation (C) will give t.

If $\phi =$ or $< 10°$, $u = v$; if $\phi > 10°$,

$$v = u \frac{\cos \phi}{\cos \theta}.$$

2. Having C, ϕ, V, and X, find the time of flight for the whole range X.

From equation (C′) we have

$$T = \frac{C}{\cos \phi} \left\{ T(u_\omega) - T(V), \right\}$$

in which u_ω, the remaining velocity at the end of the range, and T, the time to that point, are unknown.

But we have, equation (D′),

$$S(u_\omega) = z + S(V),$$

in which $z = \frac{X}{C}$ and V are known. Hence u_ω can be determined, and this value of u_ω in equation (C′) will give T.

The same remarks apply to u and V as in 1.

If $\phi =$ or $< 5°$, $\cos \phi = 1$, practically, and we have

$$t = C \left\{ T(u) - T(V) \right\};$$

$$T = C \left\{ T(u_\omega) - T(V) \right\}.$$

For Problem 3 we have, then,

Given.	Required.
C, ϕ, V, x	t
C, ϕ, V, X	T

213. Problem 4—Angle of Inclination.

1. Having C, ϕ, V, and x, find the value of θ, the inclination of the tangent at the point x.

We have, from (314),

$$\tan \theta = \tan \phi \left\{ 1 - \frac{mC}{\sin 2\phi} \right\},$$

in which θ and m are unknown. We have

$$m = f(zV),$$

in which m and z are unknown. But from (321)

$$z = \frac{x}{C},$$

from which z can be found, and hence m by formula (322), using auxiliary table m.

This value of m in (314) will give θ. The value of θ thus found, when substituted in equation (F), will give v whenever $\phi > 10°$.

The value of θ may also be calculated from equation (319),

$$\tan \theta = \tan \phi \left[1 - \frac{m}{A} \right],$$

in which m is found as above, and

$$A = \frac{\sin 2\phi}{C} \quad \text{(equation (316))}.$$

2. Having C, ϕ, V, and X, find the value of ω, the angle of fall.

We have from (317)

$$\tan \omega = \frac{BC}{2 \cos^2 \phi},$$

in which B and ω are unknown. But we have

$$B = f(zV),$$

in which B and z are unknown. From (321)

$$z = \frac{X}{C},$$

from which z can be found. We have then z and V given, from which $B = f(zV)$ can be found, using auxiliary table B and formula (322), and this value of B in (317) will give ω.

If $\phi =$ or $< 5°$, we have, equation (318),

$$\sin 2\omega = B\,C.$$

214. Problem 5—Height of Trajectory—Maximum Height.

HEIGHT OF TRAJECTORY.—Having C, ϕ, V, and x, find the height of the trajectory at the range x.

We have from (320)

$$y = x \tan\phi\left[1 - \frac{a}{A}\right],$$

in which y, a, and A are unknown. From (316) we have

$$A = \frac{\sin 2\phi}{C},$$

from which A can be determined. We have also

$$a = f(zV),$$

in which a and z are unknown. But

$$z = \frac{x}{C},$$

from which z can be determined, and we have then z and V given, from which we can find $a = f(zV)$ by the use of auxiliary table A, and formula (322). These values of A and a in (320) will give y. Equation (315) may also be used.

MAXIMUM HEIGHT.—Having C, ϕ, and V, find the maximum height of the trajectory.

This will be at the summit of the trajectory, and for this point $\theta = 0$.

We have from (320)

$$y = x \tan\phi\left[1 - \frac{a}{A}\right],$$

in which y, x, a, and A are unknown.

For the summit of the trajectory make $y = y_0$, and $x = x_0$.

To find x_0, we have, (321),

$$x_0 = C\,z,$$

in which x_0 and z are unknown. Assume equation (319),

$$\tan\theta = \tan\phi\left[1 - \frac{m}{A}\right].$$

Since $\theta = 0$ at the summit, we have

$$m = A,$$

and from (316)

$$m = A = \frac{\sin 2\phi}{C}, \quad . \quad . \quad . \quad . \quad . \quad . \quad (326)$$

from which m can be determined. Then $m = f(zV)$, in which m and V are known and z can be found by auxiliary table m and formula (322). This value of z in (321) above will give x_0, the range corresponding to the summit.

In equation (320), since $m = A$ by (326), we have

$$y_0 = x_0 \tan \phi \left[1 - \frac{a}{m}\right];$$

but

$$m = a + b.$$

Hence

$$y_0 = x_0 \tan \phi \left[\frac{a + b - a}{m}\right],$$

or

$$y_0 = x_0 \tan \phi \frac{b}{m}. \quad . \quad . \quad . \quad . \quad . \quad . \quad (327)$$

In this equation y_0 and b are unknown. But we have $b = f(zV)$, in which z and V are known, and hence b can be calculated by auxiliary table B and formula (322). This value in (327) will give y_0.

215. Problem 6—To Determine the Dangerous Space—Rule of Double Position.

DANGEROUS SPACE.—The dangerous space is the horizontal distance over which an object of a given height will be struck. Suppose the height of the object is 6 feet. If we find first the whole range for a given elevation, initial velocity, etc., and then find the range at which the height of the trajectory is 6 feet, it is evident that for every point beyond this latter range, in the descending branch of the trajectory, the height will be less than 6 feet, and the object will be struck. The dangerous space, then, is the difference between the whole range, and the range corresponding to the given height. It is also evident that in general there will be two points of the trajectory whose heights are the same

—one point in the ascending branch, and one in the descending branch. The point in the descending branch is alone considered.

The problem then resolves itself into computing first the whole range, and then the range whose ordinate is y, and taking their difference.

DATA.—The data are C, ϕ, and V.

METHOD.—To find the whole range X, use the method of Problem 2.

To find the abscissa of the point whose ordinate is y. This problem is apparently the inverse of Problem 5, for which equation (320) is used; but on examining that equation it will be found that a is a function of x, and hence we have two unknown quantities, and the equation cannot be solved. The same is true of all the equations into which y and x enter; there is no direct and simple relation between them. Hence the problem must be solved by approximation.

For this purpose combine equations (D), (E), and (303), and we have

$$I(u_0)\,S(u) - A(u) = \frac{2\cos^2\phi}{C^2}y + I(u_0)S(V) - A(V). \quad (328)$$

In this equation we can compute $I(u_0)$ by (303), and hence all the quantities which enter the second member are known.

Represent this known quantity by k. Then we have

$$k = I(u_0)\,S(u) - A(u),$$

in which u_0 is known, and we have to determine the value of u by approximation.

RULE OF "DOUBLE POSITION."—For this purpose we make use of a method called the rule of "double position." Suppose we have an unknown quantity u whose value is sought. Let u_1 represent a quantity slightly greater than u, and u_2 a quantity slightly smaller.

Suppose u_1 substituted for u in the given equation, and the latter solved. A certain value will be obtained which will be erroneous. Denote the difference between this erroneous value and the true value by ϵ_1. Similarly, substitute u_2 for u, and denote the difference between the erroneous value and the true value by ϵ_2.

Then the hypothesis upon which the rule of double position is based is, that the errors ϵ_1 and ϵ_2 in the results are proportional to the errors made in assuming the values of u_1 and u_2.

The errors in assuming u_1 and u_2 are

$$u - u_1$$

and

$$u - u_2;$$

and from the above hypothesis we have

$$\epsilon_1 : \epsilon_2 :: u - u_1 : u - u_2,$$

and by division

$$\epsilon_1 - \epsilon_2 : \epsilon_2 :: u_2 - u_1 : u - u_2;$$

$$\epsilon_1 - \epsilon_2 : \epsilon_1 :: u_2 - u_1 : u - u_1,$$

which expresses the rule of "Double Position."

216. Example.

The above is best illustrated by a numerical example.

Suppose $k = 17666.1$, and $I(u_0) = 1.55658$.

Then we have

$$17666.1 = 1.55658\, S(u) - A(u).$$

Suppose

$$u_1 = 430 \text{ ft.-secs.};$$

$$S(u) = 21579.4;$$

$$A(u) = 15797.3;$$

$$17666.1 = 1.55658 \times 21579.4 - 15797.3;$$

$$\epsilon_1 = +126.6.$$

Again, suppose

$$u_2 = 420 \text{ ft.-secs.};$$

$$S(u) = 21978.7;$$

$$A(u) = 16861.3;$$

$$17666.1 = 1.55658 \times 21978.7 - 16861.3;$$

$$\epsilon_2 = -315.9;$$

$$\epsilon_1 - \epsilon_2 = 126.6 + 315.9 = 442.5;$$

$$\epsilon_2 = -315.9;$$

$$u_2 - u_1 = -10;$$

$$u - u_2 = u - 420.$$

Then

$$442.5 : -315.9 :: -10 : u - 420.$$

$$u = 427.139 \text{ ft.-secs.}$$

This value of u in the equation containing $S(u)$ and $A(u)$ gives

$$\epsilon_1 = +6.6.$$

It is necessary therefore to make a second trial. Assuming $u_2 = 426.8$ ft.-secs., and proceeding as before, we find

$$\epsilon_2 = -8.2;$$

and forming the same proportion as before, we find

$$u = 426.9878 \text{ ft.-secs.},$$

and this value of u will satisfy the original equation. There is also another value of u which will satisfy the equation, but it will readily be seen that it belongs to the ascending branch of the trajectory, and is not used.

Having the value of u for the point whose ordinate is y we find x by equation (D),

$$x = C\left\{S(u) - S(V)\right\},$$

and the dangerous space is

$$S = X - x.$$

As an approximate value for u in making these suppositions, the value u_ω for the end of the range may be calculated and used.

217. Rigidity of Trajectory—Drift.

RIGIDITY OF TRAJECTORY.—In the previous problems it has been assumed that the point of fall of the projectile is in the horizontal plane passing through the centre of the muzzle, or that the right line drawn from the centre of the muzzle to the end of the range, or the chord of the trajectory, is horizontal.

Suppose, however, as is generally the case in practice, that the object aimed at is above or below the level of the gun, the angle of elevation or depression being α.

Then it has been proved analytically that the relations existing between the elements of the trajectory, and the chord which represents the extreme range, are the same within certain limits, whether the chord is horizontal or inclined. In other words, the whole trajectory, with its chord, may be revolved a certain distance about a horizontal axis passing through the centre of the muzzle, without changing the relations between the trajectory and its chord.

This principle is called the "Rigidity of the Trajectory," and its practical use is as follows:

Suppose we fire at an object whose elevation is α. Calculate the angle of elevation ϕ for the given range, as usual, and aim directly at the target with the rear sight set at the elevation ϕ. The act of aiming at the target gives the actual elevation $(\phi + \alpha)$. If α is depression, it is affected with a minus sign. This subject is discussed later.

DRIFT.—Mayevski's formula for drift is (see Ingalls' Hand-book)

$$D = \frac{\pi u}{n} \frac{\lambda}{h} \frac{gCV}{\cos^3 \phi} \left\{ \frac{B(u) - B(V)}{S(u) - S(V)} - M(V) \right\} \frac{X}{10{,}000}, \quad (329)$$

in which

D is the drift in feet;

$u = 0.53$ for cored shot;

$u = 0.64$ for shell;

$n =$ the twist of the rifling in calibres, at the muzzle

$\frac{\lambda}{h} = 0.41$ for projectiles 2.5 calibres long;

$\frac{\lambda}{h} = 0.37$ for projectiles 2.8 calibres long;

$\frac{\lambda}{h} = 0.32$ for projectiles 3.4 calibres long;

π 3.1416;

g 32.2 feet;

C, ϕ, and V as in other ballistic problems;

$B(u)$, $B(V)$, $M(V)$ are drift functions whose values are found from Table I, like those of $S(u)$, $A(u)$, etc.;
X, the range in feet.

The drift will be more or less affected by the wind, according to its direction and velocity, and its effects will be further explained under the subject of Pointing.

218. Problem 7—Mortar Fire—Modified Equations—Calculation.

MODIFIED EQUATIONS.—The formulas for direct fire were obtained from the differential equations (A) by assuming that the inclination θ, of the tangent, at every point of the trajectory, is relatively small, and hence its cosine or secant constant, and approximately unity. For high angle or mortar fire, however, such an assumption is manifestly incorrect, since the angle θ varies greatly throughout the trajectory.

For mortar fire, therefore, Siacci assumes that there is a mean value of co. θ, which will satisfy the differential equations, and make their second members exact integrals. This mean value is denoted by α, and its value is shown analytically to be $\alpha = \frac{(\phi) - (\theta)}{\tan \phi - \tan \theta}$, (ϕ) representing $\int \frac{d\phi}{\cos^{n+1} \phi}$, and θ, $\int \frac{d\theta}{\cos^{n+1} \theta}$, and their numerical values being given in Table IV, together with the values of tan ϕ and tan θ.

This is applied as follows: In the integration of equations (A) in the case of direct fire, the second step consisted (see page 360) in substituting for $\sec^{n-1} \theta$, the constant value $\sec^{n-2} \phi$. But for mortar fire, α must be substituted for $\sec^{n-1} \theta$, wherever the latter occurs, instead of $\sec^{n-2} \phi$.

To show the effect of this substitution, take the second of equations (285), $dt = -\frac{C}{A \sec^{n-1} \theta} \frac{dv_1}{v_{1n}}$. Writing for sec θ its mean value α, we have $dt = -\frac{C}{A\alpha^{n-1}} \frac{dv_1}{v_1^{\,n}}$. Multiplying numerator and denominator by α, $dt = -\frac{C}{A} \frac{d(\alpha v_1)}{(\alpha v_1)^n}$. Representing αv_1 by u, we have $u = \alpha v_1 = \alpha v \cos \theta$. Making the same substitutions in the remaining equations (285), and integrating, we have the following formulas for mortar fire:

$$S = C \left\{ S(u) - S(U) \right\}; \quad . \quad . \quad . \quad . \quad (330)$$

$$t = C \left\{ T(u) - T(U) \right\}; \quad . \quad . \quad . \quad . \quad (331)$$

$$x = \frac{C}{\alpha} \left\{ S(u) - S(U) \right\}; \quad . \quad . \quad . \quad . \quad (332)$$

$$\frac{y}{x} = \tan \phi - \frac{\alpha C}{2} \left\{ \frac{A(u) - A(U)}{S(u) - S(U)} - I(U) \right\}, \quad . \quad (333)$$

in which S is the length of any arc of the trajectory, measured from the origin; $U = V\alpha \cos \phi$; $u = \alpha v \cos \theta$; $v =$ velocity at the point S; $\theta =$ the inclination of the tangent at the same point.

The values of the functions $A(u)$, $S(u)$, etc., can be taken from Table I, u and U being first calculated as explained in the nomenclature.

Calculation.—The most important problems in mortar fire are to find the whole range X, and the time of flight, T, for that range. For this purpose the given data are generally C, ϕ, and V. It is evident, however, that with the given data, equations (331) and (332) cannot be solved, and the solution is obtained as follows:

For the end of the range $y = 0$, and from equation (333) we have

$$\frac{2 \tan \phi}{\alpha C} + I(U) = \frac{A(u_\omega) - A(U)}{S(u_\omega) - S(U)}. \quad . \quad . \quad (334)$$

For mortar fire the angle of fall is very nearly equal to the angle of elevation, and under this supposition we have, since $-\theta = \omega$,

$$\alpha = \frac{(\phi) - (\theta)}{\tan \phi - \tan \theta} = \frac{(\phi) + (\omega)}{\tan \phi + \tan \omega} = \frac{(\phi)}{\tan \phi},$$

from which α is known. The first member of equation (334) is therefore known, and also $A(U)$ and $S(U)$ in the second member. u_ω is therefore found by "Double Position," as previously explained. This value of u_ω in (330), (331), and (332) will give the remaining values sought.

SUMMARY TABLE.

Given.	Required.	Problem.
C, u, V	x	1
C, V, x	u	1
C, x, u	V	1
C, u_ω, V	X	1
C, V, X	u_ω	1
C, X, u_ω	V	1
C, ϕ, V	X	2
C, ϕ, X	V	2
C, V, X	ϕ	2
C, ϕ, V, x	t	3
C, ϕ, V, X	T	3
C, ϕ, V, x	θ	4
C, ϕ, V, X	ω	4
C, ϕ, V, x	y	5
C, ϕ, V	y_0	5
C, ϕ, V	x_0	5
C, ϕ, V	Dangerous space	6
C, ϕ, V	x	7
C, ϕ, V	X	7
C, ϕ, V	S	7
C, ϕ, V	t	7
C, ϕ, V	T	7
C, ϕ, V	y	7

NOTE.—Ingalls' Ballistic Tables are to be used in these problems, and the methods of Capt. Ingalls have been followed in deducing the equations.

CHAPTER VII.

ARTILLERY CARRIAGES—THEORY OF RECOIL.

ARTILLERY CARRIAGES.

219. Classification—Principal Parts of Field and Siege Gun Carriages—The Axle.

CLASSIFICATION.—Artillery carriages may be classified according to the service for which they are intended, into field, siege, and sea-coast carriages.

Field and siege carriages are generally wheeled, and are intended to support the guns in firing, and to transport them from place to place, with their ammunition and necessary supplies.

Sea-coast carriages are intended only to support the guns in firing, and hence their construction differs materially from that of field and siege carriages.

PRINCIPAL PARTS OF FIELD AND SIEGE GUN CARRIAGES.—In the field and siege services, the carriage which supports the piece, and from which it is fired, is called the gun-carriage.

Its principal parts are:

1. The axle;
2. The wheels;
3. The stock or flasks;
4. The brakes;
5. The elevating device.

THE AXLE.—The principal parts are the body, the reinforce, and the arms.

The body is the middle part of the axle, between the arms, upon which the heads of the cheeks rest, and which bears the weight of the piece and the force of recoil. It is generally made of steel, and is solid, as this construction is necessary to resist the force of recoil in these carriages. Its length is governed by the requirement that the track of the wheels shall be the same as that of ordinary vehicles, so that it can be used on the same roads.

Reinforce.—To increase the strength of the axle and its resistance to bending under the force of recoil, and also to furnish a support for the cheeks of the carriage, the axle is generally reinforced. In the old carriages the axle-body was enclosed in wood; in the new field-carriages it is enclosed between two steel plates riveted together and fitting the exterior of the body accurately. For larger carriages, or for those in which the recoil is taken up by hydraulic buffers, this is not necessary.

The Axle-arms.—These form the supports for the wheels, and are the axes about which they revolve. The arms are made solid, terminating the axle-body. They are conical in shape, as this gives stiffness with small weight, enables the wheel to be put on easily, insures a good fit between wheel and axle-arm, and enables any wear to be taken up by means of washers.

The axis of each arm is inclined slightly downward so as to make the lower element nearly horizontal. This causes the lower spoke of the wheel to stand vertical, and relieves it from cross-strain, and also prevents a thrust upon the linchpin. The axis of the arm is also inclined slightly to the front, so that when the wheel meets any obstacle in that direction it will be free from cross-strain. These two inclinations of the axle-arm are called the "set." The wheel is secured on the arm by a linchpin which passes in a vertical direction through a hole in the end of the arm, and is held in place by a semicircular catch passing under the latter.

A shoulder on the inside, next the body, holds the wheel in place.

220. The Wheels—Parts.

The principal parts are, Fig. 203, the central part or

nave N, the spokes S, the rim R, and the tire T. *The nave* receives the pressure of the axle arm and transmits it to the spokes. Formerly naves were made of wood, and lined

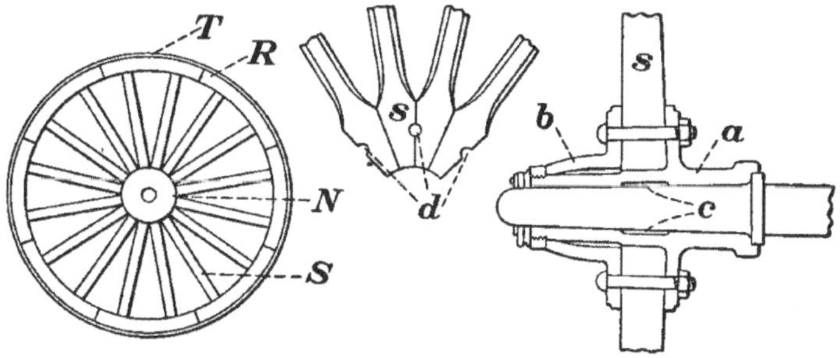

FIG. 203.

with a metal box, called the nave-box, which diminished the wear. Now they are made of malleable cast iron or bronze, in two parts, one (a) forming the nave-box and the other (b) forming a support for the spokes in front, which are inserted between these parts, pressed into place by a strong radial pressure, and bolted as shown at d, so as not to weaken them.

By this arrangement a spoke can be readily removed and replaced. This construction is used in the Archibald wheel, which is adopted in the U. S. service.

An enlargement c is sometimes made in the middle of the nave-box to contain the lubricant.

The spokes s, receive the pressure from the nave and transmit it to the rim. In our service they are made of hickory, as this gives great stiffness and elasticity for a given weight.

The stiffness is required to resist the thrust in firing, and strength is also required to enable the wheel to be used on rough ground, where the spokes are liable to be broken by contact with obstacles. The spokes are set at a slight angle with the axis of the nave, thus forming a conical surface. This is called the dish, and its object is as follows:

When the ground is inclined, the weight of gun and carriage produces a thrust on the lower wheel in the direction of the arrow. If the spokes were perpendicular to the axis of the nave, this thrust would cause a cross-strain on them, and its effect would be to loosen them in the nave, or cause them to work. The dish enables the spokes to resist this lateral thrust, and it is converted into a strain of compression. The whole structure thus acts as a circular truss, the rim being the tie.

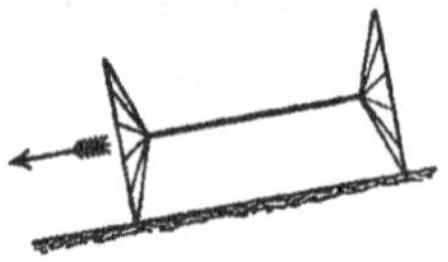

FIG. 204.

The Rim.—This distributes the weight which it receives from the spokes, to the ground. It is generally made of wood for the same reasons as in case of the spokes, and in several segments, called felloes. The object of this is to avoid cutting across the grain of the wood, and consequent weakness.

The Tire.—The segments of the rim and the spokes are held in place by the steel tire *T*, Fig. 203, which is shrunk on, and binds all the parts together. It also protects the rim from wear, and when any of the parts become loose, it can be removed, shortened, rewelded, and reset. For this purpose it is made of low steel. It is held in place on the rim by countersunk bolts passing through both.

221. Object of Wheel—The Stock.

OBJECT OF WHEEL.—The object of the wheel is to transfer the resistance to motion from the ground, where it is great and irregular, to the surface of the axle arm, which is lubricated, and the resistance of which is consequently small and regular.

The power being applied with a lever-arm, whose length is the radius of the wheel, while that of the resistance or friction is the radius of the axle-arm, the advantage of the wheel as a mechanical power increases with the radius of the wheel, and decreases with that of the axle-arm. On this account the radius of the wheel should be as great as possible and that of the axle as small as possible.

The radius of the axle-arm is fixed by the requirement of strength to support the shock of recoil; and that of the wheel by considerations of weight, draught, and facility of turning. A high wheel also is unstable. These considerations have fixed the diameters of wheels in the field and siege services as follows: field service, $57\frac{3}{4}$ inches; siege service, 60 inches. The siege wheel is much stronger and heavier than that for the field service.

An increase in width of rim also distributes the weight over a greater area and enables the wheel to better overcome the resistance offered by soft ground to traction; but it increases the weight of the wheel and decreases the facility of turning.

THE STOCK.—This consists of two pieces, called the flasks, which are separated at the upper ends, forming the cheeks, and which gradually converge at the lower ends, and are united there by a solid piece called the trail-plate or lunette. The cheeks rest upon the axle body or reinforcing plates, and have on their upper surfaces two trunnion-beds, in which the trunnions of the gun rest. The trunnions are held in place by two caps, called cap squares, which fit over bolts projecting from the cheeks at the extremities of the trunnion-beds, and are fastened by keys or bolts. The flasks are also united by various transoms, to give stiffness to the structure. The supports for the elevating screw or other device are generally attached to the stock.

The distance between the flasks varies with the size of the gun, and should be sufficient to allow the breech to be depressed to the maximum extent required in service. By this separation, also, the strain due to recoil is distributed over a greater length of axle body, and thus the resistance to bending is increased. The stock is subjected to a strong transverse stress in firing, and hence must be designed to resist this. It also acts to couple the gun-carriage to the limber when the gun is to be transported, and it gives the necessary third point of support in firing, and enables the piece to be pointed. To it are attached the supports for the sponges and rammers, and in general, if possible, no parts are allowed to project below the plane of its lower edges, to avoid striking obstacles. When not required to resist the shock of firing, its use is simply to connect the carriage and limber, and its construction then differs materially from that described, being much simpler and lighter.

222. The Brakes—Friction-brakes—Shoe—Hotchkiss—Lemoine—Nordenfelt.

BRAKES.—The object of a brake in the field and siege service is to limit the recoil, so that the piece may be kept approximately in its firing position, and thus avoid the fatigue to the cannoneers of running the piece back over a considerable distance to that position after discharge, and the consequent delay in loading.

The principles of brakes will be explained under the subject of recoil.

For the field and siege services they may be divided into—

1. Friction-brakes.
2. Elastic brakes.
3. Hydraulic brakes.

FRICTION-BRAKES.—These, as will be seen later, do not give the best results, but are sometimes preferred on account of their simplicity, and as being less liable to get out of order.

SHOE.—The simplest friction-brake is the shoe, which consists (Fig. 205) of a strong piece of iron, *a*, fitting the

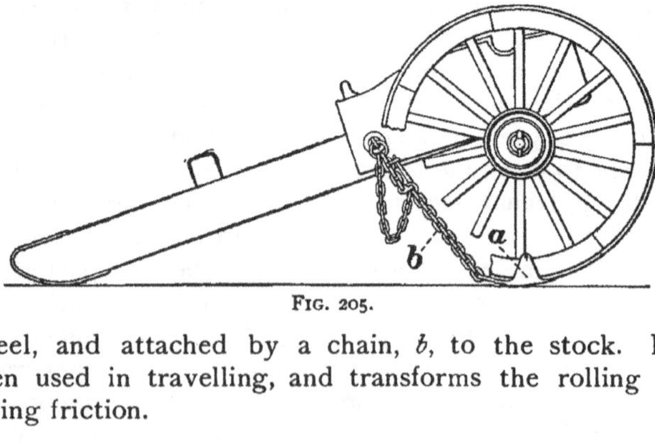

FIG. 205.

wheel, and attached by a chain, *b*, to the stock. It is often used in travelling, and transforms the rolling into sliding friction.

THE HOTCHKISS BRAKE (Fig. 206) consists of a conical box, *a*, working in screw-threads on the axle-body *b*. The nave of the wheel is also made conical at *c*.

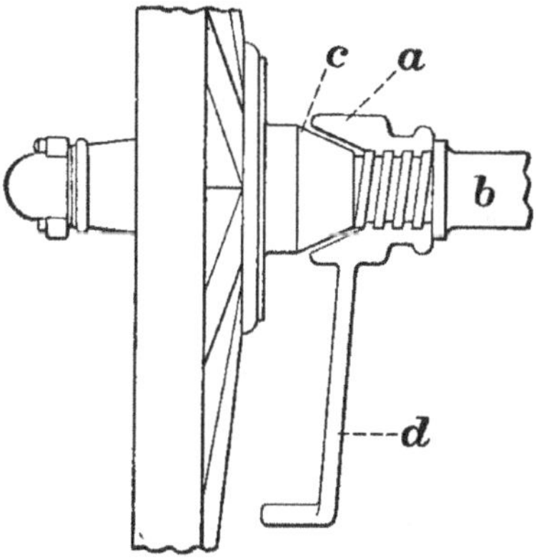

FIG. 206.

By turning the handle *d* attached to the brake, it is screwed up till the conical surfaces are in close contact. The friction between these surfaces, when the wheel rotates, tends to tighten the brake, and thus increase the resistance to rotation, while if the moment of rotation becomes too great, the surfaces will slip, and thus prevent destruction of the parts.

THE LEMOINE BRAKE is used in the French service. It consists (Fig. 207) of a rope, *a*, attached to the brake-beam at *b*, and wound loosely around the nave of the wheel.

This rope is tapering, being larger at *b*, and gradually decreasing in size. It is attached in front to a cross-bar, *c*, and this is connected to the rod *d*, which moves freely in the direction of its length, and carries a heavy mass, *e*. The action of the brake is as follows: When the piece is fired, the carriage recoils in the direction of the arrow, while the rod *d*, on account of the mass *e*, moves relatively forward. It is held in this position by the notches on *d* bearing against the edges of the plate through which it slides. This tightens the cord around the nave of the wheel, and causes it to be wound up as the wheel turns. Owing to the increase in diameter of the rope, it is wound more rapidly as the length of recoil increases and its velocity decreases, so that

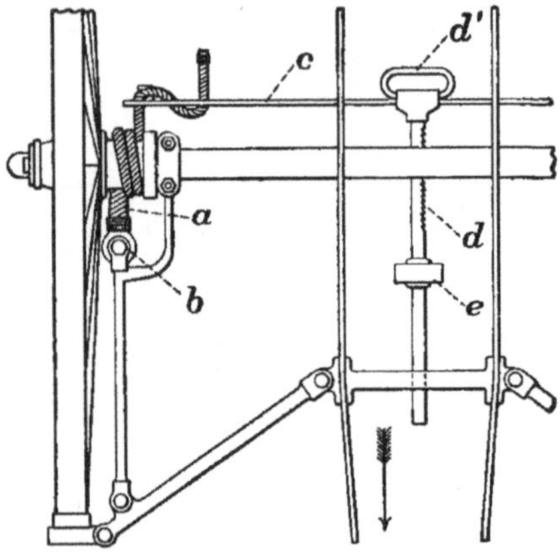

FIG. 207.

the brake is applied gradually. It may also be applied by hand, in travelling, by pulling out the rod *d* by the handle *d'*.

NORDENFELT BRAKE.—This is found on the carriage of the Nordenfelt rapid-fire gun, and also on the Hotchkiss carriage. It consists of a frame, one side of which is shown in Fig. 208, attached to the axle above its centre at the points *aa*; *bb* are the brakes, *c* the rod connecting them, *dd* rubber washers through which the brake-rods *ee* pass. As the points of support *a* are eccentric with reference to the axle, when the brakes are lowered, they come in contact with the wheels, and any rotation in recoil binds them still more tightly. When not in use, they are hooked up to the cheeks of the carriage. This brake is elastic also.

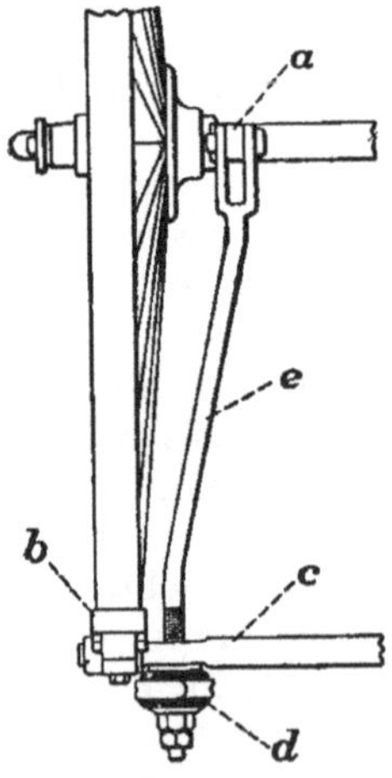

FIG. 208.

223. Elastic Brakes—Buffington—Englehardt—Belleville Springs.

ELASTIC BRAKES.—These check and moderate recoil by transmitting the first shock to some elastic body, which is thereby deformed, and when this body resumes its original form, due to its elasticity, the shock is gradually transmitted to the parts of the carriage. This relieves the carriage from the sudden shock, and thereby enables it better to sustain recoil.

THE BUFFINGTON BRAKE.—This was designed by Colonel Buffington of the Ordnance Department, and is used with the field carriages.

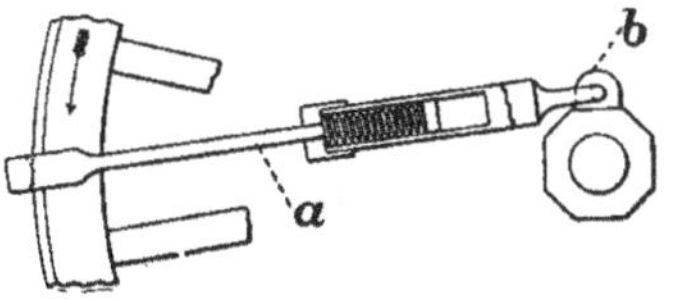

FIG. 209.

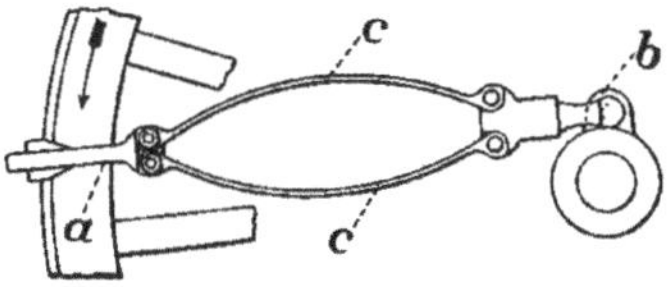

FIG. 210.

The older form consists of a rod, *a*, Fig. 209, surrounded by a spiral spring in a casing. The outer end of this rod is formed into a hook, which fits over the tire of the wheel. The casing which carried the rod and spring is attached to a hook, *b*, above the centre of the axle. When the rod and casing are lowered, the hook rests against the tire, being eccentric to the wheel. Any rotation of the wheel in the direction of recoil draws the rod out of the casing, and compresses the spiral spring. The brake is thus gradually applied. Various defects in this brake have caused the adoption of the later form shown in Fig. 210.

Later form shown in Fig. 210. Instead of the casing and spiral spring, the rod is attached to a bow-spring, *c*, which is elongated when the wheel recoils. It is held vertically when not in use.

THE ENGLEHARDT BUFFER.—This is used on some of the English carriages. It consists (Fig. 211) of an elastic buffer, *a*, of cork, rubber, or springs, which rests against a transom, *b*, attached to the cheeks of the carriage.

These cheeks have a bracket, *c*, in front, in which the axle *d* rests, and which allows them to move backward independently of the axle, and they are notched in rear at *e*, to allow a motion independent of the cross-bar *f*. The axle is attached to the cross-bar *f* by the brace *g*, and this attachment is made as near the axle-arm as possible, to

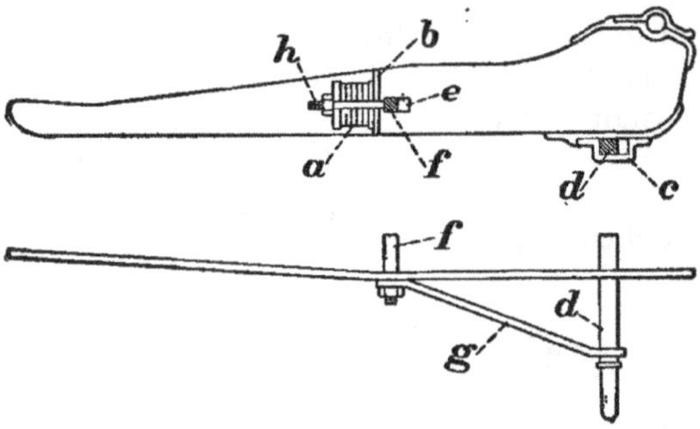

FIG. 211.

avoid bending. A bolt, *h*, passes through the buffer *a*, and through a hole in the transom *b*, and is attached rigidly to the cross-bar *f*. The action is as follows: When the piece is fired, the cheeks and transom *b* recoil together, the axle and cross bar sliding in their notches *c* and *e*. This motion compresses the buffer *a*, and as it recovers its shape, the force of recoil is gradually transferred to the wheels and axle, through the cross-bar *f* and brace *g*.

BELLEVILLE SPRINGS. — These are saucer-shaped disks of steel, *s*, Fig. 212, fitted edge to edge, and kept in place

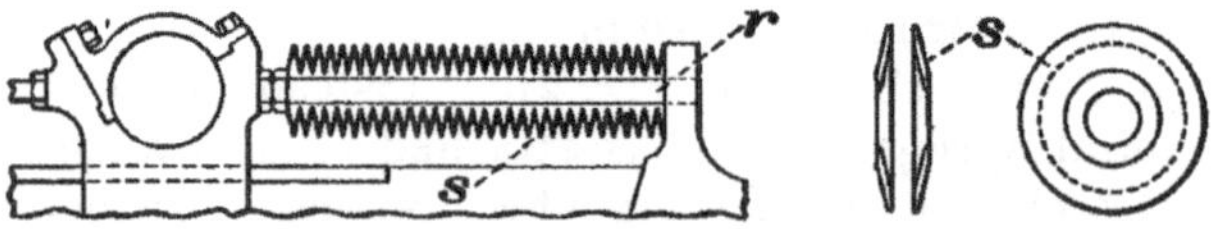

FIG. 212.

by an axial rod, *r*, for which purpose a hole is pierced in the centre of each disk. Since they occupy a relatively small space, a large number of them may be employed, and the compression of each is small. They are, however, expensive, and spiral springs are often used in place of them.

224. Hydraulic Brakes—Elevating Devices—The Elevating Screw.

HYDRAULIC BRAKES.—These are not used in the field service, owing to their weight, and liability to get out of order when subjected to rough usage, but they are used in the siege carriages, and in general wherever the recoil is great, and it is necessary to regulate it very exactly. They will be considered under Sea-coast Carriages, where they are always used.

ELEVATING DEVICES.—These are used to give the proper elevation to the piece, and may consist of—

A screw;
A toothed sector;
A combination of levers.

THE ELEVATING SCREW is generally double, and consists (Fig. 213) of an exterior hollow screw, *a*, working in a fixed nut, *b*.

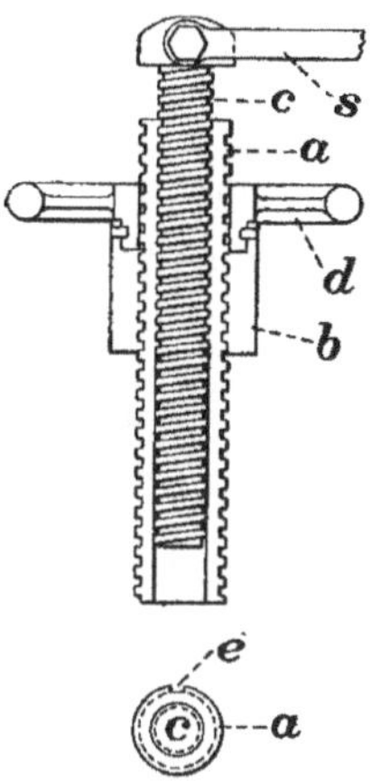

FIG. 213.

The exterior of the screw *a* has a left-hand thread, its interior a right-hand one. A second screw, *c*, works in the interior thread of *a*. *d* is a hand-wheel, which is free to rotate, but is fixed to the nut *b*, so that it has no motion of translation. A longitudinal channel or spline, *e*, is cut on the exterior of *a*, and a key on *d* fits this. The action is as follows: When *d* is turned it causes *a* to turn with it, on account of the spline and key, and at the same time *a* working in the fixed nut *b* moves parallel to its own axis. The head of *c* being fixed by a strap, *s*, to be described, cannot turn, and *c* is forced to move parallel to its own axis by the rotation of *a*, and the action of its interior screw-thread. The resultant motion is, for each turn of *d*, equal to the sum of the pitches of the two screws.

The advantage gained is that we are enabled to use an elevating screw, which is short ordinarily, but which can be lengthened to give any elevation or depression desired.

Strap.—To cause the blow on the head of the elevating

screw, upon firing, to be normal to its axis, and thus avoid bending, the nut *b*, in which the screw works, is arranged on trunnions between the cheeks of the carriage, and a strap, *s*, Fig. 214, is attached at one end to the head of the

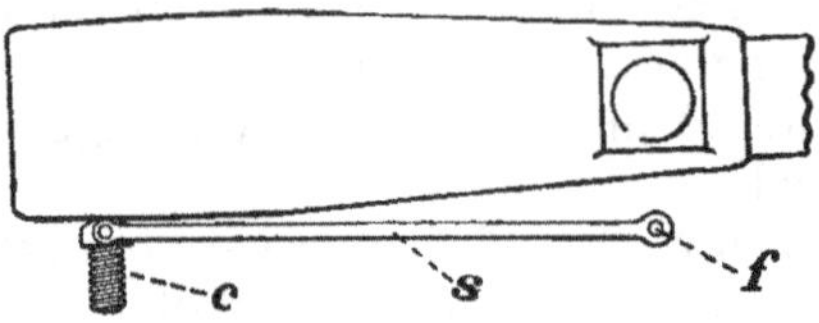

FIG. 214.

screw *c*, and at the other to an axis, *f*, parallel and near to the axis of the trunnions of the gun. In this way the axis of the screw, *c*, is kept nearly normal to the axis of the gun for all elevations.

225. Elevating Devices—The Toothed Sector—The Levers.

THE TOOTHED SECTOR.—This is used generally in combination with gearing, on the larger guns. It consists (Fig. 215) of a toothed arc, *a*, bolted to the gun, and acted

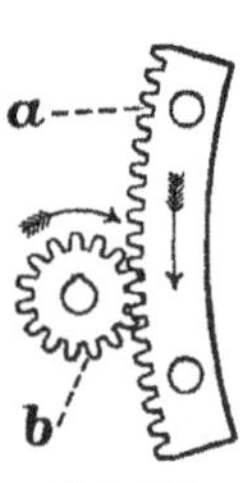

FIG. 215.

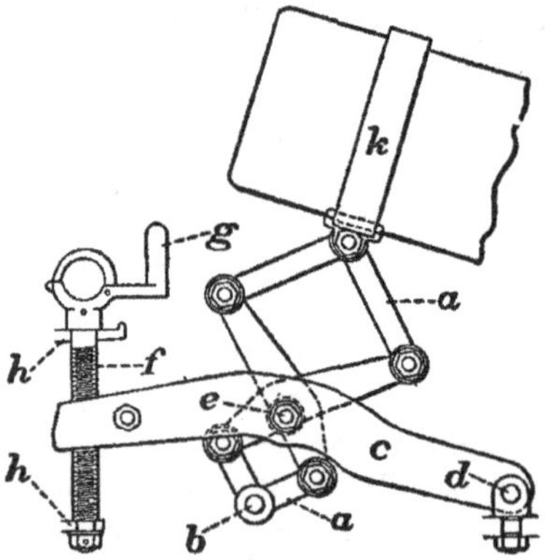

FIG. 216.

on by a gear, *b*. This gear may be worked directly by a hand-wheel, or more frequently by intermediate gearing. To gain power, and secure small motions, a worm-gear is frequently used.

THE LEVERS.—A combination of levers, called a "lazy-tongs," is used as an elevating device on the light 3.20 carriage. It consists (Fig. 216) of the arms *a*, jointed as shown and attached at *b* by a fixed axis to the carriage. Two side levers, *c*, are attached to a fixed axis at *d* on the carriage, and to the arms *a* at *e*. A screw, *f*, passes through the other extremity of the side levers *c*, and works in two collars, *hh*, attached to the carriage. When the screw *f* is rotated by the handle *g*, the side levers *c* are raised or lowered, and acting on the arms *a* through its connection *e*, it causes the structure to elongate or contract, and thus to elevate or depress the gun. The device is connected with the breech of the gun by a leather strap, *k*, passing over the breech.

226. Draught—Modes of Work of Horse—Pack-horse—Draught-horse—Angle of Traces.

DRAUGHT.—Field and siege carriages are intended not only to support their pieces during firing, but also to transport them from place to place. For this purpose the two-wheeled gun-carriage must be converted into a four-wheeled one, by the attachment of a limber. This leads to a consideration of the load which can be carried by the horse, and the best method of attaching him to the carriage.

MODES OF WORK OF HORSE.—A horse may carry his load on his back, in which case he acts as a pack animal; or he may draw this load by being attached to a carriage, as a draught animal; or these two methods may be combined.

PACK-HORSE.—This method is only used in the mountain service, when the roads are impassable for wheeled vehicles. Under such circumstances the load for a horse is from 200 to 250 lbs., and, if moving at a walk, he can carry this load 25 miles in a day. If at a trot, the load or the distance, or both, must be reduced. In this case he can carry the same load about 17 miles in a day.

The daily work of a pack-horse is considered equal to that of five men. The mule is a better pack animal than the horse, as he can carry more, is more sure-footed, and eats less. He is therefore generally used for this purpose.

DRAUGHT-HORSE.—A horse can, by the aid of the wheel, draw much more than he can carry, and hence it is always advantageous to use him as a draught animal.

In considering the draught of a horse, his effort may be divided into two parts: first, that necessary to start the carriage, and second, that necessary to keep it in motion.

The first being only temporary, may approximate the maximum strength of the horse, but it is important to know it, as upon it is based the strength of the harness.

Experiment shows that this effort varies with different horses from 600 to 1000 lbs., as measured with a spring dynamometer.

The second part of the effort varies with circumstances, such as the load, nature of the roads, etc., from $\frac{1}{10}$ to $\frac{1}{40}$ of the load. When horses are used together in a team they will do less work than the same number singly, owing to their interference with each other.

ANGLE OF TRACES.—The angle made by the traces with the ground also influences the amount of work done by a draught-horse.

If the traces be attached to the carriage at a point higher than that at which they are attached to his shoulders, it is evident that a component of the pull acts upward and decreases the hold of the horse on the ground, and, consequently, his power to pull. On the other hand, if this point is below the point of attachment at the shoulder, the vertical component acts in the opposite direction, and increases his hold on the ground. Experiment has shown that when the horse carries no load, the best result is obtained when the traces make an angle of from 10° to 12° with the ground. The tangent of 12° being about $\frac{1}{5}$, this shows that the horse pulls best when $\frac{1}{5}$ of his load is transferred to his shoulders. On the other hand, if a horse carry a load of 150 to 200 lbs., and pull at the same time, experiment shows that this angle should be 6° or 7°.

Making allowance for bad roads and rough usage, artillery horses draw less than those of commerce, and the loads allowed per horse are about as follows:

Horse artillery, 650 lbs.;
Field artillery, 700 to 850 lbs.;
Siege artillery, 1000 lbs.

227. Modes of Attachment of Horses—Attachment of Traces.

MODES OF ATTACHMENT.—Horses may be attached to a carriage in three ways:

1. In single file, with the wheel-horse in shafts;
2. In double file, with one of the wheel-horses in shafts;
3. In double file, with the two wheel-horses on opposite sides of a pole.

The team is ordinarily composed of six horses, arranged in pairs. The horses nearest the carriage are called the wheel-horses, those next in front the swing-horses, or swing-team, and those in front the lead-horses or leaders.

In each team the left horse is the *near*, and the right horse the *off* horse. The near horse carries the driver.

Single File.—The objections to this method of attachment are that much of the tractile force is lost, owing to the curving and turning of ordinary roads; it is difficult to make all the horses pull together; and the shaft-horse, being subjected to the irregular action of the other horses of the team, is soon worn out. For heavy loads over good straight roads at a slow pace, it has the advantage of a direct pull.

Double File, Wheel-horse in Shafts.—This method obviates the defects of the long line of traction on ordinary roads, and it controls the movements of the carriage well; but it subjects the shaft-horse to excessive fatigue, and hence is not generally used.

Double File, with Pole.—This method is generally adopted, as it gives the advantages of a short line of traction, with comparatively little fatigue to the wheel-horses, the control of the carriage being effected by two horses instead of one.

ATTACHMENT OF TRACES.—The traces may be attached to the carriage by a single fixed bar, *a*, called the splinter-bar, as in the old carriages, or by a double tree, *b*, which is pivoted at its middle point to the pole, and the traces attached to each end by a movable single tree, *c*. Fig. 217 shows the old and the new methods.

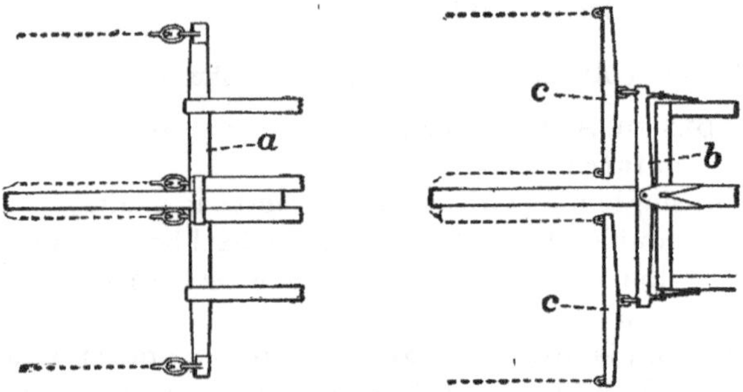

FIG. 217.

The advantage of the old method is that it is simple and strong. The disadvantages are that it throws an unequal amount of work upon the two horses, so that a willing horse will do most of the work, and there is no method by which the driver can detect this. Consequently, the resultant of the traction in this case will not coincide with the pole or middle of the axle.

The advantage of the new method is that it obviates the above difficulty, and forces an unwilling horse to do his share of the work. The single trees also prevent chafing, by yielding to the motion of the horse as he advances his shoulders in pulling.

228. Support of Pole—Line of Draught of Traces.

SUPPORT OF POLE.—The weight of the pole may be supported—

1. By counterbalancing it in rear of the axle;
2. By suspending it from the necks of the horses;
3. By a combination of these methods.

Counterbalancing Weight of Pole.—The advantage of this method is, that it frees the necks of the horses from the weight of the pole, and they are consequently not fatigued by it.

The disadvantages are, that if it is done by placing the trunnions of the gun well to the rear on the gun-carriage, it causes difficulty in limbering, because of the extra weight lifted. If it is done by allowing the trail to project over the pintle hook and rest upon a circular sweep-bar, *a*, Fig. 218, it is difficult to limber, as the trail must be raised suffi-

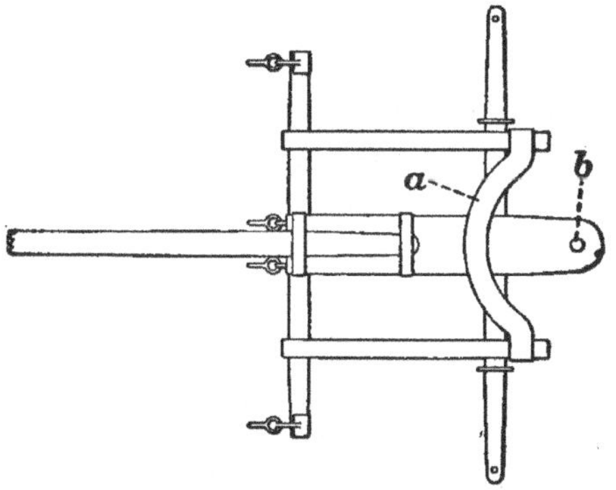

FIG. 218.

ciently to pass over the pintle *b*. As quickness of limbering is of importance with field artillery, this method is not used.

In the siege service, where quickness is not required, and where the weights are relatively great, it *is* used, especially the method shown in Fig. 218.

Suspending Weight from Necks of Horses.—The advantage of this method is that the attachment of the trail to the limber can be made at the most convenient place for limbering, and the weight of the gun can be thrown so far forward as to render the operation of limbering very easy. The great objection, however, is that it fatigues the horses too greatly, and cannot be used.

Hence a combination of these two methods has been adopted in the field service. To diminish the weight resting on the necks of the horses, the ammunition-chest of the limber has been placed, so that its centre of gravity when loaded is directly over the axis of the axle, and the weight upon the trail when the gun is limbered is regulated to partly counterbalance that of the pole.

LINE OF DRAUGHT OF TRACES. — This is so arranged that the line of traction shall be continuous from the lead-horses to the carriage, and is accomplished by attaching the traces of the swing-team directly to those of the wheelers, and those of the lead-team directly to the traces of the swing-team. By this means each horse pulls independently of all the others, and there is no interference.

229. Turning Angle—Direction of Carriage—Backing.

TURNING ANGLE.—The angle required to turn the carriage in, is called the turning angle, and is measured by one half the horizontal angle through which the pole sweeps. In practice an angle of 60° is sufficient. It varies with the arrangement of the horses, the height of the front wheels, the length of stock, the position of the pintle, and the thickness of the stock at the point where the front wheels strike it. Other considerations determine that the height of the front wheels shall be the same as those of the rear ones for interchangeability, that the length of stock shall be governed by considerations relating to recoil, and that the position of the pintle shall be considered with reference to ease of limbering and weight of pole on the horses' necks, as before explained. For carriages that do not withstand the shock of recoil, the length of stock is adjusted with reference to the turning angle.

DIRECTION OF CARRIAGE.—This is given by means of the pole. The latter being attached to the necks of the wheel-horses, the direction may readily be changed by directing the wheel-team to the right or left.

BACKING.—The pole is also used to stop and back the carriage. The wheel-horses are attached to the front of the pole, as will be explained when the harness is described.

In stopping or backing the backward thrust of the horses is applied at the end of the pole, and this thrust transmitted to the rear along the pole to the carriage.

230. The Harness.

The harness at present in use for the field artillery was designed by Major Williston of the artillery. That for the wheel-team differs slightly from the swing and lead harness. For the wheel-team it consists of—

1. The head gear to guide and hold the horses;

2. The saddle to transport the driver;

3. The draught harness, by which the carriage is moved forward;

4. The breeching, by which the carriage is stopped, or moved to the rear;

5. The breast straps, by which direction is given to the carriage, and the weight of the pole supported.

For the swing and lead teams the breeching and breast straps are omitted, and the traces are supported by a single hip strap.

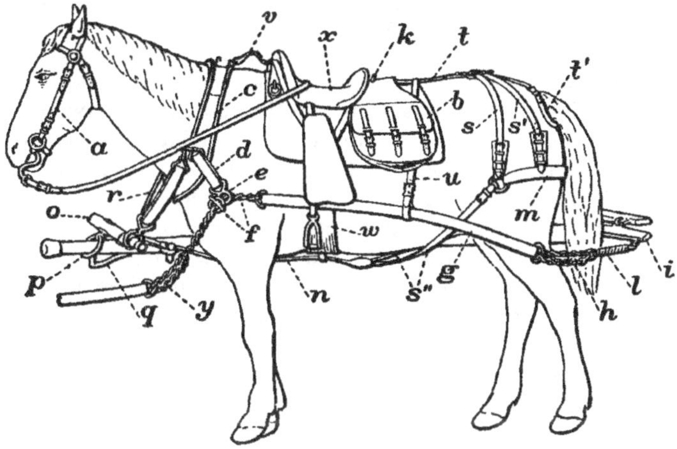

FIG. 219.

Head Gear.—The head gear consists (Fig. 219) of the bridle *a*, by which the horse is guided, and the halter for holding him when not in the carriage.

The bridle and halter are the same as those used in the cavalry service. The bridle rein of the off horse passes through a pulley on the front of his saddle, so that there is a direct instead of an oblique pull, in stopping and backing.

Saddle.—The saddle *x* is the same as that used in the cavalry service. Each horse is saddled, the saddle being held in place and motion to the front prevented by the back strap *t* and crupper *t'*, while the collar is secured to the saddle in front by the strap *v*. The girth strap or cincha *w* prevents motion of the saddle around the horse. The near horse carries the driver, and the off horse may carry an extra cannoneer. Saddle-bags, *b*, are used, as in the cavalry service, to carry the clothing, etc., of the drivers.

Draught Harness.—This consists of a collar, *c*, made of U-shaped steel. It is hinged at the top, and closes at the bottom with a spring catch. It rests against the shoulders of the horse, and is intended to distribute the force of traction over a greater area, and thus prevent chafing.

A strong leather tug, *d*, is attached to each branch of the collar, and at the outer end of the tug is an iron ring, *e*, through which the front trace-chain *f* passes.

The trace *g* is a stout leather strap, terminated at the front end by a chain and toggle, *f*, and at the rear end by a ring, through which passes the rear trace-chain *h*, having at one end a hook and at the other end the spring *l*. The rear ends of the trace-chains *y*, of the swing horses, are attached directly to the front trace-chains *f* of the wheel-horses, thus giving a continuous line of traction throughout the teams. The loin strap *u* supports the traces at their middle point. The trace-chains of the wheel-horses are attached to the single trees *i*, and in unharnessing, these are detached from the double tree, and hooked to the rear of the saddle, for which purpose a hook, *k*, is provided. The spring *l* is used to attach the traces of the wheel-horses to the single trees *i*. This allows a gradual starting of the carriage, and thus diminishes the fatigue of the horses, and the strain on the harness.

Breeching.—This consists of a breech strap *m*, hip straps *ss'*, two side straps *s''*, and a broad flat strap or martingale, *n*. The breech strap *m* passes around the hind-quarters of the horse, and is supported by the hip straps *ss'*, two on each side. The breech strap is joined to the martingale *n* by the two side straps *s''*. The martingale *n* passes along under the horse, and between his forelegs to the front, where it is connected to a transverse bar, *o*, on the front end of the pole, called the neck yoke. This yoke, *o*, is of wood, and has a ring, *p*, attached at its middle point, which slips over the end of the pole, and rests against a stop, *q*, on the under side. The collars of the wheel-horses are attached to the ends of the neck yoke *o* by the breast straps *r*.

In backing, the pressure is exerted by the horse against the breech strap *m*, and this pressure is transmitted through the side straps *s''* and the martingale *n*, to the neck yoke *o*, and thence to the pole.

The action of the harness may be understood from Figs. 220 and 221. For draught and direction *t* represents

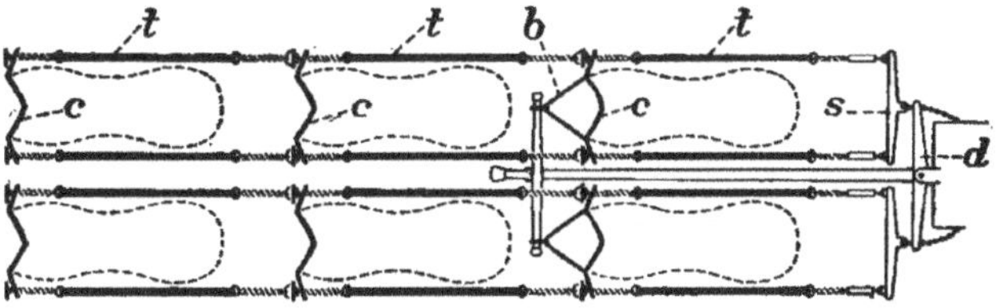

FIG. 220.

the traces, *s* the single trees, *d* the double tree, *b* the breast straps, *c* the collar. For breeching *b'* is the breech strap, *s* the side straps, *m* the martingale, *n* the neck yoke.

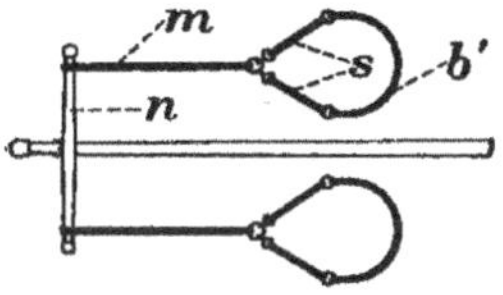

FIG. 221.

The great advantage of the present harness over the old is the change in the breeching, by which the horse has a direct instead of an oblique thrust to the rear. Many other improvements are embodied in it.

CARRIAGES FOR MOUNTAIN AND FIELD ARTILLERY.

231. Carriage for Hotchkiss Mountain Gun—For 3.6 Field Mortar.

HOTCHKISS MOUNTAIN CARRIAGE (Fig. 222).—The flasks *a* are made of steel strengthened with angle-irons, *b*, and with three transoms, *c*, *d*, *e*, and a trail-plate, *g*. The axle is solid, and the wheels have bronze naves. Recoil is checked when necessary, by a rope tied around the spokes of the wheels, and passing over the stock.

The elevating screw passes through the transom *e*. For

draught, a pair of shafts is attached to the trail by the hook *h* and a pin, and the gun and carriage drawn by one mule. The gun and carriage are generally packed on two mules,

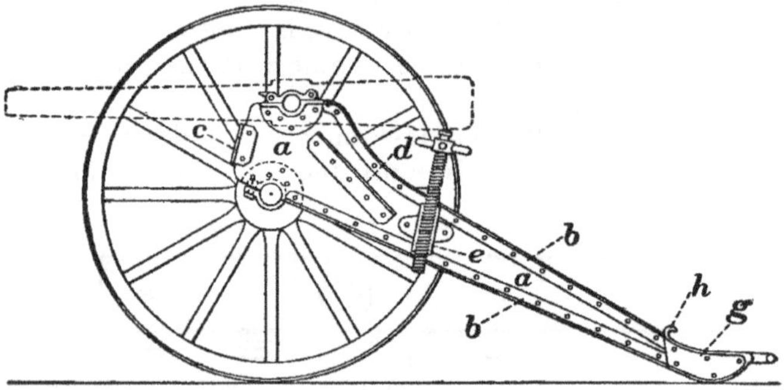

FIG. 222.

and the ammunition carried in boxes on mules also. Weight of carriage, 220 lbs. The carriage for the 3-inch gun is similar to this.

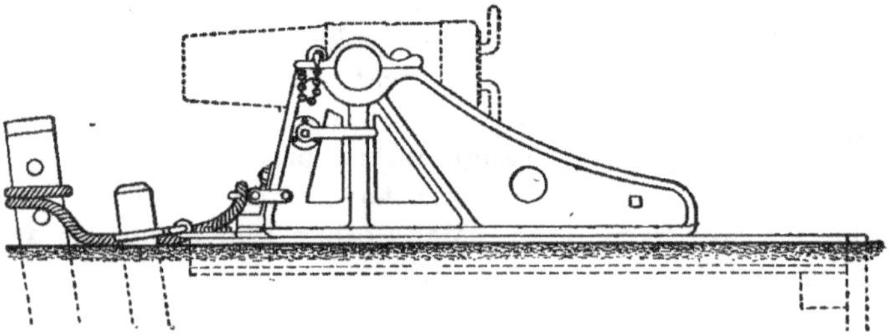

FIG. 223.

3.6 MORTAR CARRIAGE (Fig. 223).—This carriage is made of cast steel, in a single piece, provided with a clamping device in front, which bears against a steel arc attached to the mortar.

Elevation is given by the quadrant, and the mortar clamped in position by the clamping device. When in use the carriage rests on a wooden platform, and recoil is checked by a heavy rope attached to stakes in front.

232. 3.2-inch Field-gun Carriage.

This carriage was designed by Colonel Buffington of the Ordnance Department. Its principal features are—

1. The method of reinforcing the axle;
2. The formation of the flasks;
3. The elevating device;
4. The brake.

Reinforcing Axle. — To stiffen the axle against recoil, the body is enclosed (Fig. 224) between two plates of steel,

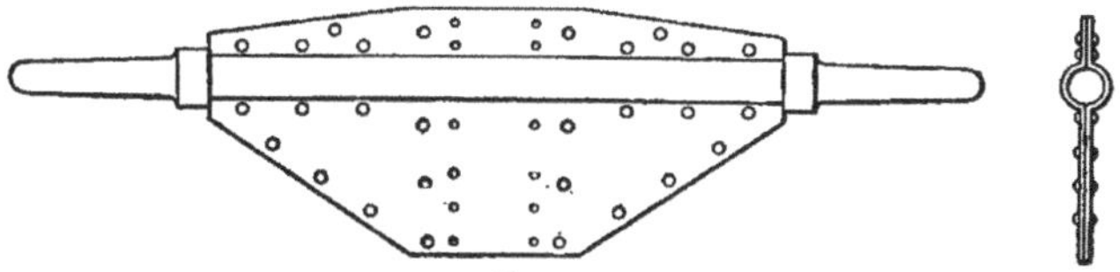

FIG. 224.

which are riveted together temporarily, bored out to a diameter slightly less than that of the exterior of the axle body, and the plates then riveted tightly together. The width of these plates is in the plane of the lower edges of the flasks, and hence they resist the force of recoil. They also serve as supports for the flasks which are bolted to them.

Formation of Flasks.—Each flask (Fig. 225) is formed of two pieces of sheet steel, stamped while hot between dies into the shape shown, and riveted together through the flanges. The lower edge *a* of the outer piece of each flask projects inward, forming a flange, to which the transoms are riveted, and by which the flasks are bolted to the axle plates. The flasks are connected by transoms, three of which form a tool-box, with a hinged lid, for carrying tools, oil-can, and loose primers.

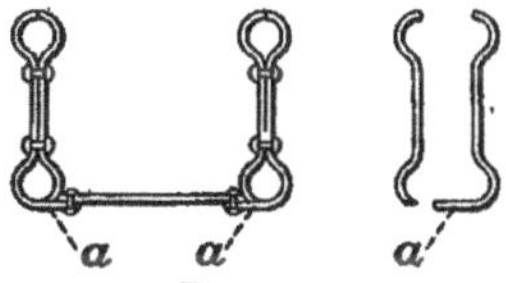

FIG. 225.

The above construction gives great lateral and vertical stiffness. The lower ends of the flasks converge to a trail-plate, at the extremity of which is the lunette ring, by which the carriage is hooked to the limber.

Elevating Device.—This is an assemblage of jointed levers, previously described, called a "lazy tongs."

Brake.—The bowspring brake is used.

Minor Parts.—The wheels are of the Archibald pattern. The trail handspike is made of two pieces of wood split axially, and having a sheet of steel between them, the whole bound together by a series of rings. The handspike is hinged to the trail-plate, and when not in use is folded against the trail, and held in place by a spring catch.

There are two seats for cannoneers on the axle. The flasks also carry the supports for the sponge and rammer, etc.

The complete gun-carriage is shown in Fig. 226.

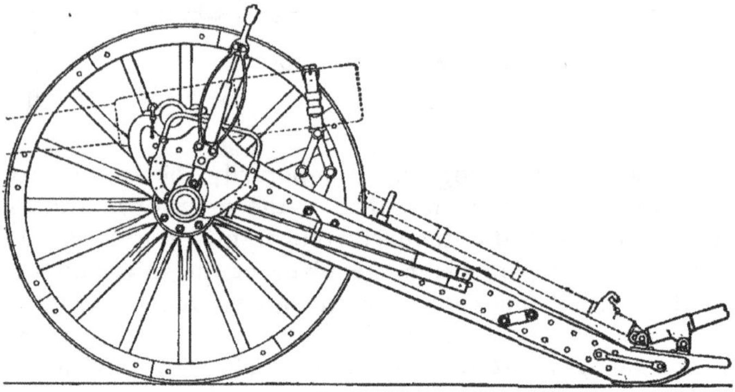

FIG. 226.

233. The Limber.

The principal parts of the limber are—

1. The wheels and axle;
2. The pole;
3. The supports for the ammunition-chest;
4. The ammunition-chest.

Wheels and Axle.—The wheels are the same as those of the gun-carriage. The axle is of steel, but as it does not withstand the shock of recoil, it is not reinforced.

The Pole.—This consists of two parts, one of which is permanently attached to the limber, and is called the fork, *a* (Fig. 227), and the other the pole proper, *b*.

The fork is of steel, of this section, ⊓, and is attached to the axle-body. It is prolonged to the rear, and carries upon its rear end the pintle-hook *c* and key *d*. The pole *b* is made of light elastic wood, and as it may be broken, is held in the fork by a bolt, *e*, and can be readily removed and replaced. Its outer end carries a pad, *f*, to prevent injury to the swing-horses, and a stop, *g*, against which the neck yoke rests; *h* is the double tree held by a bolt, *i*.

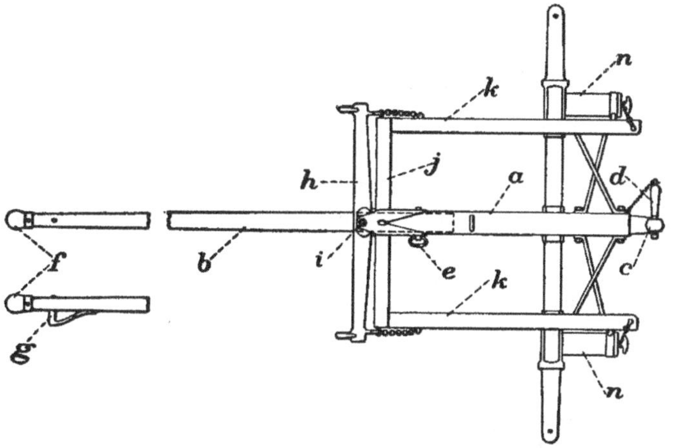

FIG. 227.

Supports for Ammunition-chests.—The ammunition-chest *l* (Fig. 229) is supported by the fork, and by the two hounds *k*. These hounds are braced to the fork in rear, and are connected together in front, and also to the fork, by a cross-bar *j*.

The hounds not only support the chest, but they strengthen all the parts, and assist in transmitting the force of traction to the axle. The chest is bolted to the hounds front and rear. The hounds and fork also support the foot-boards *m* (Fig. 229), upon which the feet of the cannoneers rest.

Ammunition-chest.—This is made of wood for lightness. It carries the ammunition for the immediate supply of the gun, and is of the same size as the chests of the caisson for interchangeability. It also furnishes seats for the cannoneers.

The lid opens on top. The advantage of this is that the chest can be made waterproof until the water reaches the lid. Its disadvantage is that the ammunition is less accessible than if the lid opened on the rear side. The advantage in case of field artillery outweighs the disadvantage, since the cartridges for this service are carried in bags, and hence the ammunition would be spoiled by access of water. For metallic ammunition, as with machine guns and the revolving cannon, as they are not liable to damage by water, the lid opens on the side for accessibility.

The interior of the chest is divided into three parts by two partitions (Fig. 228). The projectiles are placed upright in the end divisions *a*, and the cartridges in the middle division *b*. The cartridges are thus in a measure protected from fire by the projectile. The chest is low, so that a man of ordinary height can easily get at the ammunition.

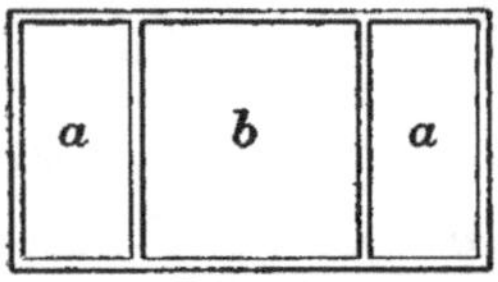

FIG. 228.

Each chest carries 42 rounds. To avoid accident, no primers are carried in the chest.

Packages of primers are carried in the cylindrical boxes, with screw tops (*n*, Figs. 227 and 229), and loose primers in the tool-box of the gun-carriage.

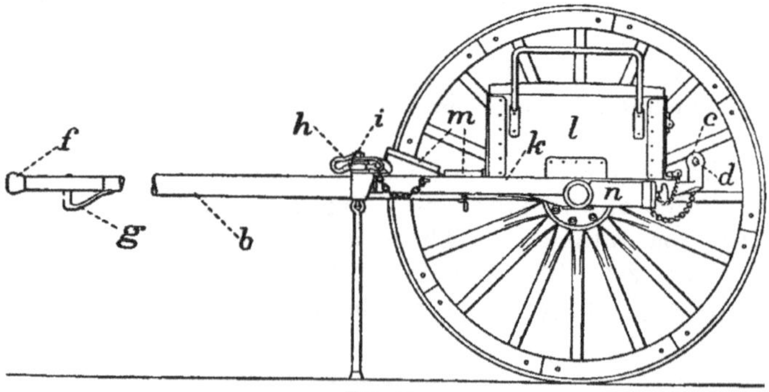

FIG. 229.

The limber complete is shown in Fig. 229. The descriptions of the caisson, forge and battery wagon, and artillery store-wagon, are omitted. The carriage and limber for the 3.6-inch field-gun resemble those for the 3.2-inch gun.

CARRIAGES FOR SIEGE ARTILLERY.

234. 5-inch Siege-gun Carriage.

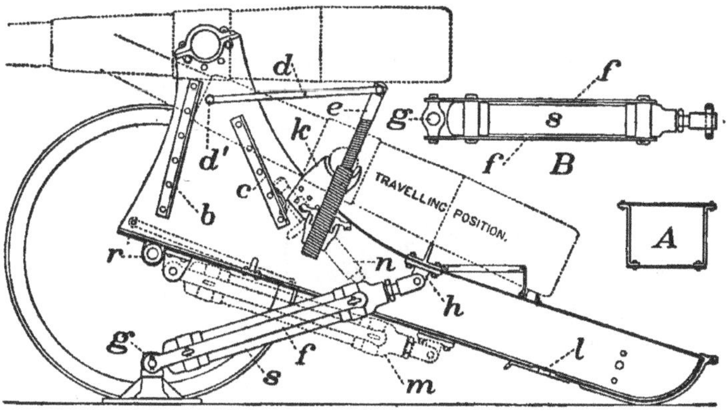

FIG. 230.

This carriage (Fig. 230) is made of steel plate ½ inch thick, the cross-section of the flasks being as shown at *A*.

The cheeks are united by two transoms, *b*, *c*, in front, and connected in rear of *c*, as shown in figure A. The axle *r* is of steel and hollow. The elevating device is a double screw *e*, connected by a strap *d* with an axis *d'*, parallel to and under the trunnions. The object of the strap has been explained.

Recoil is checked by a hydraulic buffer, *s*, which, when the gun is in the firing position, is connected by straps *f* to a bolt, *g*, on the platform. The piston-rod of this buffer is attached to the carriage by a lug, *h*. Figure B shows the arrangement of buffer and straps for attachment to bolt on platform; *k* is the travelling trunnion-bed, only one being shown; *l* is the lunette plate and lunette. When arranged for travelling, the pintle of the limber passes through *l*, the gun is moved back into the travelling trunnion-beds *k*, the hydraulic buffer occupies the position *m*, and the elevating screw the position *n*. The principal characteristic of the carriage is the height of the trunnion-beds, which are 72.25 inches, or 6 ft. ¼ in. from the ground. This is to enable the gun to be fired over a parapet of sufficient height to shelter the gunners.

235. 7-inch Siege-howitzer Carriage.

This carriage (Fig. 231) is made of steel plate ½ inch thick, the cross-section of the flasks being similar to that of the siege carriage. The cheeks are held together by transoms, and the axle is solid. The carriage differs from the 5-inch in the following points:

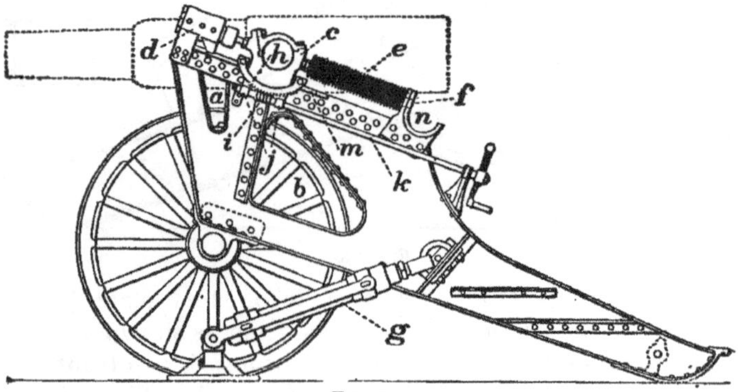

FIG. 231.

The cheeks are cut out at *ab* to decrease the weight. The piece is supported on sliding trunnion-pieces, *c*. In front are two hydraulic buffers, *d*, which limit the recoil of the trunnion-pieces *c* to about six inches. In rear of the sliding trunnion-pieces *c* are two sets of Belleville or spiral springs *e*, which return the piece to its firing position upon the carriage. The rod upon which the springs are strung passes through a hole, *f*, in the travelling trunnion beds *n*.

The recoil of the carriage is checked by the buffer *g*, attached as in the 5-inch siege-carriage. The elevating device consists of a rack, *h*, bolted to the howitzer, in which works a worm, *i*, mounted between two lugs, *j*, on the sliding trunnion-piece *c*.

A splined or square shaft, *k*, passes through this worm and its lugs, *j*, and fits loosely, so that the worm may slide along the shaft. When recoil occurs, the trunnion-carriages slide to the rear along the upper surface, *m*, of the cheeks, carrying with them the piece and the elevating-gears *h* and *i*. The springs *e* then act to force the gun and elevating gear back into position. With this carriage the first shock of recoil is taken up by the upper buffers, *d*, and the strain gradually transmitted to *g*. The carriage can thus be made lighter and stronger.

236. 7-inch Siege-mortar Carriage.

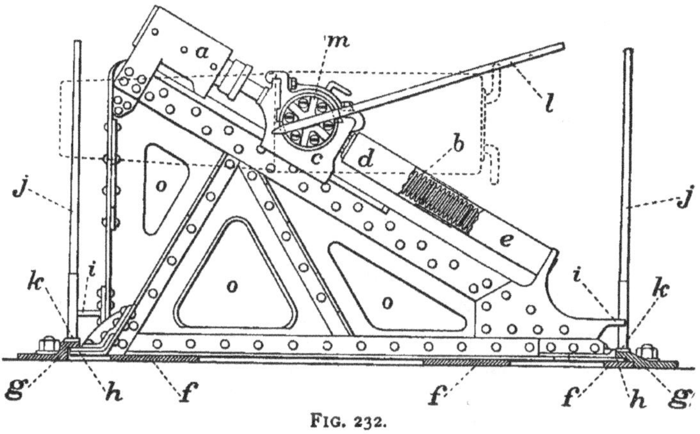

FIG. 232.

This carriage (Fig. 232) is made of steel plate as in the case of the 5 and 7-inch wheeled carriages, and in its method of checking recoil and restoring the piece to the firing position it resembles the 7-inch howitzer carriage. It differs, however, in many particulars.

It is not a wheeled carriage, but is intended to rest upon a platform when the piece is fired, like the old smooth-bore mortar carriages. Two hydraulic buffers, *a*, in front, check the recoil, while the coiled springs *b* in rear of the sliding trunnion-pieces *c*, return the piece to the firing position. These coiled springs are enclosed in a telescopic or sliding case, *de*, the part *d* sliding over *e* in recoil.

The platform has three traverse-circles, *f*, bolted to it, and also two clamping-circles *g*. Flanges, *h*, on the mortar carriage fit under these clamping-circles, and retain the carriage in place, preventing its recoil. Lugs, *i*, are attached to the carriage, against which handspikes, *j*, rest. The lower ends of these handspikes are shod, and fit into teeth, *k*, on the clamping-circles. By moving the handspikes, the mortar carriage may be traversed in azimuth for pointing. Elevation is given by a bar, *l*, which is inserted in radial grooves formed in a piece of wrought iron, *m*, bolted to the trunnion. The cheeks are connected by transoms to strengthen them, and are cut out at *o* for lightness.

CARRIAGES FOR SEACOAST ARTILLERY.

237. Classification — Barbette Carriages — Barbette Carriage for 8-inch Rifle — Principal Parts — Base-plate — Rollers and Distance-rings.

CLASSIFICATION.—Seacoast carriages are classed into—

1. Barbette;
2. Casemate or turret;
3. Disappearing;

according as the piece is fired over the parapet; through a port or embrasure; or over the parapet, the gun recoiling below it on discharge.

BARBETTE CARRIAGES.—The barbette carriages for 8, 10 and 12-inch guns resemble each other in general, differing only in certain details of construction on account of the varying weight of the guns. The 8 and 10-inch carriages are made principally of cast iron, the 12-inch of cast steel. The 8-inch carriage may be taken as a type of the others, and will be described.

CARRIAGE FOR 8-INCH RIFLE—PRINCIPAL PARTS.—The principal parts of the carriage (Figs. 233 and 234) are:

1. The base-plate or lower roller path, *A*;
2. The rollers and distance-rings, *B*;
3. The chassis, *C*;
4. The top carriage, *D*.

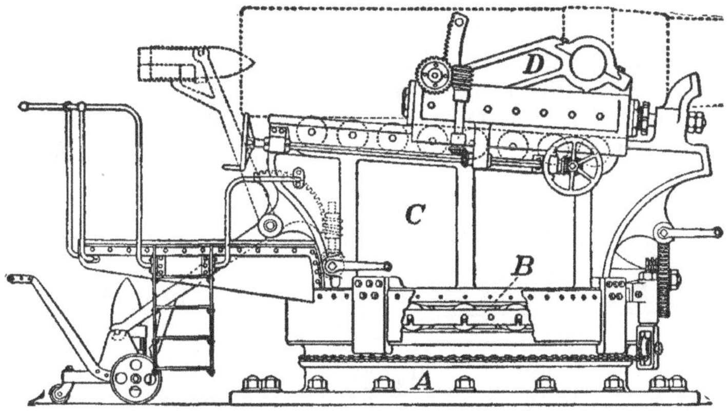

FIG. 233.

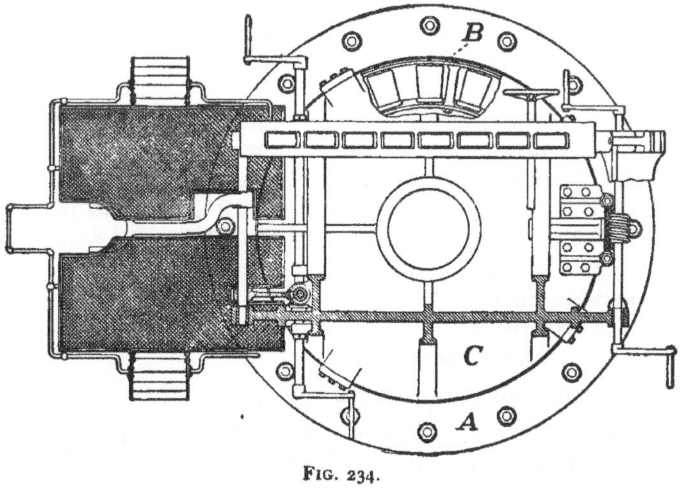

FIG. 234.

BASE-PLATE.—This consists of a heavy casting, *A*, shown in plan, section, and elevation, Fig. 235. It rests upon a bed of concrete, to which it is bolted by the anchor-bolts *a*; *b* is the roller-path upon which rests a series of conical forged steel rollers, *E*. The central portion, *c*, corresponds to the pintle in the old carriages, and around it fits a collar, *d*, Fig. 236, upon the chassis, so that rotation in azimuth occurs about this central projection.

ROLLERS AND DISTANCE-RINGS.—A ring of conical forged steel rollers, *E*, Fig. 235, rests upon the roller-path

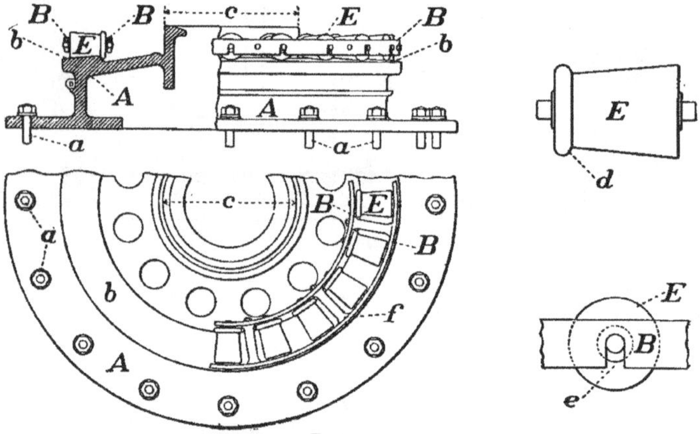

FIG. 235.

b of the base-plate, and upon these rests the corresponding upper roller-path, *c*, Fig. 236, of the chassis. These rollers are shaped as shown at *E*, the object of the flange *d* being to keep the rollers in place by its bearing on the inner edge of the roller-path. For the 8-inch carriage there are twenty of these rollers. They are held in place by two distance-rings, *B*, which are slotted for the axis of the rollers as shown at *e*. The distance-rings are kept in place and braced by the braces *f*.

238. 8-inch Barbette Carriage—The Chassis.

This consists (Figs. 234 and 236) of the circular horizontal part *a* and the two vertical cheeks *b*. The circular part, *a*, supports the cheeks, and carries on its lower side the upper roller path *c*, and the central collar *d*, which fits over the corresponding central projection, Fig. 235, in the base-plate. The upper surfaces *e* of the cheeks *b* are inclined to the front, and carry at their forward ends the lugs *f* which hold the piston-rods of the hydraulic buffers.

In modern carriages, the irregularities due to sliding friction are avoided by placing the top carriage on rollers, and throwing all the work of checking the recoil upon the buffers, which can be very accurately regulated.

These rollers are shown at *g* inserted in recesses in the

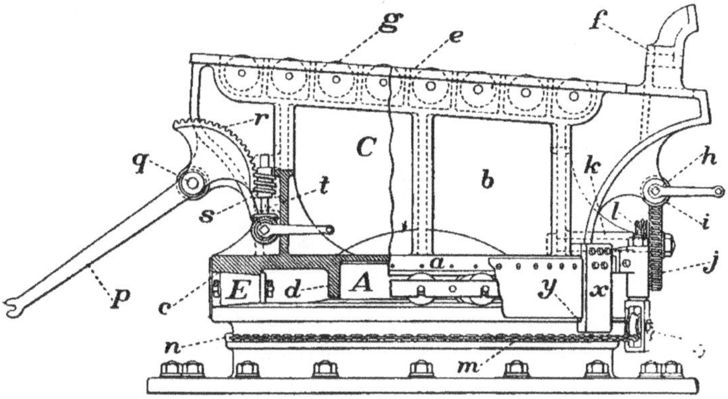

FIG. 236.

chassis-rail, and rotating on journals, so that the exterior of the roller is just above the chassis-rail.

The device for traversing in azimuth is shown in front and in plan in Fig. 234. It consists of a cross-shaft, *h*, with cranks. This shaft carries a worm, *i*, gearing into a worm-wheel, *j*, which works upon an axis, *k*, attached to the chassis.

In rear of the worm-wheel is a sprocket-wheel, *l*, on the same shaft with *j* and attached to it, so that one cannot turn independently of the other. A chain, *m*, is attached at one end to the bed-plate at *n*, and the other end of the chain at a corresponding point near the first, not shown in drawing.

This chain passes under a small wheel in a fork at *o* and

thence over the sprocket-wheel *l*. When motion is given to *l* by the shaft, worm, and cranks, the chain will pass over the sprocket-wheel and the chassis turn on its rollers, *E*. Vertical motion of the chassis is prevented by clips, *x*, bolted to it and embracing a flange, *y*, on the bed-plate.

The device for hoisting the ammunition is shown in rear of the chassis. The projectile is run up on a truck (see Fig. 233), and is then lifted together with the loading-tray by the lever *p*. This lever is on a horizontal shaft, *q*, which carries a worm-gear, *r*, acted on by the worm *s* on the shaft *t*. Its action is evident.

239. 8-inch Barbette Carriage—The Top Carriage and Buffers—Elevating Device.

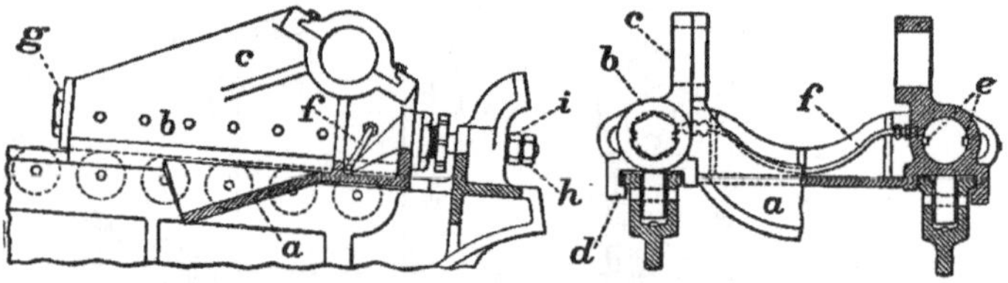

FIG. 237.

THE TOP CARRIAGE AND BUFFERS.—These are shown in Fig. 237 in section and elevation. The top carriage carries the gun, and consists of a single casting, comprising the buffers *b*, and their connecting transom, *a*. On the top of each buffer is cast the bracket *c*, carrying the trunnions of the gun.

This top carriage rests on the rollers of the chassis-rail, as shown in section, and is held in place, and prevented from lifting at discharge, by the flanges *d*. In the section are also shown the ribs or throttling-bars, *e*, which regulate the flow of liquid in the buffers, there being two in each cylinder, held in place by bolts passing through the walls of the cylinders. The action of these buffers will be explained under the subject of Recoil.

A cross-pipe, *f*, called an equalizing pipe, connects the liquid in the two cylinders, and insures their uniform resistance.

The pistons and rods are removed, by unscrewing the nuts, *g*, which close the rear ends of the cylinders, and then by removing the locking and piston nuts *h*, *i*, the piston and rod can be pushed out to the rear.

The recoil is limited to 40 inches.

ELEVATING DEVICE.—This is shown in Fig. 238. It consists of a square shaft, *a*, attached to the right side of the chassis, and working in fixed bearings at *b* and *c*; *d* is a sliding bearing attached to the top carriage. In this bearing works a bevel-gear, *e*, gearing into a second bevel-wheel, *f*, on the vertical shaft, *g*, attached to the top carriage. A worm, *h*, on this shaft gears into the worm-wheel *i*, on the horizontal shaft *j*, and on this same shaft *j* is a second gear-

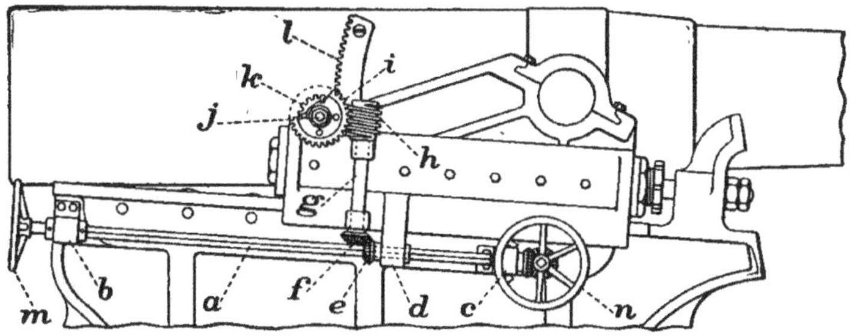

FIG. 238.

wheel, *k*, engaging with the rack *l*, on the gun. When recoil occurs, the sliding bearing *d*, moves along the square shaft *a*, carrying with it the bevel-gear *e*, so that the gears are constantly engaged, and the gun can be elevated in any position. The return to battery carries the gear *e* along the shaft *a*. By means of the hand-wheels *m* and *n*, the gun may be elevated from front or rear. The return of the piece to the firing position is due to gravity. Carriages of this kind are called gravity return carriages.

240. The 12-inch Mortar Carriage—General Features—Springs.

GENERAL FEATURES.—This carriage consists of—

1. The bed-plate or lower roller-path *A*;
2. The rollers and distance-rings *B*;
3. The upper roller path or racer *C*;

4. The cheeks *D*;

5. The spiral springs *F*, and the hydraulic cylinders *H*.

The upper roller-path is circular, and supports the cheeks, which are vertical, the two together forming the top carriage. The lower roller-path is also circular.

SPRINGS.—On the side of each cheek is cast a cylindrical recess, *E* (Figs. 239 and 240), which contains a column of spiral springs, *F*.

These springs are in ten separate lengths, and each length is composed of two coils, an inner one, *F'*, and an

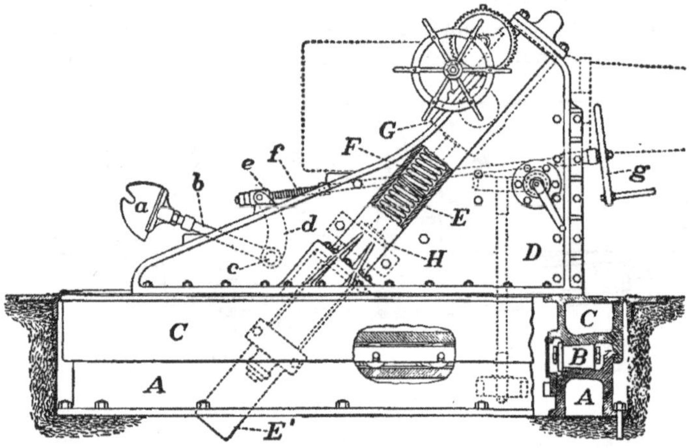

FIG. 239.

outer one, *F*. A pile of Belleville springs, *F''*, forms the upper end of the column, and upon these the mortar is supported as follows: The trunnion-carriage *G*, of cast steel, has a projecting lug, *g'*, through which passes the adjusting screw *K*.

The lower end of this screw bears on the Belleville springs, and by means of it the trunnion carriages may be adjusted till the mortar is in the proper position for loading, and it is then secured in that position by the jam-nuts *k'*.

The trunnion-carriages *G*, are two heavy blocks of cast steel, in which the trunnions rest, and which slide, under the force of recoil, along ways planed on the inner side of the cylindrical recess *E*; a slot, *m*, Fig. 241, being left in the recess for the projecting lug *g'*; a section of the trunnion-carriage and recess is shown in Fig. 241.

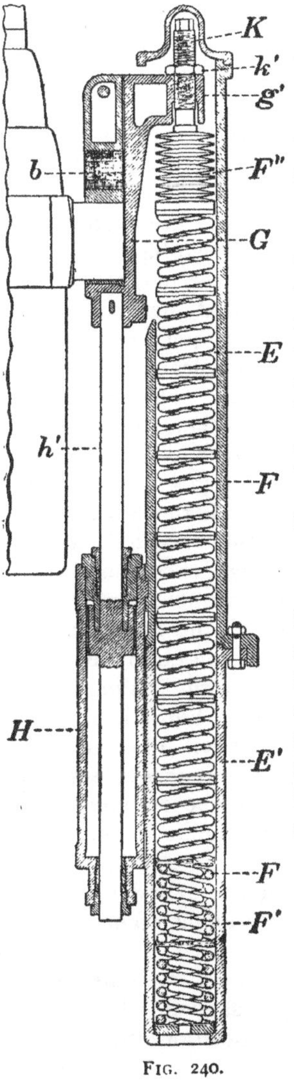

FIG. 240.

The object of the spiral springs is to return the mortar to the firing position. They are set at an angle of 50° with the horizontal, the mortar being fired between the limits 35° and 65°, so that this is a mean between them. To obtain a column of springs of sufficient length to return the piece to its proper position, the cylindrical recesses in the cheeks are lengthened by bolting a cylinder *E'*, Figs. 239 and 240, to the bottom of *E*.

241. The 12-inch Mortar Carriage—Hydraulic Buffers.

Recoil is checked by two hydraulic buffers, *H*, Figs. 239 and 240, one on each side of the carriage, bolted to the flanges of the cylindrical recesses *E*. The piston-rods *h'*, Figs. 240 and 242, of these cylinders are attached to the lower ends of the trunnion-carriages *G*. When the piece is

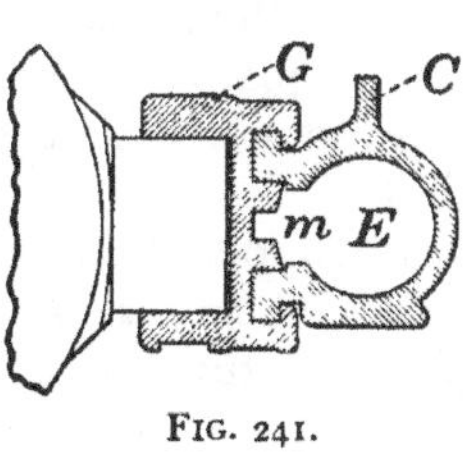

FIG. 241.

fired, the spiral springs return it to the firing position, and hence this is a spring-return carriage.

The arrangement of the hydraulic buffer for checking recoil and maintaining a constant resistance in the cylinder, differs from that for the 8, 10, and 12-inch guns as follows:

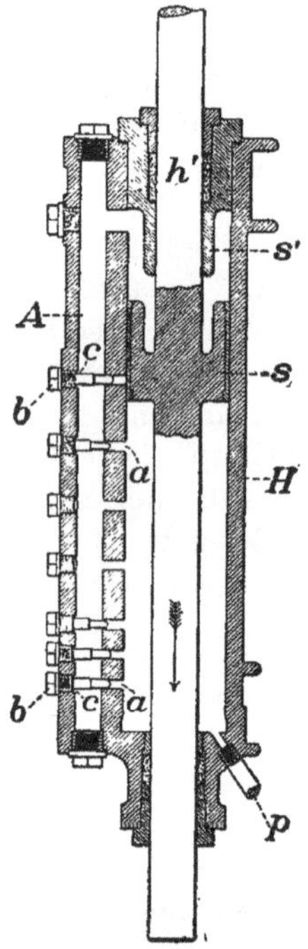

FIG. 242.

A channel, *A*, Fig. 242, is bored parallel to the axis of the cylinder *H*. Holes, *a*, are bored at different intervals along this channel, and are partially or entirely closed by screw-plugs, *b*, fitting into the holes *c*. These plugs, *b*, are of different shapes, so that they will either completely close the openings *a*, when screwed home, or will leave them partially or entirely open. They are never entirely removed. When the gun recoils, the piston moves in the direction of the arrow, Fig. 242. At the first instant of recoil, if all the holes *a* are open, it is evident that the liquid will be forced freely through these holes, and will flow along the channel *A*, and return above the piston, into the cylinder. As the motion of the piston continues, each of these holes will, in succession, be cut off, and the flow of the liquid being thus limited, its resistance will increase. By partially or entirely closing the holes *a*, it is evident that any resistance to flow, within limits, may be obtained. An equalizing-pipe, *p*, connects the two cylinders, to keep the pressure the same in both.

The piston-rod *h'*, passes through the cylinder at both ends, to equalize the volumes, and the piston-head, *s*, is solid.

Its upper side is of the shape shown, and the upper cylinder-head, *s'*, is correspondingly shaped. When the springs return the piece to the firing position, the head of the cylinder, s', enters the recess in the piston-head, s, and by gradually forcing out the liquid, the counter-recoil is checked, and the piece comes into the firing position without shock. Buffers, b, Fig. 240, on the trunnion-carriage also avoid this.

242. Remaining Parts of the 12-inch Mortar Carriage—Roller-paths—Elevating-gear—Traversing-gear—Loading-scoop.

ROLLER-PATHS.—In the 8, 10, and 12-inch carriages, horizontal motion of the parts is prevented by the central collar or pivot, as explained. In the 12-inch mortar carriage, as the recoil of the piece is downward, the central part of the carriage must be left open, and hence the central collar cannot be used. Resistance to horizontal motion is therefore obtained by forming the upper roller-path, C, so that it overlaps the lower one, A, as shown in Fig. 243, which is a section of the two. Vertical motion is prevented by the weight of the system.

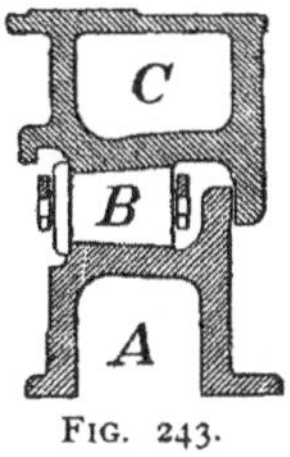

FIG. 243.

ELEVATING-GEAR.—This consists of a bronze toothed sector, a, bolted to the mortar, concentric with the axis of the trunnions, into which works a gear, b. A large gear, c, on the same shaft is driven by the gear d, and on the same shaft with d is a hand-wheel, e.

The gears b, c, d, and the hand-wheel, e, are all mounted on the trunnion-carriage, G, and as the mortar is mounted on the same carriage, the whole elevating device recoils together. Each trunnion-carriage carries its own elevating-gear.

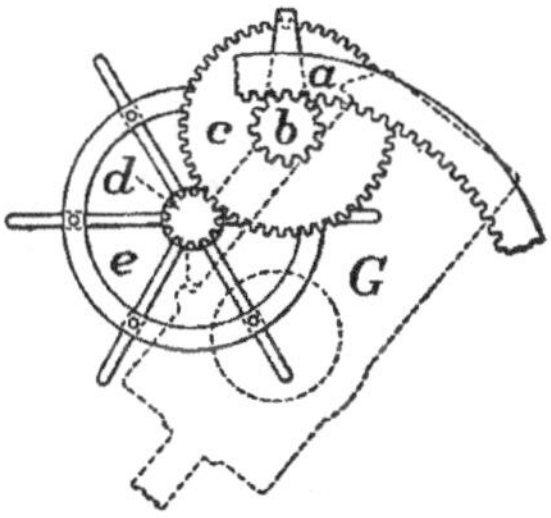

FIG. 244.

TRAVERSING-GEAR.—This consists of a vertical shaft, a, Fig. 245, attached to the upper carriage, carrying a gear, b, at its lower end, and a worm-wheel, c, at its upper end. The wheel, b, gears into a toothed ring, d, on the inside of the lower roller-path, and the shaft is rotated by a worm, *e*, driven by cranks, *f*, on a horizontal shaft, *g*, passing through the front of the cheeks of the top carriage. This device is also shown in Fig. 239.

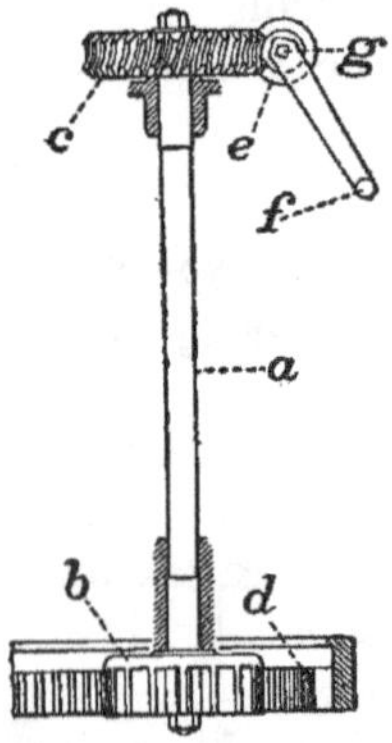

FIG. 245.

For pointing in azimuth, a cast-iron circle, graduated in degrees, is fixed around the upper roller-path, and a pointer attached to this path, indicates the direction.

LOADING-SCOOP.—This consists of a scoop or tray, *a*, Fig. 239, at the end of a lever, *b*. This lever is pivoted to the rear of the chassis on a shaft, *c*, which carries also the bent lever *d*. The outer end of this lever carries a nut, *e*, in which works the screw, *f*. This screw is supported in bearings on the left side of the top carriage, and extends to the front, where it ends in a hand-wheel, *g*. By turning this hand-wheel, the scoop is raised or lowered, carrying the projectile and charge to the breech of the mortar.

The loading position for the mortar is an elevation of 5°.

243. Casemate or Turret Carriages—General Principles—Disappearing Carriages—General Principles.

GENERAL PRINCIPLES OF TURRET CARRIAGES. — The general object of these carriages, is to secure a minimum height, and minimum embrasure opening. Hence the centre of rotation is at the centre of the embrasure, and the chassis is simply a pair of rails, which support the buffers carrying the gun.

Elevation is given by lowering or raising the rear ends of the rails, and direction by rotating the turret. None of these carriages have as yet been designed for the land service.

GENERAL PRINCIPLES OF DISAPPEARING CARRIAGES.—Owing to the great cost of modern guns and carriages, it is important to protect them as much as possible from injury from fire. This may be done either by placing them in armored casemates or turrets, or in gun-lifts, or by using the ordinary barbette battery, and placing the gun upon a disappearing carriage. The great cost and confined space of the casemates, turrets, and gun-lifts, has caused the adoption of the disappearing type of carriages in exposed sites.

The object of a disappearing carriage is, to enable the gun to be fired over an ordinary parapet, thus giving it all the advantages of an extensive field of view and of fire, with room for manœuvre, and to utilize the force of recoil in returning the gun to cover for loading, and in storing up sufficient energy, during recoil, to return the gun to the firing position.

There are therefore two points to be especially considered:

1. The means of checking recoil, so that the gun will be covered during loading.

2. The method of storing up energy sufficient to return the piece to the firing position.

Checking Recoil. — In all these carriages, the gun is mounted at the ends of lever-arms, and these arms are pivoted, in various ways, to the chassis. The recoil is checked by hydraulic buffers, or in some cases by pneumatic buffers, which allow the lever-arms to rotate gradually to the rear, bringing the gun down to the loading position.

Return to Firing Position.—The energy necessary to return the gun to the firing position, is stored up in various ways. In the English service, a hydro-pneumatic buffer is used; that is, the liquid which is forced out of the hydraulic cylinder, by recoil, passes into an air-chamber, and compresses the air sufficiently, to give the necessary pressure for returning the gun to the firing position, as soon as a valve is opened between the air-chamber and the hydraulic cylinder.

Spiral or Belleville springs are also employed. The recoil is checked by the hydraulic buffer, and the springs restore the piece to its firing position.

Counterweights may be used, either alone, or in connection with air pressure.

244. Buffington-Crozier Disappearing Carriage.

Two successful carriages of this type have been tried in the United States, and an outline description of each will be given.

BUFFINGTON-CROZIER.—This carriage was designed by Colonel Buffington, and modified by Captain Crozier, both of the U. S. Ordnance Department.

The carriage consists of the chassis, *A*, Fig. 246, the supporting levers, *B*, carrying the gun, the hydraulic buffers, *C*, and the counterweight, *D*. The carriage is a front-pintle one. The gun is mounted on the upper ends of the supporting levers, *B*.

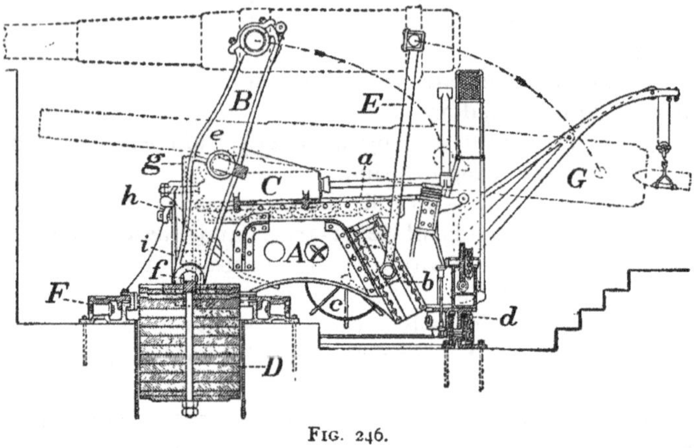

FIG. 246.

These levers have trunnions, *e*, near the middle, which are mounted on the hydraulic buffers, *C*. The lower ends of the levers are connected by a cross-head at *f*, and from this cross-head, is suspended the counterweight, *D*. This counterweight rises and falls vertically, while the trunnions, *e*, with the buffers, move horizontally along the chassis-rail, *a*. When the piece is fired, the force of recoil is taken up by the buffers, which move back as stated, while the counterweight, *D*, is raised vertically, sliding on guides, *g*. The gun in the loading position is shown at *G*. The counterweight is held in its position after firing, by a pawl, *h*, and ratchet, *i*, which being released, allows the weight to descend, and thus the gun is raised to the firing position. The trunnions of the gun describe an arc of an ellipse in their descent. The bars *E* are for giving elevation. They are attached to a straight rack, *b*, on the inside of the chassis, which is worked by the hand-wheel *c*. The elevation may be given in either the loading or the firing position. The carriage rests in front upon a ring of rollers, *F*, as previously described, and is traversed by the chain, *d*, passing over a sprocket-wheel, and worked by a crank.

245. The Gordon Disappearing Carriage.

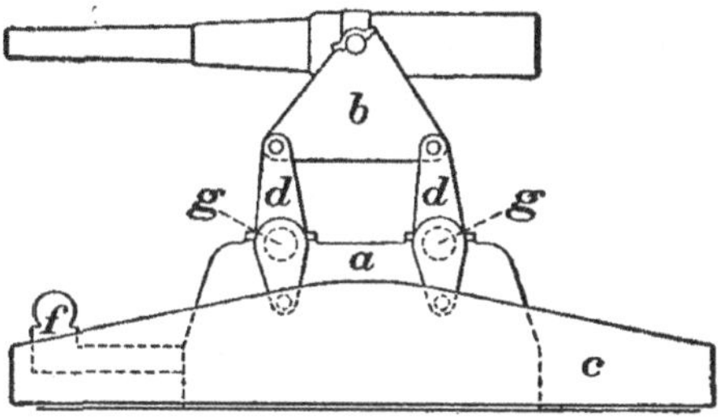

FIG. 247.

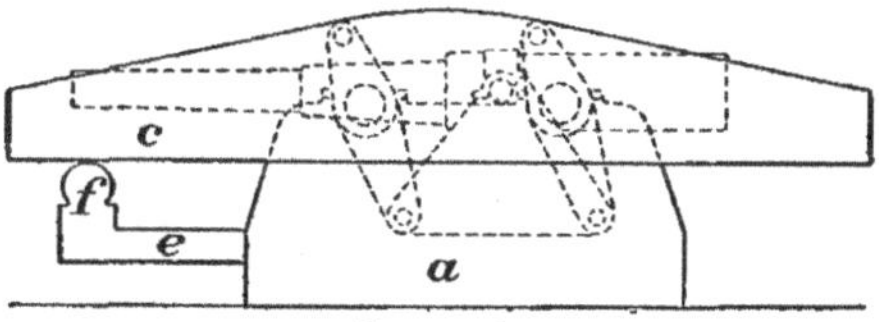

FIG. 248.

This carriage was designed by Capt. Gordon of the U. S. Ordnance Department, and consists (Figs. 247 and 248) of the chassis *a*, the top carriage *b*, the counterpoise *c*, the lever-arms *d*, the hydraulic cylinders *e*, and the air-chamber *f*.

The chassis, *a*, is a heavy casting, supporting all the parts, and it rests when in the firing position upon a platform. When the piece is to be moved in azimuth, the chassis is supported on a hydraulic pivot, not shown in the drawings, by which arrangement the traversing is effected with very little power.

On the upper side of the chassis, four levers, *d*, are mounted, two of them being shown in the drawing.

These levers rotate about the axes, *g*, and carry at their lower ends, a heavy counterweight, *c*. On the upper ends of these levers is mounted the top carriage, *b*, which supports the piece. A hydraulic cylinder, *e*, extends along the chassis.

Its piston is forced in, during recoil, and the liquid, thus forced into the air-chamber *f*, compresses the air, and stores up the energy necessary to return the piece to the firing position, when the proper valve in the air-chamber is opened. The trunnions describe an arc of about 180° during recoil, thus bringing the gun close to the parapet, and affording good cover. The elevating device is attached to the top carriage. This is a centre-pintle carriage.

Several disappearing carriages are in use abroad, as the Moncreiff, Armstrong, Canet, etc.

246. Old Seacoast Carriages in U. S. Service.

Certain old carriages are still retained in the U. S. service for the 8-inch converted rifles, and the 15-inch Rodman smooth-bore guns. They consist (Fig. 249) of a chassis, *a*, and a top carriage, *b*, made of wrought iron.

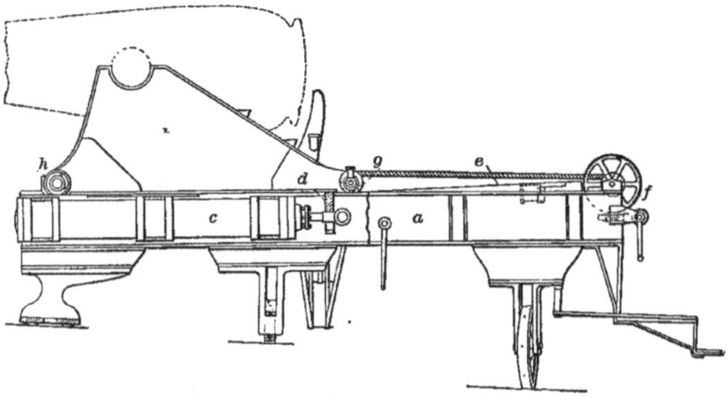

FIG. 249.

The chassis is composed of two parallel, I-shaped rails, connected by transoms, and attached to it, between the rails, is the hydraulic buffer, *c*. The piston of this buffer is attached to the top carriage by a lug, *d*, on the latter. The buffer itself,.is one of constant orifice, and variable resistance, as will be explained. Bolted to the rear end of the chassis-rail, is an inclined rail, *e*. The retraction gear is shown at *f*.

The principle of this carriage is as follows:

When the piece is fired, the top carriage rests, throughout its length, upon the chassis-rail, and hence the force of recoil is distributed over this length, and the top carriage starts to the rear on sliding friction. After a very small movement in recoil, the wheel, *g*, of the top carriage (Fig. 250), strikes the wedge-shape drail, *e*, and begins to rotate. This causes the top carriage to tip slightly forward, and brings

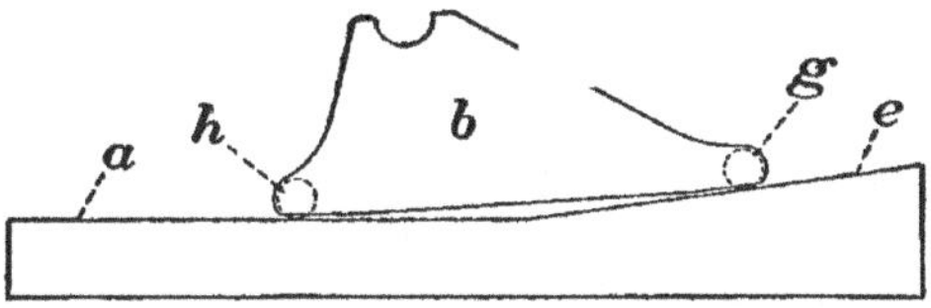

FIG. 250.

the front wheel, *h*, into bearing on the chassis-rail. The carriage then moves on rolling friction. The result of this arrangement is, that the top carriage rests, throughout the recoil, on rolling friction, as shown Fig. 250. A spring pawl and ratchet, retain the top carriage in the loading position, after recoil, and by releasing the pawl, the top carriage returns on rolling friction to its firing position, by gravity.

For drill purposes, to bring the gun from battery, for loading, the rear wheel, *g*, is mounted on an eccentric axle, and when thrown into bearing against the chassis-rail, by the action of a handspike, it tips the top carriage forward, and brings the front wheel, *h*, also into bearing. The piece, and top carriage, are then drawn to the rear, by a rope attached to the latter (Fig. 249), and wound round a drum on the shaft of the retraction gear *f*.

THEORY OF RECOIL.

247. Maximum Velocity of Recoil.

The velocity of recoil at the instant the projectile leaves the muzzle is given by equation 65, Interior Ballistics. This does not represent the *maximum* velocity of recoil, however, for the reasons stated, and a new equation is necessary to determine this velocity.

In equation 65, it is assumed, that the mean velocity of the particles of the charge, is one half that of the velocity of the projectile; that is, the equation contains the term $\left(\frac{\hat{\omega}}{2}v\right)$, in which $\hat{\omega}$ is the weight of the charge, and v the velocity of the projectile.

This is very nearly true while the projectile is in the bore, because the layer of gas next the projectile has the same velocity as the latter, and this velocity decreases to zero, for the layers toward the bottom of the bore.

But when the projectile leaves the muzzle, this condition no longer exists. The gases, which were before confined, rush out with greatly increased velocity, and this affects the recoil of the piece.

Let P denote the weight of gun, and part of the carriage which recoils;

p, the weight of the projectile;

$\hat{\omega}$, the weight of the charge;

V_m', the maximum velocity of recoil;

V, the initial velocity of the projectile;

v_m, the mean of the maximum velocities of the powder-gas upon issuing from the piece.

Then the equation

$$PV_m' = pV + \hat{\omega}\, v_m. \qquad \dots \quad (335)$$

expresses the equality of momenta of the piece, projectile, and charge at this instant.

General Sébert of the French Artillery, has determined with his velocimeter, previously described, that in order that the above equation be true, the value v_m must be about 3000 ft-secs.

Hence the maximum velocity of recoil is given by

$$V_m' = \frac{pV + 3000\tilde{\omega}}{P}, \quad . \quad . \quad . \quad . \quad . \quad (336)$$

while the velocity of recoil during the time the projectile is in the bore is (equation 65)

$$v' = \frac{pv + \frac{\tilde{\omega}}{2}v}{P}.$$

248. Periods of Recoil—Relation between Time, Velocity, and Length of Recoil in First Period.

PERIODS.—The recoil of a gun is divided into two periods:

1. From the time the gas begins to act, until the maximum velocity of recoil is attained.

2. From the end of the first period, till the piece is brought to rest.

RELATIONS BETWEEN TIME, VELOCITY, AND LENGTH OF RECOIL IN FIRST PERIOD.—In order to determine the circumstances of recoil in the first period, it is necessary to know the relations between the time, velocity, and length of recoil, and these are determined in the following manner:

If the distance recoiled by the piece, at the end of any time t, be denoted by x, the velocity at that time is

$$v' = \frac{dx}{dt},$$

and the distance x passed over is

$$x = \int v'dt.$$

Hence, considering the expression $\int v'dt$, if we construct a curve whose abscissas are the values of t, and whose ordinates are the corresponding values of v', this curve will be of the form Fig. 251, and from it we deduce the following laws:

1. The velocities of recoil increase very rapidly at first, till the point of inflection i is reached, and then more slowly,

till they cease to increase at the time corresponding to the maximum velocity V_m', which time is denoted by τ.

2. The area included between the curve, the axis T, and any ordinate v', is the distance x passed over in recoil, at the time t corresponding to that ordinate, since $x = \int v'dt$; and the total length of recoil during the first period is the area corresponding to the ordinate τ.

This curve was constructed by experiment, by obtaining with the Sébert velocimeter, the values of v' corresponding to different values of t.

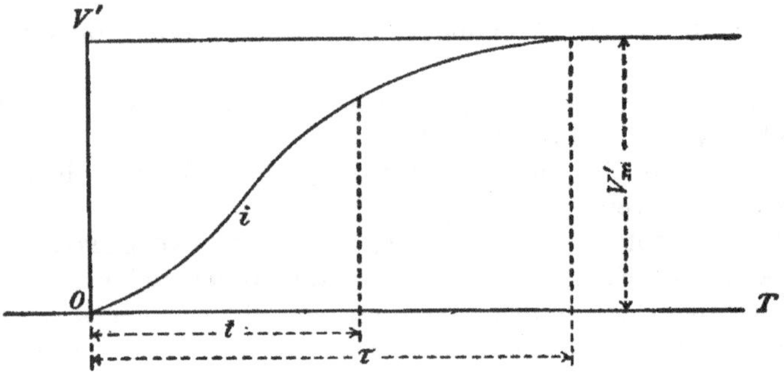

FIG. 251.

249. Ordinary Case—Steps in the Solution of the Problem.

In the case just considered, the relations between the velocities of recoil v', and the corresponding times t, were determined by experiment.

ORDINARY CASE.—Ordinarily, this relation between v' and t is not known. It may, however, be determined by a series of steps as follows:

STEPS.—We have the relation between the velocity of the projectile v and the length of its travel u in the bore, by Sarrau's monomial or binomial formulas; hence we have a relation $v = f(u)$.

1. We next determine the time t required for the projectile to pass over any length of bore u. This gives a relation $t = f(u)$.

2. Combining the curves $v = f(u)$ and $t = f(u)$, we determine the relation $v = f(t)$.

This is done by using the ordinates of the time curve $t = f(u)$ as abscissas, and those of the velocity curve $v = f(u)$ as ordinates, and constructing a curve whose equation is $v = f(t)$. This equation gives the relation between the velocity of the projectile, and the corresponding time t.

3. To pass from this curve to that of the velocity of recoil as a function of the time t, we have equations (65) and (336), giving the relation at any time t between the velocity of the projectile and that of the piece; and knowing that of the projectile we may at once find that of the piece as a function of the time, or $v' = f(t)$, which is the curve required.

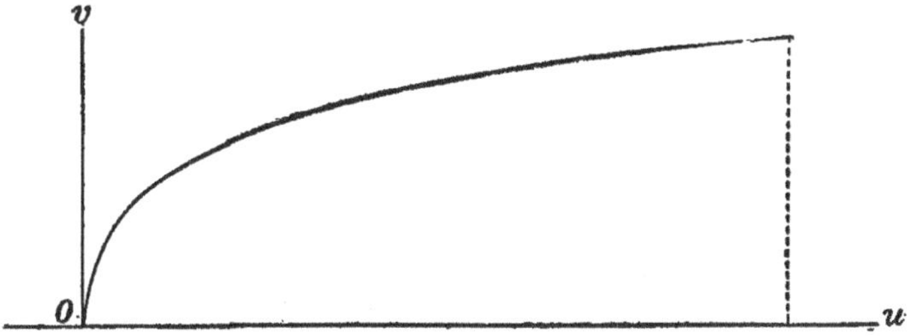

FIG. 252.

Having the curve $v' = f(t)$, we can determine, as previously shown, the time, and length of recoil, corresponding to any given velocity.

250. First Step—Time of Passage of Projectile over a Given Length of Bore—Difficulty—Remedy.

Assuming the binomial and monomial formulas (91) and (121), Interior Ballistics, we apply the one which is suitable to the particular case under consideration, and construct the curve whose abscissas are the values of u, and its ordinates the corresponding values of v. This curve will be of the form Fig. 252.

The value for the velocity at any point u is, from calculus,

$$v = \frac{du}{dt},$$

from which we have

$$\frac{1}{v} = \frac{dt}{du}.$$

Multiplying by du and integrating, we have

$$\int \frac{1}{v} du = \int \frac{dt}{du} du = \int dt = t,$$

or

$$t = \int \frac{1}{v} du.$$

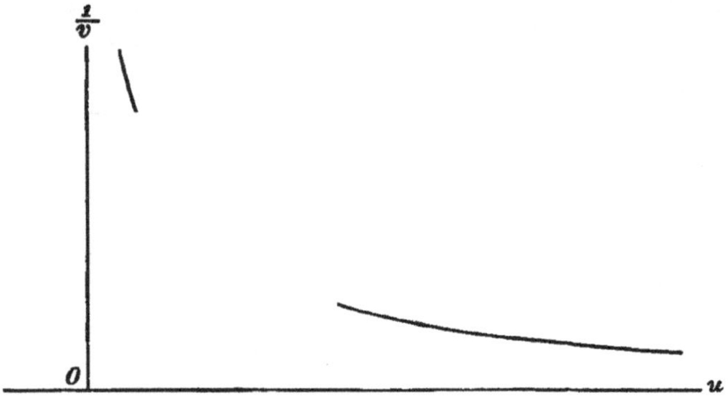

FIG. 253.

Hence if we construct a curve whose abscissas are the values of u, and its ordinates the corresponding values of $\frac{1}{v}$, the area included between this curve, the axis of u, and any ordinate $\frac{1}{v}$ will give for any value of u the time t required for the projectile to pass over this distance u in the bore.

The form of this curve is shown in Fig. 253.

DIFFICULTY.—The only difficulty in this case is that for very small values of v the ordinates $\frac{1}{v}$ will be large, and will not fall within the limits of an ordinary drawing, and hence the area under the curve cannot be accurately measured, and therefore the time corresponding to a given travel u cannot be exactly ascertained.

REMEDY.—To obviate this difficulty, we assume (as is nearly correct) that the velocity of the projectile, as a function of the time varies nearly as the abscissas and ordinates of a common parabola, whence we have

$$v = \sqrt{2pt}. \qquad (337)$$

Multiplying by dt and integrating, we have

$$\int vdt = \int \frac{du}{dt}dt = u = \int \sqrt{2pt}\,dt = \tfrac{2}{3}(2p)^{\frac{1}{2}}t^{\frac{3}{2}}. \qquad (338)$$

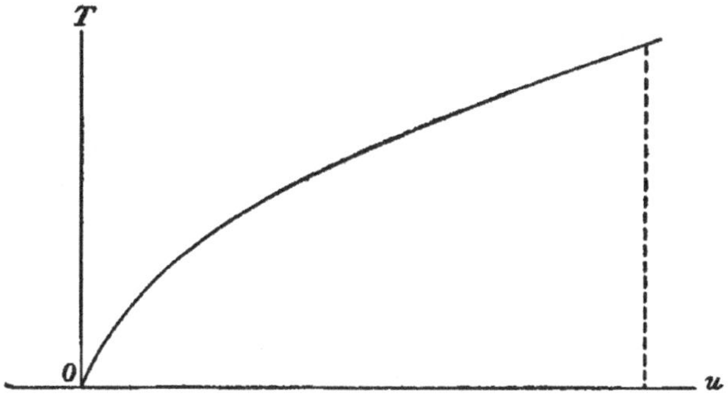

FIG. 254.

At the instant the shot leaves the bore, v in equation (337) becomes the initial velocity V. Denoting the corresponding time by t', we have, equation (337),

$$\sqrt{2p} = \frac{V}{\sqrt{t'}},$$

and this value of $\sqrt{2p}$ in equation (338) gives

$$t' = \frac{3}{2}\frac{u}{V}, \qquad (339)$$

u being the total length of travel of the projectile. Comparing this total time of passage of the projectile through the bore, with that obtained from that part of the area under the curve of reciprocals which can be measured, the value of the unmeasured portion can be ascertained very nearly. Thus the relation $t = f(u)$ is determined and the curve is given in Fig. 254.

251. Second and Third Steps in Determining the Curve of Velocity of Recoil as a Function of the Time.

SECOND STEP.—Taking the ordinates of the curve $t = f(u)$ as abscissas, and those of the velocity-curve $v = f(u)$ as ordinates, we construct a curve $v = f(t)$, showing the relations between the velocity of the projectile and the corresponding time, and this curve will be of the form shown in Fig. 255.

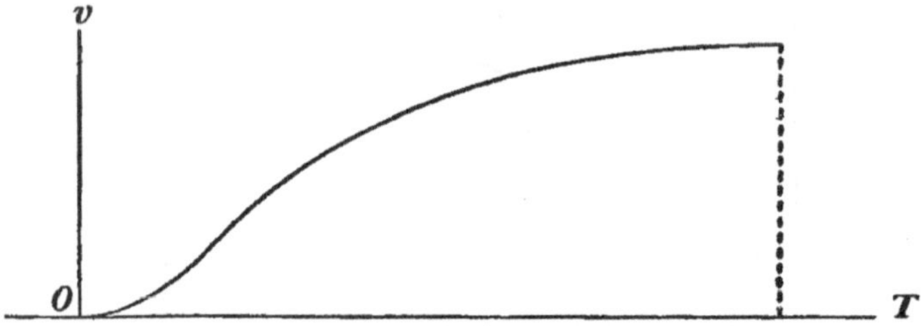

FIG. 255.

THIRD STEP—RELATION BETWEEN VELOCITY OF PROJECTILE AND THAT OF RECOIL.—We have for the velocity of recoil of the piece and carriage while the projectile is in the bore, equation (65),

$$v' = \left(\frac{p + \frac{\varpi}{2}}{P}\right) v,$$

and for the maximum velocity of recoil, equation (336),

$$V_m' = \frac{pV + 3000\varpi}{P}.$$

Since v is determined as a function of t by the second step, v' may be found for the corresponding times by equation (65), by multiplying the ordinates of the curve just determined, Fig. 255, by the ratio $\left(\frac{p + \frac{\omega}{2}}{P}\right)$. A curve can then be constructed, similar to that in Fig. 255, giving the velocities of recoil of the piece, for each instant of the passage of the projectile through the bore, and the area under this curve, bounded by the axis of T and any ordinate v' will give the corresponding space passed over, since $\int v'dt = x$.

After the projectile quits the bore, equation (65) no longer applies; but it is known that the curve becomes tangent to a line parallel to the axis of T, at a point given

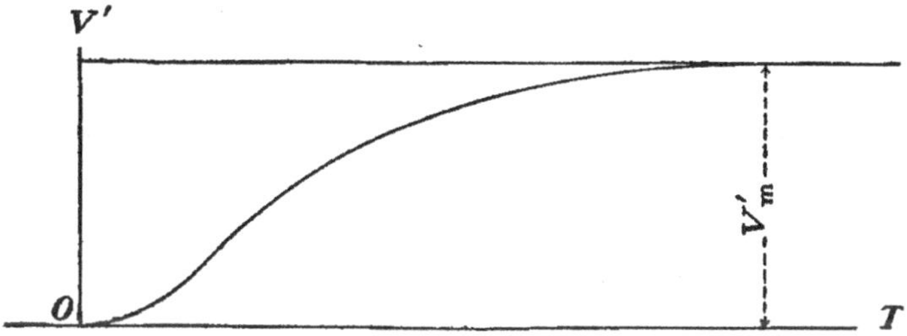

FIG. 256.

by equation (336), and it is reasonable to infer, that the rate of curvature of the curve of recoil, will continue uniform up to this point of tangency.

Hence, drawing a line parallel to the axis of T, at the distance given by equation (336),

$$V_m' = \frac{pV + 3000\omega}{P};$$

and continuing the curve already drawn, preserving its general rate of curvature up to this line, we have the curve

$v' = f(t)$, giving the time and space passed over in recoil, and this curve will be of the form Fig. 256.

252. Example—8-inch Steel B. L. Rifle—First Step—Curve $t = f(u)$.

For example take the case of the 8-inch Steel B. L. Rifle. The velocity curve for this gun, with 125 pounds brown powder, is given n Fig.i 257.

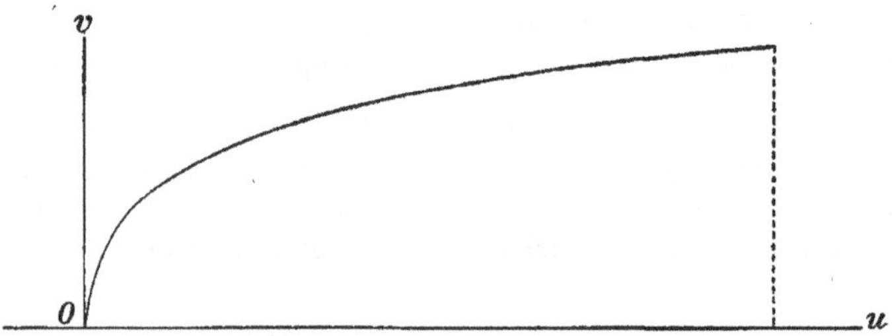

FIG. 257

From it we obtain the following abscissas and ordinates:

u (Feet).	v (Foot-seconds).
0.46	387
1.70	935
2.40	1080
3.20	1197
3.70	1259
7.30	1545
9.50	1655
11.50	1727
13.00	1787
14.30	1827
15.75	1859
17.43	1884 I. V.

From which we have the following values of $\frac{1}{v}$, and the curve of reciprocals, Fig. 258.

u (Feet).	$\frac{1}{v}$.
0.46	.002584
1.70	.001069
2.40	.0009259
3.20	.0008354
3.70	.0007943
7.30	.0006472
9.50	.0006042
11.50	.0005790
13.00	.0005596
14.30	.0005473
15.75	.0005379
17.43	.0005308

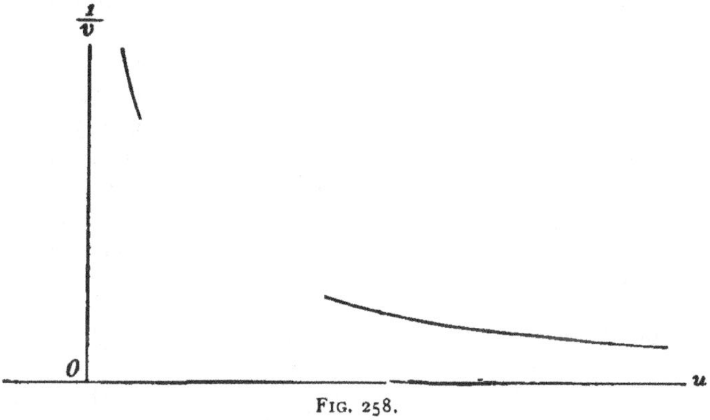

FIG. 258.

For the total time of passage of the projectile through the bore we have, using the area of the parabola (eq. 339),

$$t' = \frac{3}{2}\frac{u}{V} = \frac{3}{2} \times \frac{17.43}{1884} = .01387 \text{ sec.}$$

Hence the total area under the curve of reciprocals should be nearly .01387 second.

In the absence of a more accurate method, the area under the curve, which can be measured, may be obtained by considering each portion of the area bounded by the curve, the axis of u, and the two adjacent ordinates, as a trapezoid, and finding its area.

Thus for the first trapezoid we have

$$\text{ordinates} \begin{cases} .002584 \\ .001069 \end{cases}$$

$$\text{Value of } u = 1.24 = 1.70 - 0.46.$$

$$\text{Area } \left(\frac{.002584 + .001069}{2}\right) \times 1.24 = .00227 \text{ sec.}$$

Following this method, we have the table:

u (Feet).	Successive differences.	Successive times, seconds.	Total times, seconds.
0.46			0
1.70	1.24	.00227	.00227
2.40	0.70	.0006979	.0029679
3.20	0.80	.000704	.0036719
3.70	0.50	.0004075	.0040794
7.30	3.60	.002592	.0066714
9.50	2.20	.001386	.0080574
11.50	2.00	.001180	.0092374
13.00	1.50	.000854	.0100914
14.30	1.30	.000715	.0108064
15.75	1.45	.000787	.0115934
17.43	1.68	.000897	.0124904

From this table we can obtain the time of travel of the projectile over any distance u.

The sum of the times is .01249 second, while the total time is, as above shown, .01387 second.

Hence the difference, .00138 second, is the area that cannot be measured.

253. Examples—Second and Third Steps.

The table on page 442 gives the curve $v = f(u)$, that on this page the curve $t = f(u)$. Taking the ordinates of the latter curve as abscissas, and those of the former as ordinates, we can form the following table, whose abscissas and ordinates are those of the curve $v = f(t)$:

Total times, t.	Successive times, seconds.	Ordinates. Velocities foot-seconds, v.
0	0	
.00138	.00138	387
.00365	.00227	935
.00435	.000698	1080
.00505	.000704	1197
.00546	.0004075	1259
.00805	.002592	1545
.00944	.001386	1655
.01062	.001180	1727
.01147	.000854	1787
.01219	.000715	1827
.01297	.000787	1859
.01387	.000897	1884 I. V.,

and from this we construct the curve of velocities of the projectile as a function of the times. Passing to the consideration of the recoil of the piece and carriage, we have, equation (65),

$$v' = \frac{\left(p + \frac{\hat{\omega}}{2}\right)}{P} v.$$

$p = 300$ lbs.;
$\hat{\omega} = 125$ lbs.;
$P = 18$ tons $= 40320$ lbs.

Hence

$$v' = .00899v = .009v.$$

From this formula, having the values of v and the corresponding times t, we can form the following table and construct the curve of recoil of the piece and carriage as a function of the time.

Total times, seconds.	Velocity of recoil of Piece and Carriage, foot-seconds.
0.0000	
.00138	3.483
.00365	8.415
.00435	9.720
.00505	10.773

.00546	11.331
.00805	13.905
.00944	14.895
.01062	15.561
.01147	16.083
.01219	16.453
.01297	16.731
.01387	16.956

From equation (336) the maximum velocity of recoil is

$$V_m' = \frac{pV + 3000\tilde{\omega}}{P},$$

$$V_m' = 23.32 \text{ ft.-seconds.}$$

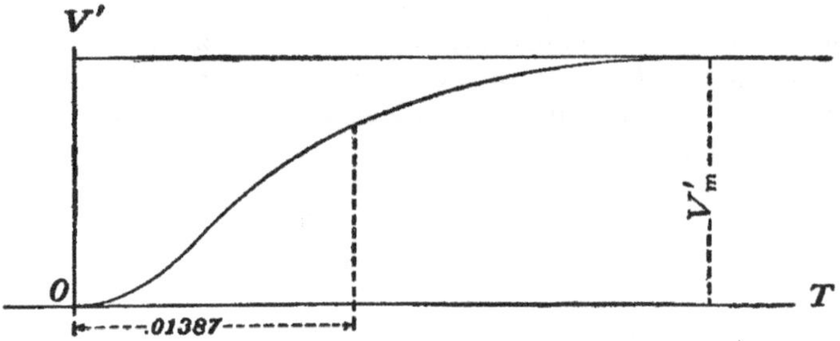

FIG. 259.

From the above data the curve of recoil, Fig. 259, may be constructed.

Drawing a line parallel to the axis of T, at the distance $V_m' = 23.32$ ft.-secs., this line will mark the limit of acceleration.

From this curve, the time corresponding to any velocity of recoil can be obtained, and the area under the curve will give the corresponding space passed over.

254. Problems.

1. Required the time at which the velocity of recoil is 13.905 ft.-sec.

The table page 446 shows it to be .00805 second.

2. Required the space passed over in recoil at the end of this time.

It will be the area under the curve, from the origin, up to the ordinate whose value is 13.905 ft.-secs., or, approximately, the sums obtained by adding the separate areas regarded as trapezoids, up to this point. As an approximation to the true result, regard the curve as a parabola. The result thus obtained will be slightly too great. The area under the curve will be two thirds the rectangle of the abscissa and ordinate of any point.

Hence for the area in question we have

$$s = \frac{2}{3} \times .00805 \times 13.905 = .0746 \text{ ft.} = .895 \text{ inches.}$$

3. Required the space passed over by the gun and carriage at the time the projectile leaves the bore.

By the same method of approximation we have

$$s = \frac{2}{3} \times .01387 \times 16.956 = .1567 \text{ ft.} = 1.88 \text{ inches.}$$

4. Required the time at which the velocity of recoil is a maximum, and the space passed over at this time.

The curve $v' = f(t)$ gives. by measurement, for the value of $t = .0395$ seconds.

The space passed over will be approximately

$$s = \frac{2}{3} \times .0359 \times 23.32 = .558 \text{ ft.} = 6.70 \text{ inches.}$$

In the same way all the circumstances of the recoil of a gun during the first period, may be obtained, having the curve $v = f(u)$, which can be obtained from Sarrau's formulas.

255. Time and Length of Recoil in Second Period.

TIME OF RECOIL.—At the beginning of the second period, the acceleration is zero, and the gun and carriage have acquired the maximum velocity of recoil V_m'. For simplicity of discussion, suppose the chassis horizontal.

The only force acting during the second period is friction, and this being practically constant, will uniformly retard the carriage and piece, till they are brought to rest.

Let f be the coefficient of friction = 0.2 about;

γ the retardation due to friction;

P the weight of the gun and carriage which is moving in recoil;

M the mass of the moving parts;

V'' the velocity of recoil at any time t during the second period.

Then from mechanics

$$\gamma = \frac{dV''}{dt} = -\frac{fP}{M}. \quad . \quad . \quad . \quad . \quad . \quad (340)$$

The velocity at any time t during the second period will be, from (340),

$$\int dV'' = \int -\frac{fP}{M}dt = -\frac{fP}{M}t + C. \quad . \quad . \quad (341)$$

When $t = 0$, or at the beginning of the second period, $V'' = V_m'$; hence in (341) we have

$$V'' = V_m' - \frac{fP}{M}t. \quad . \quad . \quad . \quad . \quad . \quad . \quad (342)$$

At the end of the second period, when recoil ceases, $V'' = 0$, and the corresponding time is, from (342),

$$T = \frac{MV_m'}{fP} = \frac{V_m'}{gf}. \quad . \quad . \quad . \quad . \quad . \quad . \quad (343)$$

LENGTH OF RECOIL.—Assuming equation (342), we have for the length of recoil, at the end of any time, t, in the second period

$$\frac{ds}{dt} = V'' = V'_m - \frac{fP}{M}t;$$

$$\int ds = \int V''dt = \int V'_m dt - \int \frac{fP}{M}t\,dt. \quad . \quad (344)$$

$$s = V_m't - \frac{1}{2}\frac{fP}{M}t^2. \quad . \quad . \quad . \quad . \quad . \quad (345)$$

At the end of recoil we have for the value of t, from (343),

$$T = \frac{V'_m}{gf}.$$

Substituting this value of t in (345) gives

$$S = \frac{V_m'^2}{gf} - \frac{1}{2}\frac{fP}{M}\frac{V_m'^2}{g^2f^2}; \quad . \quad . \quad . \quad . \quad (346)$$

but

$$\frac{fP}{M} = gf;$$

hence in (346)

$$S = \frac{V_m'^2}{2gf}. \quad . \quad . \quad . \quad . \quad . \quad . \quad . \quad (347)$$

FIG. 260.

256. Curve Representing Total Recoil—Application to 8-inch Rifle. Second Period.

CURVE OF TOTAL RECOIL.—The curve representing all the circumstances of recoil of the piece and carriage, will be obtained, by combining that for the first period, with the right line representing the second period, and will be of the form Fig. 260.

APPLICATION TO 8-INCH RIFLE.

1. Required the time, from the beginning of the second period, at which the velocity of recoil will be 10 ft.-secs., and the space passed over at that time.

From (342),

$$10 = 23.32 - .2 \times 32.2 \times t;$$

$$t = 2.068 \text{ seconds};$$

from (345),

$$s = 23.32 \times 2.068 - \tfrac{1}{2} \times .2 \times 32.2 \times (2.068)^2;$$

$$s = 34.46 \text{ feet}.$$

2. Required the time from the beginning of the second period to the end of recoil, and the total space passed over in recoil:

From (343),

$$t = \frac{23.32}{32.2 \times .2} = 3.621 \text{ seconds}.$$

From (347),

$$s = \frac{(23.32)^2}{2 \times 32.2 \times .2} = 42.22 \text{ feet}.$$

In a similar manner, all the circumstances of the recoil during the second period can be obtained, having the value of V_m' from formula (336).

257. Wheeled Carriages—Cases.

The preceding discussions relate to carriages which slide in recoil, such as those for seacoast guns.

For wheeled carriages two cases may arise:

1. The carriage may recoil, the wheels rotating, and not leaving the ground or platform upon which they rest.

2. The wheels may leave the ground or platform, depending upon the relative values of the components of the force of recoil which act to produce translation and rotation.

In the second case, the phenomenon of recoil is composed of alternate periods, during which the wheels rise, and return again to the platform.

If the carriage slides in recoil, with the wheels always in contact with the ground, the preceding discussions apply, the only change necessary being a decrease in the value of the coefficient of friction, due to the lubrication of the bearing surfaces of the nave and axle-arm. The coefficient, f, in this case is decreased to about two thirds its ordinary value.

If the wheels rise, increased pressure is produced on the trail. This increased pressure, decreases the extent of recoil as compared with that which obtains in the first case, and hence the values for time, velocity, and length of recoil deduced for the first case, will be greater than those for the second. As the calculations in the second case are somewhat complicated, those for the first case may be used instead of them, as being safe in practice.

258. To Calculate the Angle of Elevation of the Piece for which the Wheels will Rise.

The rotation of the carriage about the trail, depends on the angle of elevation at which the piece is fired. There is a limiting angle for which this rotation will occur. For all

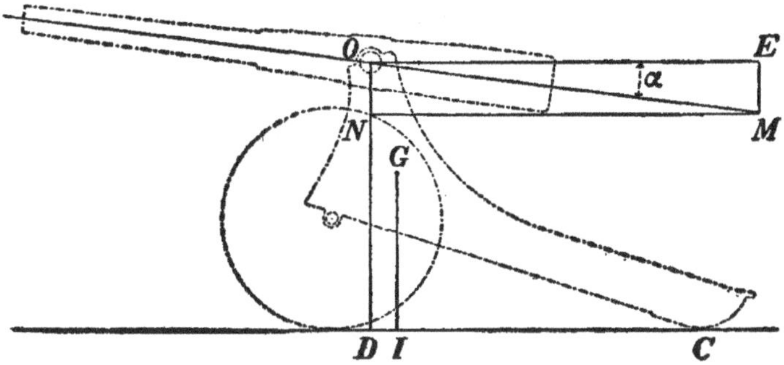

FIG. 261.

angles greater than this, rotation will not occur, and for all angles less than this, it will always occur. It is required to determine this limiting angle.

Let *OM*, Fig. 261, be the axis of the piece, and the line of action of the force P_0. Resolving this force into its components parallel and perpendicular to the ground, we have

$$OE = P_0 \cos \alpha;$$
$$ON = P_0 \sin \alpha.$$

With reference to the point *C*, the component *OE* acts with a lever-arm *OD* to raise the wheels, while *ON* acts with a lever-arm *DC* to keep them in contact with the ground. Let G be the centre of gravity of the system, composed of gun and carriage.

Then the weight P acts at G, with a lever-arm CI, to keep the wheels down. In addition to these, the system moves under the action of the force OE with an acceleration γ, and hence its force is $M\gamma$, and this force diminishes the rotative effect of the force OE, since the action of this force OE, is to produce both rotation around C, and translation along CD.

This latter force, $M\gamma$, has a lever-arm GI. We have therefore, calling those forces which tend to cause a lifting of the wheels positive, and those which oppose it negative,

$+ OE$,	lever-arm	OD;
$- ON$,	"	DC;
$- P$,	"	CI;
$- M\gamma$,	"	GI.

When the sum of the moments of these forces with respect to C is zero, the wheels will be on the point of leaving the ground.

Hence this gives the condition required.

Making

$$OD = a,$$
$$DC = d,$$
$$CI = b,$$
$$GI = h,$$

we have

$$+ P_0 \cos \alpha \, . \, a - P_0 \sin \alpha \, . \, d - P \, . \, b - M\gamma h = 0. \quad (348)$$

The value of γ is obtained as follows: Denoting the total pressure of the powder-gas by P_0, the component of this pressure causing recoil is $P_0 \cos \alpha$. This force is opposed by the friction due to the component $P_0 \sin \alpha$ and the weight of the gun and carriage P. Hence the force opposing motion is $f(P_0 \sin \alpha + P)$.

From mechanics we have therefore for the acceleration

$$\gamma = \frac{1}{M}[P_0 \cos \alpha - f(P_0 \sin \alpha + P)]. \quad . \quad . \quad (349)$$

Substituting this value for y in (348), which can be done, since the wheels are just about to leave the ground, and therefore the case is one of horizontal sliding, we have

$$P_0 \cos \alpha(a - h) - P_0 \sin \alpha(d - fh) - P(b - fh) = 0. \quad (350)$$

Now making $P = 0$ in comparison with P_0, and calling α_0 the value of α for this limit, where the wheels are just quitting the ground, we have

$$\tan \alpha_0 = \frac{a - h}{d - fh}, \quad \ldots \ldots \quad (351)$$

in which a and d can be found by direct measurement of the gun and carriage, and the centre of gravity by suspending the system in two different positions, and noting the point of intersection of the lines of suspension when produced.

BRAKES AND BUFFERS.

259. Necessity for Means of Checking Recoil—Conditions which a Good Brake should Fulfil—Classes.

NECESSITY FOR MEANS OF CHECKING RECOIL.—The above discussion, and its application in the case of the 8-inch gun, show, that unless some artificial means be employed to check recoil, its extent will be so great as to cause inconvenience.

This is especially true with modern field, siege, and seacoast guns, where the weights and initial velocities of the projectiles have greatly increased, without a corresponding increase in the weight of gun and carriage, and hence the length of recoil has been greatly increased.

In the field and siege services, this entails great fatigue upon the cannoneers in running the gun and carriage back to battery, with a consequent delay in firing, and exposure to the enemy's fire.

In the seacoast service, the length of recoil must be limited to three or four calibres, on account of cover, as the guns, if mounted in turrets, or similar places, have very limited space for working; and if mounted in barbette, a long recoil exposes the gun to hostile fire, and increases the time between shots. For these reasons, brakes or buffers are employed with modern guns.

CONDITIONS WHICH A GOOD BRAKE SHOULD FULFIL.—A good brake or buffer should fulfil the following conditions:

1. Its resistance should be constant at all times.

2. For the same piece, charge, and projectile, the length of recoil should always be the same, which is a proof of the regularity of its action.

3. It should be entirely automatic.

4. Its line of resistance should be as nearly as possible in the line of action of the force producing recoil, so as to avoid an overturning moment; and it must not interfere with the movement of the gun to and from battery, and its manœuvring.

CLASSES.—Brakes are divided into two general classes:

1. Friction brakes.
2. Hydraulic brakes.

260. Friction Brakes for Seacoast Carriages—Objections.

The various friction-brakes for field and siege carriages, have already been explained. Those for seacoast carriages consist generally of a series of plates, fixed to the chassis, between each pair of which, slides a plate attached to the top carriage. The plates are so arranged, that by means of a screw or other device they may be pressed together, and the friction due to this pressure limits the recoil.

Let P_0 represent the pressure at each surface in contact;
n, the number of surfaces;
S, the length of recoil in second period;
f, the coefficient of friction;
P, the weight of gun and carriage recoiling;
V_m', the maximum velocity of recoil with brake acting.

Suppose the chassis horizontal; then the work of friction of the plates, plus that of the piece and carriage, over the path S, will be equal to the total energy of recoil. Hence

$$f(P + nP_0) \times S = \frac{PV_m'^2}{2g}. \quad . \quad . \quad . \quad . \quad (352)$$

In this equation, for a given value of S, everything is known except P_0, and its value can be determined so as to limit the recoil to a given length S.

OBJECTIONS.—The objections to friction-brakes are:

1. They oppose to the initial motion of the system, the maximum resistance, when the velocity of recoil is greatest, and hence the resistance is not uniform during the recoil.

2. The resistance is not constant for any two consecutive shots, since it varies with the condition of the surfaces in contact. For the first shot, if the surfaces are slightly rusty, the resistance will be great; for succeeding shots, as the surfaces become polished, the resistance decreases, and if the surfaces be wet or lubricated, the resistance decreases. All these causes necessitate a regulation of the pressure for each shot.

3. The friction-brake is not automatic, as it has to be unclamped after each shot, to allow the gun to run in battery, and has then to be clamped again before firing. Accidents are liable to occur from this cause.

For these reasons the friction-brake has been abandoned for seacoast guns, and is only retained in different forms in field guns, where the weight of the hydraulic buffer, and its liability to get out of order, would be objectionable.

261. Hydraulic Brakes—General Description—Classification—Object of Discussion.

GENERAL DESCRIPTION.—This brake consists of one or more cylinders filled with non-freezing liquid, and attached either to the chassis, or to the top carriage. In modern carriages the cylinders are generally two in number, and are attached to the top carriage as near the axis of the gun as possible, to increase the mass of the system recoiling, and thus diminish its velocity; and also to bring the line of resistance of the brakes as nearly as possible coincident with the axis of the bore, and thus diminish the overturning moment.

In each cylinder moves a piston, pierced with holes parallel to the axis of the cylinder, and having a piston-rod attached to the chassis.

When the piece and top carriage recoil, the cylinders move to the rear, and the liquid is forced through the holes in the pistons. The resistance which the liquid opposes to the motion of the pistons, limits the recoil.

CLASSIFICATION.—There are two kinds of hydraulic brakes in use:

1. Those with constant orifices and variable resistance.
2. Those with variable orifices and constant resistance.

OBJECT OF DISCUSSION.—The principal points to be determined in discussing a hydraulic brake are:

1. The length of recoil;
2. The area of orifice;
3. The pressure in the hydraulic cylinder;
4. For the hydraulic brake of constant resistance, the law of variation of the areas of orifice, in order that the resistance shall be constant.

In the discussion, the friction of the liquid is neglected, as this is found in practice to be small. The flow, also, is supposed to take place through a thin partition, so that the contraction of the liquid vein may be neglected.

The velocity at the beginning of motion, is supposed to be the maximum velocity of recoil.

262. Hydraulic Brakes with Constant Orifice and Variable Resistance—Nomenclature—Value of Total Resistance Opposing Recoil.

NOMENCLATURE.—Let A be the effective area of cross-section of the piston; that is, the area of the piston minus that of the piston-rod and orifices;

a, the area of the orifices of flow;

*V_m', the maximum velocity of recoil of the system;

v', the velocity of recoil at any time t;

v, the velocity of flow of the liquid through the orifices at that time;

P, the weight of the system recoiling;

α, the angle of inclination of the chassis to the horizontal;

δ, the density of the liquid filling the cylinder (weight of unit volume):

---

* With the brake acting.

f, the coefficient of friction;

F, the total resistance which opposes the recoil.

VALUE OF F.—The total resistance, F, is composed of two parts:

1. The resistance of the brake, F'.

2. The resistance due to the friction of the moving parts, and the inclination of the chassis, F''.

We have then

$$F = F' + F''. \quad \ldots\ldots\ldots \quad (353)$$

The value of F'' is

$$F'' = P(\sin\alpha + f\cos\alpha). \quad \ldots\ldots \quad (354)$$

The value of F', the resistance of the brake, is equal to the total pressure exerted by the piston upon the liquid.

To determine this, we know from the law of continuity of the fluid, that the volume of liquid displaced by the piston, must be equal to that which passes through the orifices in it; hence we have

$$v'A = va, \quad \ldots\ldots\ldots \quad (355)$$

or

$$v = \frac{v'A}{a}. \quad \ldots\ldots\ldots \quad (356)$$

This velocity v is that due to a height of fall

$$v = \sqrt{2gH}; \quad \ldots\ldots\ldots \quad (357)$$

and if we suppose a column of liquid whose constant height is H and density δ, it will produce a pressure per unit of surface at its base, sufficient to cause the velocity of flow v. Hence this is the pressure exerted per unit of surface by the piston, upon the liquid, or it is the weight of a column of liquid whose height is H, density δ, and area of base unity. This pressure is

$$p = \delta \times H, \quad \ldots\ldots\ldots \quad (358)$$

and the total pressure on the surface A of the piston is

$$pA = F' = \delta \times H \times A, \quad \ldots\ldots \quad (359)$$

whence

$$H = \frac{F'}{\delta A}. \quad . \quad . \quad . \quad . \quad . \quad . \quad . \quad . \quad (360)$$

Substituting for v and H in (357) their values from (356) and (360), we have

$$\frac{v'A}{a} = \sqrt{\frac{2gF'}{\delta A}},$$

from which

$$F' = \frac{\delta A^3 v'^2}{2ga^2}. \quad . \quad . \quad . \quad . \quad . \quad . \quad . \quad (361)$$

Substituting the values of F' and F'' from (361) and (354) in (353), we have for the value of the total resistance to recoil

$$F = \frac{\delta A^3 v'^2}{2ga^2} + P(\sin\alpha + f\cos\alpha). \quad . \quad . \quad . \quad (362)$$

263. Length of Recoil with Constant Orifice.

Dividing both members of equation (362) by M, we have for the acceleration of recoil

$$\frac{dv'}{dt} = \frac{F}{M} = -\left[\frac{\delta A^3}{2Pa^2}v'^2 + g(\sin\alpha + f\cos\alpha)\right]. \quad (363)$$

Placing

$$\frac{\delta A^3}{2Pa^2} = B$$

and

$$g(\sin\alpha + f\cos\alpha) = K,$$

we have

$$\frac{dv'}{dt} = -(Bv'^2 + K). \quad . \quad . \quad . \quad . \quad . \quad . \quad (364)$$

But

$$dx = v'dt, \quad \therefore dt = \frac{dx}{v'},$$

and

$$\frac{dv'}{dt} = \frac{v'dv'}{dx} = -(Bv'^2 + K). \quad . \quad . \quad . \quad (365)$$

Whence

$$dx = -\frac{v'dv'}{Bv'^2 + K}. \quad . \quad . \quad . \quad . \quad . \quad (366)$$

Integrating between the limits V_m' and v',

$$\int_{v'}^{V_m'} -\frac{v'dv'}{Bv'^2 + K} = x = \frac{1}{B \log \epsilon} \log \left(\frac{BV_m'^2 + K}{Bv'^2 + K}\right)^{\frac{1}{2}}. \quad (367)$$

Replacing B and K by their values, we obtain an equation giving the relation between v' and x which is too complex for general use. To simplify it, make

$$f = 0, \quad \alpha = 0,$$

which is equivalent to supposing that the brake acts alone, without friction, and that the chassis is horizontal.

In this case we have

$$K = 0;$$

and substituting this value of K in (367), we have

$$x = \frac{2Pa^2}{\delta A^3 \log \epsilon} \log \frac{V_m'}{v'}. \quad . \quad . \quad . \quad . \quad (368)$$

When $v' = 0$, at the end of recoil, $x = \infty$ from equation (368), which shows that the recoil will continue indefinitely if the brake with constant orifices act without the aid of friction.

264. Area of Orifice for a Given Length of Recoil with Constant Orifices—Pressure in Cylinder.

Since for $v' = 0$, $x = \infty$, we cannot find directly from equation (368) the area of orifice which will limit the recoil to a fixed length l. This area can, however, for all practical purposes be determined as follows:

Suppose that when

$$v' = v_1'$$

the remaining energy of the system becomes very small, and the carriage is about to come to rest, and let l denote the length of recoil for which $v' = v_1'$. Then, from (368),

$$a^2 = \frac{\delta A^3 l \log \epsilon}{2P \log \frac{V_m'}{v_1'}} \quad . \quad . \quad . \quad . \quad . \quad (369)$$

for the area which will give the length of recoil l.

PRESSURE IN CYLINDER.—The total pressure in the cylinder at the beginning of recoil is, from (361),

$$F' = \frac{\delta A^3 V_m'^2}{2ga^2}, \quad . \quad . \quad . \quad . \quad . \quad . \quad . \quad (370)$$

and at any time at which velocity is v' it is

$$F_1' = \frac{\delta A^3 v'^2}{2ga^2}, \quad . \quad . \quad . \quad . \quad . \quad . \quad (371)$$

and the pressure per unit of area will be

$$\left.\begin{aligned} \frac{F'}{A} &= \frac{\delta A^2 V_m'^2}{2ga^2} \\ \text{and} \qquad \frac{F_1'}{A} &= \frac{\delta A^2 v'^2}{2ga^2} \end{aligned}\right\} \quad . \quad . \quad . \quad . \quad . \quad . \quad (372)$$

respectively.

265. Hydraulic Brake with Variable Orifices and Constant Resistance—Reason for Using it—Objects of Discussion—Value of Total Resistance opposing Recoil.

REASON FOR USING BRAKE WITH CONSTANT RESISTANCE.—For the brake with variable resistance, equations (372) show that the resistance is greatest when the velocity of recoil is greatest. This is contrary to the first condition imposed upon a good brake, and it evidently throws a great strain upon the carriage.

For this reason these brakes are no longer used with modern carriages, but are replaced by those whose orifice of flow is large at first, so as to allow a free flow of liquid when the velocity of recoil is greatest. As the velocity of recoil decreases, and its length increases, the orifices of flow are gradually closed automatically, and the resistance thus made constant throughout the recoil.

OBJECTS OF DISCUSSION.—The objects of the discussion are to determine—

1. The length of recoil;
2. The maximum area of the orifice of flow;
3. The law of variation of the area of orifice in order that the resistance shall be constant;
4. The constant pressure in the cylinder.

In this discussion, the first period of recoil, during which the resistance is approaching its constant value, is neglected, as being very short, and the resistance is supposed to be constant from the beginning of recoil.

TOTAL RESISTANCE OPPOSING RECOIL.—The nomenclature being the same as before, let a_0 be the maximum area of orifice of flow, and suppose it to be also the initial area.

Then we have, as in the case of the brake with constant orifice, equation (354),

$$F'' = P(\sin\alpha + f\cos\alpha). \quad . \quad . \quad . \quad . \quad (373)$$

Also, equation (361),

$$F' = \frac{\delta A^3 v'^2}{2ga^2}. \quad . \quad . \quad . \quad . \quad . \quad . \quad . \quad (374)$$

Since F' is constant by hypothesis, we must have for any other values of v' and a, as V_m' and a_0,

$$F' = \frac{\delta A^3 V_m'^2}{2ga_0^2}; \quad . \quad . \quad . \quad . \quad . \quad . \quad (375)$$

from which

$$F = F' + F'' = \frac{\delta A^3 V_m'^2}{2ga_0^2} + P(\sin\alpha + f\cos\alpha). \quad (376)$$

266. Length of Recoil with Variable Orifice.

Dividing both members of equation (376) by M, we have for the acceleration of recoil, as before (equation 363),

$$\frac{dv'}{dt} = \frac{F}{M} = -\left[\frac{\delta A^3 V_m'^2}{2Pa_0^2} + g(\sin\alpha + f\cos\alpha)\right]. \quad (377)$$

Integrating,

$$v' = -\left[\frac{\delta A^2 V_m'^2}{2Pa_0^2} + g(\sin\alpha + f\cos\alpha)\right]t + C. \quad (378)$$

When $t = 0$, $v' = V_m' = C$, hence

$$\frac{dx}{dt} = v' = V_m' - \left[\frac{\delta A^2 V_m'^2}{2Pa_0^2} + g(\sin\alpha + f\cos\alpha)\right]t. \quad (379)$$

Integrating again,

$$x = V_m't - \frac{1}{2}\left[\frac{\delta A^2 V_m'^2}{2Pa_0^2} + g(\sin\alpha + f\cos\alpha)\right]t^2, \quad . \quad (380)$$

which is of the same form as

$$h = vt - \frac{1}{2}gt^2,$$

and hence it is the equation of a uniformly retarded motion, as it should be, since the resistance is constant.

When the recoil ceases $v' = 0$, and $x = l$, the total length of recoil, and the corresponding time is, from (379),

$$T = \frac{V_m'}{\frac{\delta A^2 V_m'}{2Pa_0^2} + g(\sin\alpha + f\cos\alpha)}; \quad . \quad . \quad (381)$$

and this value of T in (380) gives for the total length of recoil

$$x = l = \frac{1}{2}\left[\frac{V_m'^2}{\frac{\delta A^2 V_m'^2}{2Pa_0^2} + g(\sin\alpha + f\cos\alpha)}\right]. \quad (382)$$

Equations (381) and (382) are similar to

$$t = \frac{v}{g}; \quad h = \frac{v^2}{2g},$$

for bodies falling freely in vacuo, under the action of gravity, as should be the case, since we have a constant force acting in both cases.

Placing equation (382) in the form

$$x = l = \frac{1}{\frac{\delta A^3}{Pa_0^2} + \frac{2g(\sin \alpha + f \cos \alpha)}{V_m'^2}}; \quad . \quad . \quad (383)$$

we see that l increases as V_m' increases and as α decreases, which should be the case.

267. Maximum Area of Orifice for the Brake with Variable Orifices.

To find the value of a_0, the maximum area of orifice in this case, solve equation (382) for a_0^2, and we have

$$a_0^2 = \frac{\delta l A^3}{P} \times \frac{1}{1 - \frac{2gl(\sin \alpha + f \cos \alpha)}{V_m'^2}} \quad . \quad . \quad (384)$$

for the maximum area of orifice.

Suppose the chassis horizontal and the top carriage mounted on rollers so as to avoid friction; then

$$\alpha = 0, \quad f = 0,$$

and (384) becomes

$$a_0^2 = \frac{\delta l A^3}{P}. \quad . \quad . \quad . \quad . \quad . \quad . \quad . \quad . \quad . \quad (385)$$

The area of orifice in this case is independent of the velocity of recoil, and hence we conclude that if the top carriage be placed on rollers and the chassis be horizontal, the length of recoil will be the same for a given area of orifice, no matter what the initial velocity of the projectile, charge of powder, or angle of fire may be.

268. Law of Variation of Areas of Orifice for Constant Pressure in Cylinder.

The energy of the system at the beginning of recoil is $\frac{PV_m'^2}{2g}$; after passing over the length of recoil x, when the velocity is v', the energy is $\frac{Pv'^2}{2g}$, and the work done over the path x is Rx, in which

$$R = M\frac{dv'}{dt} = \frac{P}{g}\frac{dv'}{dt}.$$

From mechanics, the original energy of the system is equal to its remaining energy after passing over a given path, plus the work done over that path; hence

$$\frac{PV_m'^2}{2g} = \frac{Pv'^2}{2g} + \frac{P}{g}x\frac{dv'}{dt}, \quad . \quad . \quad . \quad . \quad (386)$$

or

$$v' = V_m'\sqrt{1 - \frac{2x\frac{dv'}{dt}}{V_m'^2}}. \quad . \quad . \quad . \quad . \quad (387)$$

Since the resistance is constant and the recoil uniformly retarded, we have from the laws of constant forces

$$V_m'^2 = 2gh = 2\frac{dv'}{dt}l. \quad . \quad . \quad . \quad . \quad . \quad (388)$$

Substituting this value of $V_m'^2$ under the radical sign in (387), we have

$$v' = V_m'\sqrt{1 - \frac{x}{l}}. \quad . \quad . \quad . \quad . \quad . \quad (389)$$

Since F' is constant, we have, from (374) and (375),

$$\frac{\delta A^3 v'^2}{2ga^2} = \frac{\delta A^3 V_m'^2}{2ga_0^2}, \quad . \quad . \quad . \quad . \quad . \quad (390)$$

from which

$$\frac{v'}{V_m'} = \frac{a}{a_0}. \quad . \quad . \quad . \quad . \quad . \quad . \quad . \quad . \quad (391)$$

Dividing through by V_m' in (389) and substituting for $\frac{v'}{V_m'}$ its value from (391), we have

$$a = a_0\sqrt{1 - \frac{x}{l}}; \quad . \quad . \quad . \quad . \quad . \quad . \quad (392)$$

that is, the areas vary as the ordinates of a parabola.

These variable areas of orifice are obtained in different ways, one of the methods adopted in our service being to cut rectangular notches in the piston-head, and have bars bolted to the sides of the hydraulic cylinders, parallel to the axis, whose profile is such that at the origin the maximum opening will be a_0.

269. Profile of Rib for Constant Pressure—Maximum Pressure in Cylinder.

The profile of the rib or throttling-bar will then be a parabola. As the piston moves down the cylinder, the areas of orifice will be gradually diminished so that the pressure shall remain constant.

The equation of the parabola giving the profile of the rib is determined as follows:

Suppose there are n similar notches in the piston-head. The area of each one will be $\frac{a}{n}$. Let b and d be the breadth and depth of each notch. The rib in the cylinder must have the same breadth, b, and a variable depth, y.

Then for the area of an orifice at any time we have

$$\frac{a}{n} = b(d - y), \quad \therefore a = nb(d - y), \quad . \quad . \quad . \quad (393)$$

and this value of a in (392) gives

$$y = d - \frac{a_0}{nb}\sqrt{1 - \frac{x}{l}} \quad . \quad . \quad . \quad . \quad (394)$$

for the equation of the curve of the profile of the rib.

PRESSURE IN CYLINDER.—This is given by equation (375).

$$F' = \frac{\delta A^3 V_m'^2}{2ga_0^2} = C = \text{a constant}, \quad . \quad . \quad . \quad (395)$$

and per unit of area

$$\frac{F'}{A} = \frac{\delta A^2 V_m'^2}{2ga_0^2} = C' = \text{a constant}. \quad . \quad . \quad (396)$$

CHAPTER VIII.

POINTING; PROBABILITY OF FIRE.

POINTING.

270. Definitions.

POINTING.—To point a piece is to give it such a direction and elevation that the projectile, when fired, will hit the object aimed at.

OPERATIONS.—The pointing of a piece includes two distinct operations:

1. Giving the axis of the gun an elevation such that the projectile shall strike at the proper distance, or range, from the muzzle.

2. Giving the axis of the gun such a direction that the projectile shall strike a given point or object at that distance.

SIGHTS.—The instruments used in pointing are called sights, and there are two of these for each piece: the front sight and the rear sight.

Front Sight.—This is fixed to the muzzle of the gun, or to one of the rimbases, usually the right, and consists generally of a fixed point, or thin edge, or of cross-wires in a tube, the point, edge, or cross-wires being at a definite distance above the axis of the bore. and if on the rimbase, at a fixed distance to the right or left of that axis.

Rear Sight.—This generally consists of a rod, bar, or standard, graduated in degrees or ranges, and fixed in a socket on the breech of the gun. This standard carries a sliding notch or small hole, which is capable of being adjusted to any height on the rod, corresponding to a given range or elevation, within service limits. The notch or hole has also a motion at right angles to the axis of the bore, to correct for wind, drift, and other causes of lateral deflection.

In Fig. 262 let *A* represent the front sight;
OC, the rear-sight standard;
EC, the sliding piece, which is supposed to be at right angles to *OC*;
E, the rear-sight notch.

Triangle of Sight.—Then the triangle *OCE* is called the triangle of sight, and *ECO* is a right angle.

Zero of the Rear Sight.—The point *O*, where the line *OA*, drawn parallel to the axis of the piece through the top of the front sight, intersects the axis of the rear sight.

Natural Line of Sight.—The line *OAB*, parallel to the axis of the piece, and passing through the zero of the rear sight and the top of the front sight.

Sight Radius.—The length of the line *OA*.

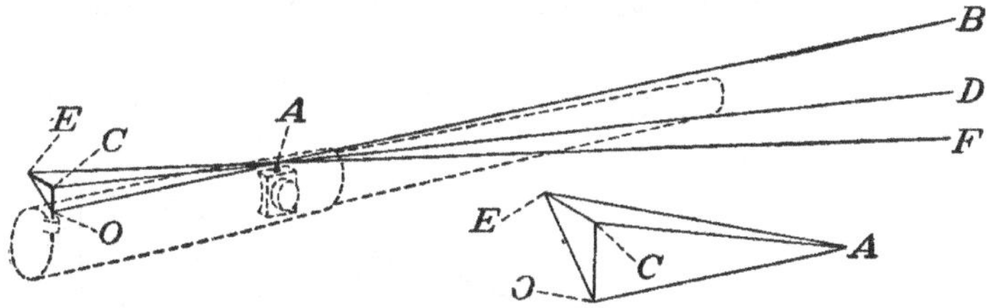

FIG. 262.

Artificial Line of Sight.—Any line, such as *EAF*, passing through the notch of the rear sight and the top of the front sight.

Natural and Artificial Planes of Sight.—The vertical planes containing the natural and artificial lines of sight respectively.

Plane of the Rear Sight.—The plane perpendicular to the axis of the piece, containing the triangle *ECO*.

271. Cases which may occur in Pointing—First Case—Height of Rear Sight—Corrections for Drift.

CASES.—The following cases may occur in pointing:

1. The axis of the trunnions may be horizontal, and the target situated in the horizontal plane passing through the centre of the muzzle.

2. The axis of the trunnions may be horizontal, and the target situated above or below the horizontal plane through the muzzle.

3. The axis of the trunnions may be inclined, and the target situated in the horizontal plane.

4. The axis of the trunnions may be inclined, and the target situated above or below the horizontal plane.

The second case is the usual one for sea-coast guns, and the fourth for field guns.

FIRST CASE.—In Fig. 263 let

HB represent the line of fire projected on the vertical plane;

H'B', the same line projected on the horizontal plane;

D and *D'*, the target;

CAD and *F'A'D'*, the projections of the artificial line of sight on the vertical and horizontal planes respectively.

Then *CO* is the height of the rear sight necessary to hit the point *D*, and *F'C'* the correction for drift.

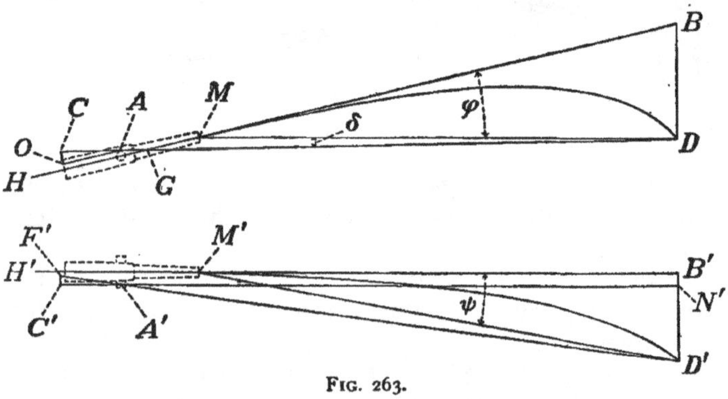

FIG. 263.

If these distances be known for a given range *MD*, it is evident that by fixing the rear sight with the proper elevation and drift, and giving the axis of the gun the direction shown, the target will be struck.

TO CALCULATE HEIGHT OF REAR SIGHT.—In Fig. 263 let h be the height of the rear sight CO;

l, the length of the sight radius OA;

ϕ, the angle of elevation BMD, the line MD being horizontal;

θ, the angle CAO made by the natural and artificial lines of sight;

δ, the angle MDG.

Then in the triangle MDG we have

$$MGD = BMD - MDG.$$

But the angle $MGD = CAO = \theta$, hence

$$\theta = \phi - \delta.$$

The angle δ is always very small, being the angle subtended at the target by the vertical projection of the chase of the gun from front sight to muzzle, and hence we have

$$\theta = \phi,$$

or

$$\tan \theta = \tan \phi;$$

but

$$\tan \theta = \frac{h}{l};$$

hence

$$\frac{h}{l} = \tan \phi; \quad \therefore\ h = l \tan \phi. \quad . \quad . \quad . \quad (397)$$

This value of h is laid off on the rear sight from the zero, and gives the graduation corresponding to the angle ϕ.

Correction for Drift.—In Fig. 263 let

D denote the drift $B'D'$;

d, the distance $F'C'$, or the correction for drift;

ψ, the angle of drift, $B'M'D'$.

In the similar triangles $F'A'C'$ and $N'A'D'$ we have

$$\frac{F'C'}{C'A'} = \frac{N'D'}{A'N'}; \quad \therefore\ F'C' = d = C'A'\frac{N'D'}{A'N'}. \quad . \quad (398)$$

But

$$C'A' = CA, \text{ nearly, } = \frac{OA}{\cos CAO} = \frac{l}{\cos \theta};$$

or, since $\theta = \phi$,

$$C'A' = \frac{l}{\cos \phi}.$$

The distance $N'D'$ is very nearly equal to the drift $B'D'$, and $A'N' = M'B'$, or the range R, very nearly.

Making these substitutions in (398), we have

$$d = \frac{lD}{R \cos \phi} \quad \ldots \ldots \ldots \quad (399)$$

272. Second Case.

In this case the axis of the trunnions is horizontal, and the target above or below the horizontal plane through the muzzle.

In Fig. 264 let D be the target situated above the horizontal plane MD'';

ϕ, the angle BMD;

α, the angle of elevation of the target above the horizontal plane.

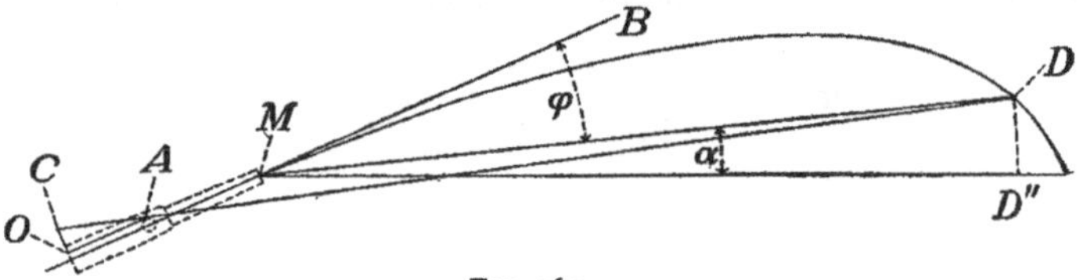

FIG. 264.

In the deduction of the equations of Exterior Ballistics, the horizontal plane MD'' through the centre of the muzzle, is alone considered, and all functions of the trajectory referred to this plane.

By the principle of the rigidity of the trajectory, the curve may be revolved through a certain angle about a horizontal axis passing through the point M, without changing the relations between the curve, its ordinates and angles, and the chord MD''.

For all practical purposes the points M and A may be considered as coinciding, and the revolution as taking place about the latter point.

Hence if the proper elevation OC, and correction for drift, be obtained from equations (397) and (399) for the range $MD'' = MD$, on the supposition that the target is situated on the horizontal plane, and then the gun be revolved about the axis of the trunnions through the angle α, till the line of sight CA passes through the point D, the projectile will hit the target, since this is equivalent to revolving the whole trajectory for the given range MD'' through the angle α. The same discussion applies to the drift, which is not altered by the elevation of the target, but is the same for the same range.

Hence, for this case, give the rear sight the same elevation and correction for drift as if the target were situated on the horizontal plane through the muzzle, and then elevate the gun till the artificial line of sight passes through the target.

273. Third Case—Errors—Fourth Case.

THIRD CASE.—In this case the axis of the trunnions is inclined, and the target is in the horizontal plane through centre of muzzle.

Let Fig. 265 be a section of the gun through the plane of the rear sight, the axis of the bore being horizontal. The zero of the rear sight, and top of front sight, are projected at O.

Let OC be the correct elevation, and FC the correction for drift, which will cause the projectile to strike the target.

Suppose that, due to inequalities of the ground or other causes, the gun is rotated to the right, about the axis of the bore, through the angle θ. The zero of the rear sight, and top of front sight, will now be projected at O', and the new position of the rear sight will be $O'C'F'$. The axis of the bore or line of fire has not changed its position, and the gun, if fired, will hit the target. But if the gun be resighted before firing, using the sights in their

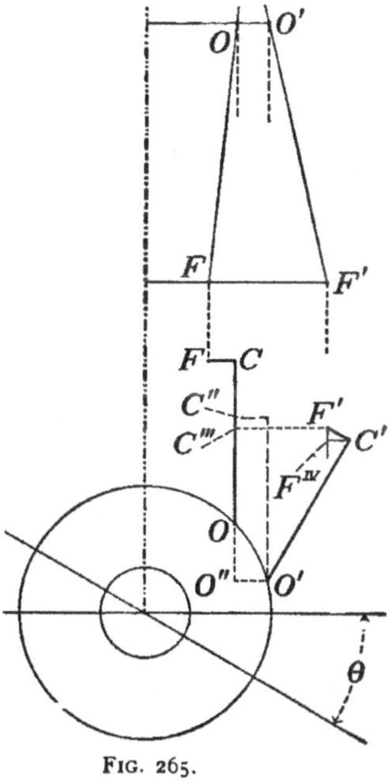

FIG. 265.

revolved position, the artificial line of sight will no longer pass through the target, and if it be made to do so, this will change the position of the line of fire, or axis of bore, and the projectile will no longer strike the target.

ERRORS.—By this rotation of the gun, the following errors have been introduced:

1. Instead of having the proper vertical height of rear sight equal to OC or $O''C''$, it is equal to $O''C'''$, or too small.

This may be explained as follows: The front sight, and zero of rear sight projected at O, have been lowered vertically by the rotation a distance $O''O$; and since the height of rear sight above the zero should remain unchanged, this height should now be $O''C''$, equal to OC. But the rear-sight notch is actually at a height $O''C'''$, and hence is too low by the distance $C''C'''$.

2. Instead of having the proper correction for drift FC, we have $F^{iv}C'$, which is too small.

3. The artificial line of sight, before rotation occurs, passes through the point F, and the top of the front sight projected at O. Hence, looking from the rear, the line of sight FO is oblique to the axis of the gun and diverges to the right.

After rotation the artificial line of sight passes through F', and the top of the front sight projected at O'. Hence, looking from the rear, this new line of sight $F'O'$ is oblique to the axis of the gun and diverges to the left. If, therefore the gun were correctly pointed before rotation, and be re-pointed after rotation, the projectile will deviate to the right.

This latter error is shown in plan, the same letters being used. It is of very frequent occurrence in small-arms firing, when the rear sight is not held vertically, the bullet deviating to the side toward which the sight is inclined.

To avoid these errors it is necessary to construct the rear sight so that its standard or upright will rotate about the point O. By means of a spirit-level fixed at right angles to the axis OC, this axis can always be kept in a vertical plane. With this arrangement, whenever the front sight and the zero of the rear sight are lowered vertically through a distance $O''O$, by rotation due to the inequality of the ground or other causes, the notch of the rear sight will be lowered the same amount, the vertical heights OC and $O'C''$ and the correction for drift remaining unchanged. This arrangement is made in all field-gun sights.

With this arrangement of the sights the pieces are pointed as in the first case.

FOURTH CASE.—In this case the axis of the trunnions is inclined, and the target above or below the horizontal plane.

If the standard of the rear sight rotates about the zero, as explained, the pointing is executed as in case two.

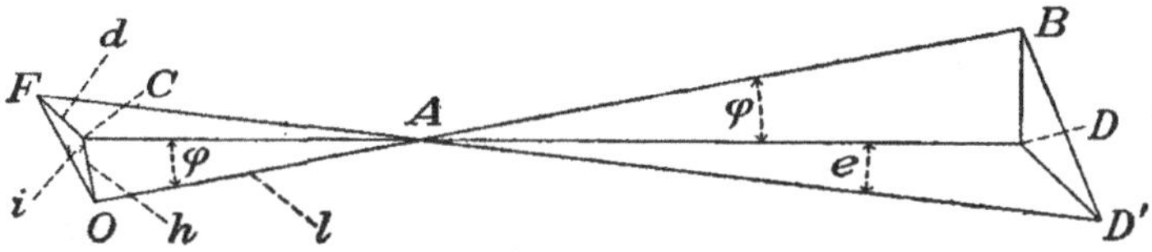

FIG. 266.

274. Permanent Angle of Drift.

Referring to Fig. 263, it is evident that the drift increases more rapidly than the range, and hence each range requires a special correction for drift. It is found, however, that within certain limits of error, a permanent correction may be made for drift, by giving the rear-sight standard an inclination to the left at a certain angle i with the vertical. This applies only to small arms where the barrel can be held with the front sight vertical, and to guns with fixed platforms, where the axes of the trunnions are horizontal. For field-guns, as before explained, the standard of the rear sight is kept vertical.

To determine this permanent angle, in Fig. 266

let $i =$ the angle FOC required;
$h = OC$, the height of rear sight;
$d = FC$, the correction for drift,
$l = OA$, the length of the sight radius;

ϕ = the angle of elevation BAD;
e = the angle DAD'.

Then in the triangle FOC we have

$$d = h \tan i;$$

but from (397),

$$h = l \tan \phi;$$

hence

$$d = l \tan \phi \tan i,$$

$$\tan i = \frac{d}{l \tan \phi}. \quad \ldots \ldots \quad (400)$$

From the triangle OAC we have

$$AC = l \sec \phi, \quad \ldots \ldots \quad (401)$$

and from FAC,

$$d = AC \tan e,$$

or from (401),

$$d = l \sec \phi \tan e; \quad \ldots \quad (401a)$$

but from the figure,

$$\tan e = \frac{DD'}{AD} = \frac{\text{drift}}{\text{range}};$$

hence in (401*a*)

$$d = l \sec \phi \frac{D}{R}, \quad \ldots \ldots \quad (402)$$

D denoting the drift and R the range.

Substituting this value of d in (400), we have

$$\tan i = \frac{D}{R \sin \phi}. \quad \ldots \ldots \quad (403)$$

It is known from Exterior Ballistics that $\sin \phi$ increases more rapidly than the range, and so also does D. Hence these variations partly correct each other and tend to make the angle i constant, and hence this constant correction for drift can be applied without great error. For long ranges, however, this correction is only partial, and an additional one must be made with the sight.

From equation (403) the amount of drift corrected for by this method is

$$D = R \sin \phi \tan i;$$

if any greater drift exists, as D', the difference,

$$D' - D = D' - R \sin \phi \tan i$$

must be corrected for by the sight.

275. Indirect Pointing.

When the target cannot be seen from the gun, the above methods must be modified.

The simplest case is that of mortar-firing, where the target is invisible from the piece, but may be seen from the parapet of the emplacement. In this case the plane of sight is established by plummets suspended from trestles, the elevation given approximately, and the mortar moved or traversed, till the planes of fire and of sight coincide.

The second case is where the target cannot be seen from the battery, but is visible from some place sufficiently near to communicate with the battery. In this case an observer watches the point of fall, and signals to the battery its position as to range and deviation. The aim is then corrected, and this is continued till the target is struck. An auxiliary mark, which is visible from the gun, is selected, and the sights directed upon this mark before each shot. When the target is struck, the piece is thereafter sighted upon this auxiliary mark, without changing the sights as adjusted for that shot. If a suitable mark cannot be obtained, the bearing of the target may be observed by a compass. Then by placing the compass in rear of the gun, the bearing of the target may be laid off by stakes and the gun directed upon them, the firing being corrected by an observer.

In some services sights are arranged so that they may be reversed in position; that is, the rear sight is placed in the position ordinarily occupied by the front sight, and the latter replaces the rear sight. (See 7-inch howitzer sight, page 493.) In this case the marking stake or stakes are placed in rear, and the sights directed upon them. Reflecting sights are also used when cover is of importance, and for turrets, the sights are placed on the exterior, and at the opposite extremity of the diameter upon which the gun is situated. While the turret is revolved 180° for loading, the sighting is effected, and when the gun is loaded, a traverse of 180° brings the gun into firing position.

276. Causes of Deviations in Pointing—Effect of Wind.

CAUSES OF DEVIATIONS.—When a gun is correctly pointed, as explained, the projectile should pass through the target. This, however, does not occur unless further corrections be made to eliminate other causes of error.

The principal of these are:

1. The effect of the wind.
2. Errors in estimating distance to target.
3. Effects of light on sights.
4. Personal errors of the eye.
5. Errors in height of front and rear sights.
6. Motion of target.
7. Rotation of the earth.
8. Variations in ammunition.
9. Jump.

EFFECT OF WIND.—The effect of wind is to increase or decrease the range, according as it is blowing from rear to front, or from front to rear, and to increase or decrease the drift of the projectile.

The velocity of the wind is generally expressed in miles per hour, and is obtained from an anemometer. A vane or other indicator gives its direction. Let W be the velocity of the wind in miles per hour, and ϕ the angle which its direction makes with the line of fire. The angle ϕ is measured from front to rear, being zero when the wind blows directly from the front along the line of fire.

Then the component which increases or decreases the range is $W \cos \phi$, and that which increases or decreases the drift is $W \sin \phi$.

The effect upon the increase or decrease of range is obtained by reducing W to feet, and regarding the projectile as having a velocity equal to $v + W \cos \phi$ or $v - W \cos \phi$.

Using these values for v in the ballistic formulas, the effect of the wind along the range can be calculated.

The component which produces deviation is $W \sin \phi$. Various formulas are given for calculating its effect, but the subject is very difficult.

When the wind is blowing from the left, its relative motion with respect to the projectile is less because the latter is moving in the same direction. When it blows from the opposite direction the reverse is the case. To correct for lateral deviation due to wind, the drift-slide is set towards the wind. That is, if the wind is from the left, the slide is moved to the left.

For small-arm firing, the direction of the wind is expressed by a clock-face notation, the clock being supposed to be held in the hand of the firer, with the XII-o'clock mark toward the target and the III-o'clock mark to the right. A wind blowing directly from the target is called a XII-o'clock wind; one directly from the left, a IX-o'clock wind, etc.

Assuming the force of the wind as unity, a table is given in the "Rifle and Carbine Firing," showing the proportions of the rectangular components of the different winds, and it is found practically, that the lateral deflections produced by them are proportional to these components.

The amount of lateral deviation produced by a wind blowing at right angles to the line of fire, with a velocity of one mile per hour, is called the coefficient of deviation. Calling this coefficient k, we have for any wind whose velocity is W, and which makes an angle ϕ with the line of fire, for any given range,

$$D = kW \sin \phi.$$

The relation between k and R must be determined by experiment.

277. Estimating Distances—By the Eye—By Sound—Le Boulengé Telemeter.

It is evident that unless the distance of the target be known, the proper elevation and correction for drift cannot be given, since these depend on the range. In actual operations the distance of the target is seldom known. For sea-coast guns, the channel or harbor is surveyed and plotted, and buoys may be planted at different known distances. In siege operations, the distance of the target may be obtained by observations with various instruments of precision; but for field artillery, where time is lacking, the distance must be obtained rapidly by estimation, by range-finders, or by trial shots.

Distances may be estimated—

1. By the eye;
2. By sound.

BY THE EYE.—This method requires considerable practice to obtain results of any accuracy. For short ranges the eye may be trained by trial, by observing carefully the appearance presented by known objects at different distances, such as the height of a man, the parts of his dress, etc., which are visible at those distances. Each individual must form a standard of comparison for himself; and since this method is only applicable for relatively short distances, it is of more importance for small-arm fire.

Objects vary in appearance according to the nature of the ground, being apparently nearer for level ground; also on a clear day, or with a distinct background, they appear nearer than under opposite conditions.

BY SOUND.—This method is based on the fact that sound travels about 1100 feet per second in air. Hence, if the time in seconds be noted between the flash and the report of a gun, or between the flash and the report of a shell fired from the battery, the distance is obtained by multiplying the time in seconds by 1100 feet. This time may be measured by a stop-watch, or by counting the number of steps taken in the interval, and knowing the number of similar steps which the observer takes per second.

LE BOULENGÉ TELEMETER.—An instrument called the Le Boulengé telemeter is used for measuring distances by sound. It consists of a glass tube filled with liquid, in which a disk is placed, whose specific gravity is slightly greater than that of the liquid. When the tube is held vertical, the disk falls through the liquid with a motion which is nearly uniform. To use the telemeter, the tube is held horizontal, with the disk at zero. When the flash is seen, it is turned quickly to a vertical position; when the report is heard, it is turned back to a horizontal position. A scale on the tube gives the range directly, corresponding to the distance passed over by the disk.

It is evident that it would be difficult in practice to observe the burst, and hear the report of any particular shell.

278. Range-Finders—Principle—Class 1.

The estimation of distances by the eye and by sound being so inaccurate, instruments called range-finders or telemeters have been devised to measure the distance to the target.

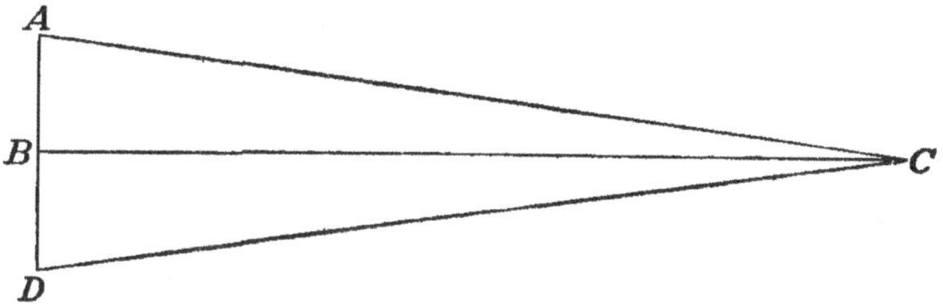

FIG. 267.

PRINCIPLE.—In Fig. 267 let C be the target.

In the isosceles triangle ADC, if the angles at A and D, and the base AD, are known, the angle ACD can be found, and we have

$$BC = \frac{\frac{1}{2}AD}{\tan \frac{1}{2}ACD};$$

or in the right-angled triangle ABC or BCD, if the angle at A or D be known, we have

$$BC = \frac{AB}{\tan ACB},$$

or

$$AC = \frac{AB}{\sin ACB}.$$

The object of these instruments is to measure rapidly and accurately the angles A and D.

To avoid calculations, the angles A and D are so chosen that the value of the tangent or sine of C shall be some simple number, as $\frac{1}{10}$, $\frac{1}{20}$, $\frac{1}{40}$, etc.; or the angles may vary and the base have a fixed value, the corresponding multipliers being inscribed on the instrument.

This divides the instruments into two general classes—

1. Those having fixed angles and variable bases;
2. Those having variable angles and fixed bases

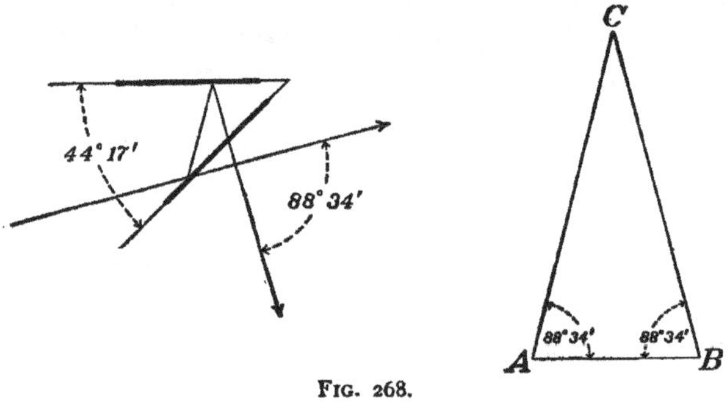

FIG. 268.

and a third class which combines the qualities of the two above, viz.,

3. Those having variable angles and variable bases.

CLASS 1.—In this class of range-finders the base is proportional to the range. This gives greater accuracy, as with a small base, a slight error in measurement of either angles or base leads to a large error in range. The general idea of this class of instruments is as follows: Two mirrors are fixed at an angle, say, of 44° 17′ (Fig. 268). A ray of light striking one of these mirrors is reflected twice, and according to a well-known principle of optics, the ray, after two reflections, makes with the original direction of the incident ray, an angle of 88° 34′, or twice the angle of the mirrors. An observer standing at A and looking toward B, sees B directly, and by reflection in the mirrors, makes the image of C coincide with B. The point A is then marked by a stake. Moving to B, which must be found by moving along AB and looking towards A, the reflection of C is made to coincide with A. The angle at C is then

$$180^\circ - 2 \times 88^\circ\, 34' = 2^\circ\, 52',$$

and measuring AB we have

$$AC = \frac{\frac{1}{2}AB}{\sin 1^\circ\, 26'} = \frac{\frac{1}{2}AB}{\frac{1}{40}};$$

$$AC = 40 \times \tfrac{1}{2}AB = 20AB.$$

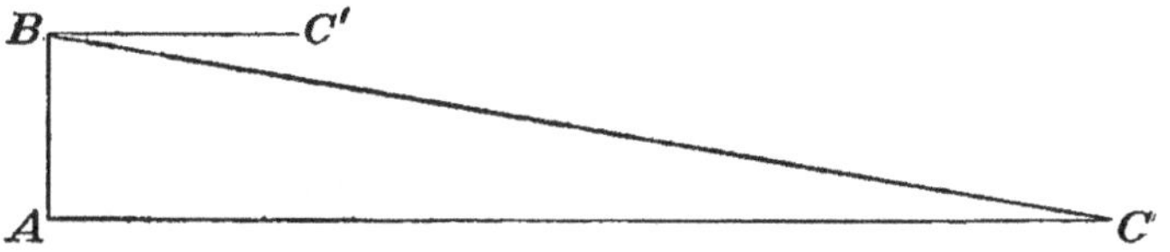

FIG. 269.

279. Range-finders—Class 2—Class 3—Depression Range-finders —Range and Position Finder.

CLASS 2.—In this class we have a fixed base and a variable angle. To save time, the instruments are generally adjusted so that the range can be read off at once. As the measurements must be very accurate, telescopes are often used to measure the angles, and this necessitates a very accurate mounting and increases the difficulty of transportation.

CLASS 3.—These instruments can be used by either method, but the variable base is generally preferred.

DEPRESSION RANGE-FINDERS.—Let AB, Fig. 269, represent the vertical height of a gun, or of a range-finder, above the surface of the water, and C an object, such as a ship, whose distance is to be determined. If the angle $C'BC$ be measured, and the height AB be known, it is evident that the distance BC can be determined as before. These instruments are called depression range-finders, the angle being measured in a vertical plane.

RANGE AND POSITION FINDERS.—In sea-coast batteries it is often necessary to fire at objects, such as ships, which cannot be seen from the guns. In this case it is necessary to find not only the range but also the *position* of the object, which is generally in motion, in order to hit it at any given time. An instrument used for this purpose is the Fiske Range and Position Finder, invented by Lieut. Fiske of the U. S. Navy.

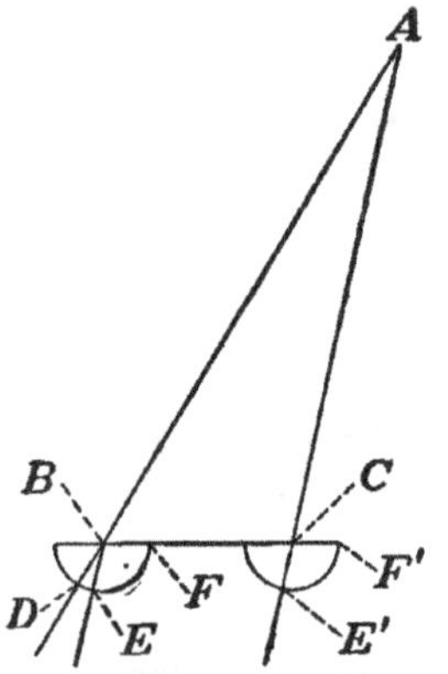

FIG. 270.

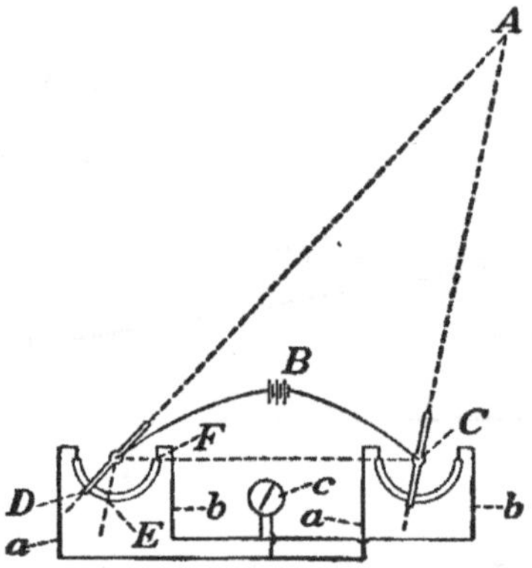

FIG. 271.

280. The Fiske Range-finder.

In Fig. 270 let A represent the target, and BC a known base. Then

$$AC : BC :: \sin ABC : \sin BAC.$$

$$AC = BC \times \frac{\sin ABC}{\sin BAC}.$$

The angle ABC can be readily measured. The angle $BAC = DBE$, the line BE being parallel to AC. The Fiske Range-finder measures the angle DBE by the use of the Wheatstone bridge, as follows:

Suppose the two semicircles in Fig. 270 replaced by two metallic arcs (Fig. 271). At the centre of each of these arcs is pivoted a telescope, the pivot of which is connected to a battery, *B*. The telescopes are in electrical contact with the arcs. These metallic arcs are connected at their extremities with a galvanometer, *c*, the whole forming a Wheatstone bridge, whose arms are *aa bb*.

When the two telescopes are pointed on the object *A*, it is evident that the arms of the bridge are unequal, and hence do not balance, and this fact is indicated by the deflection of the needle of the galvanometer. The arc *FD* is noted.

By swinging the telescope at *F*, around, till the needle of the galvanometer indicates zero, the bridge balances, the telescope being parallel to the one at *C*, and the arc or angle $DF - FE = DE$ is equal to the angle at *A*. From this the distance *AC* can be calculated, or be read off directly on a properly constructed scale.

Generally, in using the instrument, the telescopes are mounted at a distance from the battery where the view is uninterrupted, while the galvanometer is at the gun. The observers keep the telescopes constantly directed on the target, and the man at the gun balances the bridge, by introducing a variable resistance into the circuit, till the needle stands at zero. This variable resistance is graduated so as to indicate the range corresponding to the resistance introduced.

281. The Fiske Position-finder—Range by Trial Shots.

To find the position of the object, the Fiske Range-finder is modified as follows, Fig. 272.

Let *A* and *B* be the arcs with their telescopes as described, and *D* a chart drawn to scale, on which are two metallic arcs, *A′* and *B′*. The arc *A′* is connected electrically with *A* and with a galvanometer, *A″*, forming a Wheatstone bridge, and in the same way the arc *B′* is connected with *B*, and with the galvanometer, *B″*. The arc *A′* carries a metallic rule, *A′C′*, pivoted at the centre of the arc, and *B′* a rule, *B′C′*, similarly pivoted.

When the rule *A′C′* is parallel to the telescope at *A*, the galvanometer *A″* is at zero. When *B′C′* is parallel to the telescope at B, the galvanometer B'' is at zero. Hence their intersection C' marks the position of the object on the chart. Let G be the gun in battery, and G' its corresponding position on the chart. The gun has a metallic arc with which a pointer is in contact, and the arc G' has a metallic rule, $G'C'$, in contact with it. The gun G with its arc and pointer, and the metallic rule $G'C'$, are electrically connected with a galvanometer, G'', near the gun, forming a third Wheatstone bridge. It is evident that by traversing the gun in azimuth till its axis is parallel to the rule $G'C'$, the galvanometer G''

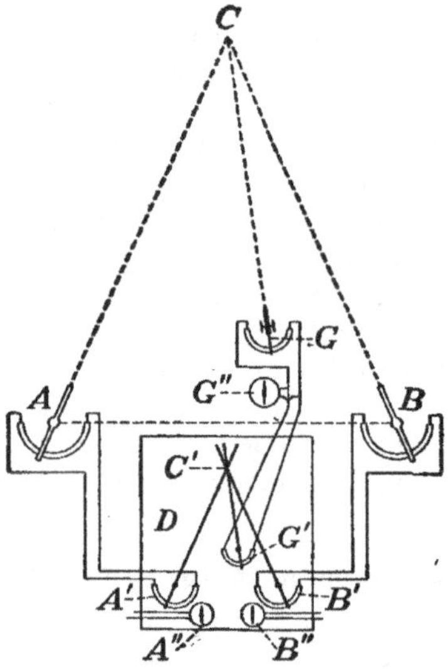

FIG. 272.

will indicate zero, and the gun will have the proper direction. The elevation may be telephoned from the observing station, or else the gunner, knowing the range and direction of the object, may take the elevation directly from a range table. Other arrangements of the same nature may be made with this instrument, using the principle of the Wheatstone bridge.

RANGE BY TRIAL SHOTS. — Owing to various causes,

the determination of distances in the field by range-finders is attended with difficulty, and the method actually adopted in all services is that by trial shots. Two plans are used.

In the first a percussion-shell is fired with an elevation which will cause it to strike short of the target. The point of fall is observed. A second shell is then fired with an elevation which will cause it to strike beyond the target. Its point of fall is also observed. The target is then enclosed in a fork.

Taking a mean of these two elevations will give a still closer approximation. If this shot falls short, a mean of this elevation and of that beyond will give another approximation, and so on.

By this means the range is soon found.

The second method is to fire the first shot short, and then increase the elevation slightly, and so on by successive increments till the proper range is attained. The difficulty is in observing accurately the point of fall for long ranges.

282. Effect of Light—Errors of the Eye—Errors in Height of Front and Rear Sights.

ERRORS DUE TO LIGHT.—In clear weather shots usually fall short, since, in a bright light, objects appear nearer, and the distance is underestimated, and, in addition, a finer sight is taken, owing to the distinctness of the front sight. The converse is true on a dark day. With regard to lateral deviation, if one side of the sight is brighter than the other the deviation will be from the light.

ERRORS OF THE EYE.—These vary with different individuals, and must be corrected by training.

ERRORS IN HEIGHT OF FRONT AND REAR SIGHTS.—In the previous discussions it has been assumed that the zero of the rear sight, and the top of the front sight, are at the same distance from the axis of the piece. If this be so, the natural line of sight is parallel to the axis of the piece in all positions, and hence the vertical plane containing this line is likewise parallel to the plane of fire in all positions.

If, however, the height of the front sight is not the same as that of the zero of the rear sight, an error is introduced. To show this, in Fig. 273 let A and B be the vertical projections of the rear and front sights, respectively, and AB the horizontal projection of the natural line of sight when the axis of the trunnions is horizontal.

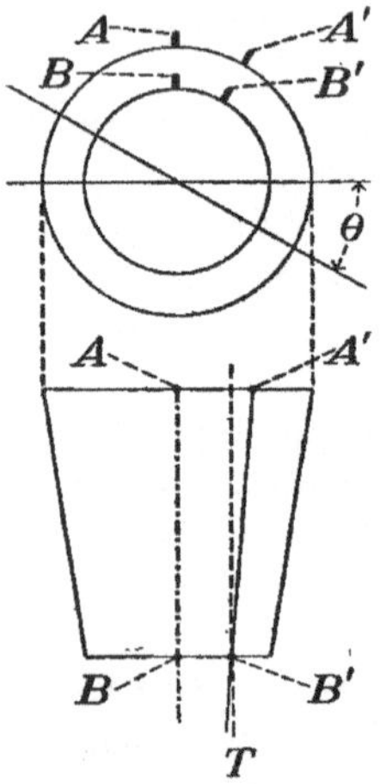

FIG. 273.

Suppose that, due to inequalities of the ground, the axis of the trunnion is revolved through the angle θ.

Then A' and B' will be the vertical projections of the rear and front sights, and $A'B'$ the horizontal projection of the natural line of sight. This line is now inclined, in its revolved position, to the axis of the piece, and hence in the revolved position the plane of sight will intersect the plane of fire. Hence, to an observer behind the gun, if the line of sight be directed on the target T, the gun will shoot to the left.

This error is similar to that discussed in subject 273. In that case it can be removed by keeping the rear sight vertical. The error in the present case can only be removed by making the heights of front sight and of zero of rear sight equal.

283. Motion of Target — Rotation of the Earth — Variations in Ammunition.

The target may move directly toward or from the gun, at right angles to the line of fire, or oblique to that line. As the last case includes both the others, it will be considered.

Let AB, Fig. 274, be the line of fire, and BC the direction of motion of the target, making the angle ϕ with the line of fire. Suppose the range AB and the rate of motion of the object known.

During the time that the projectile is moving over the distance AB the object has moved over the distance BC. Let v denote the velocity of the object and t the time of flight of the projectile; then

$$BC = vt.$$

Suppose the correction for drift to be such as to cause the projectile to strike at B. Then to hit the object the correction should be made to cover the additional distance,

$$D + CC' = D + vt \sin \phi,$$

in a direction toward the motion of the target. If this motion is not known it must be estimated.

This will still leave a small error in range $BC' = vt \cos \phi$ which must be compensated for by a slight increase in ele-

FIG. 274.

vation, or by retaining the elevation corresponding to AB and aiming beyond the target the estimated distance BC'.

ROTATION OF THE EARTH.—This is not generally taken into account. Its effect in the northern hemisphere is to cause projectiles to deviate to the right.

VARIATIONS IN AMMUNITION.—The effect of increasing the charge and density of loading is to increase the initial velocity. That of increasing the weight of the projectile is to decrease this velocity, as has already been shown in Interior Ballistics. With modern guns these variations in ammunition are very slight, and their effects may be neglected. Variations in moisture also affect the initial velocity, a damp powder giving less velocity and a dry powder greater velocity, for the reasons previously explained.

Also the heating of the bore increases this initial velocity, since less heat is lost by the gases.

284. Jump.

This error is caused by the motion of the gun upon discharge, due to the elasticity of the parts of the carriage, the lack of accurate fitting of gun to carriage, the vibration of the chase, etc. It varies with different guns and carriages, and is determined by experiment for any particular gun as follows:

In Fig. 275 let *AB* be a vertical screen or target, placed

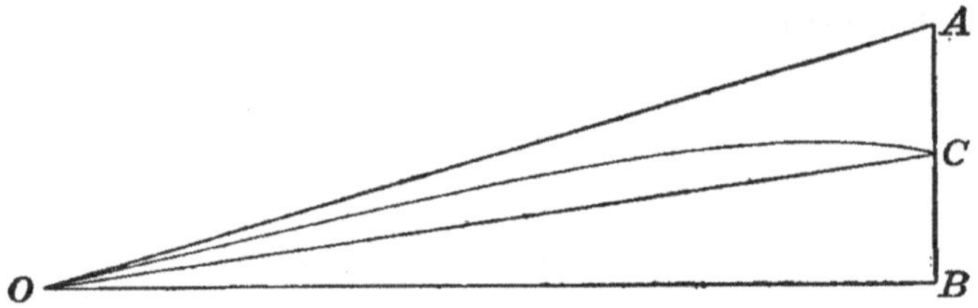

FIG. 275.

at such a distance from the muzzle of the gun *O* that it will not be affected by the blast. Let *OB* be the axis of the bore, supposed horizontal. The point *B* where the axis of the bore prolonged pierces the target is found by inserting a disk in the breech of the gun with a small peep-hole in the centre, and placing in the muzzle a pair of cross-hairs whose intersection is at the axis of the bore. Looking through the peep-hole and at the cross-hairs, the point *B* is marked on the target. When the gun is fired, suppose *OA* to be the line of departure. Then *AOB* is the angle of jump required. The projectile will strike the target at some point *C*. From the triangle *AOB* we have

$$\tan AOB = \frac{AB}{OB} = \frac{AC + CB}{OB}.$$

From the laws of falling bodies we have

$$AC = \tfrac{1}{2}gt^2. \qquad . \quad . \quad . \quad . \quad . \quad . \quad . \qquad (403a)$$

For the short distance OB we may regard the velocity of the projectile as uniform. Denoting this velocity by v, and the distance OB by a, we have

$$a = vt; \quad \therefore t = \frac{a}{v}.$$

This value of t in (403a) gives

$$AC = \frac{ga^2}{2v^2}.$$

The distance BC to the centre of shot-hole C can be measured; calling this b, we have, for tan AOB,

$$\tan AOB = \frac{ga}{2v^2} + \frac{b}{a};$$

in which the second member is known, since v can be calculated by Exterior Ballistics.

If the shot does not strike vertically above B, there will be a lateral deviation c, whose measure is $\tan^{-1}\frac{c}{a}$.

If OB is not horizontal, the same principle applies, the triangle AOB being an oblique, instead of a right-angled triangle.

285. Description of Sights for 8, 10, and 12 Inch Seacoast Guns.

These guns have two sets of sights. The first set, AA', Fig. 276, is on the middle element of the reinforce, and consists of a simple rear-sight notch, A, and a conical front sight, A'. They cannot be used for elevations, and are for catching the target readily, and giving the general direction.

The second set, BB', Fig. 276, is placed on the left side of the gun, as shown, and are on this side so as to be

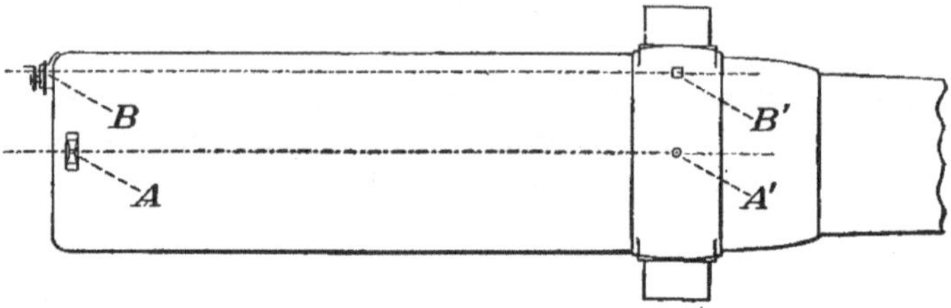

FIG. 276.

out of the way of loading, and so that the gun may be sighted while the charge and projectile are being inserted.

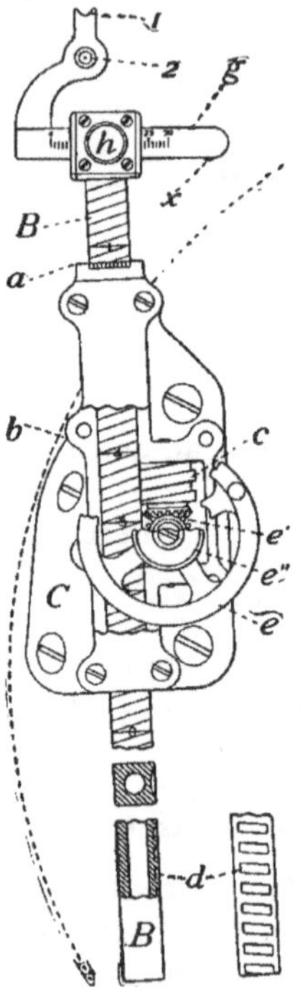

FIG. 277.

The rear sight, Fig. 277, slides through a bronze socket, C, bolted to the breech-plate. This socket is inclined to the left at the permanent drift angle, which is 2° 30′ for the 8-inch gun, 2° 45′ for the 10-inch, and 3° 00′ for the 12-inch. It is prolonged upward a distance ab, to give increased support and steadiness to the sight.

The socket carries a worm, c, which engages in a corresponding thread, d, in the right-hand edge of the sight. This worm is worked by a hand-wheel, e, and pinion, e', and the hand-wheel e is held in place, and motion prevented, by a clamping-wheel, e''. The functions of the worm are to raise and lower the rear sight, and to hold it fixed in any given position, so that it will not be moved by the shock of firing.

The sight consists of a hollow steel bar, B, one inch square, graduated in degrees, and each degree into six parts. The smallest reading on the sight is therefore 10 minutes. The top, a, of the bronze socket, C, is divided into 10 equal parts, and the divisions on the sight are diagonal, so that by means of the scale on the socket, each of these graduations on the sight, can be divided into ten equal parts, giving one minute for the least reading.

The top or head of the sight, consists of a deflection-bar, *g*, with a vertical projection, carrying a notch, 1, and a peep-sight, 2. The notch is used in connection with the top of the front sight, to catch the target quickly, and the peep-sight with the cross wires of the front sight, for final adjustment. The deflection-bar, *g*, has a horizontal sliding motion through the top of the rear-sight bar, to correct for wind, drift, and other errors, and is clamped in any position by the clamp-screw *h*. It is graduated as shown, each graduation being $\frac{1}{1000}$ of the range.

For deflection to the left, the bar is used in the position shown. For deflection to the right the small pin *x* is pushed in, the bar entirely removed from its socket, and reinserted from the right, the graduations being the same on the reverse side.

The right-hand side of the sight-bar, contains a portion, of a screw-thread *d*, into which gears the worm, *c*, for raising and lowering, as before explained.

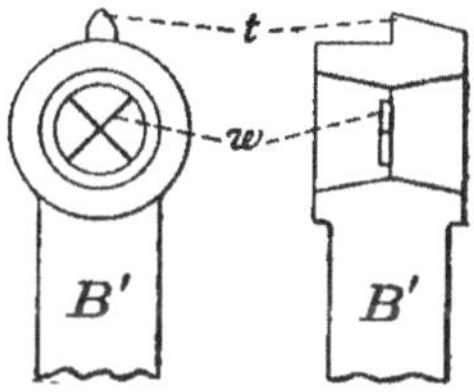

FIG. 278.

The front sight *B′*, Fig. 278, consists of two truncated cones, with the smaller bases together at the middle, and carrying two flat steel cross ribbons, *w*, halved into each other. It has also a top sight *t*, which is used with the open notch of the rear sight *B*.

286. Gunner's Quadrant for Mortars.

When the angle of elevation exceeds 15°, the rear sight above described, becomes so long as to be difficult to handle and it may bend under its own weight, so as to cause inaccuracy in aiming.

The target, also, in such cases, is not generally visible from the gun or mortar, and for these reasons the rear sight is not used. To give the necessary elevation in such cases, and for mortar-fire generally, the gunner's quadrant is used.

This consists, Fig. 279, of the body *a*, and a movable arm, *b*.

The body is made of bronze, and carries an arc graduated on one side from 0° to 44°, and on the opposite side from 45° to 89°. On the inside edge of the graduated arc are

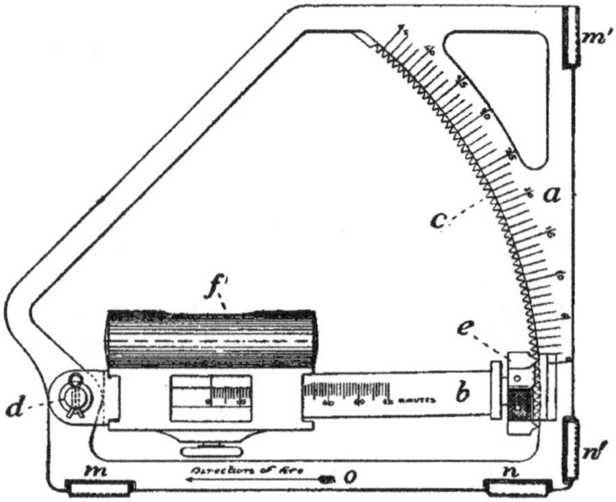

FIG. 279.

teeth *c*, each of which corresponds to one degree. An arrow is marked on each side of the body, and when the quadrant is in use, the arrow on the side on which the reading is taken must always point in the direction of the target.

The movable arm, *b*, is pivoted to the body at *d*.

This arm has a small toothed sector, *e*, which is acted on by a spiral spring, contained in the arm, and by which the sector is pressed outward, so that its teeth will remain engaged with those of the graduated arc. The upper surface of the movable arm, *b*, is the arc of a circle, and on this arc rests a level, *f*. This level bears on the arm at two points only, and is of such a length that when moved along the arc from its zero-point, to its extreme position at the other end of the arm, the angle moved over is one degree.

The arm is graduated in minutes.

Degrees are read on the graduated arc, and minutes by the scale on the movable arm.

To Use the Quadrant.—Suppose the elevation to be 20° 18′.

Press back the toothed sector, *e*, and move the arm, *b*, till its index is opposite the 20° mark on the graduated arc.

Slide the level, *f*, along the movable arm, *b*, till its index is opposite the 18′ mark on the arm. The quadrant is now set to 20° 18′.

Place the side, *mn*, on the flat surface prepared for it, near the breech of the gun, or the side, *m′n′*, against the face of the muzzle, being careful to keep the side on which the reading is taken, to the left, and the arrow, *o*, pointing toward the target. Elevate the piece till the bubble in the level comes to rest in the centre. For any elevation greater than 45°, as 60° 33′, use the graduations on the other face of the arc, and the scale on the movable arm as above.

The other side of the quadrant must now be turned to the left, and the arrow on it pointed toward the target. Elevate the gun as before.

287. Sights for Siege Artillery—For 7-inch Howitzer.

The sight for the 5-inch siege-gun, is exactly similar to that for the 3.2 field-gun to be described. That for the 7-inch mortar is the gunner's quadrant already described.

SIGHT FOR 7-INCH HOWITZER.—This piece has a comparatively low initial velocity, and curved fire. Hence the sight for this gun should give a large scale for correcting lateral deviations, as they will be greater than for the siege-gun, which has a high velocity.

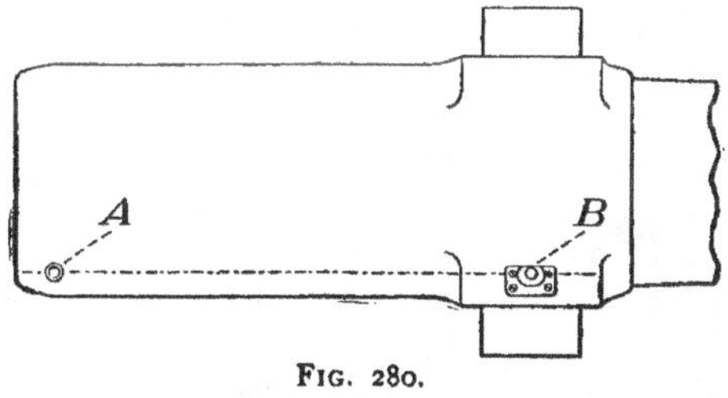

FIG. 280.

Being fired from a fixed platform, the inclination of the trunnions may be neglected, in comparison with the errors due to low velocity, and hence the standard does not rotate about its zero-point, so as to remain always in a vertical plane.

For the 5-inch siege-gun, its high velocity renders it more accurate, and although it is fired from a fixed platform, and the inclination of the trunnions is consequently small, any error due to this cause is eliminated by the rotation of the rear-sight standard.

Position.—The sights for the 7-inch Howitzer, (Fig. 280), are placed on the right side of the piece, the rear sight, *A*, in a hole drilled through the rear end of the jacket, and the front sight, *B*, on the right rim-base.

The Rear Sight.—A hole, (Fig. 281), is drilled in the jacket, in which fits a socket, *b*, held in place by a set-screw, *c*. The sight is a round steel rod, *a*, made flat on the rear side, which contains the graduations in degrees, and the usual diagonal scale.

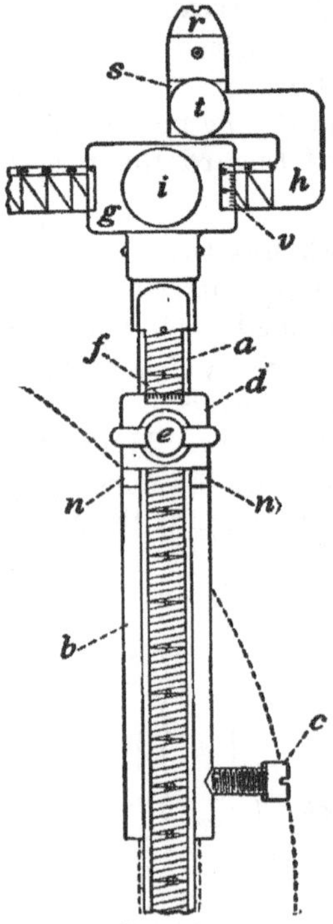

FIG. 281.

This rod fits accurately in the socket *b*, and carries a sliding collar, *d*, (Figs. 281 and 282), which may be fixed at any point along the scale by the clamp-screw, *e*. The rear upper edge *f* of this collar is bevelled, and carries a scale by means of which the diagonal divisions may be read to 1 minute.

The bottom of this collar has two projections, *n*, diametrically opposite. These fit into corresponding notches, *n'* (Fig. 282), in the top of the socket *b*.

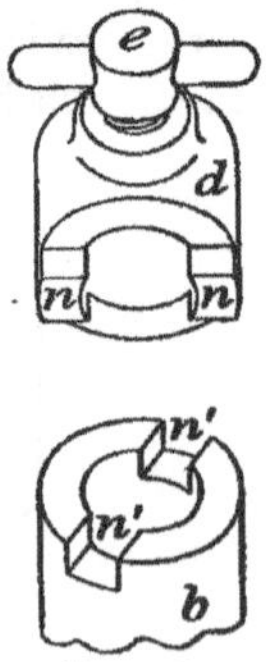

FIG. 282.

The axis of these notches is at right angles to the axis of the bore, and they are so arranged that the sight may be inserted into the socket for a reading of the deflection-bar to the right, or by lifting the sight and turning it 180° about its axis, the sight may be reinserted in the socket, and the deflection-bar can be read to the left. The correction for lateral deviation is given by a deflection-bar, *h* (Fig. 281), sliding in a socket, *g*, on the top of the rear sight-bar, and clamped in any position by the clamp-screw *i*. It is graduated similarly on both sides, and by turning the sight 180°, as previously explained, deflections may be read to the right or to the left.

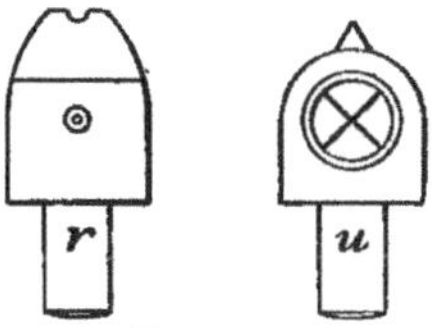

FIG. 283.

A vernier, *v*, on the right edge of the socket, *g*, enables the divisions on the deflection-bar to be read to $\frac{1}{10}$ of the scale, the vertical divisions on this bar being $\frac{1}{100}$ of the range. The deflection-bar is provided with two movable sight-pieces, *r* and *u* (Figs. 281 and 283), fitting into a socket, *s*, on the deflection-bar, and held in place by a clamp-screw, *t*. For direct pointing the sight piece, *r*, is placed in its socket in the deflection-bar, and *u* in the front-sight socket. For indirect pointing upon an object in rear, when the target is not visible, the sight-pieces *r* and *u* are interchanged.

FIG. 284.

Front Sight.—The front sight consists (Fig. 284) of a base, *a*, fastened to the rimbase by four screws.

The top is provided with a socket of the same shape and dimensions as that of the deflection-bar *h* (Fig. 281). The thumb-screw *d* clamps the sighting-piece firmly in position.

288. Sights for Field Artillery—3.6 Mortar—3.2 Field-gun.

SIGHTS FOR 3.6 MORTAR.—The sights for this piece

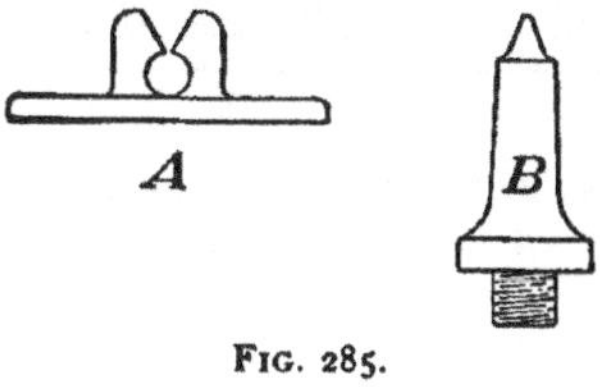

FIG. 285.

(Fig. 285) are a notch, *A*, in rear, and a point, *B*, in front, near the muzzle.

SIGHTS FOR 3.2 FIELD-GUN.—The rear sight consists of a base, *A*, Fig. 286, which fits in a corresponding socket in the gun; a pivot, *B*, Fig. 287, which fits into the bearing *b*, Fig. 286, of the base, and rotates around the bearing *b*, Fig. 287; and a standard, *C*, Fig. 288, carrying the graduations. The pivot *B*, Fig. 287, has two cuts, *c*, *c'*, of the shape shown, in which slide corresponding projections, *c*, *c'*, Fig. 288, of the standard.

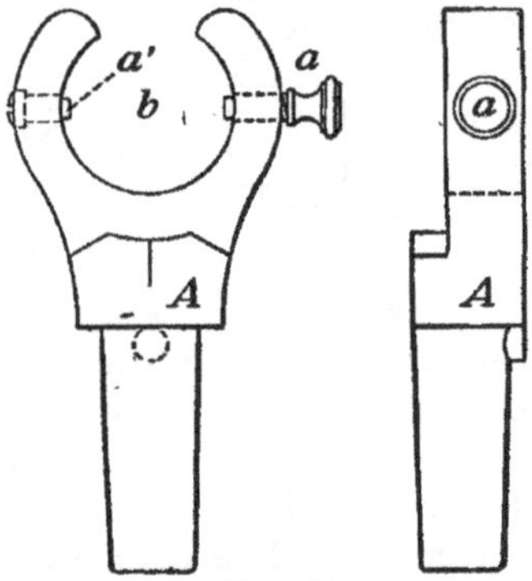

FIG. 286.

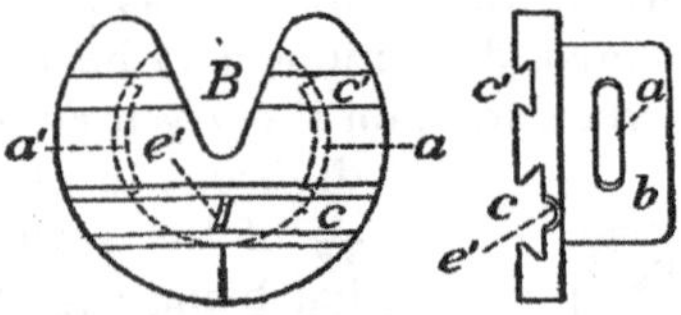

FIG. 287.

The cylindrical part, *b*, of the pivot, *B*, is held in place

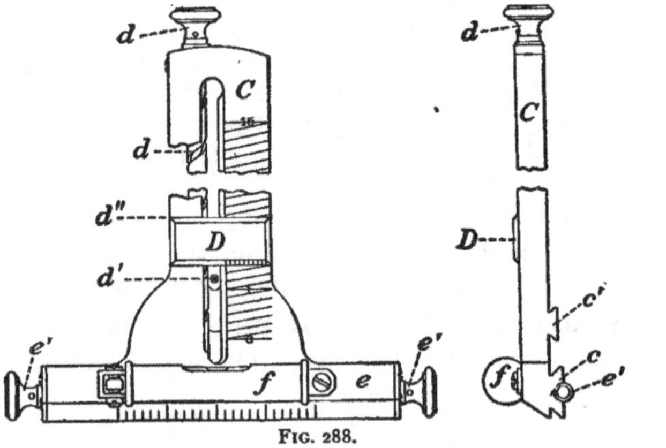

FIG. 288.

in the base *A*, Fig. 286, by two screws, *a*, *a'*, passing through the side of the base. These screws enter slots, *a*, *a'*, Fig. 287, in the pivot *B*, and allow a certain amount of rotation to the pivot.

The standard which carries the graduations consists (Fig. 288) of an upright bar, *C*; a sliding-piece, *D*, moved up and down along the bar, *C*, by a screw, *d*, and carrying the peep-sight, *d'*; a cross-bar, *e*, carrying the graduations for lateral deflection; a screw, *e'*, working in a half-nut, *e'*, in the pivot *B*, Fig. 287, by means of which the standard is moved to the right or left; a spirit-level, *f*, which indicates the vertical position of the standard; and two projections, *c*, *c'*, which fit in the corresponding cuts, *c*, *c'*, in the pivot *B*.

The axis of rotation is at the zero of the scale, and the usual diagonal scale, divided into 10-minute intervals, is read to one minute, by a scale on the sliding-piece *D*.

The assembled sight is shown in Fig. 289, and its action is evident.

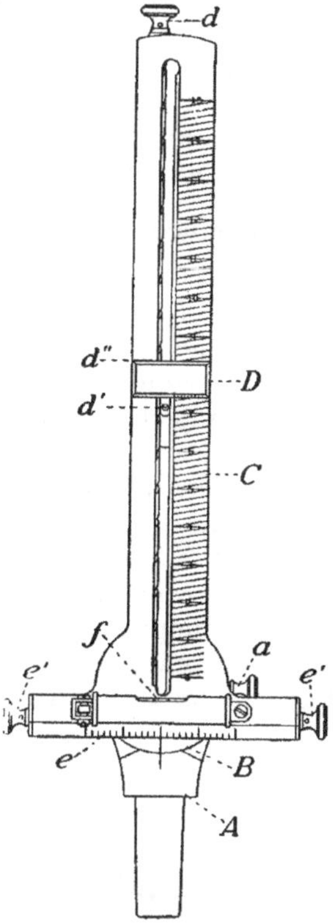

FIG. 289.

Front Sight.—This consists (Fig. 290) of a base, *a*,

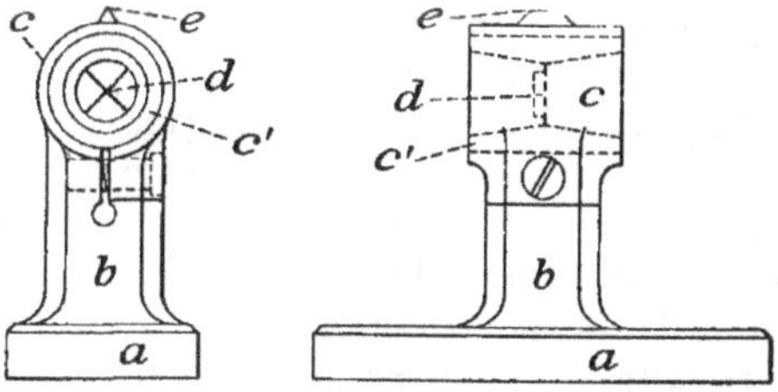

FIG. 290.

bolted to the right rim-base; a standard, *b*; and a cylinder, *c*.

These are all formed in one piece. The cylinder carries two thin cross-ribbons of steel, *d*, in an inner cone, *c'*, and a front sight-point, *e*. The point *e* and top of slide *d''*, Fig. 289, are used for coarse sighting, while the cross-ribbons and peep are for fine sighting.

The sights are on the right side of the piece. That for the 3.6-gun is similar.

289. Deviations—How Measured.

DEVIATIONS.—Owing to the causes previously explained, if a series of shots be fired at a given point of a target, they will in general not hit the point aimed at, nor will they be grouped symmetrically around this point. Each shot will have a trajectory differing from the other shots, and all these trajectories taken together will form a sheaf of trajectories, whose shape in general is that of a bent cone. The axis of this cone is called the mean trajectory, and all the others are grouped symmetrically about it. The point where this axis pierces the target is called the centre of impact, and the distance of this centre of impact from the point aimed at, is called the mean deviation. In Fig. 291 let *O* be the point aimed at; *AC* the axis of the

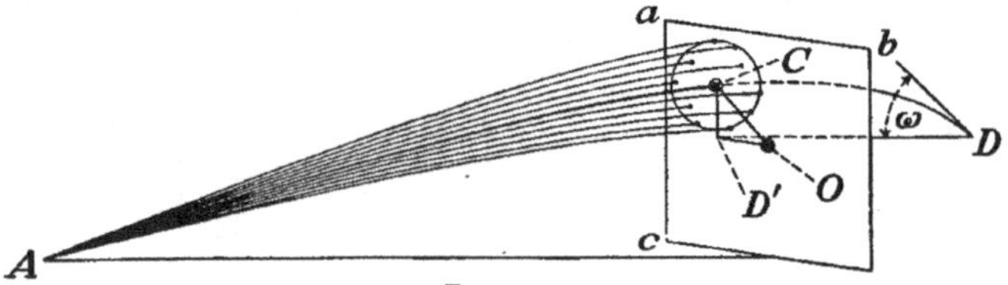

FIG. 291.

sheaf of trajectories; *C* the point where this axis or mean trajectory pierces the plane of the target. Then *C* is the centre of impact, and *OC* the mean deviation.

HOW MEASURED.—It is usual to measure deviations in three directions:

1. In the direction of the range;
2. Laterally, in the direction *ab*;
3. Vertically, in the direction *ac*.

For the mean range deviation the target is usually taken

horizontal, and the measurements made from the centre of the target, in the direction of the range. In case a horizontal target cannot be used, the mean range deviation may be obtained from the mean vertical deviation, by considering that part of the mean trajectory, CD, in rear of the target to be a straight line, making an angle ω with the horizontal equal to the angle of fall. This angle can be calculated by the formulas of Exterior Ballistics.

We have then

$$D'D = CD' \text{ cotang } \omega;$$

or, if the mean range deviation is measured on a horizontal target, we have for the mean vertical deviation

$$CD' = DD' \tan \omega.$$

The same method applies to any shot of the sheaf of trajectories. The mean lateral deviation is measured parallel to ab, and is OD' in the figure, and the mean vertical deviation is measured parallel to ac, and is CD' in the figure. The lateral and vertical deviations of any shot are measured in the same way, from the point aimed at, to the centre of the shot-hole.

290. To Find the Centre of Impact—Example.

In order to measure the mean deviations, it is necessary to determine the position of the centre of impact. For this purpose, in Fig. 292, assume an origin of co-ordinates at the

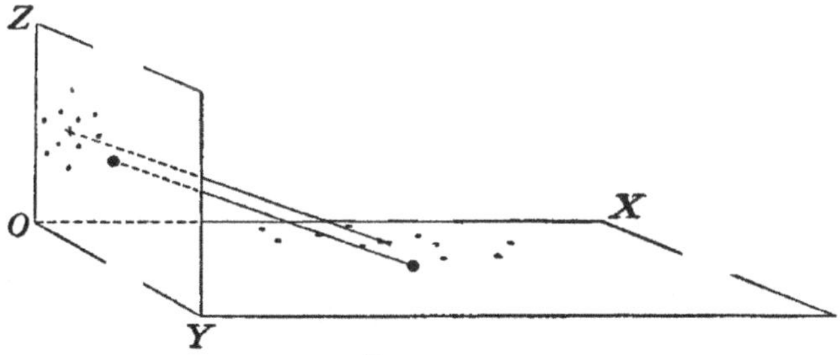

Fig. 292.

lower left-hand corner O of the target, and axes OX in the direction of the range, OY laterally, and OZ vertically.

The point O is selected as an origin, for convenience, to avoid the use of negative co-ordinates.

Let x', x'', etc., denote the distances of the shot-marks from O measured parallel to OX;

y', y'', etc., parallel to OY;

z', z'', etc., parallel to OZ;

X', Y', Z', the co-ordinates of the point aimed at;

X, Y, Z, the co-ordinates of the centre of impact;

n, the number of shots.

Then

$$X = \frac{x' + x'' + x''' + \text{etc.}}{n};$$

$$Y = \frac{y' + y'' + y''' + \text{etc.}}{n};$$

$$Z = \frac{z' + z'' + z''' + \text{etc.}}{n};$$

and the point whose co-ordinates are (XY) in the horizontal plane, and (YZ) in the vertical plane will be the centre of impact.

The mean deviations in range, laterally and vertically, will then be

In range $X - X'$;
Laterally $Y - Y'$;
Vertically $Z - Z'$.

Similarly for any shot the deviations in range, laterally and vertically, will be

In range $x' - X'$;
Laterally $y' - Y'$;
Vertically $z' - Z'$.

In these calculations the positive sign indicates distances beyond, to the right, and above the centre of impact; the negative sign distances short of, to the left, and below that centre.

EXAMPLE.—Eight shots are fired from the 3.20-inch steel field-gun at a vertical target, range 1760 yards.

Size of target 40 by 20 feet. The co-ordinates of the shots, measured from the lower left-hand corner of the target, are as given in the table.

Find the mean deviation in range, laterally and vertically, or the co-ordinates of the centre of impact.

TABLE.

No. of Shots.	Co-ordinates, feet.	
	Lateral.	Vertical.
1	9.00	9.50
2	21.68	5.00
3	14.25	5.66
4	17.00	5.00
5	11.00	8.66
6	19.00	9.82
7	17.00	10.32
8	14.83	9.00
	8)123.76	8)62.96
	$Y = 15.47$	$Z = 7.87$

The co-ordinates of the centre of the target are

$$Y' = 20, \quad Z' = 10.$$

Hence the mean lateral and mean vertical deviations are:

Mean lateral $Y - Y' = -4.53$ feet left;
Mean vertical $Z - Z' = -2.13$ feet below.

The mean deviation in range must be calculated.

In Fig. 293 OA is the vertical height of the point aimed at, 10 feet. Assume the angle of fall for this range and elevation $\omega = 3^\circ.00$, which is very nearly correct.

The centre of impact C is below the point O, 2.13 feet. Find first the point B, which is the position of the centre of the target O on the horizontal plane; then C', that of the centre of impact; and the difference $C'B$ is the mean error in range.

We have

$$AB = 10 \times \text{cotan } 3^\circ = 190.8 \text{ feet} = 63.6 \text{ yards} = X';$$
$$AC' = (10 - 2.13) \times \text{cotan } 3^\circ = 150.1 \text{ feet} = 50.03 \text{ yards} = X.$$

Mean deviation in range $X - X' = - 40.7$ feet

$= 13.56$ yards short.

Mean range $1760 - 13.56 = 1746.44$ yards.

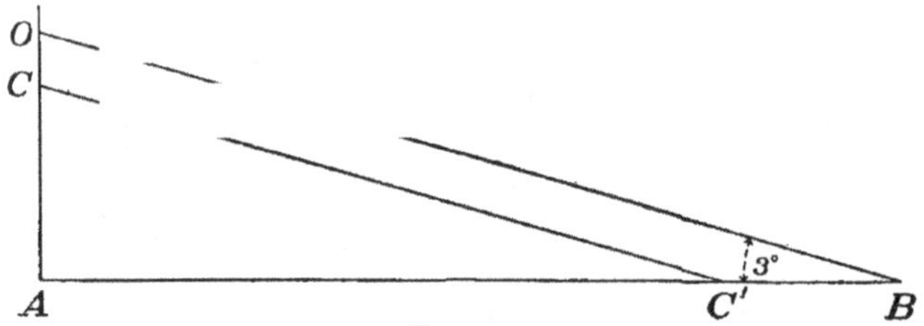

FIG. 293.

291. Errors.

The centre of impact, as its name indicates, is the centre of the group of shots fired, and all the shot are grouped symmetrically about it. Hence, if this point be taken as a new origin of co-ordinates, for every positive abscissa or ordinate, there must be a corresponding negative one, and the algebraic sums of the abscissas or ordinates measured from this point, are equal to zero.

The abscissa or ordinate of any shot measured from the centre of impact is called the error. Corresponding to the case of deviations, errors are measured in three directions: along the range, laterally, and vertically. The distinctions between deviations and errors are:

1. Deviations are measured from the point aimed at; errors, from the centre of impact.

2. Deviations are not grouped symmetrically about the point aimed at unless this point coincides with the centre of

impact, while errors *are* grouped symmetrically about the latter point.

It is theoretically possible, by carefully correcting for wind, drift, and the various other causes before enumerated, to make the centre of impact coincide with the point aimed at, and the mean trajectory pass through that point. But when this has been done the trajectories will still form a sheaf or cone about the mean trajectory as an axis. This is due to accidental errors which cannot be corrected, and whose consideration requires the application of the doctrine of probability, to be discussed later.

To find the error of any shot, and the mean errors for n shot, we have given the co-ordinates of the shot and those of the centre of impact, referred to the origin at the lower left-hand corner of the target.

Let X, Y, Z be the co-ordinates of the centre of impact;

x', y', z', those of a shot;

e_x, e_y, e_z, the errors in the directions X, Y, Z, respectively for each shot;

ϵ_x, ϵ_y, ϵ_z, the mean errors in range, laterally and vertically respectively, for n shot.

Then

$$e_x = x' - X,$$
$$e_y = y' - Y,$$
$$e_z = z' - Z,$$

and

$$\epsilon_x = \frac{e_x + e_x' + e_x'' + \text{etc.}}{n};$$

$$\epsilon_y = \frac{e_y + e_y' + e_y'' + \text{etc.}}{n};$$

$$\epsilon_z = \frac{e_z + e_z' + e_z'' + \text{etc.}}{n}.$$

The sums $e_x + e_x' +$ etc., $e_y + e_y' +$ etc., $e_z + e_z' +$ etc., if taken with their proper signs, are each equal to zero, according to the principle previously explained. Hence, in adding these, they must be taken without regard to sign, and the sum of their numerical values obtained. For instance, if two shots are fired, and one strikes 10 feet beyond and the other 10 feet short of the centre of impact, the values of e_x and e_x' will be $+10$ and -10, respectively, and this sum, considering the signs, is zero.

But the total error committed in this case is 20 feet for the two shots, and hence the mean error is $\frac{20}{2} = 10$.

The error, measured from the centre of impact directly to a shot, is called the absolute error. Denoting this error by r, we have

$$r = \sqrt{e_x^2 + e_y^2} \quad \text{or} \quad = \sqrt{e_y^2 + e_z^2},$$

and the mean absolute error is

$$\epsilon_r = \frac{r + r_1 + r_2 + \text{etc.}}{n}.$$

292. EXAMPLE.—Find the errors laterally and vertically and the mean errors for the 3.20-inch field-gun with the data in the last example.

TABLE.

No. Shots.	Co-ordinates—Feet.		Errors—Feet.		Squares of Errors.	
	Lateral.	Vertical.	Lateral.	Vertical.	Lateral.	Vertical.
1	9.00	9.50	− 6.47	+ 1.63	41.8609	2.6596
2	21.68	5.00	+ 6.21	− 2.87	38.5641	8.2369
3	14.25	5.66	− 1.22	− 2.21	1.4884	4.8841
4	17.00	5.00	+ 1.53	− 2.87	2.3409	8.2369
5	11.00	8.66	− 4.47	+ 0.79	19.9809	0.6241
6	19.00	9.82	+ 3.53	+ 1.95	12.4609	3.8025
7	17.00	10.32	+ 1.53	+ 2.45	2.3409	6.0025
8	14.83	9.00	− 0.64	+ 1.13	0.4096	1.2769
	123.76	62.96	25.60	15.90	$\Sigma e_y^2 = 119.4466$	$\Sigma e_z^2 = 35.7208$
	$Y = 15.47$	$Z = 7.87$	$\epsilon_y = 3.20$	$\epsilon_z = 1.9875$		

It will be observed that the positive and negative lateral errors balance each other, and also the positive and negative vertical errors, as they should do, for the reasons already explained. Also, that the sum of the positive or the negative errors in either column, divided by one half the number of shots, will give the correct values of ϵ_y and ϵ_z.

To calculate the range errors and the mean error in range, the same principle applies as for deviations, that is,

$$e_x = e_z \text{ cotan } \omega,$$
$$\epsilon_x = \epsilon_z \text{ cotan } \omega.$$

In our service a different mean absolute error is sometimes used as a measure of the accuracy of a gun. It is taken as the hypothenuse of a right-angled triangle of which the other two mean errors are the sides. Thus for the 3.20-inch gun in the example the mean absolute error is,

$$\epsilon_m = \sqrt{\epsilon_y^2 + \epsilon_z^2} = \sqrt{(3.20)^2 + (1.9875)^2} = 3.77 \text{ feet}.$$

This differs slightly from the true mean absolute error ϵ_r previously explained, which would be in this case 4.003 feet.

PROBABILITY OF FIRE.

293. Division of Sheaf of Trajectories—Law of Error—Probability Curve—Principles upon which Form of Curve Depends.

DIVISION OF TRAJECTORY.—Considering the errors in a given number of shots, it is found that they vary in magnitude according to a certain law. As we approach the centre of impact the shot-marks become more numerous, and as we recede from it they decrease in number. That part of the sheaf of trajectories which contains one half the whole number of shots is called the "*nucleus*"; outside of the nucleus, the surrounding part, containing 40 per cent, is called the "*envelope*"; and outside of this, the remaining 10 per cent is called the "*tailings*."

LAW OF ERROR.—Since one half the shot are grouped within a small distance of the centre of impact, it may be inferred that small errors are more apt to occur than large ones; and since only 10 per cent of the shot lie at any considerable distance from the centre of impact, it may be inferred that the chances of committing large errors are small, or that very large errors are not likely to occur.

PROBABILITY CURVE.—This law is general and applies not only to errors of shot, but to accidental errors of any kind. It may be expressed by a curve, called the probability curve, whose form is shown in Fig. 294.

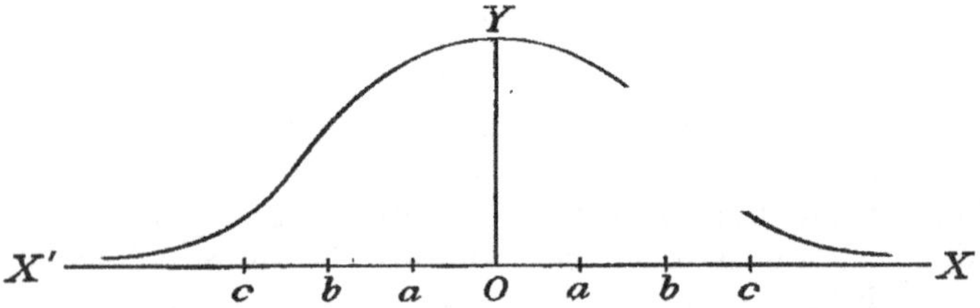

FIG. 294.

In this figure let O represent the centre of impact, and XX' the direction of the range. Let Oa, Ob, Oc represent errors in range, their magnitude being represented by the lengths of Oa, Ob, etc., measured from O. Then from the law of error it is evident that the smaller error Oa is more likely to occur than the larger one Ob, and this latter than the larger one Oc.

In a large number of shots, the error Oa will also occur more frequently than Ob, and so on.

If in 10 shots the error Oa occurs four times, Ob three, and Oc once, the fractions

$$\tfrac{4}{10},\ \tfrac{3}{10},\ \tfrac{1}{10}$$

measure the probability of the occurrence of these errors respectively.

Hence if we lay off errors along XX', measuring from O, and at the points a, b, c, etc., erect ordinates proportional to the probability of the corresponding errors, we will obtain the curve in the figure.

The same discussion applies to lateral and vertical errors, as they follow the same law.

PRINCIPLES UPON WHICH FORM OF CURVE DEPENDS.—The form of the curve depends upon the following general principles:

1. The number of shots striking at O will be greater than at any other point, or the probability of the error zero will be greater than that of any other error, and hence the maximum ordinate of the curve will be at O.

2. The number striking in the vicinity of O will be greater than for points farther to the right and left, and hence the ordinates of the curve will decrease slowly near O.

3. The number of hits will decrease rapidly as the distance to the right and left of O increases, and hence the ordinates of the curve will decrease rapidly in these directions.

4. For great distances from O, corresponding to large errors, the ordinates will be very small, since great errors are not likely to occur.

5. The only error that cannot occur is one infinitely great, and hence the ordinate of the curve becomes zero at an infinite distance, or the axis of X is an asymptote to the curve.

6. Since the shot are as likely to fall short of the point O as beyond it, the same error Oa is as likely to occur on one side of O as on the other, and hence the curve is symmetrical with respect to the axis OY.

294. Equation of the Probability Curve—Properties of the Curve—Limits.

EQUATION.—The equation of the probability curve, deduced by analytical methods, is (see Johnson, equation 1)

$$y = \frac{h}{\sqrt{\pi}} e^{-h^2x^2}, \quad . \quad . \quad . \quad . \quad . \quad . \quad . \quad (404)$$

in which y is the ordinate of the curve corresponding to the abscissa x, or the probability of the error x;

h is the modulus of precision, whose meaning and value will be explained;

$\pi = 3.1416$;

$e =$ the base of the Napierian system.

PROPERTIES OF THE CURVE.—Differentiating equation (404) twice, we have

$$\frac{dy}{dx} = -\frac{2h^3}{\sqrt{\pi}} x e^{-h^2x^2}, \quad . \quad . \quad . \quad . \quad . \quad . \quad . \quad . \quad (405)$$

$$\frac{d^2y}{dx^2} = -\frac{2h^3}{\sqrt{\pi}} e^{-h^2x^2}(1 - 2h^2x^2), \quad . \quad . \quad . \quad (406)$$

From equation (405), when $x = 0$, we have

$$\frac{dy}{dx} = 0,$$

and hence the tangent at y is parallel to the axis of x.

Placing $\frac{d^2y}{dx^2} = 0$, we have

$$1 - 2h^2x^2 = 0,$$

or

$$x = \frac{1}{h\sqrt{2}}; \quad . \quad . \quad . \quad . \quad . \quad . \quad . \quad . \quad (407)$$

hence $\frac{d^2y}{dx^2}$ passes through zero and changes its sign for the value of x given in equation (407). There is therefore a point of inflection for the curve corresponding to this abscissa.

LIMITS.—In discussing the probability of making an error, it is usual to consider this error as lying within certain limits. Hence it is necessary to consider the area bounded by the probability curve, the axis of errors, and any two ordinates whose abscissas represent the limits between which the given error lies. The area so determined represents the probability of the occurrence of the error within the given limits.

The general expression for the area of a curve is

$$P = \int y dx.$$

Replacing y by its value from (404), we have

$$P = \frac{h}{\sqrt{\pi}} \int e^{-h^2x^2} dx. \quad . \quad . \quad . \quad . \quad . \quad . \quad (408)$$

The axis of XX' extends to infinity in both directions as explained; hence the total area under the curve will be obtained by integrating equation (408) between the limits $+\infty$ and $-\infty$. That is,

$$P = \frac{h}{\sqrt{\pi}} \int_{-\infty}^{+\infty} e^{-h^2x^2} dx.$$

Place

$$hx = \alpha, \quad \therefore\ dx = \frac{d\alpha}{h};$$

hence

$$P = \frac{1}{\sqrt{\pi}} \int_{-\infty}^{+\infty} e^{-\alpha^2} d\alpha. \quad . \quad . \quad . \quad . \quad . \quad (409)$$

The value of the integral between limits is, from calculus,

$$\int_{-\infty}^{+\infty} e^{-\alpha^2} d\alpha = \sqrt{\pi}; \quad . \quad . \quad . \quad . \quad . \quad (410)$$

hence in (409)

$$P = 1,$$

or the total area under the curve is unity. This means that it is certain that the error will be contained between $+\infty$ and $-\infty$.

Similarly, for the probability that an error shall be contained between any limits $+x$ and $-x$, we have

$$P = \frac{h}{\sqrt{\pi}} \int_{-x}^{+x} e^{-h^2x^2} dx. \quad . \quad . \quad . \quad . \quad . \quad (411)$$

Since the curve is symmetrical with respect to the axis OY, we have

$$P = \frac{h}{\sqrt{\pi}} \int_{-x}^{+x} e^{-h^2x^2} dx = \frac{2h}{\sqrt{\pi}} \int_{0}^{x} e^{-h^2x^2} dx \quad . \quad . \quad (412)$$

for the probability that the error shall be less than x regardless of its sign.

295. Modulus of Precision h—Use of Table I.

Assume, equation (404),

$$y = \frac{h}{\sqrt{\pi}} e^{-h^2x^2}.$$

The smallest possible error is zero. Making $x = 0$ in the above equation, we have

$$y = \frac{h}{\sqrt{\pi}} \quad . \quad . \quad . \quad . \quad . \quad . \quad . \quad . \quad (413)$$

for the probability of the error zero. It is evident that this is the greatest value of y, and gives the maximum ordinate OY.

Suppose we have another series of shots for which the value of h differs from that for the first series, as $h' = 2h$. Then the probability of the error zero in the second case will be

$$y' = \frac{2h}{\sqrt{\pi}}, \quad . \quad . \quad . \quad . \quad . \quad . \quad . \quad . \quad (414)$$

and the ordinate will be twice that in equation (413).

That is, the probability of the error zero will be twice as great in the second series as in the first. As the accuracy

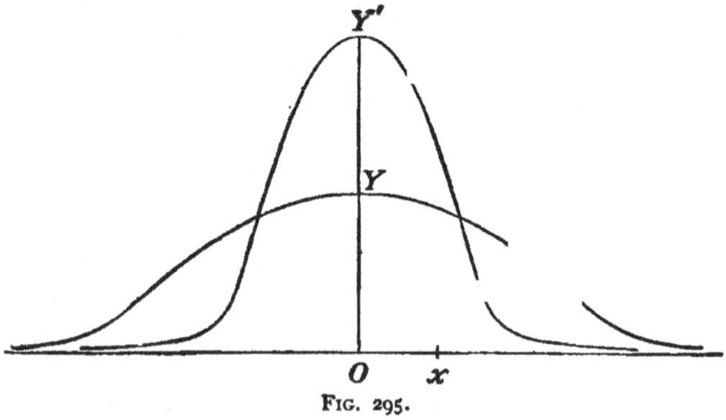

FIG. 295.

increases with the probability of making no error, we conclude that the second series of shots is more accurate than the first. The quantity h is then a measure of the precision of the shots, and hence is called the modulus of precision.

If we construct the probability curves for the two series of shots, since the areas under them are always unity, we will have those represented in Fig. 295, OY representing the maximum ordinate for the first series and OY' for the second. It is evident, from the above discussion and figure, that for the same error, x, the curve of probabilities will vary with the modulus h. Hence it is usual to change the form of the equation for probability, so as to introduce h as a factor of the error x, and hence into the limit. Equation (412) is therefore generally written

$$P = \frac{2}{\sqrt{\pi}} \int_0^{hx} e^{-h^2x^2} d(hx). \quad . \quad . \quad . \quad . \quad (415)$$

The values of P for different values of hx have been calculated and tabulated, and are given in Table I. The value of h as deduced analytically is

$$h = \sqrt{\frac{n - 1}{2\Sigma e^2}}, \quad . \quad . \quad . \quad . \quad . \quad . \quad (416)$$

in which

n is the number of shots;

Σe^2, the sum of the squares of the errors in any given direction.

USE OF TABLE I.—Let it be required to find for the 3.20 gun the probability of committing a lateral error less than 2 feet, at a range of 1 mile.

The value of Σe_y^2 for this gun is 119.45 (see example, Subject 292); hence

$$h = \sqrt{\frac{7}{2 \times 119.45}} = .1711;$$

$$hx = .1711 \times 2 = .3422;$$

$$P \text{ for } hx = .3422 = .37155;$$

$$P = \text{about } \tfrac{4}{10};$$

or about four shots in ten will make an error less than 2 feet laterally, at one mile range.

TABLE I.

PROBABILITY OF ERRORS.

hx	*P*	*hx*	*P*	*hx*	*P*	*hx*	*P*	*hx*	*P*
0.00	0.00000	0.40	0.42839	0.80	0.74210	1.20	0.91031	1.60	0.97635
.02	.02256	.42	.44747	.82	.75381	1.22	.91553	1.62	.97804
.04	.04511	.44	.46622	.84	.76514	1.24	.92050	1.64	.97962
.06	.06762	.46	.48465	.86	.77610	1.26	.92523	1.66	.98110
.08	.09008	.48	.50275	.88	.78669	1.28	.92973	1.68	.98249
.10	.11246	.50	.52050	.90	.79691	1.30	.93401	1.70	.98379
.12	.13476	.52	.53790	.92	.80677	1.32	.93806	1.72	.98500
.14	.15695	.54	.55494	.94	.81627	1.34	.94191	1.74	.98613
.16	.17901	.56	.57161	.96	.82542	1.36	.94556	1.76	.98719
.18	.20093	.58	.58792	.98	.83423	1.38	.94902	1.78	.98817
.20	.22270	.60	.60386	1.00	.84270	1.40	.95228	1.80	98909
.22	.24429	.62	.61941	1.02	.85084	1.42	.95537	1.82	.98994
.24	.26570	.64	.63458	1.04	.85865	1.44	.95830	1.84	.99073
.26	.28690	.66	.64938	1.06	.86614	1.46	.96105	1.86	.99147
.28	.30788	.68	.66378	1.08	.87333	1.48	.96365	1.88	.99216
.30	.32863	.70	.67780	1.10	.88020	1.50	.96610	1.90	.99279
.32	.34912	.72	.69143	1.12	.88679	1.52	.96841	1.92	.99338
.34	.36936	.74	.70468	1.14	.89308	1.54	.97058	1.94	.99392
.36	.38933	.76	.71754	1.16	.89910	1.56	.97263	1.96	.99443
.38	.40901	.78	.73001	1.18	.90484	1.58	.97455	1.98	.99489
								2.0	.99532
								3.0	.99998
								∞	1.00000

296. Probable Error—True Mean Error—Relation between Probable and True Mean Errors.

Among the errors which may be committed, from zero to infinity, there are two whose values are of constant use and importance.

These are the *probable error* and the *true mean error.*

PROBABLE ERROR.—In Fig. 296, the total area under the probability curve being unity, if the abscissas Op on each side of O be so taken that the area $pp'p'p$ included between the curve, the ordinates pp', and the axis XX' is equal to one half, the error Op is called the probable error. That is, it is the error whose probability is one half, or the error which is as likely to be exceeded as not. For example, if ten shots be fired, and O be the centre of impact, the probability is that five of these shots will strike within the distance Op from the centre of impact and the other five at a greater distance. It is to be noted that while the probable error is Op, it may

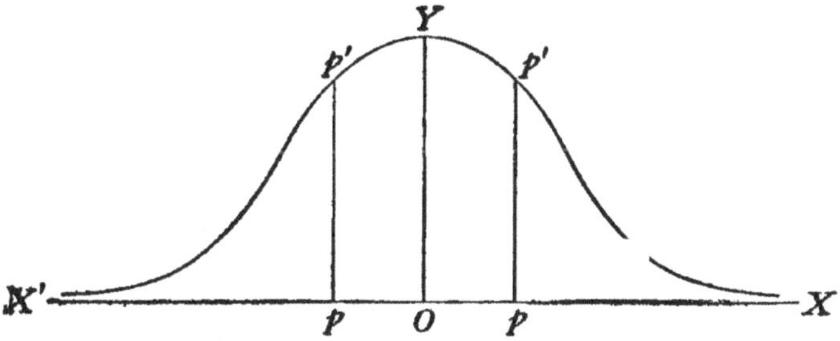

FIG. 296.

occur on either side of O, and hence it must be measured in both directions from O.

From Table I, the value of hx for $P = \frac{1}{2}$ is

$$hx = 0.4769.$$

Hence, calling this error x_p, we have

$$x_p = \frac{0.4769}{h}. \quad \ldots \ldots \ldots \ldots \quad (417)$$

Substituting for h its value from (416), we have

$$x_p = 0.6745\sqrt{\frac{\Sigma e^2}{n-1}}. \quad \ldots \ldots \quad (418)$$

TRUE MEAN ERROR.—The *mean error* has already been calculated for the 3.20-inch gun for a limited number of shots. If the number of shots be increased, a different value for the mean error would be obtained, and the true value of this mean error can only be found for an infinite number of shots: hence the name. Since it is impossible in practice to fire an infinite number of shots, the value of the true mean error must be found by analytical methods from the equation of the curve. This method gives for its value

$$x_m = \frac{1}{h\sqrt{\pi}}. \quad . \quad . \quad . \quad . \quad . \quad . \quad . \quad . \quad . \quad . \quad (419)$$

Substituting the values of π, and of h from (416), we have

$$x_m = 0.79788\sqrt{\frac{\Sigma e^2}{n-1}}. \quad . \quad . \quad . \quad . \quad (420)$$

Dividing equation (418) by (420), we have

$$\frac{x_p}{x_m} = 0.8453; \quad \therefore \; x_p = 0.8453 x_m. \quad . \quad . \quad . \quad (421)$$

297. Probable Zone—Examples—Comparison of Mean and True Mean Errors.

PROBABLE ZONE.—The probable zone is one which will probably contain 50 per cent, or one half, the total number of shots. Hence the probability for this zone is $P = \frac{1}{2}$. Now, in considering the probable error, it was shown that it measured a distance on each side of the point of impact within which one half the whole number of shots would strike; and hence if we lay off a distance on each side of the centre of impact equal to the probable error, and draw through the points thus determined two lines at right angles to the plane of fire and extending indefinitely in both dircc-tions, these lines will determine a zone which will contain 50 per cent of the shots.

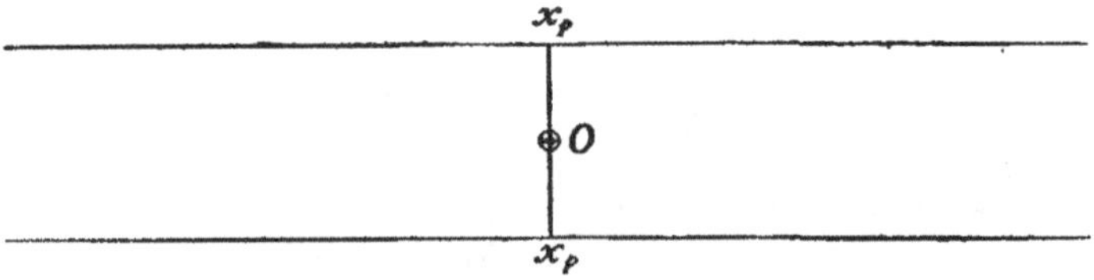

FIG. 297.

In Fig. 297 let O be the centre of impact; x_pOx_p, the direction of the range, or line of fire; Ox_p, measured in both directions from O, the probable error. Then the zone defined by the parallel lines extending to infinity in both directions is called the probable zone, and will contain one half the whole number of shots fired. The same reasoning applies to the lateral and vertical probable zones.

The width of this zone is twice the probable error, or

$$2x_p = 2 \times 0.6745\sqrt{\frac{\Sigma e^2}{n-1}} = 1.349\sqrt{\frac{\Sigma e^2}{n-1}}. \qquad (422)$$

EXAMPLES.—Find the probable error, true mean error, and probable zone vertically for the 3.20-inch gun at one mile range.

$$x_p = 0.6745\sqrt{\frac{\Sigma e_z^2}{n-1}} = 0.6745\sqrt{\frac{35.72}{7}} = 1.523 \text{ feet};$$

$$x_m = 0.79788\sqrt{\frac{\Sigma e_z^2}{n-1}} = 0.79788\sqrt{\frac{35.72}{7}} = 1.802 \text{ feet};$$

$$2x_p = 2 \times 1.523 = 3.046 \text{ feet}.$$

The same method will give the corresponding errors and zones laterally and in range, Σe^2 differing for the different directions.

COMPARISON OF MEAN AND TRUE MEAN ERRORS.—The true mean vertical error in this case is $x_m = 1.802$ feet, while the mean error as obtained from eight shots is (see table) 1.9875 feet. Hence the mean and true mean errors differ very slightly, and with a large number of shots the difference would be still less. This is true generally, and hence the calculated mean error may be used instead of the true mean error without appreciable error in the result. This leads to a simple method of calculating probable zones, as follows:

From (421) we have

$$x_p = 0.8453x_m; \quad \therefore 2x_p = 1.69x_m. \quad . \quad . \quad . \quad (423)$$

Substituting for x_m the calculated mean error ϵ_x, ϵ_y, or

ϵ_z, we have for the probable zone in any one of these directions:

$$\left.\begin{array}{ll}\text{Range} & 2x_p = 1.69 \times \epsilon_x; \\ \text{Lateral} & 2y_p = 1.69 \times \epsilon_y; \\ \text{Vertical} & 2z_p = 1.69 \times \epsilon_z.\end{array}\right\} \quad . \quad . \quad . \quad . \quad (424)$$

298 25 per-cent Rectangle—Probable Rectangle—Rectangles of any Percentage.

Let O, Fig. 298, be the centre of impact, XX' the 50 per-cent zone for the range, YY' the 50 per-cent zone laterally.

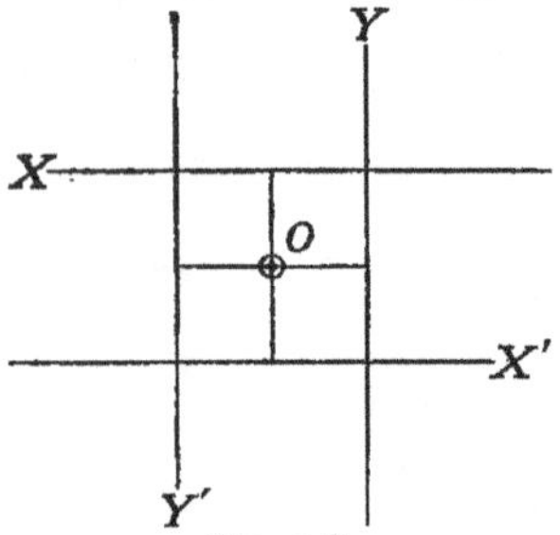

FIG. 298.

The intersection of these zones will form a rectangle about the centre of impact, and this rectangle will contain one fourth or 25 per cent of the shots, for

$$.50 \times .50 = .25 .$$

This is called the 25 per-cent rectangle. It is the rectangle formed about the centre of impact by the intersection of the two 50 per-cent zones; and since each contains 50 per cent of the shots independently of the other, by the doctrine of probability, when they intersect, their common part will contain a percentage equal to the product of the two.

PROBABLE RECTANGLE.—The probable rectangle is one which is formed by the intersection of two zones of equal probability, and which will probably contain 50 per cent of the shots. That is, its probability is $P = \frac{1}{2}$.

Now the probability of any rectangle, as illustrated in the case of the 25 per-cent rectangle, is equal to the product of the probabilities of the two zones whose intersection forms the rectangle; and denoting the probabilities of these zones by P' and P'' respectively, we have

$$P = P' \times P''.$$

It is evident, however, that there are an infinite number of zones whose intersection will give a rectangle having the probability $P = \frac{1}{2}$, since any two, the product of whose probabilities is one half, will fulfil this condition. To fix the rectangle, therefore, we impose the condition that the probabilities of the two intersecting zones shall be equal. Substituting in the above equation this condition, we have

$$\tfrac{1}{2} = \sqrt{\tfrac{1}{2}} \times \sqrt{\tfrac{1}{2}},$$

or the probability of committing an error less than one half of either side of the rectangle is $P' = P'' = \sqrt{\frac{1}{2}}$.

For

$$P' = \sqrt{\tfrac{1}{2}} = 0.7071$$

we have, from Table I,

$$hx = 0.7438;$$

hence

$$x = \frac{0.7438}{h}. \quad \text{. . . . .} \quad (425)$$

Substituting for h its value from (416), we have

$$x = 1.052\sqrt{\frac{\Sigma e^2}{n-1}}.$$

This is the value of one half the side of the rectangle. Hence

$$2x = 2.104\sqrt{\frac{\Sigma e^2}{n-1}}.$$

Calling the sides of this rectangle in the horizontal plane $\varDelta_x$ and $\varDelta_y$, and in the vertical plane $\varDelta_y$ and $\varDelta_z$, we have

$$\left.\begin{aligned} \varDelta_x &= 2.104\sqrt{\frac{\Sigma e_x^2}{n-1}}; \\ \varDelta_y &= 2.104\sqrt{\frac{\Sigma e_y^2}{n-1}}; \\ \varDelta_z &= 2.104\sqrt{\frac{\Sigma e_z^2}{n-1}}. \end{aligned}\right\} \quad \text{. . . .} \quad (426)$$

RECTANGLES OF ANY PERCENTAGE.—By similar reasoning we can find the probable rectangle which will contain any given percentage of shots. For instance, required the rectangle which will contain twelve out of twenty shots fired from the 3.2 gun at one mile range.

We have

$$\frac{12}{20} = 60 \text{ per cent.}$$

The probability of this rectangle must then be

$$P = .6,$$

and hence the probability of its sides $P' = \sqrt{.6}$, since

$$\sqrt{.6} \times \sqrt{.6} = .6 = P.$$

From Table I the value of hx corresponding to

$$P' = \sqrt{.6} = .7746$$

is

$$hx = .8572;$$

hence

$$x = \frac{.8572}{h}.$$

Substituting for h its value from (416), we have

$$x = 1.212\sqrt{\frac{\Sigma e^2}{n - 1}}$$

and

$$2x = 2.424\sqrt{\frac{\Sigma e_x^2}{n - 1}},$$

$$2y = 2.424\sqrt{\frac{\Sigma e_y^2}{n - 1}},$$

$$2z = 2.424\sqrt{\frac{\Sigma e_z^2}{n - 1}},$$

and so on for a rectangle of any percentage.

299. Examples—Measure of Accuracy of Guns—Calculation of Probable Rectangle from Mean Error.

Find the 25 per-cent rectangle, the probable rectangle, and the 60 per-cent rectangle for the 3.20 gun, in the vertical plane, at one mile range.

1. The 25 per-cent rectangle:
The probable zone vertically is (see subject 297)

$$2z_p = 3.046 \text{ feet.}$$

The probable zone laterally is

$$2y_p = 1.349\sqrt{\frac{\Sigma e_y^2}{n-1}} = 1.349\sqrt{\frac{119.45}{7}} = 5.57 \text{ feet.}$$

Hence the 25 per-cent rectangle is

$$3.046 \times 5.57 = 16.97 \text{ sq. feet.}$$

2. The 50 per-cent rectangle:

$$\Delta_y = 2.104\sqrt{\frac{\Sigma e_y^2}{n-1}} = 2.104\sqrt{\frac{119.45}{7}} = 8.69 \text{ feet;}$$

$$\Delta_z = 2.104\sqrt{\frac{\Sigma e_z^2}{n-1}} = 2.104\sqrt{\frac{35.72}{7}} = 4.75 \text{ feet.}$$

Hence the 50 per-cent rectangle is

$$8.69 \times 4.75 = 41.31 \text{ sq. feet.}$$

3. The 60 per-cent rectangle:

$$2y = 2.424\sqrt{\frac{\Sigma e_y^2}{n-1}} = 10.01 \text{ feet;}$$

$$2z = 2.424\sqrt{\frac{\Sigma e_z^2}{n-1}} = 5.48 \text{ feet.}$$

Hence the 60 per-cent rectangle is

$$10.01 \times 5.48 = 54.82 \text{ sq. feet.}$$

COMPARISON OF ACCURACY OF GUNS.—The probable or 50 per-cent rectangle, is generally used to compare the accuracy of different guns, and may be taken either in the horizontal or in the vertical plane. For small arms and high-power guns the vertical rectangle is the more accurate means of comparison. It is evident that for high-power guns, with flat trajectories, the horizontal rectangles will be larger than for guns with high-angle or curved fire. Hence if we compare guns and mortars by their horizontal rectangles, the mortar will appear the more accurate. On the other hand, for high-angle or curved fire, horizontal targets should be used as a means of comparison. The most accurate method for all guns is to take the plane of the target at right angles to the trajectory at the point of impact, but this is generally impracticable.

CALCULATION OF PROBABLE RECTANGLE FROM MEAN ERROR.—The mean error of a given number of shots may be readily obtained as previously shown.

For the side of the 50 per-cent rectangle we have, equation (425),

$$\Delta_x = 2x = \frac{2 \times 0.7438}{h}. \qquad (427)$$

From (417),

$$h = \frac{0.4769}{x_p}; \qquad (428)$$

and from (421),

$$x_p = 0.8453 x_m. \qquad (429)$$

Substituting the value of x_p from (429) in (428), we have

$$h = \frac{0.4769}{0.8453 x_m};$$

and this value of h in (427) gives

$$\Delta_x = 2x = 2.636 \times x_m. \qquad (430)$$

300. Use of Probable Error in Calculating Probabilities—Use of Table II—Example.

The probable error is generally used as a standard of comparison for other errors, since it represents an error which is as likely to be exceeded as not.

For the probable error we have, equation (417),

$$x_p = \frac{0.4769}{h}.$$

Using this as a unit of comparison, the ratio of any other error x to this is

$$\frac{x}{x_p} = \frac{hx}{0.4769}. \quad . \quad . \quad . \quad . \quad . \quad . \quad (431)$$

In Table I we have given, values of P corresponding to hx, or, conversely, values of hx corresponding to P. Dividing the values of hx in Table I by 0.4769, we have the corresponding values of $\frac{x}{x_p}$.

Finding the values of P, from Table I, corresponding to these values of $\frac{x}{x_p}$, we can form a new table, giving the values of P corresponding to $\frac{x}{x_p}$, or the values of $\frac{x}{x_p}$ corresponding to P.

This table is called Chauvenet's Table, and is given below.

TABLE II.

PROBABILITY OF ERRORS.

P	0	1	2	3	4	5	6	7	8	9
0.0	0	.02	.04	.06	.07	.09	.11	.13	.15	.17
.1	.18	.20	.22	.24	.26	.28	.30	.32	.34	.36
.2	.38	.40	.41	.43	.45	.47	.49	.51	.53	.55
.3	.57	.59	.61	.63	.65	.67	.70	.72	.74	.76
.4	.78	.80	.82	.84	.86	.89	.91	.93	.95	.98
.5	1.00	1.02	1.04	1.07	1.09	1.12	1.14	1.17	1.19	1.22
.6	1.25	1.27	1.30	1.33	1.36	1.39	1.42	1.45	1.48	1.51
.7	1.54	1.57	1.60	1.64	1.67	1.71	1.74	1.78	1.82	1.86
.8	1.90	1.94	1.98	2.03	2.08	2.13	2.18	2.24	2.30	2.37
.9	2.44	2.52	2.60	2.69	2.78	2.91	3.04	3.22	3.45	3.82

USE OF TABLE II—EXAMPLES.

1. Required the probability of committing a lateral error with the 3.2 gun, at 1 mile range, of less than 4.354 feet.

The probable error laterally, for this range, is 2.785 feet, hence

$$\frac{y}{y_p} = \frac{4.345}{2.785} = 1.56.$$

From Table II, P for $\frac{y}{y_p} = 1.56$ is

$$P = 0.7071 = \sqrt{\tfrac{1}{2}}.$$

2. The probable lateral error of the 3.2 gun at 1 mile range is 2.785 feet.

The probability of committing an error less than x, is $P = \sqrt{\tfrac{1}{2}} = 0.7071$. Find the value of x.

From Table II, for $P = 0.7071$ we have

$$\frac{y}{y_p} = 1.56, \quad \therefore\ y = 1.56 \times 2.785 = 4.345 \text{ feet.}$$

301. Probability of Hitting any Plane Figure.

By the previous methods it has been shown how to determine the sides of a rectangle which will contain any given percentage of shots. By the use of Table II we can readily determine the probability of hitting any plane figure of a given size and shape.

As the simplest case, consider first a rectangular object.

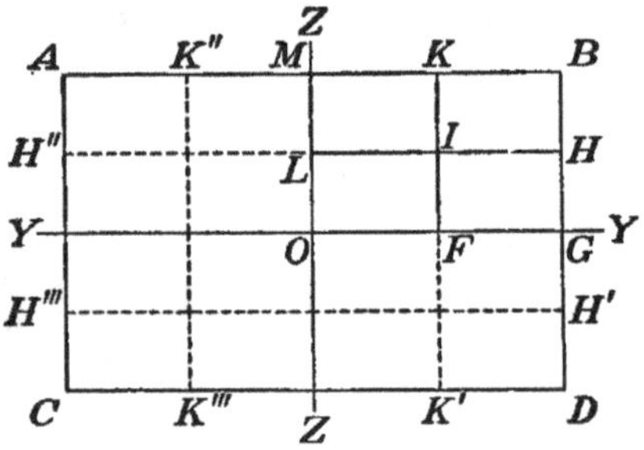

FIG. 299.

Let O, Fig. 299, be the centre of impact, OY and OZ the rectangular axes, and suppose vertical errors to be measured along OZ, and horizontal errors along OY. For the given gun and range, the probable errors horizontally and vertically will be known by firing a certain number of shots and calculating the probable errors by equation (418).

1. What is the probability of striking the rectangle $ABDC$?

From Table II we find the probability of committing the error OG by taking out the value of P corresponding to

$$\frac{OG}{y_p} = \frac{y}{y_p}.$$

From the same table we find the probability of committing the error OM by taking from this table the value of P corresponding to

$$\frac{OM}{z_p} = \frac{z}{z_p},$$

and the probability of hitting the rectangle $ABCD$ is the probability of committing these two errors simultaneously, or the product of the above separate probabilities.

2. What is the probability of striking the rectangle $OGBM$?

From the fact that the shot are grouped symmetrically about O, owing to the law of probability, it follows that the number of hits in $OGBM$ will be $\frac{1}{4}$ of those in the rectangle $ABCD$. Hence the probability of hitting the rectangle $OGBM$ is $\frac{1}{4}$ that of hitting the rectangle $ABCD$.

3. What is the probability of striking within $OMKF$?

This is found exactly as for the rectangle $OMBG$. Find from Table II the probabilities corresponding to $\frac{OF}{y_p}$ and $\frac{OM}{z_p}$, and multiply these probabilities together. The result will be the probability of striking within the rectangle $K''KK'K'''$, and $\frac{1}{4}$ of this will be the probability required.

4. What is the probability of striking $FKBG$?

It is the difference between the probabilities for $OGBM$ and $OFKM$, which have already been found.

5. What is the probability of striking $OGHL$?

Find from Table II the probabilities for $\frac{OG}{y_p}$ and $\frac{OL}{z_p}$; multiply these probabilities together and take $\frac{1}{4}$ of the product for the probability required.

6. What is the probability of striking $FGHI$?

It is the probability of striking *OGHL* minus the probability of striking *OFIL*.

7. What is the probability of striking *IKBH* ?

It is the probability of striking *OMBG* minus the sum of the probabilities of striking *OGHL* and *LIKM*.

In the same way any figure may be divided into rectangles, approximately, whose centres coincide with the centre of impact.

The probability of striking the rectangles or parts of rectangles about the centre of impact may be readily calculated by Table II, and the probability of striking those parts whose centres do not coincide with the centre of impact may be determined by subtraction.

302. Right-line Method.

The area under the probability curve being unity, and the curve being symmetrical with respect to the axis *OY*,

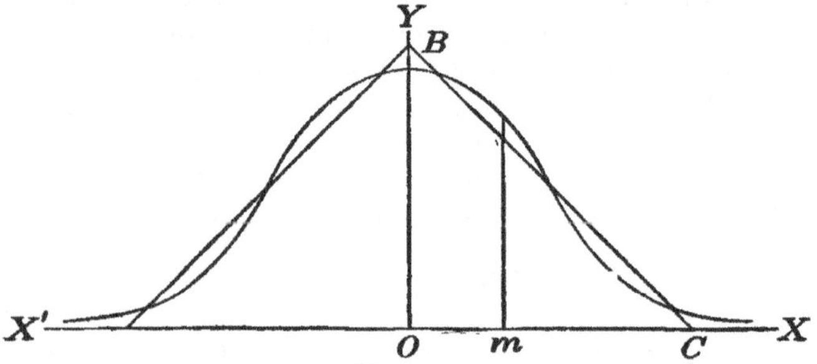

FIG. 300.

the area under each branch is ½. If a right line *BC*, Fig. 300, be drawn so that the area of the triangle *OBC* = ½, and the abscissa of its centre of gravity be at a distance *Om* from *O*, equal to the true mean error x_m, then the right line *BC* may be substituted without appreciable error for the probability curve.

In this case the greatest possible error is *OC* and the greatest possible ordinate is *OB*, and to show that the right line may be substituted for the curve it is necessary to prove:

1. That the probability of the error OC does not differ sensibly from that of the error ∞, which is the greatest possible error in the case of the probability curve.

2. That the ordinate OB does not differ sensibly from the maximum ordinate OY of the curve.

1. *Probability of the error OC.*

Since in a triangle the centre of gravity is situated at a distance from its base equal to $\frac{1}{3}$ its height, we have

$$OC = 3Om;$$

but, from (419),

$$Om = x_m = \frac{1}{h\sqrt{\pi}},$$

hence

$$OC = \frac{3}{h\sqrt{\pi}},$$

and

$$h \times OC = hx = \frac{3}{\sqrt{\pi}} = 1.6925.$$

From Table I the value of P corresponding to $hx = 1.6925$ is

$$P = .983.$$

The value of P for $hx = \infty$ is

$$P = 1.00,$$

hence the probabilities of the extreme errors in the two cases are as

$$.983 : 1.00.$$

That is, out of 100 shots 98 will make an error less than OC.

2. *Value of the ordinate OB, as compared with OY.* The maximum ordinate OY of the probability curve is found by making $x = 0$ in equation (404). The value thus obtained is

$$OY = y = \frac{h}{\sqrt{\pi}} = \frac{h\sqrt{\pi}}{\pi};$$

or since

$$x_m = \frac{1}{h\sqrt{\pi}}, \therefore h\sqrt{\pi} = \frac{1}{x_m},$$

we have

$$OY = \frac{1}{\pi \times x_m}. \quad . \quad . \quad . \quad . \quad . \quad . \quad (432)$$

From the triangle OBC, since its area is $\frac{1}{2}$ and its base $3x_m$, we have

$$\frac{1}{2} = \frac{(OB \times 3x_m)}{2};$$

hence

$$OB = \frac{1}{3 \times x_m}. \quad . \quad . \quad . \quad . \quad . \quad . \quad (433)$$

Comparing (432) and (433), we see that the numerators of the values of OY and OB are the same, and the denominators differ but slightly, and hence OB may be taken for OY, or the right line may be substituted for the curve without appreciable error.

303. Value for Probability by the Right-line Method.

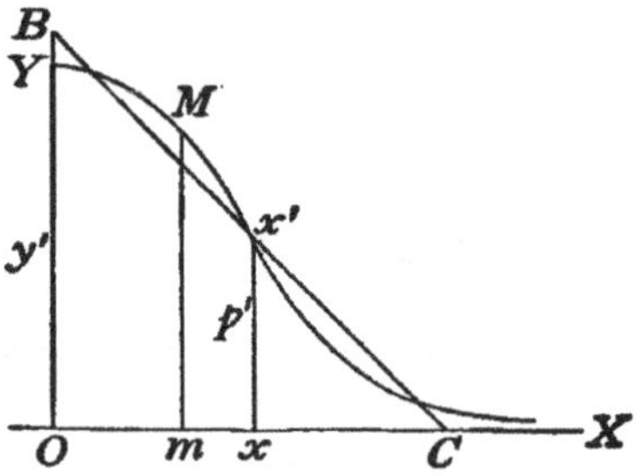

FIG. 301.

In Fig. 301 make

$$OB = y';$$
$$OC = x'' = 3x_m;$$
$$Ox = x;$$
$$Om = x_m;$$
$$xx' = p'.$$

Then the area $OBC = \frac{y'x''}{2} = \frac{3x_m y'}{2}$.

$$\text{Area } OBx'x = \left(\frac{OB + xx'}{2}\right)Ox = \left(\frac{y' + p'}{2}\right)x. \quad (434)$$

From the similar triangles OBC and $xx'C$ we have

$OB : OC :: xx' : OC - Ox$, or $y' : x'' :: p' : x'' - x$;

hence

$$p' = \frac{y'(x'' - x)}{x''};$$

but

$$x'' = 3x_m,$$

hence

$$p' = \frac{y'(3x_m - x)}{3x_m}. \quad . \quad . \quad . \quad . \quad . \quad (435)$$

Substituting in (434) for p' its value from (435), we have

$$\text{area } OBx'x = \frac{(6x_m y' - xy')}{6x_m}x. \quad . \quad . \quad . \quad (436)$$

Now the probability of an error less than $Ox = x$ is the ratio of the area of the triangle OBC to that of the trapezoid $OBxx'$; hence, dividing (436) by the area of the triangle $\frac{3x_m y'}{2}$, we have

$$P = \frac{2}{3}\frac{x}{x_m} - \frac{1}{9}\frac{x^2}{x_m^2}. \quad . \quad . \quad . \quad . \quad . \quad (437)$$

In this equation, having the value of the true mean error given by the equation (420), or that of the mean error obtained as explained from a number of shots, we can find the probability of any error x without using the probability tables.

This discussion of probability may be extended to include the methods for hitting circles or ellipses, and also for determining the number of shots necessary to produce a given result, such as to make a breach in a wall, etc., but the discussion is too extensive for the present course.

CHAPTER IX.

PORTABLE ARMS.

304. Division — Hand Arms — Cutting Arms — Principles—Light Artillery Sabre.

DIVISION.—Portable arms are those which are carried by the individual soldier, and are divided into—

1. Hand arms.
2. Small arms.

HAND ARMS are those which are used for attack and defence at very short distances, and are divided according to their mode of action into—

1. Cutting arms.
2. Thrusting arms.
3. Thrusting and cutting arms.

CUTTING ARMS—PRINCIPLES.—A cutting arm is one which acts by its edge, and, being used entirely against animate objects, is based upon the following general principles:

1. Since the object to be cut is elastic and fibrous, the blow must be struck so that only a few points of the cutting edge at a time will come in contact with the body, and in order to prevent the fibres or the muscles from mutually supporting each other, they must be cut one at a time.

For these reasons the edge of a cutting weapon should be curved, and the blow oblique rather than direct.

The kind of curvature of the edge (convex or concave) will depend on the direction in which the weapon is moving at the time of the blow. If moving toward the object, the edge should be convex; if from it, concave.

Extreme examples are seen in the Turkish sabre, *a*, and the Arab yataghan, *b*, Fig. 302.

2. In order to give force to the blow, the centre of

gravity should be well forward; an example is seen in the axe.

3. For facility of handling, the centre of gravity should be near the hilt.

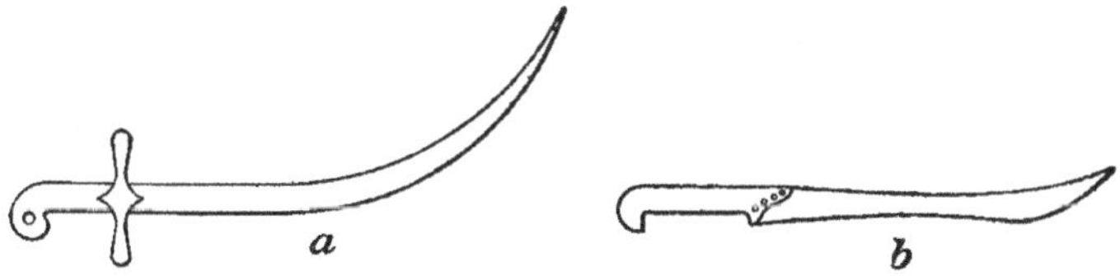

FIG. 302.

As these two principles are conflicting, a compromise is generally effected by throwing the centre of gravity well forward in a cutting weapon, and well to the rear in a thrusting one, and giving it an intermediate position where, as in the cavalry sabre, the two functions are combined.

LIGHT ARTILLERY SABRE.—This is the only distinct cutting weapon in service, and it has a short curved blade with a comparatively light hilt (Fig. 303), the centre of gravity being well forward. The cross-section is grooved for lightness and strength.

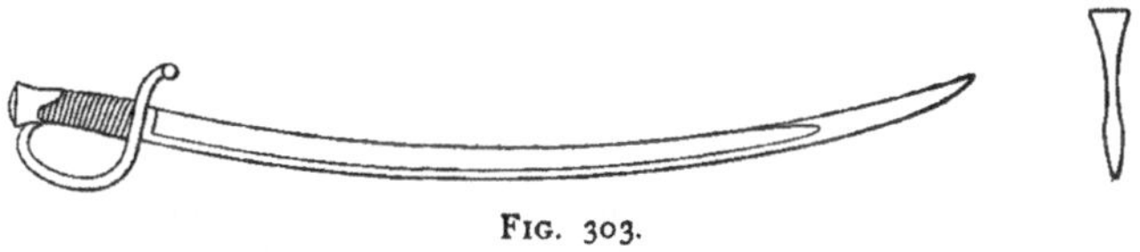

FIG. 303.

305. Thrusting Arms—Principles—Straight Sword—Bayonet—Lance—Cutting and Thrusting Arms—Cavalry Sabre.

A thrusting arm is one which acts by its point, and is based upon the following principles:

1. Its penetration depends on the power of the wedge at its point, and hence this point or wedge should be as sharp as possible consistent with strength.

2. For a given power of wedge, the penetration also depends on the position of the axis of the wedge with reference to the thrusting force. Hence the blade of a thrusting

weapon should be straight, to prevent the turning aside of the point by the oblique component of this force.

3. For facility of handling, the centre of gravity should be well to the rear, and the blade should be light.

The principal thrusting weapons are

The straight sword;
The bayonet;
The lance or pike.

THE STRAIGHT SWORD (Fig. 304), as its name indicates, has a straight blade and sharp point, and the centre of gravity well to the rear in accordance with these principles.

FIG. 304.

THE BAYONET.—This is intended to convert the gun into a pike. It was formerly employed very extensively, but its use has gradually decreased as ranges and velocities have increased. It is still supplied with the latest model guns, and is shown in Fig. 305.

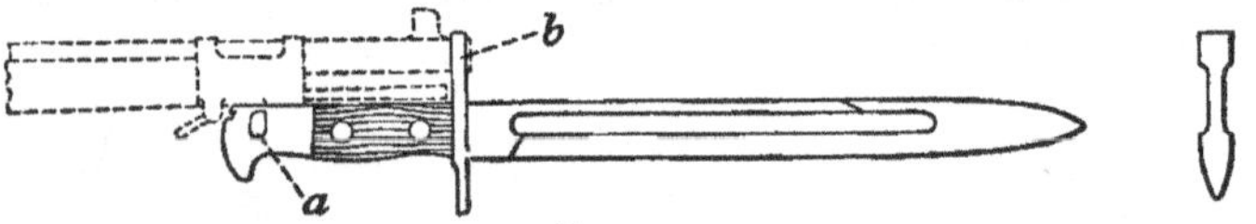

FIG. 305.

It is fixed to the muzzle of the gun by a spring clasp, *a*, engaging over a stud on the upper band, and by a ring, *b*, which encircles the muzzle.

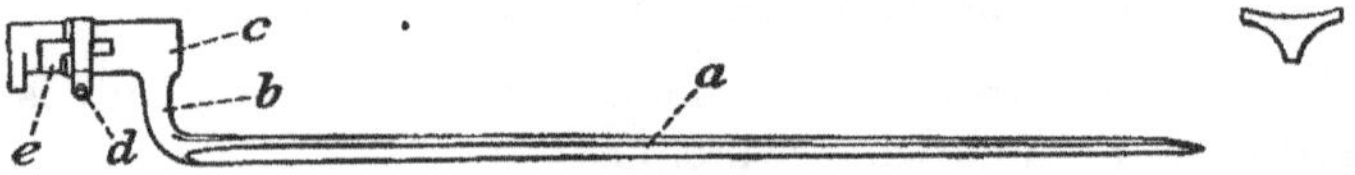

FIG. 306.

The older form of bayonet in use on the Springfield Rifle cal. .45 is shown in Fig. 306. Its cross-section is shown in the figure, and is such as to give lightness and stiffness.

The parts are: the blade *a*, neck or shank *b*, socket *c*, clasp *d*, and groove *e*. Its method of attachment to the gun is well known.

THE LANCE OR PIKE.—This is still used in some foreign services, and is a sharp steel blade fixed to the end of a long wood handle. This handle is provided with a loop at the centre of gravity, for convenience in carrying and guiding. It has the advantage of greater length than the other thrusting weapons, but is inconvenient to carry and handle.

CUTTING AND THRUSTING ARMS—CAVALRY SABRE.—These weapons combine the functions of the other two classes and hence exhibit features common to each class.

THE CAVALRY SABRE (Fig. 307) is the only weapon of this

FIG. 307.

class in service, and the following points may be noted. As it is used both for cutting and thrusting, its blade is longer and less curved than that of the light artillery sabre; the hilt is heavier, to bring the centre of gravity further to the rear, and the hand is better protected by the guard.

SMALL ARMS.

306. Principal Parts—The Barrel—Calibre—Recoil.

PRINCIPAL PARTS.—The essential parts of all breech-loading small arms are:

The barrel;
The receiver;
The breech mechanism;
The firing mechanism;
The sights;
The stock and mountings;

and for magazine arms

The repeating mechanism.

THE BARREL—CALIBRE.—The determination of the calibre of a small arm involves the consideration of recoil, initial velocity, and various other questions which will be discussed in detail.

RECOIL.—Experience has shown that a certain amount of recoil can be borne by the soldier without fatigue. The fatigue caused by recoil will vary not only with the weight of the arm and the velocity of recoil, but also with the nature of the powder, the inclination of the small of the stock, the area of the stock resting against the shoulder, etc.

For convenience of carrying and to avoid fatigue the weight of a small arm should not greatly exceed 9 lbs. This fixes the weight of the barrel, and for a given weight of barrel, or of gun, we conclude generally that the fatigue due to recoil increases with the velocity of recoil. We have for the velocity of recoil while the projectile is in the bore, equation (65), Interior Ballistics,

$$v' = \frac{pv}{P}\left(1 + \frac{\omega}{2p}\right).$$

Since P, the weight of the gun, is fixed by other considerations, as above explained, the velocity of recoil can be reduced only by decreasing the initial velocity v or the weight of the bullet p.

Objections to Decreasing Initial Velocity. — These are obvious. The object of all improvements in modern guns is to obtain as great an initial velocity as possible, keeping the maximum pressure within safe limits, as this increase of velocity gives greater energy, longer ranges, flatter trajectories, etc., as will be explained. It is evident, therefore, that the fatigue due to recoil can only be reduced and kept within proper limits by decreasing the weight of the bullet.

Advantages of Decreasing Weight of Bullet.—Considering the equation

$$v' = \frac{pv}{P}\left(1 + \frac{\omega}{2p}\right),$$

it is evident that for an allowable value of v', since P is constant, a decrease in the weight of the bullet p, will increase the initial velocity v.

Therefore a decrease in weight of bullet gives a value for the recoil which can be easily supported by the soldier, and it also increases the initial velocity of the projectile, which is the object sought. Whether this increase in initial velocity will be advantageous at different ranges depends on the manner in which the weight is reduced, and it is necessary therefore to consider the best method of doing this.

307. Reduction of Weight of Bullet—First Method—Decreasing the Length, keeping the Diameter Constant.

The weight of the bullet may be decreased:

1. By decreasing its length, keeping the diameter constant.

2. By decreasing the diameter, keeping the length constant.

3. By changing both length and diameter.

To determine which of these methods is best, assume the equations

$$\frac{dv}{dt} = \frac{P}{M}; \qquad (438)$$

$$v = \int \frac{dv}{dt} dt. \qquad (439)$$

$$R = A \frac{\delta}{\delta_1} \frac{cd^2}{W} f(v); \qquad (440)$$

$$\frac{v^2}{\rho} = g \cos \theta. \qquad (441)$$

Equations (438) and (439) are from Mechanics. In (438) P is the total pressure acting to produce acceleration, and M the mass of the projectile. In (439) v is the velocity of the projectile at any time t, and in the present case is the initial velocity. Equations (440) and (441) are from Exterior Ballistics. In (440) R is the retardation of the projectile due to the resistance of the air, and the quantities in the second member are all defined in Exterior Ballistics, d and

W being the diameter and weight of the projectile. In (441) v is the velocity of the projectile at any point of its trajectory, ρ the radius of curvature at that point, and θ the inclination of the tangent, g being 32.2 ft.-seconds.

DECREASING LENGTH OF PROJECTILE, DIAMETER CONSTANT.—In equation (438),

$$\frac{dv}{dt} = \frac{P}{M},$$

the total pressure $P = p' \times \frac{1}{4}\pi d^2$, p' being the pressure of the powder per unit of area of base of projectile.

For constant values of p' and d, P will remain constant. Hence if the length of the projectile be decreased, the diameter being constant, M will decrease, and from equation (438) $\frac{dv}{dt}$, or the acceleration, will increase.

In equation (439),

$$v = \int \frac{dv}{dt} dt,$$

since $\frac{dv}{dt}$ increases, v, or the initial velocity, will increase.

In equation (440),

$$R = A \frac{\delta}{\delta_1} \frac{cd^2}{W} f(v),$$

since W decreases while d remains constant, R will increase.

This will cause v to decrease for all points of the trajectory, and hence in equation (441),

$$\frac{v^2}{\rho} = g \cos \theta, \quad \therefore \rho = \frac{v^2}{g \cos \theta},$$

ρ will decrease, or the trajectory will be more curved.

If p', the pressure of the powder per square inch, be increased, P in equation (438) will increase, and hence also $\frac{dv}{dt}$. This, in equation (439), will cause an increase in v, but since from (440) the retardation is still great, the velocity will fall off rapidly, and from (441) the trajectory will be very much curved.

The results obtained by decrease of weight of bullet, by the method of shortening it, and keeping the diameter constant, are, therefore:

1. The velocity of recoil is decreased;
2. The initial velocity is increased;
3. The remaining velocity at different points falls off very rapidly;
4. The curvature of the trajectory is increased.

From the 3d and 4th results we conclude that this method of reducing the weight of the bullet should not be adopted.

308. Reduction of Weight of Bullet—Second Method—Decreasing Diameter, keeping Length Constant.

In equation (438),

$$\frac{dv}{dt} = \frac{P}{M},$$

we have as before

$$P = p' \times \tfrac{1}{4}\pi d^2.$$

Suppose p' fixed and d decreased. Then, since the area of cross-section of the projectile decreases in this case, P will decrease directly with it, and the mass M will also decrease directly with the same area, the length being constant, and hence the ratio $\frac{P}{M}$ will not change. The same may be shown for an *increase* in diameter, the length being constant.

The sectional density of a projectile is $\frac{W}{d^2}$ (see Projectiles, subject 164). Substituting for W its value, we have

$$\frac{W}{d^2} = \frac{\tfrac{1}{4}\pi d^2 l\delta}{d^2} = c'l,$$

or the sectional density varies with the length. Hence when the length is constant, the sectional density is constant, and from the above we conclude that for the same pressure per square inch, and the same sectional density of projectile, no increase of velocity is obtained by reducing the weight, assuming the same pressure curve in the two cases.

In equation (440),

$$R = A \frac{\delta}{\delta_1} \frac{cd^2}{W} f(v),$$

since $\frac{d^2}{W}$ and v do not change, there is no change in retardation, and consequently there is no change in curvature, equation (441).

Therefore the only effect of reducing the weight of the projectile by decreasing the diameter and keeping the length constant, or, in other words, keeping the sectional density constant, when the pressure per square inch p' remains constant, is to diminish the velocity of recoil.

Suppose, however, that p' is increased.

Then in equation (438),

$$\frac{dv}{dt} = \frac{P}{M},$$

M, as before, will decrease directly with the area of cross-section, but P will increase, and hence the ratio $\frac{P}{M}$ will increase. This will cause $\frac{dv}{dt}$ to increase.

In equation (439),

$$v = \int \frac{dv}{dt} dt,$$

v will increase. In equation (440),

$$R = A \frac{\delta}{\delta_1} \frac{cd^2}{W} f(v),$$

R remains constant, since $\frac{d^2}{W}$ does not change; or it may even decrease, owing to the increase in v, and the consequent change in the exponent of $f(v)$ from 3 to 2 (see Mayevski's experiments, Exterior Ballistics). v therefore will be greater for all points of the trajectory, and in equation (441),

$$\rho = \frac{v^2}{g \cos \theta},$$

ρ will be greater, and hence the curvature of the trajectory will be less, or it will be flatter.

Hence by decreasing the weight of the bullet by the second method, that is, by reducing the diameter and keeping the length constant, and at the same time increasing the pressure per square inch of the powder-gas, we obtain:

1. A decrease in velocity of recoil;
2. An increase in initial velocity;
3. No increase in retardation, and perhaps a reduction;
4. A flatter trajectory.

309. Reduction of Weight of Bullet—Third Method—Changing Length and Diameter—Smokeless Powder—Advantages of Reduction of Calibre—Flatness of Trajectory.

The method at present adopted is to vary the pressure per square inch, the length, diameter, and weight of projectile, so as to obtain the best ballistic results. This has led to a reduction of the calibre from 0.45 to 0.30 inch, a decrease in the weight of the bullet from 500 to 220 grains, the length being very slightly changed, and an increase of pressure per square inch from a maximum of 30,000 lbs. to a maximum of 45,000 lbs. per square inch, an increase of initial velocity from 1300 to 2000 ft.-seconds, with a reduction of velocity of recoil from 14 to 9.6 ft.-seconds, and of energy of recoil from 27 to 11 foot-pounds.

SMOKELESS POWDER.—It is evident that to obtain any ballistic advantage from a reduction of calibre, the pressure per square inch on the projectile must be increased. When the calibre of small arms was first reduced, various attempts were made to obtain this increase of pressure by the use of the old black powder in various forms, such as larger charges compressed, slower burning, etc., but the results were unfavorable, giving high and irregular pressures, increase of fouling, etc. The effort to overcome these difficulties led to the introduction of smokeless powder. Its advantages have been explained in High Explosives, one advantage of great importance being that, as the smokeless powder burns more slowly and regularly, it acts upon the projectile like the slow-burning powders already described in large guns, and hence for a given initial value of p' we obtain a greater initial velocity than would be produced by the same initial value of p' with the old black powders.

ADVANTAGES OF REDUCTION OF CALIBRE.—The principal of these are:

1. Flatness of trajectory, and increase of range;
2. Decrease in weight of cartridges;
3. Increase of accuracy of fire;
4. Decrease of recoil;
5. Increased penetration.

FLATNESS OF TRAJECTORY.—The advantage of this may be illustrated as follows: Assume, equation (441),

$$\rho = \frac{v^2}{g \cos \theta}.$$

FIG. 308.

Let H, Fig. 308, be the height of a man, and suppose this height to be the maximum height of the trajectory. The total range in this case is called the maximum continuous dangerous space, and is frequently used in comparing the ballistic qualities of guns.

The value of $\cos \theta = 1$ at the summit of the trajectory; hence

$$\rho = \frac{v^2}{g}$$

for this point. It is evident that ρ increases rapidly with v.

Hence for a low velocity we will have the trajectory AB, and for a high velocity, $A'B'$, the maximum continuous dangerous spaces being the horizontal distances AB and $A'B'$ respectively. The flat trajectory, then, gives a greater continuous dangerous space $A'B'$, and this is true when the dangerous space is not continuous, as in the case of the object H'; the dangerous spaces being $H'B$ and BB' respectively.

An error in estimating distance is also of less importance with a flat trajectory; as in the figure, an error $H'B$ for the curved trajectory, and BB' for the flat one, may be committed, and the target will still be struck. The distances AB and $A'B'$ for the calibres .45 and .30 are 418 and 600 yards respectively.

310. Advantages of Reduction of Calibre—Decrease in Weight of Cartridges—Increase of Accuracy of Fire—Increased Penetration.

DECREASE IN WEIGHT.—The number of rounds carried is limited by the physical endurance of the soldier, just as the weight and recoil of his piece are fixed by the same conditions. A reduction in the weight of the cartridge increases the number of rounds that can be carried, and this increase is very important owing to the great increase in rapidity of fire with modern breech-loaders, and the difficulty of supplying the fighting-line with fresh ammunition. This reduction in weight is due not only to the reduction in calibre, but also to the introduction of smokeless powder, by which the weight of the charge has been reduced nearly one half.

INCREASE OF ACCURACY OF FIRE.—Owing to the greater velocities at all points of the trajectory, the small-calibre projectile is less affected by the wind and other deviating causes, and the drift is not greater than with the old projectile. Hence the horizontal deviations are less than with the old projectile. As has already been shown, the flatness of trajectory makes it more accurate in a vertical direction, and hence its absolute accuracy, which is taken to be the radius of the circle containing one half the whole number of shots, is greater than with the old bullet, the radius of the circle being less.

There is, however, one exception to this. Owing to the relative increase in length of the new bullet, it is necessary to give it greater velocity of rotation about its longer axis to insure stability, and hence the pitch of the rifling is more rapid for the new small calibre.

This increases the passive resistances in the bore, and, with the greater pressure per square inch on the projectile, causes increased vibration of the barrel. The result is that for short ranges the accuracy of the small calibre is slightly less than that of the old arm.

Beyond these ranges the small calibre is more accurate.

INCREASED PENETRATION.—This is due to increase of velocity, and also to the fact that the exterior of the bullet is covered with a jacket of harder metal, such as copper, German silver, or nickeled steel. This jacket holds the projectile together and prevents deformation on striking. It is stated that the bullet of the 8-mm. rifle has pierced a tree 17 inches in diameter, and afterwards passed through the bodies of five men. The penetration of the cal. .30 bullet with steel jacket is in sand 14 inches, and in oak from 16 to 24 inches, the target being 3 ft. from the muzzle. The penetration of the cal. .45 bullet under the same circumstances is 3.3 inches in oak.

Numerous experiments have been made upon human bodies to test the effect of the small-calibre bullet, with the general result that the wounds are less serious and the shattering effect on the bones less than with the old projectile. The shock or stopping power is also less as the calibre decreases, unless the bullet acts explosively, and hence it has been proposed for the very small calibres to remove the jacket from the point of the bullet, thus causing it to spread out in front on striking.

311. Disadvantages of Reducing the Calibre.

The principal of these are:

1. The decrease in weight of bullet, and hence the relative increase in its length, necessitates a more rapid twist of rifling to give it stability in flight, and this increase of twist increases the passive resistances in the bore and gives rise

to greater vibrations of the barrel. These vibrations, as stated, decrease the accuracy for short ranges. A test with the barrel confined in a fixed rest showed greater inaccuracy at 500 yards than with the cal. .45. A heavy barrel cal. .30, made expressly for the purpose, was then tried in the fixed rest under similar conditions, and remarkable accuracy, greater than ever before recorded, was obtained. This shows that the inaccuracy is due to vibrations of the barrel, and it is probable that when the gun is fired from the shoulder in the ordinary manner the targets will be much better than when a fixed rest is used, as the barrel in this case will not be rigidly held, and consequently its vibrations will be less.

The increase in twist also renders the projectile more apt to strip in the bore; that is, to be forced across the lands without taking the rifled motion, with the result that the bore is scored or fouled by the metal, and the projectile rotates about its shorter axis in flight, or tumbles. This has been remedied by the use of a harder metal jacket.

2. The cleaning of the bore is more difficult. Since the introduction of smokeless powder this objection has less weight.

3. The manufacture is more difficult.

This has been a serious objection, as it is a very difficult operation to bore and rifle accurately such a small calibre, and any inaccuracy here is fatal to the accuracy of fire. This difficulty has also been overcome, and guns below 0.30 calibre are now successfully made.

4. The pressure in the bore is greater.

The necessity for this has been shown, and it has been difficult to provide steel of sufficiently high elastic qualities to withstand this pressure. It has also caused the abandonment of nearly all the old forms of breech mechanism, in order to obtain a secure fermeture.

In spite of these objections, the great advantages of a reduction of calibre have led to its universal adoption in all countries, and the tendency now is to go below the .30 calibre. This has been done in some countries. One of the points still in doubt is the effect of the small-calibre bullet upon the nervous system, and whether a wound from this bullet, when not fatal, will stop a man.

This can only be solved in actual war, and hence in our service it has been thought best not to go below the cal. .30 at present.

312. Rifling—Pitch—Number of Grooves and Lands—Width—Depth—Direction of Twist.

PITCH.—The pitch of the rifling in small arms is always uniform, because, when fired, the bullet is molded accurately into the grooves and lands, and the length of the surface of the bullet in contact with the bore is great. If the pitch be uniform, no change of form of the molded surfaces takes place during the passage of the projectile from breech to muzzle; if the pitch be increasing, a change of form is constantly occurring, resulting in increased resistance, deformation of projectile, and inaccuracy.

It has been found necessary in practice to increase the twist as the calibre decreases, as already explained (see subject 163).

In the Springfield cal. .45 the twist is one turn in 48.9 calibres, in the new cal. .30 it is one turn in $33\frac{1}{3}$ calibres, or one turn in 22 and 10 inches respectively.

NUMBER OF GROOVES AND LANDS.—The number of grooves has no effect apparently upon the accuracy of fire, and hence for convenience of manufacture, cleaning, and strength, these are as few as possible. The cal. .45 has three grooves and lands, the cal. .30 four. As a general rule the number has varied from three to seven.

WIDTH OF GROOVES AND LANDS.—The width depends on the kind of bullet. When of hardened lead, the bullet is slightly upset by the shock of discharge and forced into the grooves, the lands cutting into the projectile. As this metal offers comparatively little resistance, and the twist is not rapid, the grooves and lands in the cal. .45 are of equal width.

With the jacketed bullet, the resistance to deformation being much greater, the grooves are wider and the lands narrower, since these latter do the work of cutting into the projectile. In the U. S. cal. .30 the grooves are three times the width of the lands.

DEPTH OF GROOVES AND LANDS.—If the depth of groove is too great, there is too much work lost in forcing the projectile, and the forcing may not be perfectly accomplished. This latter will cause erosion and inaccuracy. If the depth is too small, the groove may be easily filled by fouling. The depth also varies with the kind of bullet. The depth of groove in the Springfield cal. .45 is .005 inch, and in the cal. .30 it is .004 inch.

The exterior diameter of the lead bullet is 0.457 inch, that of bore at bottom of grooves .455 inch (see Fig. 309).

Hence with the Springfield rifle, in addition to the work done by the lands in cutting into the projectile, the latter exceeds the diameter of the bore at bottom of grooves by .002 inch. This, added to the upsetting action of the powder, gives a very energetic forcing, and insures its accomplishment, but without great strain on the gun. The cannelures or grooves in the bullet also assist in reducing the work. With the cal. .30 the exterior diameter of the bullet is 0.308 inch, and that of the bore at the bottom of grooves the same. Hence the bullet exactly fills the bore from groove to groove, and there is no forcing in the grooves, aside from what may be due to upsetting of the metal by the action of the powder and the pressure of the lands.

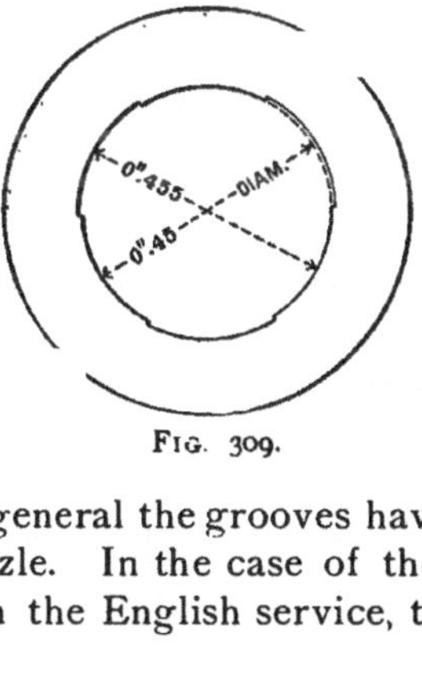

FIG. 309.

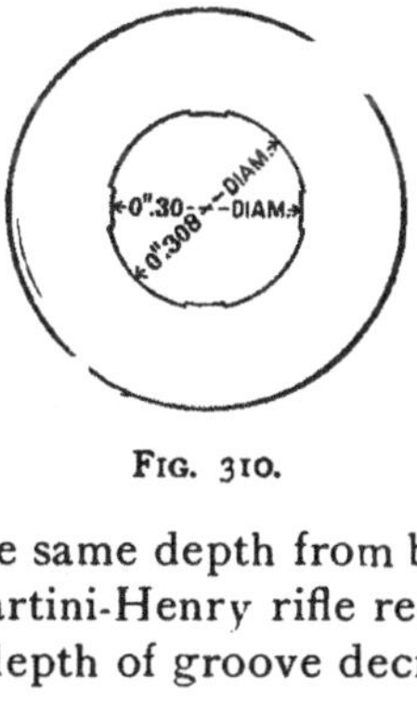

FIG. 310.

In general the grooves have the same depth from breech to muzzle. In the case of the Martini-Henry rifle recently used in the English service, the depth of groove decreased from breech to muzzle, to make the forcing more gradual, and thus decrease the pressure at the origin, and consequently the vibrations.

Figs. 309 and 310 show the cal. .45 and cal. .30 grooves in section.

DIRECTION OF TWIST.—This has no influence upon the accuracy of fire, as it produces "drift," which can be allowed for. All small arms are rifled with a right-hand twist, and the resulting drift is to the right as already explained.

A case occurs in the French service, where the vibrations of the barrel, owing to the peculiarity of the breech mechanism, caused the bullet to deviate to the right, and to correct this the gun was rifled with a left-hand twist.

313. Profile of Chamber—Thickness and Length of Barrel.

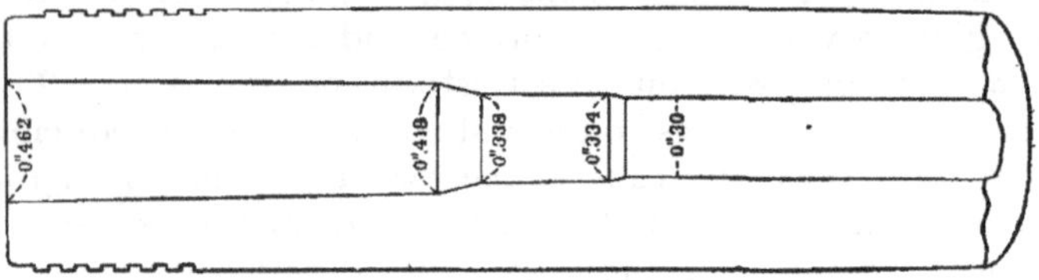

FIG. 311.

PROFILE OF CHAMBER.—The chamber is made slightly conical to facilitate the extraction of the cartridge-case. That for the cal. .30 rifle is shown in Fig. 311. The chamber must be free from all cuts or scratches, since the cartridge-case will be forced into them on firing, and will either stick or rupture. All dimensions must be exact, and very little variation can be allowed.

THICKNESS OF BARREL.—The case is that of a single cylinder under extension, the exterior pressure being zero (see "Elastic Strength of Guns"). For the thickness of the cal. .30 rifle-barrel just in front of the powder-chamber, assume, equation (205),

$$R_1 - R_0 = R_0\left(\sqrt{\frac{3\theta_0 + 2P_0}{3\theta_0 - 4P_0}} - 1\right).$$

We have

$$R_0 = \frac{.30}{2} = .15,$$

$$\theta_0 = 61,500 \text{ lbs.},$$

$$P_{0\theta} = 40,000 \text{ lbs.},$$

which values substituted in the above equation give

$$R_1 - R_0 = .3429 \text{ inch} = 1.14 \text{ calibres.}$$

The actual thickness is 0.34 inch = 0.49 − 0.15 = 0.34.

For the thickness at various points along the bore the pressure curve must first be calculated, but other considerations, such as stiffness to resist vibrations and to prevent bending in service, etc., enter, and the exterior is given the general form of a conical frustum, the thickness at the muzzle being 0.53 calibres, 0.16 inch.

LENGTH OF BARREL.—This is so adjusted that the rear-rank man can fire over the shoulder of the man in front without danger to the latter, and for the small-calibre rifle this length is fixed at 30 inches (100 calibres).

Experiment shows that increasing this length gives very little increase of initial velocity, while it increases weight and difficulty of manufacture. The length of the cal. .45 barrel is 32.6 inches. The length of travel of the projectile in the bore for the cal. .30 is 28.19 inches (94 calibres), and for the cal. .45, 30.445 inches (67.6 calibres).

314. The Receiver—General Features—Receiver for Springfield Rifle.

THE RECEIVER is a distinctive feature of breech-loading small arms, and forms an extension of the barrel, for the purpose of receiving the cartridges and breech mechanism.

GENERAL FEATURES.—The shape of the receiver depends on the breech mechanism, and also upon whether the gun is a single-loader or a magazine arm.

In general it must have the following features:

1. A method of attachment to the barrel.
2. A method of attachment to the stock.

3. An opening through which cartridges are inserted, empty shells extracted, and in which the breech-block or bolt works.

4. An axis about which the block rotates, or guide-grooves for regulating the motion of the bolt.

5. A recess or groove for locking the block or bolt.

6. An arrangement for ejecting empty cartridge-shells; and for a magazine arm in addition to the above—

7. An opening for the admission of cartridges from the magazine.

8. A "cut-off" by which this opening may be used or not at will.

RECEIVER FOR SPRINGFIELD RIFLE.—Fig. 312 shows the receiver for the Springfield rifle.

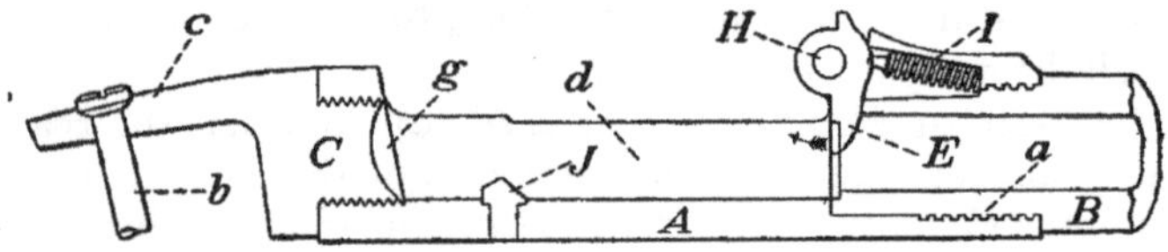

FIG. 312.

It is attached to the barrel *B* by the screw-threads *a*; to the stock, by a screw, *b*, passing through the tang *c*; *d* is the opening through which the cartridges are fed, and in which the breech-block works; *H* is the axis about which the breech-block rotates; *g*, the recess into which the breech-block is locked by its cam-latch, to be described; *I* is the ejector spring and spindle. The cartridge-case is loosened in its seat by the positive action of the extractor *E*, which rotates in the direction of the arrow. The axis of the spindle of the spring *I* is at first above the axis of rotation, *H*, of the block and extractor. After a small rotation of *E*, the axis of the spindle is carried below the axis *H*, and the spring *I* then acts to rotate *E* quickly, and throw out or eject the empty case. As the case moves backward, it strikes the inclined stud *J*, and is, by it, deflected upward out of the receiver.

315. Receiver for Cal. .30.

This is shown in Fig. 313.

It is attached to the barrel by a screw-thread, and to the stock by the screws *X* and *Y* passing through the trigger-guard

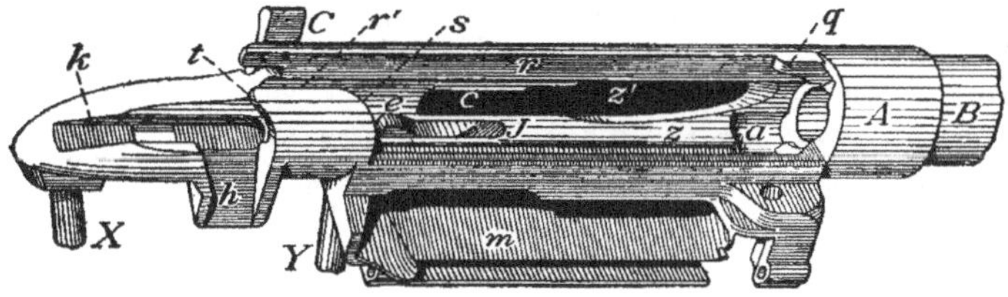

FIG. 313.

into it from below (see Fig. 336); *z* is the opening through which the cartridges are fed when the gun is used as a single-loader, and *z′* when used as a magazine arm. The left side, *r*, of the opening *z*, is parallel to the axis of the bore and, together with the surface, *r′*, on the right, forms a guide for the bolt when moving forward or back. A second groove, *h*, forms a recess for the operating-handle of the bolt to rest in, when this handle is rotated to the right in closing the breech. The forward shoulder or cam, *s*, in front of the groove, *h*, is so shaped as to give a screwlike motion to the bolt in closing, thus moving it slowly forward to its seat against the breech. The rear shoulder, *t*, arrests the forward motion of the bolt in closing. A third groove, *k*, prevents the firing mechanism from turning with the bolt in closing the breech.

The groove *a* locks the bolt, a lug on the latter entering it. When in the firing position, the pressure of the gas is transmitted to the surface of the groove *a*; the surface *s*, and the rear surface of the groove *h*, acting as safety-supports.

The empty shell is ejected as follows: The bolt is drawn slowly backward at first, by the action of the inclined surface, *t*, of the groove *h*, against the operating-handle. The extractor, which is on the bolt, and engaged with the rim of the cartridge, draws the case back slowly, due to this motion of the bolt. When the bolt is free to move along the axis of the receiver, it moves quickly, drawing back the empty case.

At the end of the travel of the bolt, the short arm, *e*, of the ejector-lever, in the bottom of the receiver, is struck by a shoulder at the end of a groove in the bolt, and the long arm, *f*, is thrown up, striking the empty case and ejecting it. The opening *m* is the magazine, which will be explained later.

The cut-off for the magazine, Fig. 314, is a pin or rod, the rear part, *a*, of which is round, and the front part, *c*, is cut away partly, as shown. The cut-off is inserted in the left-hand side of the receiver, parallel to its axis (see Fig. 313, *C*), the cut-away portion, *c*, projecting over the opening *z'* of the magazine. When the magazine is in use, the flat part of the cut-off forms a portion of the surface *z'* of the magazine opening. When the magazine is to be cut off, the rod is rotated by turning the handle *C*. This brings the rounded part of *c* into such a position that it projects into the opening *z'* and forces the cartridges down slightly, so that the bolt will pass over without touching them.

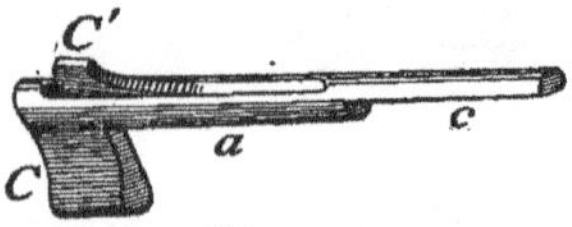

FIG. 314.

The cut-off is held in the open or closed position by the spring *C'*, which works in a groove in the receiver.

316. Breech Mechanism—General Classification—Sliding Mechanism.

The functions of the breech mechanism are to open, close, and lock the breech, extract the empty cartridge-case, and for magazine arms, in addition, to operate the repeating mechanism, and insert the cartridge.

GENERAL CLASSIFICATION.—Breech mechanisms may be classified generally into:

1. Those which operate by sliding.
2. Those which operate by rotation.

SLIDING MECHANISM.—The sliding may take place—

1. By the motion of the barrel parallel to its axis.

This arrangement is now obsolete, and is unsuitable for a military weapon on account of the weight of the barrel.

2. By the motion of the breech-block parallel to its axis. Guns with this mechanism are called bolt-guns, and the mechanism resembles in its action the bolt of a door, whence the name. All magazine arms at present in use belong to this system. It presents the following advantages:

a. Extreme simplicity, great strength, and small number of parts.

b. Secure locking against the effects of discharge.

c. Ease of extraction of empty case.

d. Better adapted to magazine arms than any other system.

The objections to the system formerly were:

a. Danger of blowing out the bolt by premature discharge, before the breech was securely locked.

b. Liability to explode the cartridge when pushing it home, either by striking it a blow, or by the projection of the firing-pin striking the primer in the cartridge-case.

These objections have been overcome.

3. The block may slide at right angles to the axis of the barrel. An example in seen in the Krupp fermeture.

The advantage of this system is that it is not liable to blow out, as the direction of the pressure is normal to the bearing surfaces of the block; the disadvantages are that it tends to guillotine the cartridge, does not push it home, and renders extraction of the empty case difficult.

317. Rotating Mechanism.

The rotation may take place—

1. Around an axis parallel to the axis of the gun, and at one side. This is now obsolete.

2. Around an axis parallel to the axis of the gun, and below it; example, revolvers.

This system is objectionable for a military arm, on account of the weight of the revolving cylinder, and also because of the break in the barrel at the junction of the cylinder and barrel proper, through which gas may escape.

3. Around an axis at right angles to the axis of the gun, above that axis, and at the front of the block; example, the Springfield rifle cal. .45.

This system has the following advantages:

a. The block acts to push the cartridge home in closing.

b. It forms, in connection with the extractor, a strong lever for extracting the empty case.

c. It is simple and has comparatively few parts.

Its disadvantages are:

a. It does not securely and positively lock the block, and the tendency of the pressure is to force it strongly against the breech-recess; hence for high pressures, as in the present small calibre, it is difficult to open after firing.

b. It is not adapted for a magazine arm.

4. Around an axis at right angles to the axis of the bore, above that axis, and in rear of the block; example, the Martini-Henry recently used in the English service (Fig. 315).

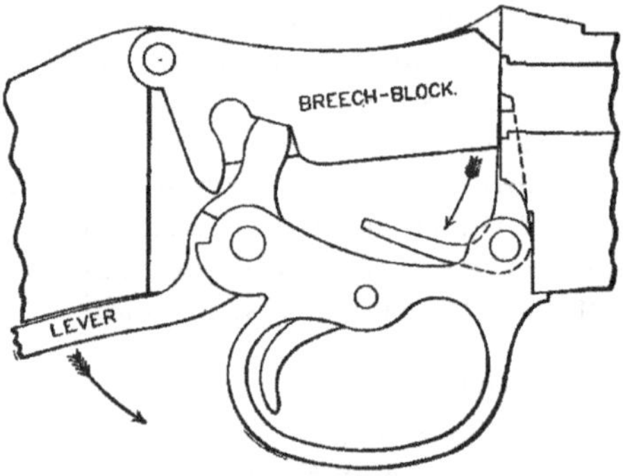

FIG. 315.

The advantages of this system are:

a. It is simple and solid, and the block is well protected against accident.

b. The pressure does not tend to blow open the block.

Its disadvantages are:

a. A space must be left between the front of the block

and the rear of the chamber to allow for rotation of the block, and hence the chamber cannot be tightly closed.

b. The extraction of the empty case is difficult.

c. The block is liable to guillotine the cartridge, unless the latter is forced completely home before closing the breech.

5. Around an axis at right angles to the axis of the gun, below that axis, and in front of the block; example, the Remington, Fig. 316.

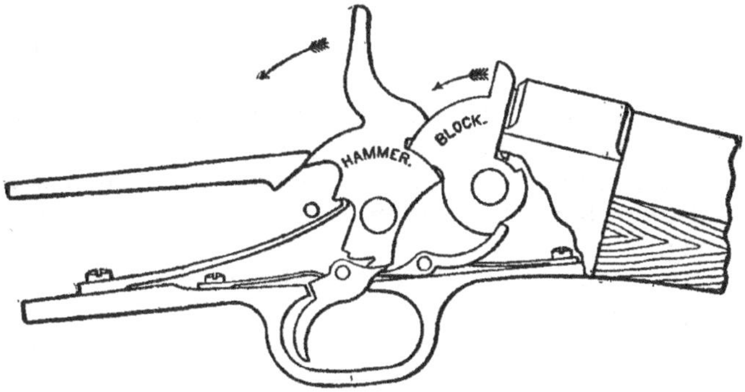

FIG. 316.

This system is simple, but requires an exact adjustment of all the parts, especially of the hammer, as in addition to its ordinary functions it locks the breech-block.

A system is also used in which the barrel rotates about an axis at right angles to the bore and below that axis; example, shot-guns. This is not used for military arms, on account of the weight of the barrel.

318. Requirements of a Good Breech Mechanism.

A good breech mechanism should fulfil the following requirements:

1. It should be simple, strong, and safe in action, and should work freely under all conditions which are liable to prevail in active service, even when rusty or covered with dust.

2. It should be easy to clean, take apart and assemble,

and should be composed of few pieces, which are not liable to break or work loose, and which are interchangeable. Screws are objectionable, as they are liable to work loose.

3. The motions in loading should be as few as possible, and executed in regular order, and it should be impossible to fire the gun till the breech is securely locked.

4. To increase the rapidity of fire, the motion of opening and closing the breech should cock the firing mechanism. In general it is preferable to cock the firing mechanism by the motion of opening the breech, as this withdraws the firing-pin so that it will not strike the primer of the cartridge when the latter is pushed home, and in addition, if the mechanism is cocked in closing, a slip of the hand before the bolt is home will cause the latter to spring back, and either throw out the cartridge which is partly introduced, or, in case of a magazine arm, it may cause the introduction of a second cartridge before the mechanism, and thus produce jamming.

5. The opening of the breech should automatically eject the empty case.

6. The working of the mechanism should cause as little fatigue as possible to the firer.

7. A safety-device should be provided which can be readily seen and operated, and by which the mechanism can be locked in place and accidental discharge rendered impossible.

8. It should be impossible to strike a blow on the cartridge, either by the bolt, or by the firing-pin, while the breech is being closed.

319. Breech Mechanism of Springfield Rifle, Cal. .45.

This mechanism belongs to the system in which the block rotates about an axis perpendicular to the axis of the gun, above that axis, and in front of the block.

Although it is to be replaced by the cal. .30, the arm is still (1895) in service, and is likely to be used in any emergency arising within the next few years, and hence its mechanism will be explained.

The principal parts are (Fig. 317):

The breech-block *D*;
The hinge-pin *H*;
The cam-latch *G*;
The extractor *E*;
The ejector spring and spindle *I*.

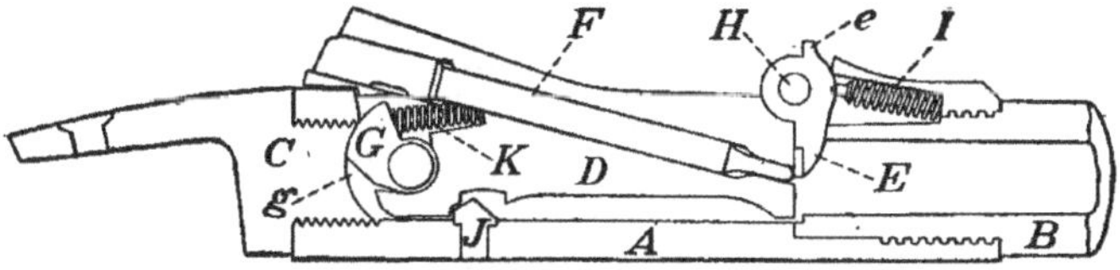

FIG. 317.

The receiver and ejector spring, spindle, and stud have been previously explained.

THE BREECH-BLOCK.—This has an oblique hole, *f*, Fig. 318, through it for the firing-pin. In front is the hinge-pin hole *h*, which is elongated parallel to the axis of the bore, and through which passes the hinge-pin *H*, Fig. 317, around

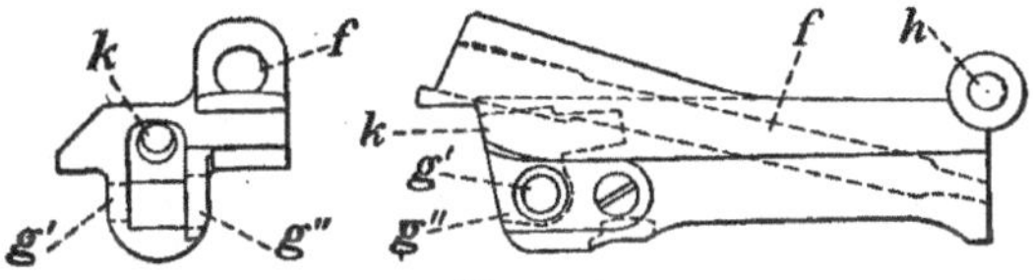

FIG. 318.

which the block rotates. In rear is a recess, *k*, called the cam-latch recess, for the cam-latch *G* and its spring *K*, Fig. 317. The shaft of the cam-latch passes through the hole *g'*.

THE HINGE-PIN.—This forms the axis about which the block rotates. It passes through two holes in the lugs of the receiver, and is kept from turning by an arm with a stud which fits in a hole on the side of the receiver.

THE CAM-LATCH.—This locks the breech-block in firing by entering a circular recess, *g*, in the breech-screw *C*, Fig. 317. It is fixed to a shaft one end of which passes through the hole *g'*, Fig. 318, in the block, and the other end is supported by the breech-block cap *g''*, which is removable. The axis of the cam-latch shaft projects on the right-hand side, and to it is attached a thumb-piece by which the cam-latch is operated. This axis fits loosely in the hole *g'* in the breech-block, and also in the corresponding hole in the breech-block cap *g''*.

The cam-latch is pressed to the rear into its recess in the breech-screw by the cam-latch spring *K*.

THE EXTRACTOR.—This is mounted on the hinge-pin, on the left side of the chamber. Part of its lower extremity is cut into such a shape as to form, when in place, a part of the counterbore of the chamber, in which the rim of the cartridge rests. It has also in front, and slightly above the axis of the hinge-pin, a recess for the reception of the head of the ejector-spindle; and a lug, *e*, Fig. 317, projects beyond the upper surface of the receiver, against which the breech-block bears in opening.

ACTION OF MECHANISM.—When the piece is fired, the breech-block slides bodily to the rear, owing to the elongation of the hinge-pin hole *h*, Fig. 318. Owing to this motion of the block, and to the loose fit of the cam-latch shaft in its holes in the block, the pressure of the powder-gas is transmitted directly, through the breech-block and the body of the cam-latch, to the breech-screw *C*, and there is no strain upon the hinge-pin *H* or the cam-latch shaft. The block is opened by pressing the thumb-piece forward, which disengages the cam-latch from its recess. When the block has nearly completed its rotation upward, it strikes against the projecting lug, *e*, of the extractor *E*, Fig. 317, and rotates the latter slowly, thus extracting the empty case. The ejector then acts as before explained, and throws the case out of the receiver.

320. Breech Mechanism of the Cal. .30 Rifle—The Bolt.

This mechanism belongs to the system in which the block slides parallel to the axis of the bore, and the gun is a bolt-gun.

The principal parts are:

The bolt *D*, Fig. 319;
The sleeve *I*, Fig. 320;
The extractor *E*, Fig. 321.

THE BOLT, Fig. 319, is a hollow cylinder, closed at the front end except a small opening, *f'*, in the centre, for the

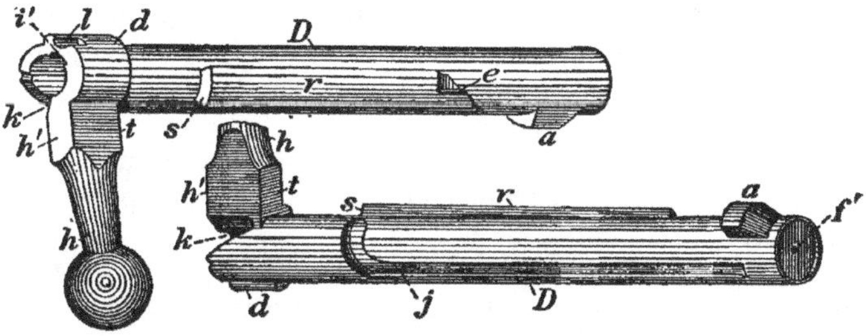

FIG. 319.

passage of the point of the firing-pin. Its interior shape is shown in section, Fig. 336.

The head of the cartridge rests against the front of the bolt, which is hollowed to receive it, and which supports the pressure of the powder-gas. *a* is the locking-lug, which engages in the locking-recess *a*, Fig. 313, in the front part of the receiver; *r* is the guide-rib which rests against the left side, *r*, of the guide-groove, Fig. 313, when the bolt is unlocked and rotated, and guides the motion of the bolt. It also forms a stop to limit the motion when the bolt is rotated in opening. The guide-rib has a shoulder, *e*, in front, against which a corresponding shoulder on the extractor rests.

The rear end, *s*, of this rib rests in front of a corresponding shoulder, *s*, Fig. 313, on the receiver, when the block is locked for firing, and forms a safety-support to resist the pressure, the lug *a* being the first support. *D* is the body of the bolt, in one piece with the operating-handle *h*. This handle terminates at the bolt, in a collar, *d*, Fig. 319, which only partly encircles the rear end of the bolt.

This collar serves to connect the bolt with the other parts of the mechanism.

The rear side, *h'*, of the operating-handle *h*, rests in front of a corresponding shoulder, *h*, Fig. 313, in the receiver in the firing position, and forms a second safety-support to resist the pressure.

At the rear end of the bolt is a notch, *k*, one side being straight and the other inclined. This notch cocks the firing-pin when the bolt is rotated in opening, and also allows the cocking-piece, carrying forward the firing-pin and striker, to move down into it, when the piece is discharged, as will be explained

There is also a longitudinal groove, *j*, its rear end turning to one side, and its front end terminating abruptly in a shoulder.

This groove works the ejector-lever *J*, Fig. 313, in the bottom of the receiver. The notch *i'* admits a stud, *i'*, Fig. 320, on the sleeve, and the notch *l* is for the safety-lock *l*, Fig. 320.

321. Breech Mechanism Cal. .30—The Sleeve—The Extractor.

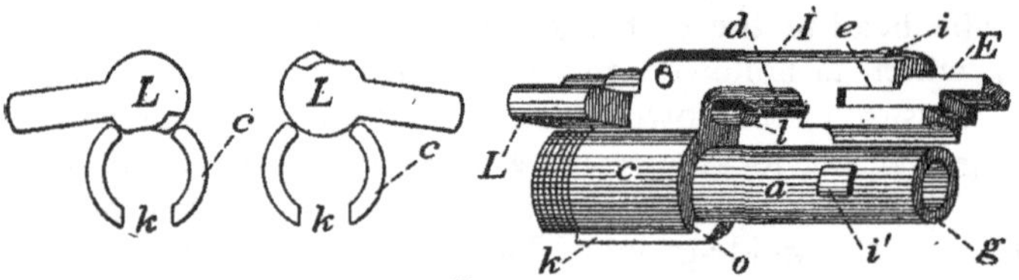

FIG. 320.

THE SLEEVE (Fig. 320) serves to connect the firing mechanism and the bolt, and carries the safety-lock and the extractor. It consists of a single piece of metal, the lower parts, *a* and *c*, of which are hollow cylinders, and the upper part, *I*, an arm of the shape indicated. The firing-pin and cocking-piece *Fk*, Figs. 323 and 336, pass through the cylinders *ac*, and the slot *k*, in *c*, is to allow the hammer to move forward and back in firing and cocking. The stud *i'* enters the notch *i'*, Fig. 319, in the interior of the bolt, and locks the bolt and sleeve on the interior. The arm *I* has a cut, *d*, which embraces the circular collar *d*, Fig. 319, on the bolt, and locks the sleeve and bolt together on the exterior.

This locking on the interior and exterior, allows the bolt to turn, while the sleeve remains fixed, but does not allow longitudinal motion of the bolt without the sleeve. The fork *e* in front carries the extractor *E*, which is secured in it by a screw *i*. The shoulder *g* forms a seat for the spiral main spring *G*, Fig. 323, which surrounds the firing-pin *F*. The safety-lock is shown in rear. It consists of a thumb-piece, *L*, and spindle, *l*. Its object is, first, to lock the bolt in the firing position, so that the breech cannot open accidentally; and second, to lock the firing-pin in the full-cock position, so that the piece cannot be accidentally discharged. Both these operations are performed at the same time, as follows:

The spindle *l* is half cut away, as explained in the case of the magazine cut-off. The thumb-piece *L* is cut away also, so that when turned to the left the cut-away portion forms part of the interior surface of the cylinder *c*, through which the cocking-piece can pass freely. When in this position also, the cut-away part of the spindle *l* forms part of the interior surface of the cut *d*, and the collar *d*, Fig. 319, on the bolt can rotate freely in this cut. When the thumb-piece *L* is turned to the right, the cut-away part no longer forms a portion of the interior surface of the cylinder *c*, and hence the cocking-piece cannot enter this cylinder to move forward. At the same time, the rounded part of the spindle *l*, turns down into the cut *d*, and its front end enters the notch *l*, Fig. 319, in the bolt, thus preventing the latter from rotating. The rear end of the bolt fits against the shoulder *o*, so that the exterior of the cylinder *c* and the exterior of the bolt form one continuous surface.

The cylinder *a* enters the interior of the bolt.

THE EXTRACTOR (Fig. 321) is a long bar, *E*, having a hook, *o*, at its extremity which engages over the rim of the cartridge. It is attached at the other extremity to the sleeve *I*, as already explained. *e* is a projection which rests against a corresponding shoulder, *e*, on the guide-rib *r*, of the bolt, Fig. 319. *q* is a recess fitting against a shoulder, *r*, in the receiver, Fig. 313, in the locked position, and *p* a spring which acts against the lower surface of *q* on the receiver, to force the extractor down over the rim of the cartridge. The extractor has a slight motion around the screw *i*, which is necessary in dismounting the mechanism.

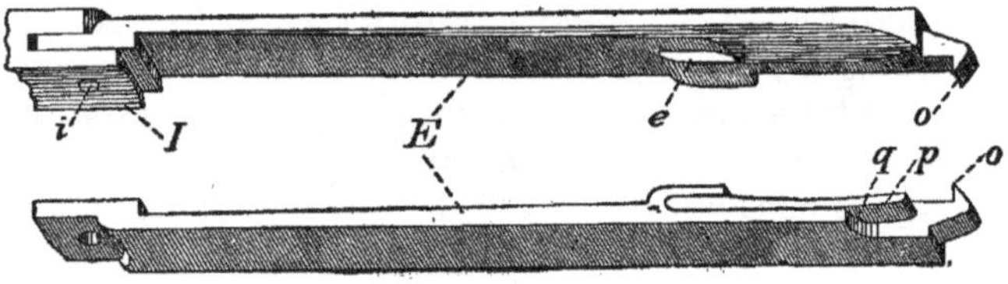

FIG. 321.

322. Firing Mechanism—General Principles—Conditions for Good Firing Mechanism—Firing Mechanism of Springfield Rifle.

The ammunition used with all modern small arms contains a central primer of mercuric fulminate, which is ignited by a blow from the firing mechanism.

The method generally adopted is to transmit this blow through the medium of a firing-pin passing through the breech mechanism. The pin may be acted on directly by a spring which forces it forward when the trigger is pulled, or it may be acted on by a hammer which is itself acted on directly by a spring. The first method is that now generally adopted for bolt-guns, the second being used in the Springfield and some older forms of breech-loaders.

CONDITIONS TO BE FULFILLED BY A GOOD FIRING MECHANISM.—A good firing mechanism should fulfil the following conditions:

1. It should ignite the primer with certainty and without piercing it.

2. It should not be hard to operate, as this causes loss of aim; nor too easy, as this leads to accidents.

3. Its parts should be simple, strong, few in number, easily dismounted and assembled, and interchangeable.

4. It should be cocked automatically by the opening or closing of the breech. The reasons why cocking on opening is preferred have been given.

5. It should have a safety-device to prevent accidental

discharge when the piece is carried loaded, and should show clearly whether it is cocked or not.

FIRING MECHANISM OF SPRINGFIELD RIFLE.—The principal parts of this mechanism are (Fig. 322):

The firing pin *F*;
The hammer *b*;
The tumbler *c*;
The main spring *d*;
The sear *e* and sear-spring *e'*;
The trigger *f*.

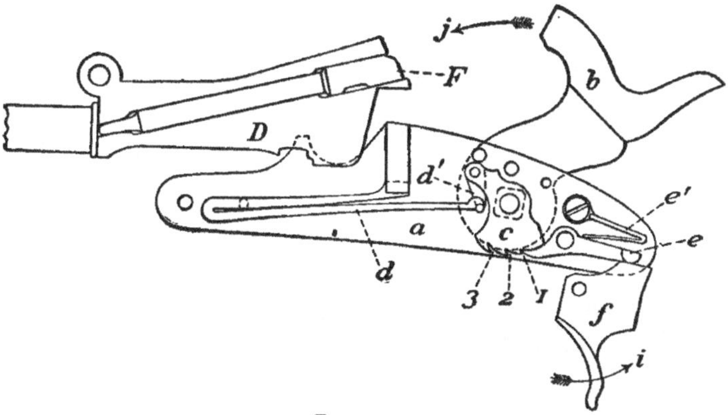

FIG. 322.

The firing-pin *F* passes through the breech-block *D*, and projects to the rear. The hammer *b* is fastened to the tumbler *c* by the tumbler-screw, and fits on a square arbor or shaft, so that the hammer and tumbler must rotate together. The tumbler has three notches: a full-cock, 1; half-cock, 2; and safety-notch, 3. The main spring *d* is attached to the tumbler by a swivel, *d'*. The sear *e* is a pivoted lever, and is acted on by the sear-spring *e'*, which forces it against the tumbler, and hence it is always ready to catch in one of the notches 1, 2, or 3. The trigger *f* is a pivoted lever, and acts against a projection on the long arm of the sear. The tumbler and sear are held in place, and supported on the inside, by a piece called the bridle, not shown in the figure.

All the parts except the firing-pin and trigger are assembled to a flat plate, *a*, called the lock-plate, which is secured to the right side of the gun by two screws.

ACTION OF THE MECHANISM.—When the trigger *f* is pulled in the direction of the arrow *i*, the sear is withdrawn from its notch in the tumbler, and the action of the main spring causes the hammer and tumbler to rotate in the direction of the arrow *j*, striking a blow upon the firing-pin, which is thus driven forward against the primer, exploding it.

323. Firing Mechanism of the Cal. .30.

The principal parts of this mechanism are (Fig. 323):

The firing-pin *F* and striker *F'*;
The main spring *G*;
The cocking-piece *K*;
The sear *H* and sear-spring *H'*;
The trigger *T*.

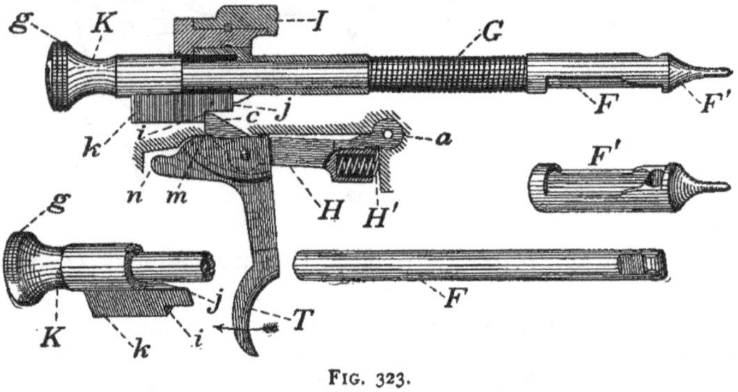

FIG. 323.

The firing-pin is composed of two parts, the body *F* and the striker *F'*, the method of connection of the two being indicated in the figure. The striker can thus be readily replaced when broken, or removed to permit the replacing of a broken main spring. The firing-pin passes through the sleeve *I* and the bolt, as already explained, and the main spring *G* rests between the rear shoulder of the striker *F'* and the front shoulder *g* on the sleeve, Fig. 320. It is evident that when the firing-pin is drawn back, the main spring will be compressed, since the sleeve *I* is fixed with reference to the pin. The cocking-piece *K* is screwed to the rear end of the firing-pin. The part *g* is roughened to give a firm hold to the fingers in cocking. The part *k* carries the full-cock notch *i* and the wedge-shaped cocking-nose *j*, all these being in one piece. The cocking-nose *j* engages in the notch *k*, Fig. 319, in the rear end of the bolt.

The sear *H* is a piece of metal of the shape shown, hinged at *a* to the receiver, and its nose *c*, passing through a cut in the bottom of the receiver, engages in the full-cock notch *i*. It is constantly pressed upward into this notch by the spiral sear-spring *H'*, one end of which bears against the receiver, and the other against the sear. The trigger *T* is pivoted to the sear. At the rear it rests against the bottom of the receiver, at the point *m*, and after the trigger is pulled slightly the point *n* comes into bearing against the bottom of the receiver, the point *m* losing contact.

ACTION OF MECHANISM. — Suppose the piece fired. When the bolt is rotated to the left by its operating-handle, the inclined side of the notch *k*, Fig. 319, in the bolt, presses against the corresponding side of *j*, Fig. 323, and forces the cocking-piece, firing-pin, and striker backward, till the end of *j* rests against a notch on the rear end of the bolt. The firing-pin is thus drawn back and cocked, the main spring *G* being compressed.

After the introduction of the cartridge into the receiver, the bolt is pushed forward and rotated to the right, to lock it.

In moving forward, the full-cock notch catches against the sear *H*, and the firing-pin is now held back by the sear. When the bolt is rotated to the right for locking, the slight forward motion completes the compression of the main spring, and the rotation brings the nose *j* and part *k* of the cocking-piece opposite the notch *k*, Fig. 319, in the bolt. When the trigger is pulled in the direction of the arrow, the nose *c* of the sear is lowered slowly at first out of the full-cock notch *i*. As the pull of the trigger continues, the point *m* loses its bearing against the bottom of the receiver, and the point *n* comes into bearing. The lever-arm being thus increased, the nose of the sear at the last moment, moves quickly out of its notch *i*; the firing-pin is forced forward under the action of the main spring *G*, and the cartridge is fired. If the bolt is not properly locked, the notch *k*, Fig. 319, on the bolt will not be opposite the nose *j*, Fig. 323, of the cocking-piece, and the latter either cannot move forward sufficiently far to allow the firing-pin to strike the primer, or, if the bolt is nearly locked, the forward motion of the cocking-piece will cause *j* to strike the inclined side of the notch *k*, Fig. 319, and thus cause the bolt to rotate to the right, and completely lock it. This is an additional safety-device.

324. Sights—General Principles—Position.

There are two sights for small arms:

1. An adjustable rear sight·
2. A fixed front sight.

REAR SIGHTS.—A good rear sight for a military arm should be simple, solid, easy of repair, graduated so that the marks can be readily seen, and so arranged that when the rear-sight notch is set to any particular graduation it will not be displaced by the shock of firing, or by any other means, except when changed by the firer. The form of the notch should be such as to enable the target to be seen easily.

It should be out of the way and well protected when not in use, to avoid being broken, and it should contain all the graduations required up to the extreme effective range of the arm. The requisite of simplicity, excludes peep and telescope sights, except for selected marksmen, and the flat trajectories of the small-calibre rifles have greatly simplified the rear sights by reducing their heights, and doing away with corrections for wind, and to some extent for drift.

Elevations are marked in ranges and not in degrees, as the ammunition is invariable.

The rear sight generally consists of a leaf which is hinged to a base, the latter being screwed to the barrel of the gun.

The base carries a flat spring which bears against the lower edge of the leaf and keeps it upright when in use, or folded down against the base when not required.

The leaf is graduated in ranges (yards) and carries a slide which has a notch cut in it forming the rear-sight notch.

This slide moves along the leaf, and is clamped at any graduation and held firmly in place.

FRONT SIGHT.—The front sight is generally a stud set at the muzzle, and terminates in a thin edge parallel to the axis of the bore. It should be sufficiently strong, to prevent injury by the rough usage of service.

POSITION.—The front and rear sights are generally so placed, that the notch of the rear sight, and top of the front sight, shall be in a plane passing through the upper element of the barrel and the axis of the bore, and at as great a distance apart as possible, so as to give the longest sight-radius attainable, consistently with distinct vision of the target and the two sights. In some arms, as the Springfield, the rear sight has an arrangement for correcting for wind, and the slide is set with an inclination to the left equal to the permanent angle of drift.

325. Sights for Springfield Rifle—Sights for the Cal. .30.

SIGHTS FOR SPRINGFIELD RIFLE.—*Rear Sight.*—The principal parts are (Fig. 324):

The fixed base *A*;
The movable base and spring *B*;
The sight-leaf *C*;
The sight-leaf slide *D*.

The fixed base *A* is screwed to the barrel. The movable base *B* carries a flat spring, which bears against the lower edge of the sight-leaf *C* and keeps it vertical or folded down. This movable base rotates about the pivot *E*, and is moved by the screw *F* working in a worm on the end of *B*. The sight-leaf is thus moved to right or left, and corrections made for wind. The sight-leaf *C* carries the graduations, and is hinged to the movable base at *G*. It also carries the binding-screw *H*, which can be made to bear against the sight-leaf slide, and thus clamp it in any position.

The sight-leaf slide *D* carries the rear-sight notches 1, 2, 3, 4, and 5. No. 5 is used for ranges up to 200 yards with the leaf down.

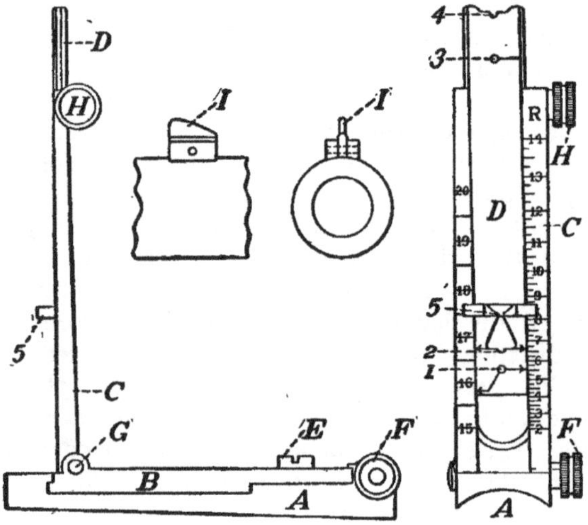

FIG. 324.

For distinction Nos. 1 and 3 will be called peep-sights, and Nos. 2 and 4 open sights.

If the peep-sights 1 and 3 are to be used, No. 1 is employed from 200 to 1350 yards, the right-hand arrow on No. 1 coinciding with the graduations.

For 1400 yards the leaf-slide is pushed down, and No. 3 is used, its mark coinciding with the graduation 14.

From this to 2000 yards the left-hand arrow on No. 1 coincides with the left-hand graduations, the sight being taken through No. 3.

If the open sights 2 and 4 are to be used, No. 2 is employed from 200 to 1400 yards. The leaf-slide is then pushed down, and No. 4 is used, the left-hand arrow on No. 2 coinciding with the left-hand graduations. The corrections for wind are marked on the fixed base, and the leaf-slide is set at an inclination to the left equal to the permanent angle of drift.

Front Sight.—The front sight, Fig. 324, consists of a thin blade, *I*, set in a stud.

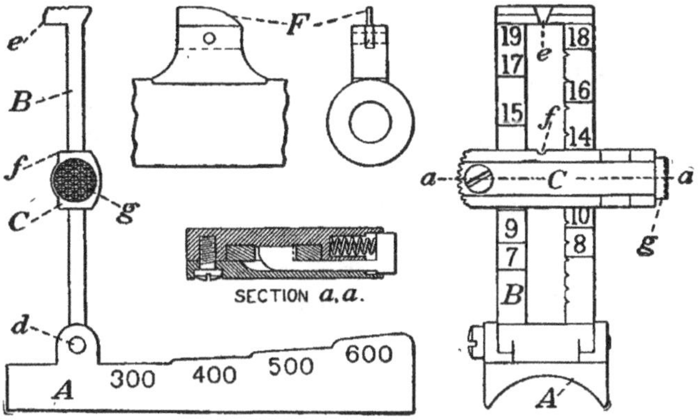

FIG. 325.

SIGHTS FOR CAL. .30.—*Rear Sight.*—The principal parts are (Fig. 325):

The fixed base and spring *A*;
The leaf *B*;
The leaf-slide *C*.

The fixed base is screwed to the barrel, and carries the flat spring which bears against the lower edge of the leaf *B*, and keeps it upright or folded. The leaf *B* is hinged to the base at *d* and is graduated on both sides, beginning with 700 yards. From 300 to 600 yards the fixed base is cut in steps, and the steps marked as in the figure. For ranges up to 600 yards, the leaf is folded down, and the slide *C* rests upon the corresponding step, the sight being taken through the notch *e*. Beyond 600 yards the leaf is upright, the top surface of the slide coinciding with the graduation, and the sight is taken through the notch *f*. The slide *C* is clamped in place by a serrated piece contained in the slide, and acted on by a spiral spring which presses it constantly against notches on the right-hand inner edge of the leaf.

A pressure on the button *g* releases this catch, and the slide may be moved up or down. The arrangement is shown in section in the figure.

The notch *e* is set slightly to the left of the axis of the bore and corrects for drift at 500 yards. For distances less than 500 yards this correction is too great, and for those greater than 500 yards too small.

The notch *f* is similarly set to the left of the axis of the bore and corrects for drift at 1000 yards. For distances less than this the correction is too great, and for distances greater than 1000 yards, too small. The notches on the leaf and slide are exactly similar to those on the sight of the Springfield rifle.

FRONT SIGHT.—This is shown at *F*. It resembles the Springfield front sight, and differs from it principally in being higher.

326. The Stock and Mountings.

THE STOCK.—The stock is that part of the arm to which all the other parts are assembled, and it serves—

1. To facilitate the handling and pointing, to diminish the shock of recoil by distributing it over a greater area at the shoulder, and to stiffen and protect the barrel.

2. In some magazine arms to carry the supply of ammunition required for rapid fire, and in all arms to carry certain parts necessary for the service, security, or preservation of the piece.

For lightness it is made of wood, and for strength this wood should be of close grain, and it should be well seasoned to prevent warping; walnut is generally used.

It is widened at the butt, *a*, Fig. 326, to distribute the

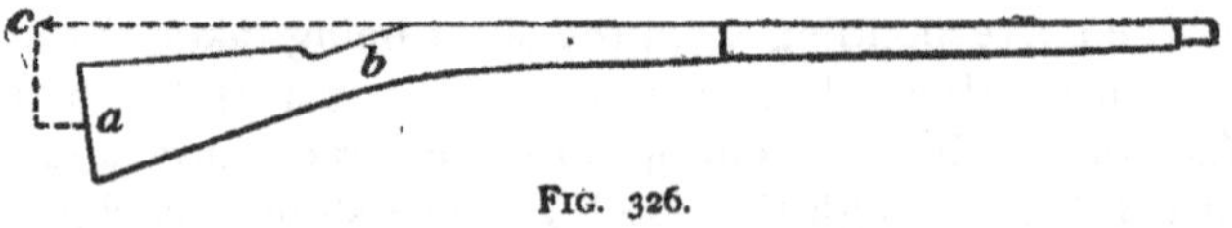

FIG. 326.

pressure of recoil over the shoulder, and is crooked at the small of the stock, *b*, for convenience of aiming. This crook

must not be excessive, as it causes rotation about the shoulder, with a lever-arm, *ac*, which increases with the crook, and may cause inconvenience. It also weakens the stock, since the wood is cut across the grain at this point.

In some cases, to avoid this weakening and give room for the mechanism, the stock is made in two distinct parts, called respectively the butt-stock and the tip-stock. With smokeless powders, the barrel becomes excessively heated; and to prevent contact with the hand, the upper part of the barrel, at the rear, is also covered with wood.

Under the head of mountings are included the wiping-rod, the bands and tip, the butt-plate, trigger-guard, swivels, and the various pins and screws by which these parts are secured to the gun.

The wiping-rod is screwed into its seat for a short distance, to avoid displacement in firing.

The bands assemble the barrel to the stock, and are not continuous, but split, Fig. 327, and are assembled by a screw, *a*. They can thus be readily adjusted to the stock and barrel, and any undue binding prevented, as this might cause vibration in firing.

The butt-plate and trigger-guard preserve the butt and trigger respectively from wear and accident, and the swivels are used for stacking and to support the gun-sling.

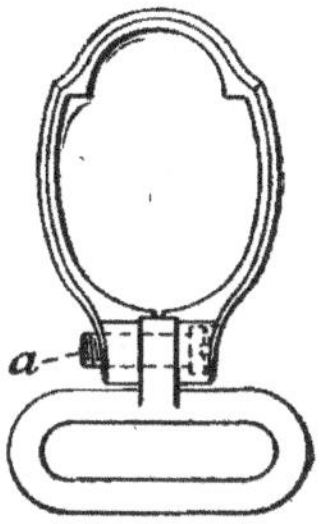

FIG. 327.

REPEATING OR MAGAZINE ARMS.

327. Advantages of Magazine Arms—Definition—Conditions to be fulfilled by a Good Magazine Arm.

ADVANTAGES.—In ordinary breech-loaders three operations are necessary to prepare for firing:

1. Open the breech;
2. Insert the cartridge;
3. Close the breech.

The longest of these is the time required to take the cartridge from the box or belt and insert it in the gun.

The rapidity of fire is therefore greatly increased if the cartridges can be automatically introduced, and the three operations reduced to two, viz., opening and closing the breech.

As, however, the cartridges so introduced must be carried by the piece in some convenient receptacle, it is evident that the number so carried is limited, and hence automatic introduction of the cartridges cannot be continuous beyond a few shots.

The advantage of a magazine arm is, then, that it can furnish a certain number of shots in a very small interval of time; and in order to make use of this advantage it is necessary to be able to reserve this supply till needed, and ordinarily to use the arm as a single-loader.

This leads to the conclusion that a good magazine arm should be also a good single-loader, and should fire as rapidly, when used as such, as any good single-loader, since the arm is used habitually as such, and only in emergencies as a magazine gun.

DEFINITION.—A magazine or repeating arm may then be defined as one in which a certain number of cartridges are introduced in succession, automatically and rapidly, into the receiver.

CONDITIONS TO BE FULFILLED BY A GOOD MAGAZINE GUN.—A good magazine gun should fulfil the following conditions:

1. When used as a single-loader it should fire as rapidly as any ordinary single-loader.

2. When used as a magazine arm it should give the greatest possible rapidity of fire, and the mechanism should work well and regularly when rapidly used.

3. It should allow the change from single-loader to magazine fire to be readily and quickly made, and the device for making this change should be readily seen, so that no mistake can be made; and so placed that it cannot be accidentally operated.

4. It should afford an easy and rapid method of recharging the magazine.

5. The cartridges in the magazine must not be damaged

or deformed by firing or by handling the piece, or be liable to explode by the shock of discharge.

6. The weight of the piece with magazine and cartridges must not exceed that usually allowed for small arms.

7. It should afford a ready view of the number of cartridges in the magazine at all times, so that the supply may not be exhausted before they are needed.

328. Classification of Repeating Mechanism.—The Detachable Magazine—Lee Magazine—Advantages and Disadvantages of Detachable Magazines.

CLASSIFICATION.—The repeating mechanism includes the magazine in which the supply of cartridges is carried, and the means by which the supply is fed to the receiver. As these are generally combined, it is customary to classify the mechanisms according to the magazines used.

Magazines are classified into—

1. Detachable;
2. Fixed.

DETACHABLE MAGAZINES.—The detachable magazines are generally box-shaped, and are placed in rear of the barrel and below the receiver. They are called detachable because they may be readily detached from the gun. They

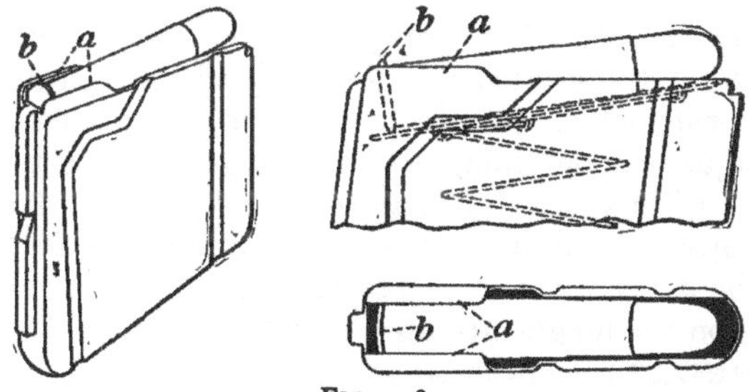

FIG. 328.

are generally made of thin sheet steel, and contain a spring or some device by which the cartridges are constantly pressed upward toward the receiver. The top of the magazine is folded over for a short distance at the rear, *a*, Fig. 328, and these folds hold the cartridges in place against the action of the spring. When the bolt is drawn to the rear over the cartridge, a portion, *b*, of its rim projects beyond the folds. As the bolt is pushed forward, it strikes the rim *b* and pushes the cartridge forward beyond the folds, out of the magazine and into the receiver and chamber. This device in some similar form is found in all box magazines.

LEE MAGAZINE.—The Lee magazine is a good example of this system, and is shown in Fig. 329 with its method of attachment to the gun.

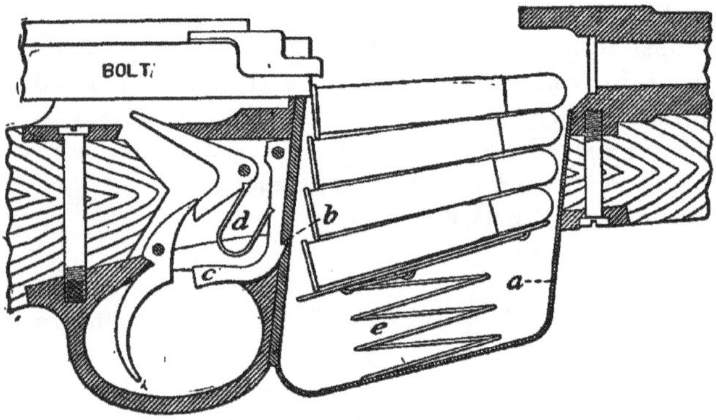

FIG. 329.

a is the magazine, *b* a projection on its rear end, *c* the magazine-catch operated by the U-shaped sear-spring *d*, *e* a folded spring which pushes the cartridges upward. The filled magazine, containing five cartridges, is inserted from below, in a cut made for it in the stock and receiver, and pushed upward till the magazine-catch *c* snaps under the projection *b*. When the magazine is empty it is released by pressing on the magazine-catch *c*, and withdrawn.

ADVANTAGES.—These are:

1. Since they can be used only when fixed in place, it is always evident whether or not the magazine supply is being employed. This does not, however, apply to those which are lowered vertically to cut off the supply.

2. A number of these can be carried loaded, and as they

can be inserted quickly, the rapid fire can be kept up con tinuously for some time.

DISADVANTAGES.—1. The magazine has considerable weight and adds to the burden carried by the soldier. This additional weight could otherwise be utilized to increase the number of cartridges carried.

2. The magazines are apt to be thrown away or lost when empty, and when lost the gun cannot be used as a magazine arm.

3. The cut through the bottom of the receiver is inconvenient when the gun is used as a single-loader, and when the magazine is attached it must generally be used; that is, the gun cannot be used with facility as a single-loader.

To remedy the inconvenience of the cut in the receiver, the Lee gun has a spring slide which closes this cut as soon as the magazine is withdrawn, and the insertion of the magazine pushes this slide out of the way. In some guns of this type a cut-off is arranged by which the magazine is lowered vertically, so that the cartridges will be out of the way of the bolt when the gun is to be used as a single-loader.

329. Fixed Magazines—Classification—Description of the Jarmann Magazine.

CLASSIFICATION.—Fixed magazines may be classified according to their shape into—

1. Tubular;
2. Box.

Tubular magazines may be placed either under and parallel to the barrel, in the front part of the stock; or in the butt, in rear of the barrel.

Box magazines are placed in rear of the barrel, and directly in front of the trigger-guard.

TUBULAR MAGAZINE UNDER BARREL—JARMANN MAGAZINE.—The Jarmann magazine-gun, formerly used in Norway, may be taken as an example of the tubular magazine under the barrel. In Fig. 330, *a* is the barrel; *b* the magazine; *c* the spiral spring which forces the cartridges to the rear; *d* the piston attached to the end of the spring; *e* the carrier which lifts the cartridges from the mouth of the magazine to the receiver; *f* the carrier-spring, which is fork-shaped and rests on two pins, *g*, pressing the carrier down; *h* the pin by which the carrier is attached to the receiver, and around

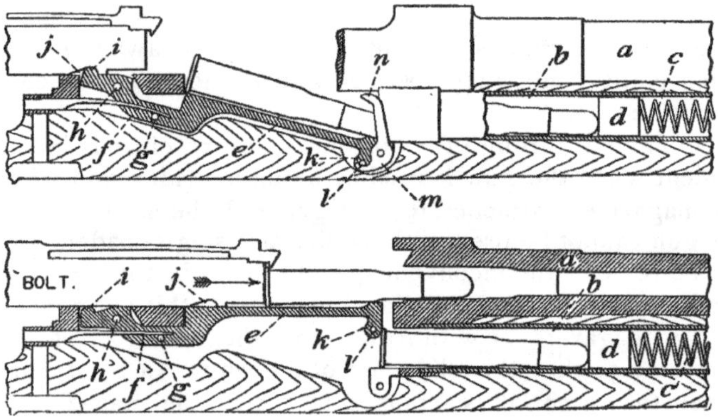

FIG. 330.

which it rotates; *i* a shoulder on the rear end of the carrier, projecting above the bottom of the receiver when the carrier is down; *j* a corresponding shoulder and recess on the lower side of the front of the bolt, which, as the bolt is drawn back allows the carrier first to drop under the action of the spring *f*, and immediately afterward, as the bolt moves further back, strikes against *i* and raises the carrier; *k* is a projection from the lower front end of the carrier, whose object is to work the cartridge-stop and hold back the next cartridge in the magazine. This projection *k*, carries a pin, *l*, which works the cartridge-stop.

ACTION OF MECHANISM.—Suppose the piece fired. The breech is then closed by the bolt, the carrier *e* is held up in the position shown in the lower figure by the bearing of the lug *i* on the bottom of the bolt. The carrier thus forms a part of the bottom of the receiver. The cartridges are held back in the magazine against the action of the spring *c*, by the projection *k* of the carrier, bearing on the head of the rear cartridge.

As the bolt is withdrawn, the carrier *e* remains in the position just described, because its lug *i* bears continually against the bottom of the bolt. When the bolt, in its backward motion, reaches a position such that the cut *j* comes over the lug *i* of the carrier, the latter is free to rotate, and moves downward under the action of its spring *f*. During its downward motion the cartridges are kept in place by the bearing of the front of the carrier *e* against the head of the rear cartridge. At the last moment of the rotation of the carrier, when it occupies the position shown in the upper figure, this bearing is removed, and the rear cartridge is forced by the spring *c*, out of the magazine, and on the carrier. As soon as this is done, the downward rotation of the carrier being completed, the pin *l*, on the projection *k*, strikes the cartridge-stop *m*, and causes it to rise and partly close the mouth of the magazine, thus preventing the other cartridges from being forced out. All this occurs while the cut *j* in the bolt is over the lug *i* of the carrier. As the bolt is pulled backward still further, the shoulder of *j* strikes the shoulder of *i* and raises the carrier *e* with its cartridge quickly to the mouth of the chamber. The bolt is then pushed forward and the cartridge inserted. It will be remembered that at this time the cartridges are held back in the magazine by the cartridge-stop. To release this stop, the bolt, in moving forward, strikes the long lever *n* of the cartridge-stop, which is situated on the right-hand side of the receiver. This pushes the lever *n* forward, lowers the stop, and frees the mouth of the magazine, and under the action of the spiral spring *c*, the cartridges move forward till the head of the rear one comes into bearing against the projection *k*, which is now in the position shown in the lower figure, the bolt being closed.

The principles explained here are found in modified forms in all magazines of this type. The magazine has a cut-off by which the carrier is locked in its upward position and the gun may then be used as a single-loader.

330. Objections to Tubular Magazines under the Barrel—Advantage.

OBJECTIONS.—The objections to tubular magazines under the barrel are:

1. The cartridges lie with the primer of one against the bullet of the next, and hence the shock of discharge is liable to explode the primer, or to upset and deform the point of the bullet. With modern smokeless powders, although the bullet is not so liable to be deformed, owing to its harder jacket, it is more liable to be driven down into its case, since any excessive crimping of the bullet to the case, which would tend to prevent this, increases the pressure in the gun, by increasing the resistance to motion at the origin. If the bullet be forced down into the case, the density of loading of the charge is increased, and hence also the pressure.

2. The spiral spring which forces the cartridges into the carrier, must be long, as it has to act over a great distance. Hence it is tightly compressed at first, and its action becomes very slight on the last cartridge, and is therefore irregular.

3. When the magazine is full, the centre of gravity of the system is carried forward, and as it is emptied this centre changes.

4. The weight of the arm increases considerably when the magazine is loaded.

5. The magazine is difficult to load, as the cartridges must generally be inserted singly.

6. The state of supply of the magazine cannot be seen.

7. Unless the bolt is drawn back to its full extent, and quickly, the carrier will not work properly.

8. As the magazine-tube is thin, a slight damage to the stock may close up the tube so that it will not feed.

Its greatest advantage is the number of cartridges carried.

331. Tubular Magazine in Butt — Fixed-box Magazines — The Mannlicher Magazine.

TUBULAR MAGAZINE IN BUTT.—This was the earliest form of magazine, as seen in the Spencer rifle, which was used during the Civil War. It has been abandoned, however, because it has nearly all the disadvantages belonging to the tubular magazine under the stock, and in addition it weakens the small of the stock and does not carry a large number of cartridges. The Hotchkiss is probably the best example of this type.

FIXED-BOX MAGAZINE.—This type of magazine has been adopted by many of the foreign powers and by the United States.

THE MANNLICHER MAGAZINE.—The Mannlicher magazine may be taken as a type of this system used abroad.

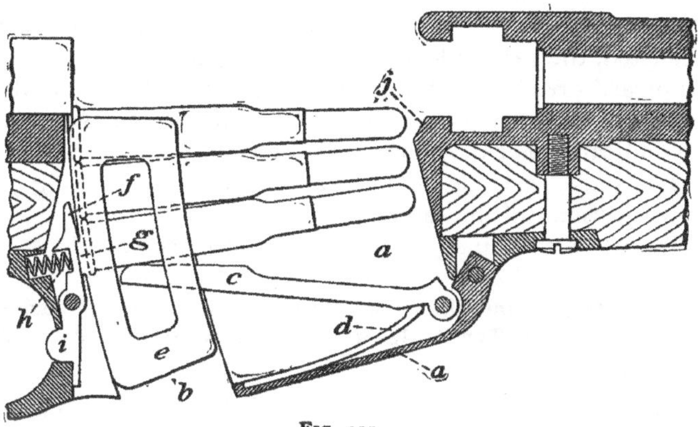

FIG. 331.

In Fig. 331, *a* is the fixed box, having the bottom open at *b*. This box is fixed in rear of the barrel and in front of the trigger-guard, and projects below the stock as in the Lee magazine. *c* is the carrier-lever; *d* the magazine-spring, which pushes the carrier-lever upward against the cartridges. The cartridges are carried in a packet, *e*, made of tin, the top and bottom edges being slightly folded over, as shown in Fig. 328.

This packet carries five cartridges, and is inserted with its cartridges, from above, through the cut in the bottom of the receiver, into the magazine *a*. It is held in place in the magazine, by the upward pressure of the carrier-lever *c* on the cartridges, which is transmitted to the packet *e* by the folded edges, and as this would push the packet out of the magazine, the catch *f*, acted on by the spiral-spring *h*, engages against a lug, *g*, on the rear end of the packet, and prevents it from rising.

ACTION OF THE MAGAZINE.—When the filled packet *e* is introduced into the magazine, it compresses the magazine-spring *d*. The packet is pushed down till the catch *f* snaps over the lug *g*. The cartridges being constantly pressed upward by *c* and *d*, when the bolt is pushed forward it strikes the exposed part of the base of the upper cartridge, and pushes it forward beyond the folded edges of the case, the surface *j* of the receiver guiding the point of the bullet upward and into the chamber. When all the cartridges are exhausted, the carrier-lever *c* and spring *d* no longer exert an upward pressure on the packet, and hence the latter falls through the opening *b* in the bottom of the receiver, and thus indicates that the supply is exhausted.

The packet may be removed at any time by pressing on the projection *i* of the catch.

To cut off the supply the packet must be removed.

332. Advantages and Disadvantages of the Fixed-box Magazine, Mannlicher Type—General Principles of the Cal. .30 Magazine.

ADVANTAGES.—The Advantages of the fixed-box magazine, Mannlicher type, are:

1. In common with all box magazines, the cartridges lie so that the spring which moves them acts in the direction of their least dimension, and therefore the great length and irregularity of its action, as in the tubular magazine, are avoided.

2. The cartridges are not liable to explode, or to be deformed in handling and firing.

3. The centre of gravity of the system is not changed.

4. The magazine is easily charged.

5. The packets are light, and hence do not add much useless weight to the soldier's burden, and they are cheap and may be thrown away.

6. The exhaustion of the magazine is automatically indicated.

7. The magazine cannot be lost, and is not liable to damage.

DISADVANTAGES.—The objections are:

1. When the packet is in place the arm cannot be used as a single-loader without great care; and when the packet is withdrawn the bottom of the receiver is not solid, which is an inconvenience.

2. The cartridges must be carried in packets, and cannot be placed in the magazine without them. The packet therefore becomes a necessary part of the mechanism, just as the magazine in the Lee gun.

GENERAL PRINCIPLES OF THE CAL. .30 MAGAZINE.—The magazine of the cal. .30 remedies the last two defects.

In this gun the magazine is a fixed box, but, instead of projecting vertically below the receiver, it is partly horizontal and partly inclined at the left side, where it opens into the receiver. This gives a solid bottom to the receiver, so that no inconvenience results from using the gun as a single-loader, and the cartridges may be inserted into the magazine either singly by hand or quickly from a packet carrying five cartridges. This latter arrangement is called a quick-loader, and is used in many other box-magazine guns, in which the packet does not form an essential part of the mechanism, as with the Lee, and the Lee-Speed or English gun.

333. Description of the Magazine for the Cal. .30.

This magazine is situated under the receiver, in front of the trigger-guard and in rear of the barrel. It consists, Figs. 332 and 333, of the horizontal part *m* (see also Fig. 313) and the curved part *O*.

The horizontal part is in one piece with the receiver, and the curved part is formed by the separate piece *O*, of the proper shape, secured to the left side of the receiver.

The opening *z'*, through which the cartridges pass to the receiver, is narrowed at the rear (see Fig. 313) corresponding to the folding down of the sides of the magazines in the

Lee and Mannlicher, and for the same purpose—that is, to hold the cartridges in the magazine, against the action of

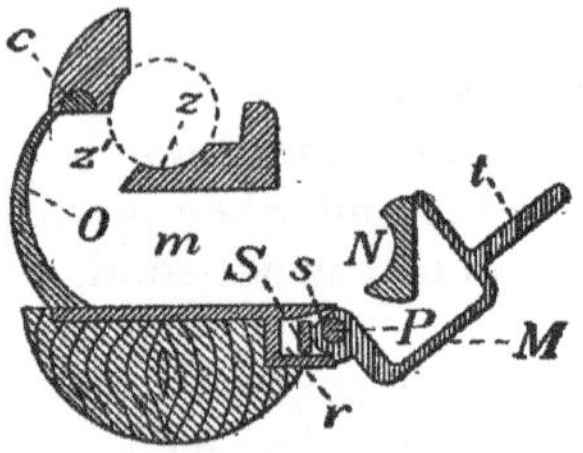

FIG. 332.

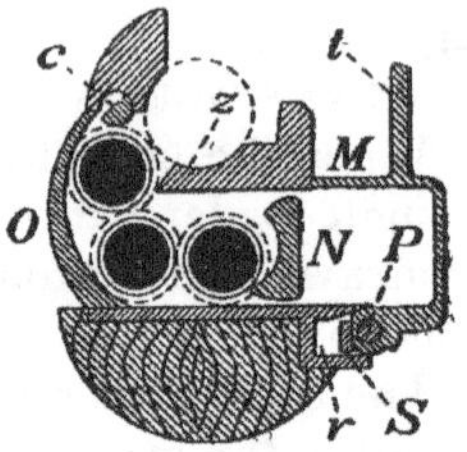

FIG. 333.

the carrier-lever and spring; and as with other box-magazines, the cartridge must be pushed forward by the bolt, beyond this narrow part, before it can rise into the receiver. The bottom of the receiver at *z* is left solid, with the advantages noted.

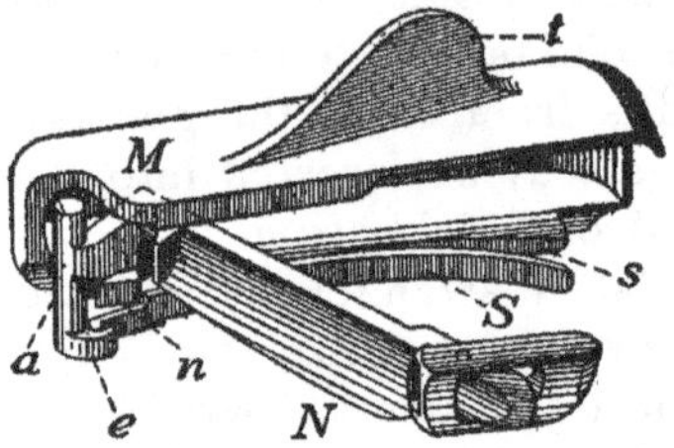

FIG. 334.

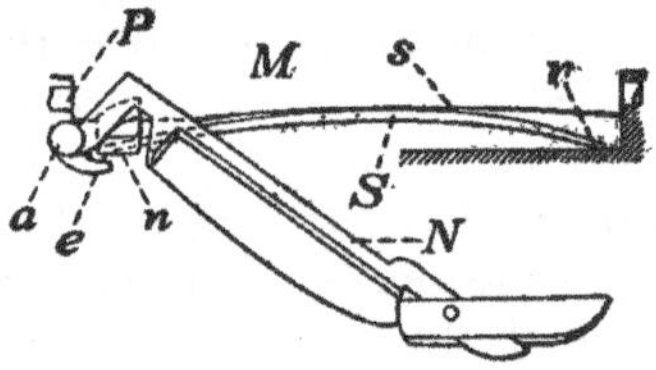

FIG. 335.

The cartridges are pushed to the left, and into the receiver, by the carrier-lever *N*, Figs. 334 and 335. This lever has a spindle, *a*, and lug, *e*, at its forward end and below, against which rests a flat bow-spring, *S*. This spring *S* is carried in a small recess, *r*, below the receiver (see also Figs. 332 and 333). The rear end of *S* bears against the side of this recess, *r*, the front end against the lug *e* on the carrier-lever, and against the back of the spring rests the lower edge, *s*, of the gate *M* which opens and closes the mouth of the magazine.

The spring *S* is thus under constant compression, due to the action of the gate, and it forces the carrier-lever to the left in the magazine. When the gate is opened, as in Fig. 332, a lug, *n*, attached to it, Figs. 334, 335, presses against the carrier-lever and forces it to the right against the action of the spring *S*, thus leaving the magazine clear for loading. The spring *S* acts also to keep the gate *M* open or closed, just as the flat spring on the rear sight keeps the sight-leaf up or down.

The gate *M*, Fig. 334, has a thumb-piece, *t*, by which it is opened and closed, and it is assembled to the side of the receiver by the pin *P*, Figs. 332 and 333. The cut-off is shown at *c*, same figure.

ACTION OF MECHANISM.—To fill the magazine the gate *M* is opened by the thumb-piece *t*, and the five cartridges inserted by hand singly, or all at once from a quick-loader, the carrier-lever *N* being held back as explained.

When the gate is closed the carrier-lever comes into action, and forces the cartridges to the left and upward. The first four cartridges, by their shape, act to push each other upward as soon as they reach the curved part of the receiver. The fifth cartridge is pushed upward by the shape of the upper side of the follower-lever. If the cut-off

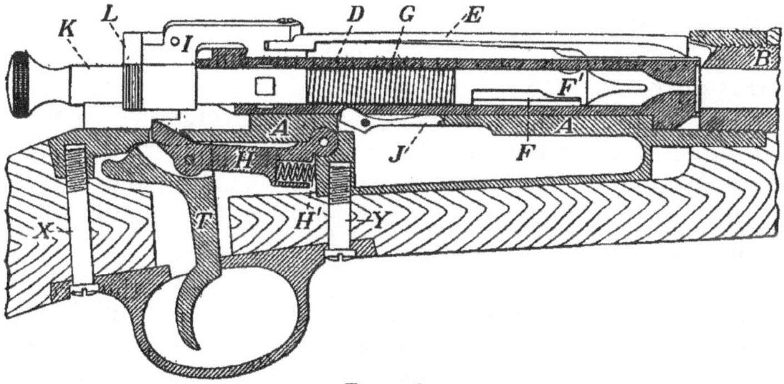

FIG. 336.

c is used, it projects as explained into the opening *z'* of the magazine, Fig. 333, and forces the upper cartridge down sufficiently far to be out of the way of the bolt.

The assembled mechanism of the cal. .30 rifle is shown in the firing position in Fig. 336.

334. Revolvers—Classification—Conditions to be fulfilled by a Good Service Revolver—Remarks.

CLASSIFICATION.—The revolver is a weapon for personal defence at short distances, not exceeding 50 or 60 yards, and is employed principally by mounted troops and by officers.

They are divided into three classes:

1. Single-action revolvers, or those which must be cocked by hand before each fire.

2. Self-cocking revolvers, in which by pulling the trigger the cocking and firing are accomplished, till all the chambers are emptied.

3. Double-action revolvers, which act as single-action or as self-cocking at the will of the firer.

CONDITIONS TO BE FULFILLED BY A GOOD SERVICE REVOLVER.—1. Its mechanism should be simple, strong, easy to dismount and assemble, and interchangeable.

2. Each chamber which is to be fired should stop exactly in the prolongation of the barrel.

3. The mechanism should work well whether the revolver be fired rapidly or slowly; this rapid or slow fire being readily employed at will.

4. The bullet should possess sufficient energy to stop a man at 50 or 60 yards.

5. It should be easy to load, and the empty cases should be readily extracted.

REMARKS.—The principal points with reference to the working of a revolver are, to insure the stoppage of rotation of the cylinder in the proper position, to obtain rapidity of fire when needed and slow fire at other times, and to be able to load and extract easily.

The stoppage of rotation of the cylinder at the proper time has been successfully accomplished. The rapid and slow firing at will requires a revolver of the third class, or a double-action revolver. The single-action revolver gives the slow fire, but will not fire rapidly, while the self-cocking revolver, although giving a rapid fire, does not give an accurate slow fire, because of the prolonged pull upon the trigger, which is apt to derange the aim. The loading and extraction are readily accomplished in the service revolver, all the empty cases being ejected automatically at the same instant, and the chambers can be loaded from a quick-loader. The condition of certainly stopping a man at 50 yards has caused the retention of larger calibres for the revolver than for the rifle, those of the revolver being 0.38 and 0.45 inch. The revolvers adopted for the U. S. service are the Colt's double-action cal. .38 and the cal. .45. The mechanism of the revolver is best explained from a model.

AMMUNITION.

335. History—Advantages and Disadvantages of Metallic Cartridge-cases—Folded-head Cartridges.

HISTORY.—Small arms were originally loaded by pouring in the powder and then inserting the ball, each of these being carried separately and loose.

The powder-charge was next wrapped in paper, and hence the name cartridges, from *charta*, paper; later the powder and ball were united in one package, and the operation of loading was preceded by the tearing open of the paper containing the powder, pouring it into the barrel, and then inserting the ball. These arrangements were used with muzzle-loaders, and continued up to the Civil War.

With the introduction of breech-loaders, a change in the cartridge became necessary. The gas from the powder escaped through the opening of the breech, occasioning loss of force, and it also clogged the firing mechanism.

To obviate the defect of the escape of gas through the opening of the breech, various devices were provided, such as the De Bange pad in the French Chassepot rifle, a rubber packing-ring, etc. These devices prevented the escape in the direction indicated as long as they were uninjured by the gas, but did not prevent it from penetrating into the firing mechanism, which was soon clogged.

To avoid the expense of manufacture, and the increase of weight, which the use of the metallic case entailed, and also to avoid the difficulties of extraction, combustible cartridge-cases were used with the early breech-loaders.

But the objections already stated caused them to be abandoned and led to the adoption of the metallic case.

ADVANTAGES AND DISADVANTAGES OF THE METALLIC CASE.—The metallic case presents the following advantages: the escape of gas is entirely prevented; the powder is well protected against shock and moisture; the components of the cartridge—powder, primer, and bullet—are complete and invariable; the dimensions of the cartridge-case are exact, and there is no difficulty in loading.

The disadvantages are, the increase of weight of the cartridge and the expense of fabrication. The first is greatly reduced by the use of smokeless powder and the reduction of calibre, and the second by improved processes of manufacture, by which all the parts are rapidly and cheaply made by machinery.

FOLDED-HEAD CARTRIDGES.—The earliest metallic cartridges were made of copper, with a folded head (Fig. 337), the fulminate by which the charge was fired being contained in the fold *a*.

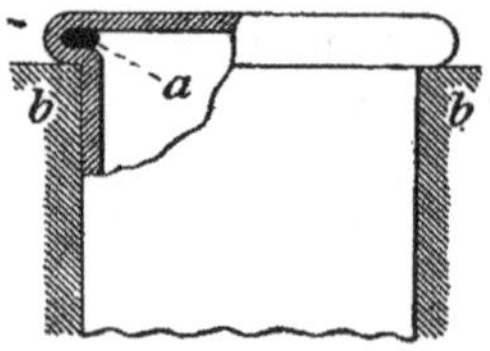

FIG. 337.

These are called rim-fire cartridges. The objections to them are:

1. The fulminate is exposed to shocks, which may cause accidental discharge in handling.

2. The charge of fulminate is larger than necessary to produce discharge, and hence tends to rupture the head of the shell at the fold.

3. The fulminate is not evenly distributed; and as the firing was produced by a blow of the hammer on the rim, if this blow fell where there was no composition a miss-fire would result.

4. The head of the case is not supported by the walls of the chamber *b* at the fold, and hence, due to this cause and to the excess of fulminate, the head was liable to shear off.

The principal advantage is that, as it was generally used in arms with tubular magazines, there was little danger of explosion by the shock of firing, since the point of the bullet did not rest against the primer in the magazine.

336. Folded-head Cup-anvil Cartridge—Solid-head Cartridge.

FOLDED-HEAD CUP-ANVIL CARTRIDGE.—To remedy the shearing of the head of the cartridge, due to non-support by the walls of the chamber, and also the defects of the rim fire, the folded-head cup-anvil cartridge, with the fulminate at the centre of the head, was devised. Cartridges with central primers are called centre-fire cartridges.

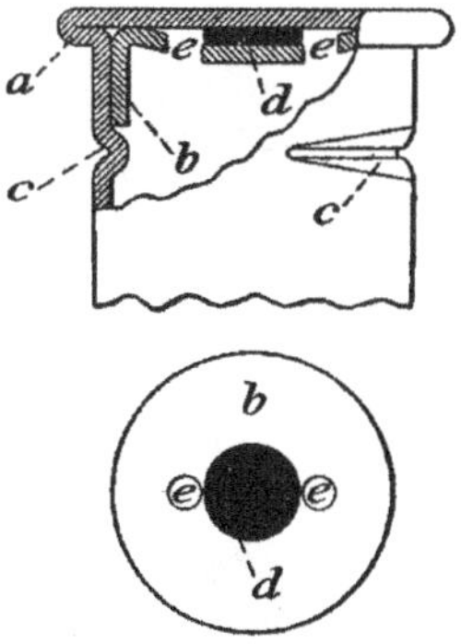

FIG. 338.

This cartridge is shown in Fig. 338. In order to prevent the action of the gas upon the fold *a*, a gas-check cup *b* was inserted in the head of the case. When the gas expanded, its pressure was exerted upon the cup, and the fold *a* protected.

In the rim-fire cartridge, the blow of the hammer upon the fold of the head was resisted by the wall of the chamber, which thus prevented the fold from yielding, and the effect of the blow was transmitted to the fulminate. In this case the wall of the chamber acted as an anvil.

When the fulminate was placed at the centre of the head it was necessary to provide an anvil as before, to resist the blow of the firing-pin, and this anvil was furnished by the cup, *b*. As this cup performed both functions, as above explained, it was called a cup anvil. The anvil is a common feature of all primers, and is necessary for the reason stated.

The cup anvil was held in place by two crimps, *c*, in the case. The fulminate is at *d*, and *ee* are the two vents through the cup anvil, by which the flame from the fulminate escapes to the charge, the fulminate being fired by the blow from the firing-pin.

The copper of which the folded-head cup-anvil cartridge is made is objectionable, as it is too soft, and the extractor frequently cuts through it and fails to withdraw the case, and also, owing to its lack of elasticity, it is apt to stick in the chamber after firing. For these reasons brass is preferable, as it is harder and more elastic, but owing to its hardness the head cannot be folded.

SOLID-HEAD CARTRIDGES.—For these and other reasons the folded-head cup-anvil cartridge of copper was abandoned, and the solid-head brass cartridge adopted in its place.

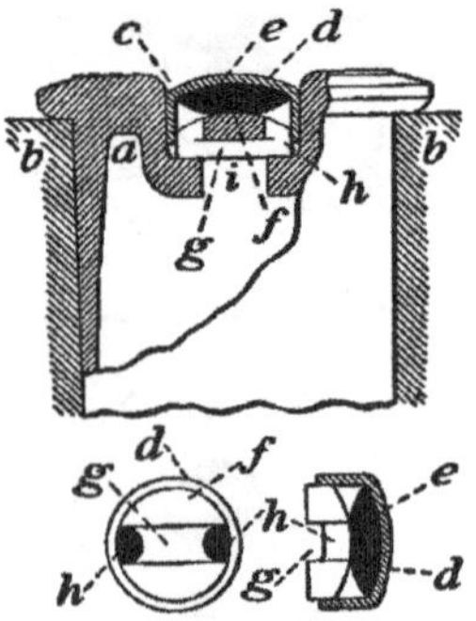

FIG. 339.

In this cartridge (Fig. 339) the head is formed by pressure, causing the metal to flow into the shape shown.

The danger of shearing at the head is avoided, since the bottom of the case, *a*, inside, is in front of the shearing plane, *b*.

The primer is inserted from the outside in a pocket, *c*, in the base of the cartridge, and consists of the cup *d*, the fulminate *e*, and the anvil *f*. The anvil is made of copper, and has a cut, *g*, across the bottom, and two vertical holes, *h*, at the sides, communicating with *g*, through which the flame from the fulminate passes to the charge by the vent *i* in the primer pocket.

The principal defect of this cartridge is that its walls are thinner in front than in rear, and when the cartridge is fired the front part expands more than that in rear, and is liable to stick. Hence if there is any movement of the case to the rear, it is apt to tear apart.

337. Components of the Cartridge—The Bullet—The Powder.

THE BULLET.—For the older arms the bullet was made of pure lead cast in a mold. As improvements were made, the soft lead was found to shear in the grooves and cause "leading."

Hence the lead was hardened by alloying it with some other metal, such as tin or antimony. The Springfield bullet is an alloy of lead and tin. With the introduction of small calibres, high velocities, and rapid twist, the hardened lead did not present sufficient resistance to shearing, and the jacketed bullet was adopted. The jacket at present used is cupro-nickelled steel.

The casting of the bullet also was objectionable, since the density was not uniform, and the centre of gravity frequently did not coincide with the longer axis, giving rise to irregularity in flight. For this reason the bullet was formed by compression between dies, and more uniform density thus obtained. In the Springfield bullet (Fig. 340) three grooves or cannelures are formed at the rear end, and these are filled with vegetable wax for lubrication of the bore. With the cal. .30 bullet (Fig. 341) it is found that these are not necessary, and they have been abandoned.

The shape of the bullet is cylindro-ogival for the Springfield and cal. .30. The two bullets are shown (Figs. 340 and 341). The weights are: Springfield, 500 grains; cal. .30, 220 grains.

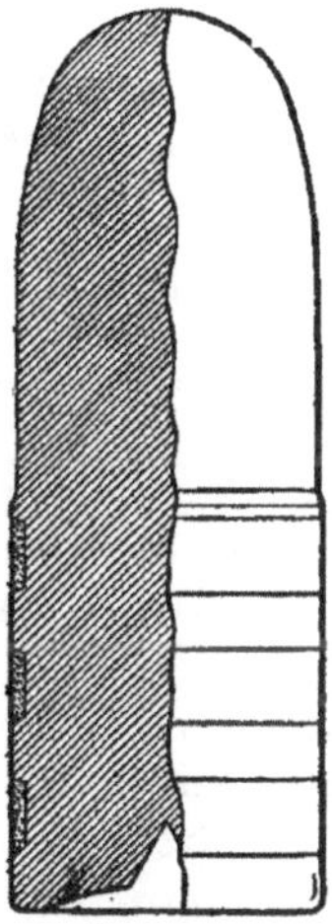

FIG. 340.

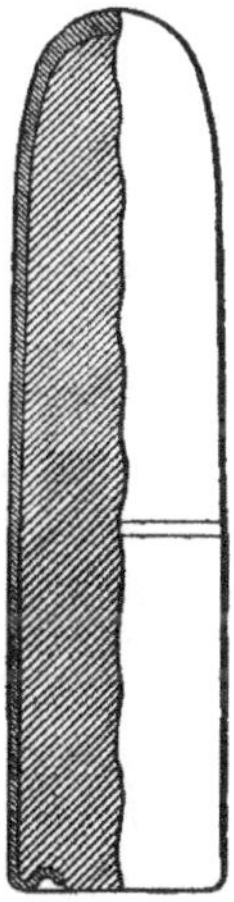

FIG. 341.

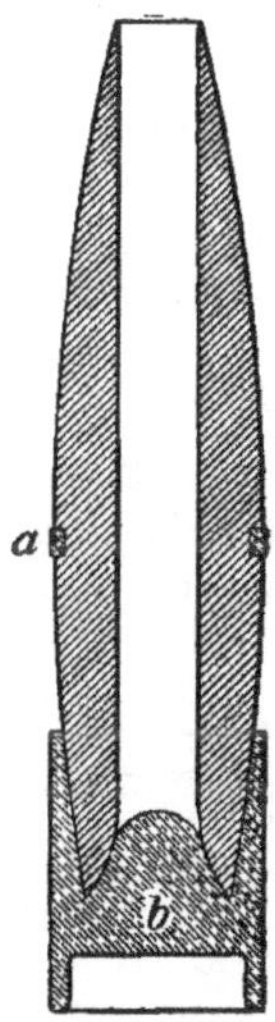

FIG. 342.

Recent experiments have been made with a tubular steel bullet, the Krnka-Hebler.

This bullet (Fig. 342) is made entirely of steel except the narrow copper rotating band, *a*, around the middle. On the rear end is a sabot, *b*, made of vulcanized fibre and weighing only a few grains; its object being to receive the pressure of the powder-gas over a greater extent of surface, and to act as a gas-check, preventing the escape of gas along the sides of the projectile. When the projectile leaves the bore, the pressure of the air upon the front surface of the sabot causes it to drop off. The central hole allows the air to pass through freely in flight, and thus diminishes the retardation owing to the decreased surface presented. An initial velocity of 3000 ft.-secs. has been obtained with this projectile, with a pressure of 46,000 lbs. per square inch in the gun.

THE POWDER.—Small-arms powder is used in the Springfield rifle, weight 70 grains. It is measured automatically in a loading-machine, and after insertion in the case is slightly compressed before the bullet is put in. The charge of smokeless powder varies from 32 to 43 grains, 37 of Wetteren or 43 of Peyton powder being at present used. As it is important with smokeless powders to secure the same amount for each charge in order to regulate the pressure, these charges are weighed, and to insure greater regularity the powder is sieved before loading.

338. Components of the Cartridge—The Case—The Primer.

THE CASE.—The general features of the cartridge-case have already been described. The rim is for the purpose of extraction, limits the forward motion of the cartridge in loading, and fixes its position in the chamber. In certain box magazines the rim occasions some difficulty if care is not exercised in placing the cartridge in the magazine. For example, in Fig. 343, if the cartridges occupy the position there shown, it is evident that the top cartridge is held by the rim of the one next below, and consequently the bolt cannot without difficulty push it out of the magazine. To remedy this it has been proposed to make rimless cartridges, as in Fig. 344, the notch *a* being for the purpose of extraction. The ob-

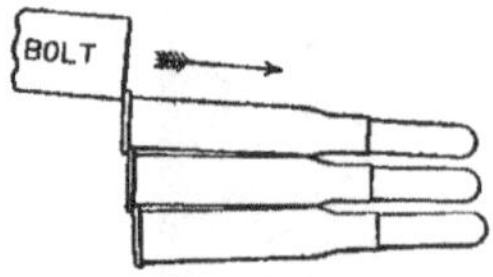

FIG. 343.

jections to these cartridges are that their position in the chamber is regulated by the bearing of the shoulder *b* against a corresponding shoulder in the forward part of the chamber, and as it is impossible to make the length *cd* exactly the same for all cartridges and all chambers, short cartridge will have too much play, and the head of the case, moving to the rear on firing, while the front sticks in the chamber, for the reasons already explained, will cause rupture of the case and fouling of the mechanism. In addition, the operation of the extractor is not always certain.

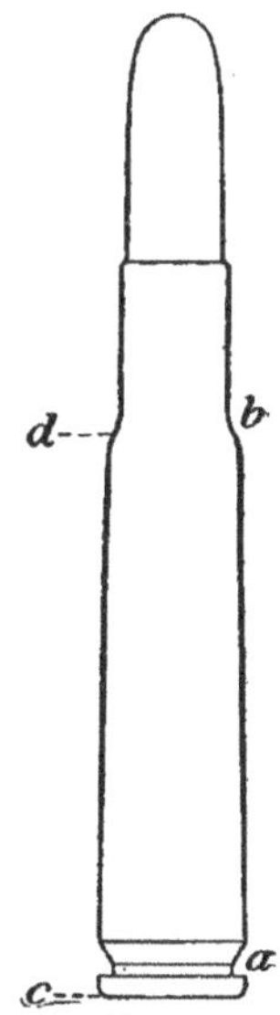

Fig. 344.

The case of the cal. .30 cartridge is made bottle-shaped to reduce its length as much as possible in order to give a longer path for the gas to work over and to diminish wave action, and the exterior is conical to facilitate extraction in both the Springfield and the cal. .30.

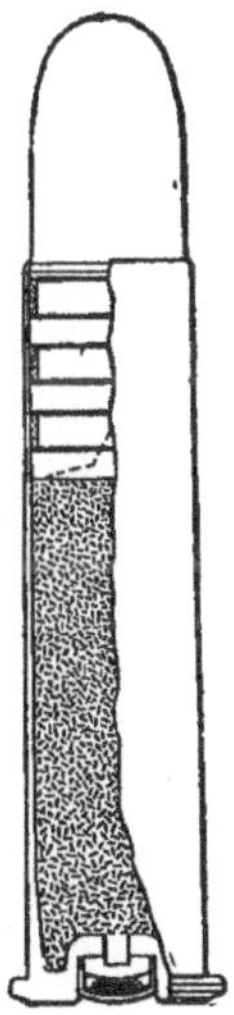
Fig. 345.

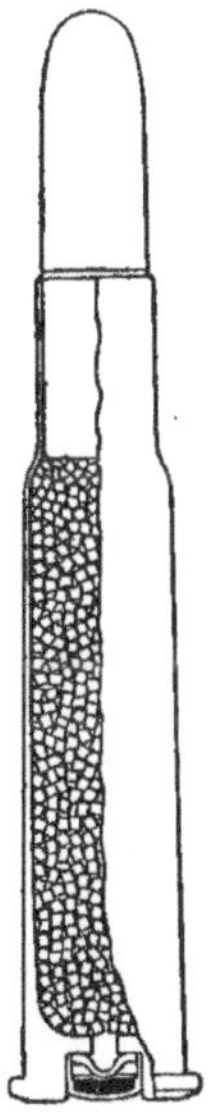
Fig. 346.

The contact of the old nitrate powders with the brass case caused deterioration of the latter, and it was tinned to prevent this.

The effect of the new smokeless powders on the case is not known, but the cases are tinned as with the old powders.

The Primer.—Its composition has already been explained.

With the new smokeless powders some difficulty has occurred in igniting the charge, and the strength of the primer has been increased, with successful results as regards ignition. The primer is, for safety, sunken below the level of the head of the cartridge. The old Springfield cartridge can be reloaded; the new smokeless-powder cartridge cannot be, except at the arsenals, on account of the danger from excessive crimping, and the high pressures that result from an error in weight of charge, or from inserting the bullet too far into the case, and also because of difficulty in providing reloading-tools for the small calibre.

The complete cartridges for the Springfield and the cal. .30 rifles are shown in Figs. 345 and 346.

CHAPTER X.

MACHINE AND RAPID-FIRE GUNS.

MACHINE GUNS.

339. Definition — Object — Advantages — Disadvantages — Requirements — Kinds of Machine Guns

DEFINITION.—A machine gun is one that is loaded and fired by machinery.

OBJECT.—Its object is to deliver a rapid and continuous fire, and thereby enable a few men to produce the same effect as a larger number armed with the ordinary rifle.

ADVANTAGES.—Owing to the great volume of fire delivered by them, they may be employed at decisive moments of the attack, and to defend defiles, ditches of permanent works, and for their moral effect against mobs and in street-fighting. In the Naval Service they are mounted in the tops, to sweep the enemy's decks, drive the cannoneers from their guns, and to repel boarders.

DISADVANTAGES.—These guns are mounted on wheeled carriages, and transported like artillery. They therefore appear naturally to belong to that arm of the service. But as they generally fire small-arm ammunition, they are unable to cope with field-artillery at the fighting range of the latter.

This limits the use of machine guns in the attack to the infantry arm, and it is generally considered that for purposes of attack they are inferior to infantry, as they do not possess its mobility.

For defence the guns are very useful in holding positions where they may be permanently mounted, and fired in a fixed direction.

REQUIREMENTS.—In order that a machine gun may fulfil its functions, it should, when once pointed in a given direction, retain that direction unchanged by the shock of firing; which requires that there shall be no recoil, and that the mechanical operations of loading and firing shall not interfere with the aim or working of the gun. These conditions are very difficult to fulfil, and are perhaps more nearly attained in the Maxim automatic gun than in any other. The gun must also be capable of being rapidly directed upon any particular object, and of having this direction quickly changed. This is accomplished in most of them by mounting the gun on a fork placed upon the carriage, by which means a quick motion in azimuth, and also around the axis of the trunnions, can be given.

The gun must be managed by a small number of men, who should be well protected by shields from the enemy's fire. The ammunition should in general be the same as that used by the infantry, to avoid complication, and a large supply must be carried by the gun in a condition ready for feeding, in order to insure rapid and continuous fire. The mechanism should not be liable to jam or get out of order when the gun is fired rapidly; it should be simple and easily repaired, and the gun should not become heated to such an extent as to interfere with firing.

KINDS OF MACHINE GUNS.—The principal machine guns which have been tried in the United States are

The Gatling;
The Gardner;
The Maxim;
The Hotchkiss revolving cannon.

Of these, the Gatling and Hotchkiss revolving cannon have been adopted for service.

340. The Gatling Gun—Parts—Barrels—Cylinders—Casing.

PARTS.—The gun consists, Fig. 347, of a number of breech-loading rifled barrels, *B*, usually ten, placed around and parallel to a central shaft, *S*. These barrels are held in place by two barrel-plates, *P*, *P'*, called respectively the front and rear barrel-plates. The barrel-plates are circular disks assembled to the central shaft *S*, and having holes in them through which the barrels pass. The barrels and central shaft thus form a cylinder, of which the barrels are the elements and the central shaft the axis.

In rear of the barrels is the carrier-block *C*, which is a metal cylinder attached to the central shaft *S*. On the surface of this cylinder are grooves forming extensions of the barrels. These grooves receive the cartridges from the

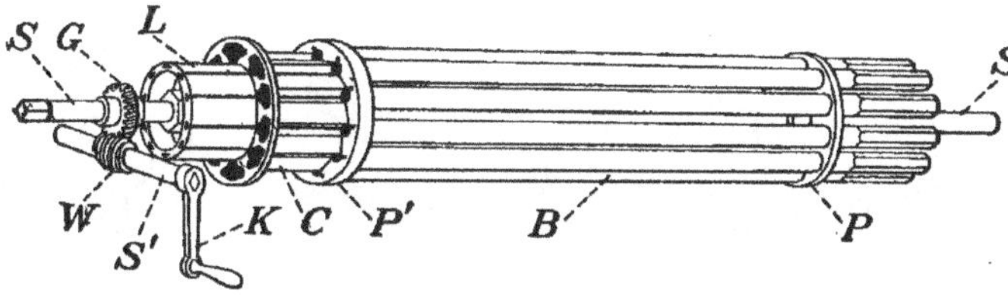

FIG. 347.

feed, and guide them while they are being pushed into the barrels by the bolts, and they also guide the empty shells while they are being withdrawn from the barrels after firing.

The outer edges of these grooves have projections which act to feed the cartridges, as will be explained.

In rear of the carrier-block is the lock-cylinder *L*, a second metal cylinder attached to the central shaft *S*, the surface of which forms guides in which slide backward and forward, the bolts by which the breech is opened and closed, and the cartridges fired.

On the rear end of the central shaft is a worm-gear, *G*, in which works a worm, *W*, on the transverse crank-shaft *S'*. By attaching the crank *K* directly to the rear end of the central shaft *S*, a rapid fire is obtained; when attached as shown, the fire is comparatively slow.

CASING.—The central shaft, with barrels and mechanism, is mounted in a frame, the mechanism being covered by a bronze casing which protects it from dust. The shaft *S* is journalled in this frame and casing in front and rear, so that the shaft, barrels, carrier-block, and lock-cylinder revolve independently of them.

The trunnions are attached to the exterior of the frame, and the gun is mounted on a fork attached to the carriage.

The fork has a motion in azimuth, and hence the direction may be quickly changed without moving the carriage as before explained.

341. The Gatling Gun—Parts—The Bolts—The Cam-groove—Action of Mechanism.

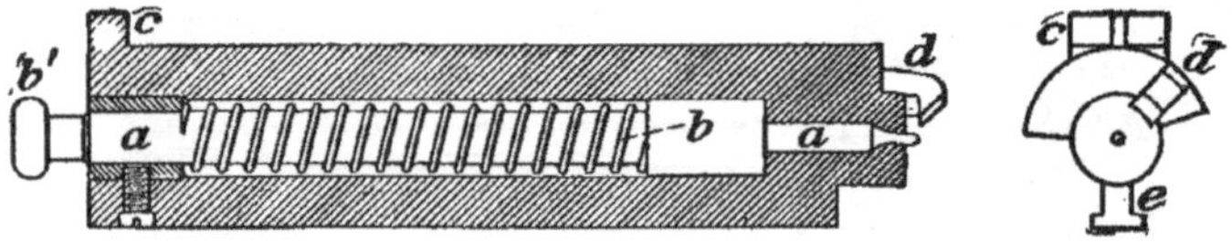

FIG. 348.

THE BOLTS.—There is one bolt for each barrel. Each bolt consists, Fig. 348, of a hollow cylinder, through which passes the firing-pin *a*, surrounded by its spiral main spring, *b*. The firing-pin terminates in rear in a head, *b'*, which is used in cocking and firing. Each bolt has a lug, *c*, projecting from its rear end. This lug fits into a groove in the casing, and is the means by which the forward and backward motion is communicated to the bolts during the rotation of the barrels. Each bolt acts with reference to its own barrel, like the bolt in the cal. .30 rifle, opening, closing, and locking the breech. The extractor, *d*, engages over the rim of the cartridge before firing, and by the backward motion of the bolt extracts the empty case from its barrel. *e* is the guide-rib which fits in a corresponding groove in the lock-cylinder, and guides the bolt in its forward-and-back motion.

THE CAM-GROOVE.—The rear part of the cylindrical bronze casing surrounding the lock-cylinder contains a groove, called the cam-groove, which may be regarded as formed by the intersection of the interior of the cylindrical casing by a plane, *cd*, oblique to the axis, as in Fig. 349.

This gives an ellipse, the upper and lower ends of which,

at *c* and *d*, are cut off by two planes perpendicular to the axis of the cylinder. Hence the sides *cd* of the groove are arcs of an ellipse, and the ends *a* and *b*, arcs of circles, with their planes perpendicular to the axis of the cylinder. The arc *b* is at a distance in rear of the barrels equal to the length of a bolt, and the arc *a* at a distance equal to the length of the bolt plus that of the cartridge, with a small allowance for play added.

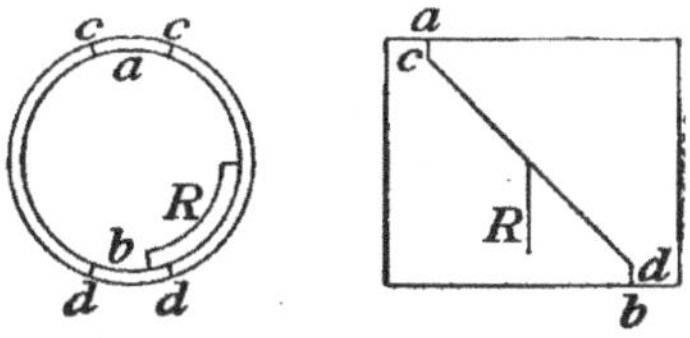

FIG. 349.

ACTION OF THE MECHANISM.—When the crank *K*, Fig. 347, is rotated, it causes the central shaft, with the barrels, carrier-block, and lock-cylinder, to rotate in the casing.

The bolts, being held by the guides in the surface of the lock-cylinder, also rotate with the barrels and other parts. But by the bearing of the lugs *c*, Fig. 348, of the bolts, in the elliptical groove *cd*, Fig. 349, in the breech-casing, the bolts on the right-hand side are forced to move forward toward the barrels, and those on the left to move backward.

Fig. 350 shows a development of the cam-groove, barrels. and firing mechanism; *cd* being the development of the right-hand side of the elliptical groove *cd*, Fig. 349, and *c'd'* that of the left-hand side of the same groove, while *cc'* and *dd'* are the developments of the circular arcs *b* and *a*, Fig. 349, respectively.

When the lugs *c* of the bolts, Fig. 348, in this rotation, reach the part *dd'*, called the "loading flat," the cartridges drop from the feed into the grooves in the carrier-block, in front of the bolts; as the rotation continues, each right-hand bolt is forced forward by the inclined groove *cd*, pushing its cartridge into the barrel. When the cartridge is completely inserted, the lug *c* of its bolt has reached the part *cc'*, called the "firing-flat," and the bolt thus closes the barrel, just as the bolt of the cal. .30 rifle.

While the bolts are thus moving forward, a groove *R* on the right-hand side of the casing, catches the head of the firing-pin and retains it, thus compressing the spiral main spring and cocking the firing pin.

This groove *R* is called the cocking-rib, and is essentially a short arc of a circle whose plane is parallel to those of *cc'* and *dd'*. This arc ends abruptly, so that when the firing-pin

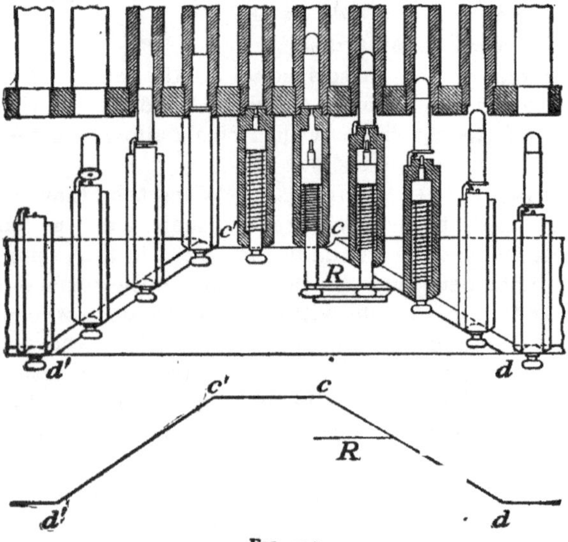

FIG. 350.

is cocked and the barrel closed, as in the figure, a continuation of the rotation causes the head of the firing-pin to pass out of the cocking-rib. The firing-pin then moves forward under the action of the spiral main spring and fires the cartridge. The rotation still continuing, the bolts are withdrawn by the left-hand groove *c'd'*, and as they move back, the empty cases are drawn out by the extractors on the bolts.

342. The Gatling Gun—Feeds—Tin Feed-case—Objections—Bruce Feed—Objections.

FEED.—The feed is the method of supplying the cartridges to the gun. Various feeds have been used with the Gatling gun, and changes have been made in them to correct defects as they developed.

TIN FEED-CASE.—The first feed consisted of a tin case, *A*, Fig. 351, of trapezoidal cross-section, containing 40 cartridges.

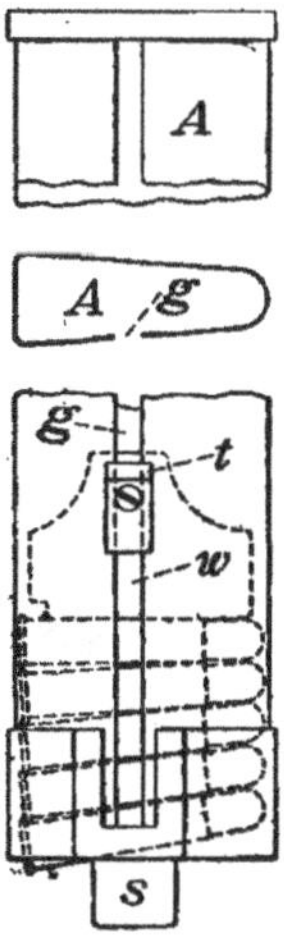

FIG. 351.

The cartridges were placed horizontally in this case, lying one above the other, and were held in the case by a spring, *s*, at the lower end, the upper end being closed.

A weight, *w*, at the upper end rested on the column of cartridges, and was provided with a projecting thumb-piece, *t*, the whole sliding along the case in a groove, *g*, cut in the side. When in use, the lower end was placed in an opening over the carrier-block, the case being in a vertical plane, and the spring *s* which closed the lower end being forced aside by the operation of inserting it.

The cartridges then fell of their own weight into the grooves in the carrier-block, and were pushed forward by the bolts. The sliding weight *w*, and thumb-piece *t*, were intended to aid the fall of the cartridges, especially at high angles of elevation.

OBJECTIONS.—The objections to this feed were, that it did not work regularly for different angles of elevation, since the component of gravity parallel to the case varied with the angle of elevation. Also, the cartridges did not always fall parallel to the guide-grooves, and hence jamming was liable to occur, and in very rapid firing the cartridges did not fall quickly enough to supply the barrels. For these reasons a second feed was introduced.

THE BRUCE FEED.—This is a gravity feed, but is intended to force the cartridges to fall parallel to the guide-grooves and hence avoid jamming. It consists, Fig. 352, of an upright bronze standard, *a*, to which is pivoted a swinging piece, *b*, having two grooves in it. Below the grooves is a fixed mouth, *c*, and below this a wheel, *d*, turning freely on its axis. When in use the feed is inserted in an opening in the breech-casing directly over the carrier-block *e*. The paper box containing the cartridges, the top being removed, is placed in the fixed standard *a*, with the heads of the cartridges to the rear. The heads of the cartridges engage in the grooves of the swinging-piece *b*, and the paper box may then be pulled off. In the position shown in the figure, the left-hand column of cartridges passes at once directly into the fixed mouth *c*, and as each cartridge strikes the wheel *d*, its weight causes the latter to revolve and present a new groove for the reception of a cartridge. The cartridges thus delivered to the wheel *d* are in turn carried round by it and deposited in the grooves in the carrier-block *e* in the proper position. As soon as the left-hand column of cartridges is exhausted, the weight of the right-hand column causes the swinging-piece *b* to rotate to the left, and thus brings the right-hand column over the fixed mouth *c*. This operation is repeated as long as the supply of cartridges is kept up.

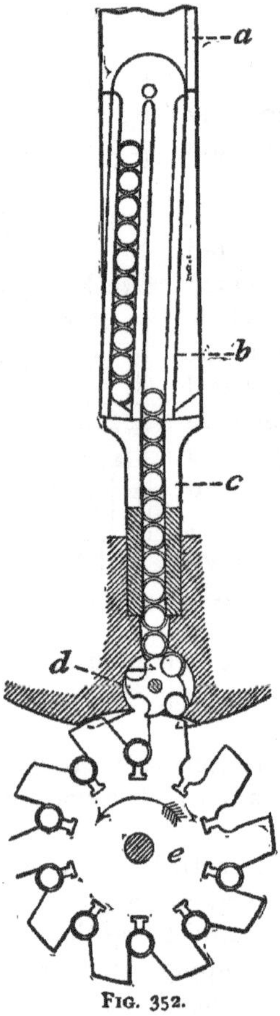

Fig. 352.

Objections.—This feed delivers the cartridges parallel to the barrels, and thus avoids jamming; but as it depends on gravity, its action is variable for different angles of elevation, as with the old tin case, and this objection has been overcome by the introduction of the Accles feed-drum.

343. The Gatling Gun—The Accles Feed—Advantages and Objections.

This feed consists (Fig. 353) of a drum, with two heads of brass, connected by a sheet-brass casing. The distance apart of the two heads is equal to the length of a cartridge.

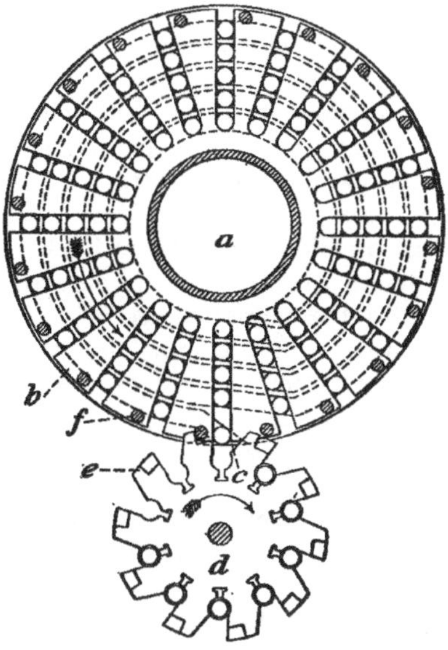

FIG. 353.

The inside of each head is grooved in a spiral form, the spiral beginning at the centre and ending at the mouth or opening of the drum. The central part, *a*, of the spiral is removed, and its place occupied by the axis or pivot of a set of radial arms, *b*, which rotate about this axis. The cartridges are inserted through the mouth *c* into the drum, the heads of the cartridges entering the spiral of one of the drum-heads, and the point of the bullet the corresponding spiral on the opposite drum-head.

The cartridges thus rest in the spirals and between the radial arms. When in use, the feed-drum is inserted in an opening in the breech-casing, directly over the carrier-block *d*, the opening *c* of the drum being down, and over the grooves of the block, and the planes of its heads at right angles to the axis of the barrels. Projections, *e*, are formed on the outer edges of the grooves of the carrier-block which engage with pins, *f*, joining the outer extremities of the radial arms of the drum, like the teeth of gear-wheels.

When the crank *k*, Fig. 347, of the gun is rotated, the lock-cylinder, barrels, etc., revolve, and the projections on the grooves of the carrier-block cause the radial arms of the drum to rotate. These arms bearing against the cartridges in the drum force them along the spirals toward the opening *c*, from which they are delivered to the grooves of the carrier-block parallel to the latter.

ADVANTAGES AND OBJECTIONS.—This drum feeds the cartridges without the aid of gravity and is hence a *positive* feed, and is independent of the angle of elevation.

As it is driven by the carrier-block, it supplies the cartridges as fast as they are needed, and thus the feed is perfectly regulated; and as the cartridges are guided by the spirals, they are delivered in the proper position to the carrier-block at all angles of elevation, and jamming is avoided.

The objections are the weight of the drum, and the extent of its surface exposed to hostile fire. A bullet striking the drum would render it useless. For these reasons a new feed has recently been introduced.

344. The Gatling Gun—Latest Improved Feed.

The latest feed introduced has a small surface exposed to fire, is independent of gravity, and can therefore be used with equal facility at any angle of elevation, and it is cheap and light.

Long strips of tin or any cheap flexible metal, Fig. 354, have tongues or slits, *a*, punched in them, one end of the tongue being left attached to the strips, and the other separated.

These tongues surround the cartridge and hold it in place on the strip. The small rectangular slots *b*, are punched completely through, and in these slots fit the rims of the cartridge-cases, thus preventing any side or longitudinal motion of the cartridges with respect to each other.

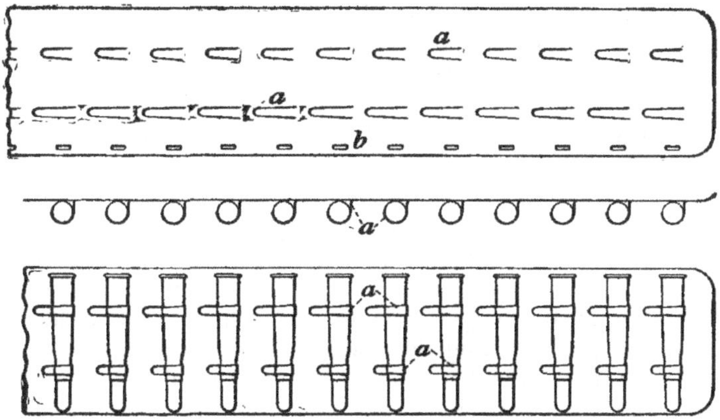

FIG. 354.

A hopper, *a*, Fig. 355, is hinged to the frame which supports the gun, just over the carrier-block, and this hopper has an opening, *b*, on the left side through which the strips holding the cartridges are fed. This opening is narrow in front and wide in rear, in order to prevent the cartridges being introduced with the wrong end to the front. Below

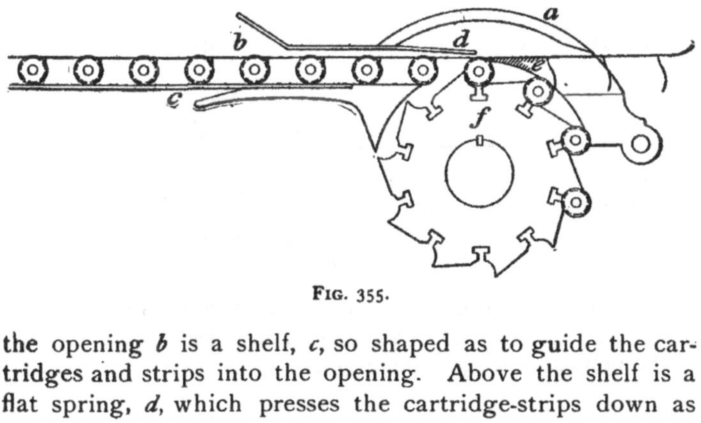

FIG. 355.

the opening *b* is a shelf, *c*, so shaped as to guide the cartridges and strips into the opening. Above the shelf is a flat spring, *d*, which presses the cartridge-strips down as they pass through the opening. A wedge, *e*, projects from the opposite side of the hopper and, acting on each cartridge in turn, forces it out of the strip, the tongues *a*, Fig. 354, bending downward into recesses provided for them. *f* is the carrier-block, provided with projections which act like the teeth of a wheel upon the cartridges, forcing the strip to the right.

When in use the strip containing the cartridges is pushed into the opening *b* of the hopper. The crank is then rotated, which causes the projections on the grooves of the carrier-block to act upon the cartridges, forcing the strip to the right through the hopper. This action brings each cartridge in succession against the point of the wedge *e*, and the action of the wedge forces the cartridge out of its hold on the strip by bending downwards the tongues *a*, Fig. 354, and the cartridge is deposited in the groove of the carrier-block, the empty strips passing out at the right.

345. The Gardner Gun—Parts—The Barrels—The Casing—The Bolts.

PARTS.—The parts of the Gardner gun are
The barrels;
The casing;
The bolts;
The firing and extracting mechanism;
The cams;
The feed-valve and guide.

BARRELS.—There are two barrels, *a*, Fig. 356, which are

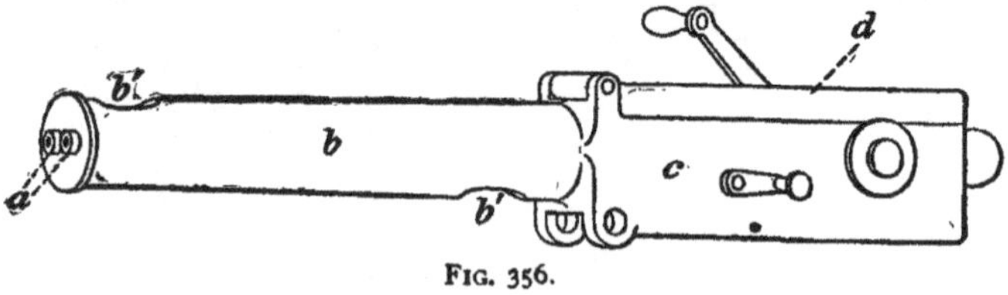

FIG. 356.

parallel and have their axes in the same horizontal plane. They have no motion, and are loaded and fired by the action of the bolts and firing mechanism.

THE CASING.—This is of bronze, the front part, *b*, being cylindrical and forming a support and protection for the barrels. Two openings, *b'*, are made in the top and bottom, to permit a current of air to circulate around the barrels and keep them cool in firing. The rear part, *c*, of the casing is box-shaped and contains the mechanism. It is closed at the top by a cover, *d*, which is hinged to the forward part of the casing, and secured by a screw on the neck of the cascable, and may be raised, thus allowing the mechanism to be seen and readily removed.

THE BOLTS.—There are two bolts of U shape, Fig. 357,

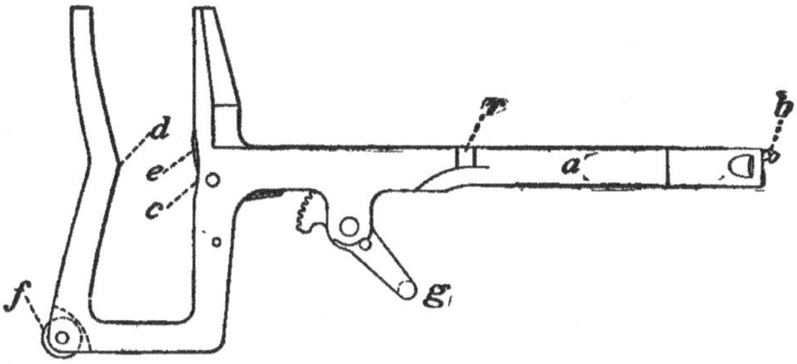

FIG. 357.

one for each barrel. One side of the U has an arm, *a*, extending at right angles to its length, and this arm forms the bolt proper, and carries the firing mechanism, and the extractor, *b*. The U-shaped part of the bolt has a recess, *c*, into which the surface of the driving cam (*a*, Fig. 360) fits, at a certain period of its rotation, and it has also a projection, *d*, which at the proper period in the rotation of the cam, bears against its exterior surface. The sear *e*, projects in the recess *c*, and is acted on by the cam, when the latter enters that recess. The bolt as a whole has a backward and forward motion in the casing, running on the truck-wheel *f*. *g* is the cocking-lever, whose action will be explained.

346. The Gardner Gun—The Firing Mechanism—Action—The Extracting Mechanism.

THE FIRING MECHANISM.—This consists (Fig. 358) of a firing-pin, *h*; spiral main spring, *i*; cocking-lever, *g*; sear, *e*; and sear-spring, *j*. The firing-pin has a collar, *k*, in front, and a toothed sleeve, *l*, in rear; the latter sliding longitudinally along the firing-pin. The firing-pin terminates in rear

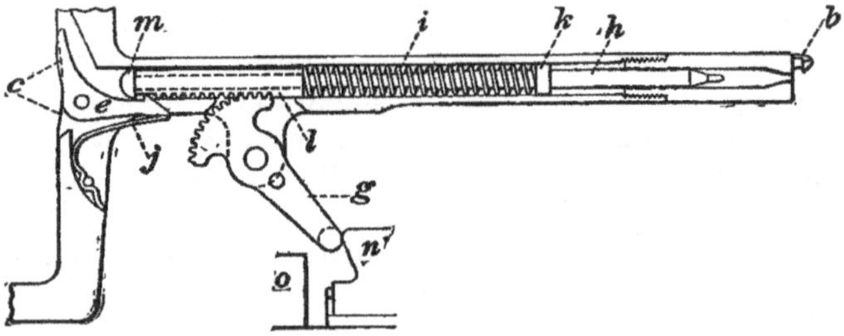

FIG. 358.

in a head, *m*, fixed to the pin; *n* is the main-spring compressor, and *o* the cocking-cam.

ACTION OF FIRING MECHANISM.—In the position represented in the figure, the firing-pin is cocked, but the main-spring is not compressed. The head, *m*, of the firing-pin, is engaged with the sear, *e*. As the bolt is moved forward by the cam, the cocking-lever, *g*, moves with it, and the lower end of this lever bears against the main-spring compressor *n*, thus causing *g* to rotate, and acting by its teeth on those of the sleeve *l*, the latter is forced forward, compressing the main spring *h*, since the firing-pin is held by the sear *e*. When the main spring, *h*, is fully compressed, the cam *a*, Fig. 360, enters the recess *c*, Fig. 358, in the bolt, and pressing on the sear *e*, releases it. The firing-pin then moves forward through the sleeve, under the action of the main spring, and fires the cartridge.

As the bolt moves backward under the action of the cam, the lower end of the cocking-lever, *g*, bears against the cocking-cam *o*, and the hring-pin, by the action of the teeth of the cocking-lever on those of the sleeve, is forced to the rear till its head, *m*, catches over the sear.

THE EXTRACTING MECHANISM.—This consists (Fig. 359) of a hook-shaped extractor, *b*, on the end of the bolt *a*, which rides over the rim of the cartridge-case *s*, as tne latter is forced home, and withdraws the empty shell as the bolt moves backward. The ejectors are two levers, *p*, pivoted to the

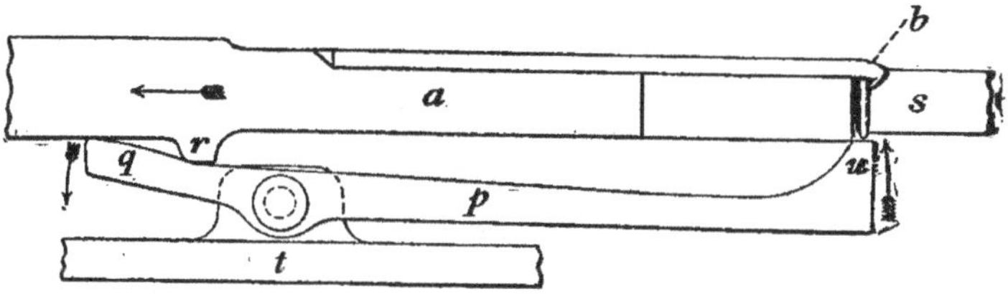

FIG. 359.

sides *t* of the casing, the rear or bent ends, *q*, of which are struck by lugs, *r*, on the bolts as they move backwards. The cartridge-case being held by the extractor *b*, the end, *u*, of the lever, strikes the case, disengages it from the extractor, and throws it out of the casing. The ejectors also act as stops, to prevent the cartridges from dropping through the openings in rear of the barrels, when fed down by the valve.

347. The Gardner Gun—The Cams—The Feed-valve and Guide.

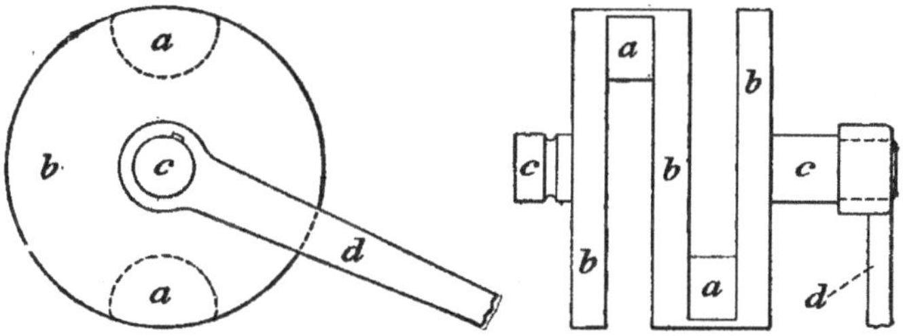

FIG. 360.

THE CAMS.—Motion is given to all the parts by two cams, *a*, Fig. 360. These cams are attached to three steel disks, *b*, at opposite extremities of a diameter, and the whole caused to rotate around the axis *c* by the crank *d*. As rotation continues, each cam acts against the U-shaped portion of its bolt, pushing it forward, and holding it motionless while firing occurs; then moving it backwards, and holding it motionless while loading occurs. Firing and loading take place when the cams are in prolongation of the axis of the arm of the bolt carrying the firing-pin, at which time the direction of the force of recoil passes through the axis *c* of the cam-disks, and hence there is no tendency to rotate. The bolts are motionless for about $\frac{1}{8}$ of a revolution of the cams, to allow for hang-fires.

THE FEED-VALVE AND GUIDE.—The feed is arranged as follows: A vertical bronze guide, *g*, Fig. 361, resembling

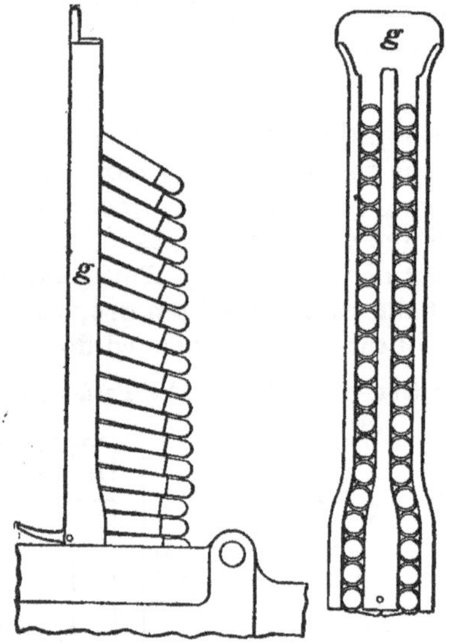

FIG. 361.

the Bruce guide already explained for the Gatling gun, but without the wheel, is fixed to the casing in rear of the barrels, and holds the cartridges as previously explained. Below this feed-guide, the casing is perforated with two holes, for the passage of the cartridges, and below these holes is the feed-valve, *v*, Figs. 362 and 363, which is a flat plate having two holes corresponding to those in the casing. This valve slides at right angles to the barrels, and is driven by a fork-shaped lever, *l*, which receives its motion from the bolts *d*, as they move forward. By this arrangement the cartridges drop from the feed-guide *g*, through the holes in the casing, and these holes are alternately opened and closed by the feed-valve *v*, as it moves to the right and left. When in the proper position, one of the holes in the feed-valve *v*, is in prolongation of the corresponding hole in the casing the other hole being closed, and the cartridge drops through, and is forced by the bolt into the chamber. The hole over the other barrel is then opened, and a cartridge drops, and is forced forward into that barrel. The details are best explained from the gun.

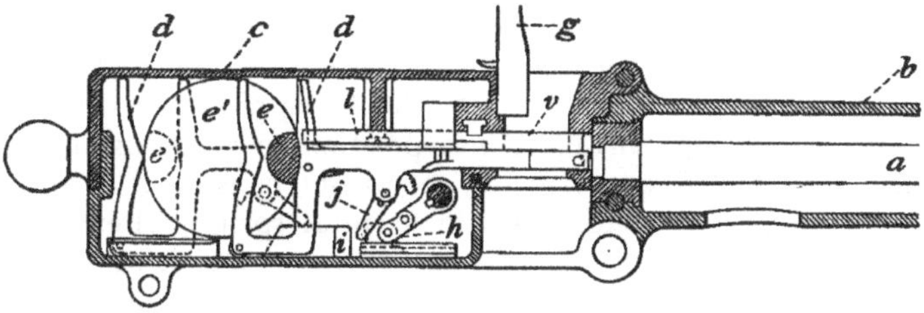

FIG. 362.

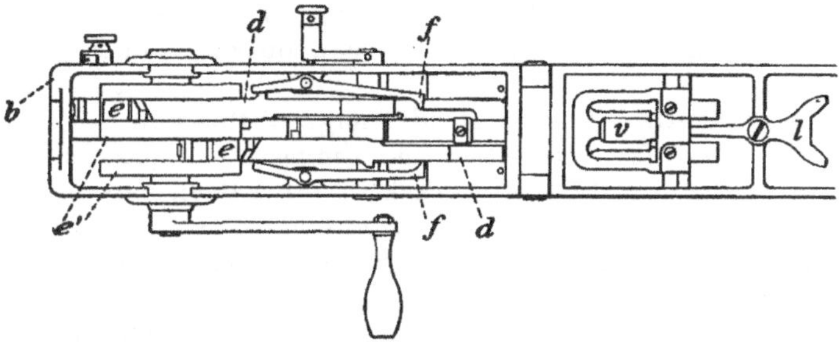

FIG. 363.

The assembled mechanism is shown in Figs. 362 and 363. *a* the barrels, *b* the casing, *c* the breech-cover, *d* the bolts, *e* the cams, *e'* the disks, *v* the feed-valve, *l* the feed-valve lever, *f* the ejectors, *h* main-spring compressor, *i* cocking-cam, *g* feed, *j* cocking-lever.

348. The Maxim Automatic Machine Gun—General Principles—Action of Mechanism—Advantage—Parts.

GENERAL PRINCIPLES.—The Maxim automatic machine gun is so constructed, that on firing a single shot, the force of the recoil is utilized for opening the breech, extracting the empty case, and effecting the various operations necessary to reload and again fire the arm, or prepare it for firing; so that after the gun has been once fired, all these operations are performed automatically, and the gun continues firing with great rapidity so long as the trigger remains pulled, and the supply of cartridges lasts.

ACTION OF MECHANISM.—The breech mechanism is operated by hand to insert the first cartridge in the barrel, and the trigger is then pulled. The pressure of the powder-gas on the breech-block, causes the latter, with the barrel, to recoil. During this recoil, the breech is opened, the empty cartridge case extracted, the firing-pin cocked, and a loaded cartridge brought into position to be thrust into the chamber. The energy of recoil not consumed in the above operation is stored up in a spiral spring, which by its reaction causes the barrel to return to the firing position, forces the loaded cartridge into the chamber, and closes the breech. The moment the breech is closed, the gun is fired automatically, if the trigger be held in the pulled position. The rate of fire is about 660 rounds per minute.

ADVANTAGE.—The great advantage of this gun is, that being automatic in its action, the aiming is not interfered with by the operation of a crank or other device to work the mechanism, and hence it can be pointed readily in any direction, and the direction changed with great facility.

PARTS.—The gun consists practically of two parts—a recoiling, and a non-recoiling part. The recoiling part embraces the barrel, the lock, the crank, the breech-block, and an inner frame with guides and bearings, on which these parts move. The recoiling part may be considered the gun proper.

The non-recoiling part consists of a casing and two side frames, in which the recoiling part moves.

349. The Maxim Automatic Machine Gun—The Barrel and Frame —The Breech Mechanism.

THE BARREL AND FRAME.—The gun has a single barrel,

a, Fig. 364, attached to the inner frame *b*, and is an ordinary rifled one of the desired calibre. It has bearings at *c* and *d*, which rest in corresponding supports in the bronze casing, and on these bearings the barrel slides back and forth in action. The frame *b* is open at the top and bottom, and resembles a box. This frame carries the breech mechanism, and hence the latter moves back and forth with the barrel and frame, and has also motion with respect to the frame, as will be explained. Near the rear end of the frame, the crank-shaft *e* passes through both sides, and has a motion of rotation in the frame. On the right hand side, this crank-shaft projects, and upon it is fixed a bent lever, *ff'*, of the shape shown. On the left side of the frame *b* the crank-shaft *e* projects also, and to it is attached the short crank *g*. The strong spiral spring *h* by which the counter-recoil is produced, is attached at *h'* to the fixed casing, and at *h''* to the short crank *g*. Any rotation of the crank-shaft *e* in the direction of the arrows, will therefore increase the tension of the spiral spring *h*, which will be wound up around *e*. Also any backward movement of barrel and frame, with reference to the fixed casing, will increase this tension, since the spiral spring is fixed to the casing at *h'*.

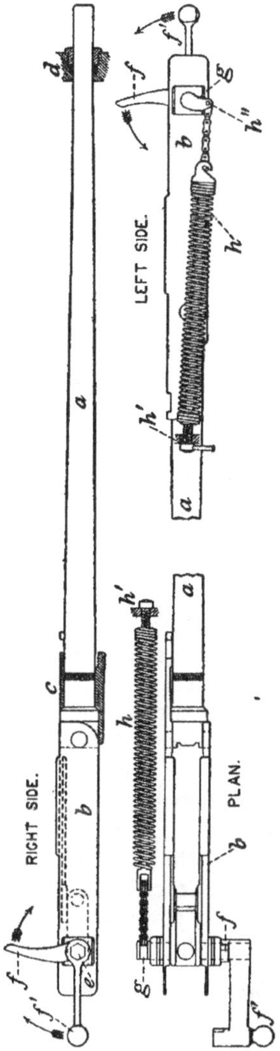

FIG. 364.

THE BREECH MECHANISM.—This is contained between the sides of the inner frame *b*. In Fig. 365 let *e*, as before,

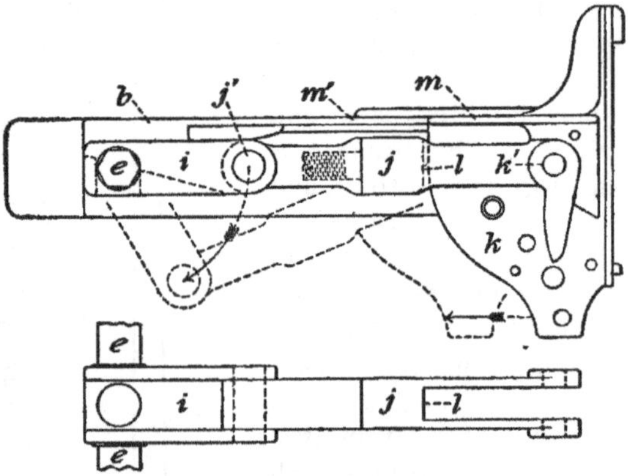

FIG. 365.

represent the crank-shaft. Upon this shaft, and between the sides of the inner frame *b*, is fixed an arm or crank, *i*, so that any rotation of *e*, causes *i* to rotate also.

At the end of *i*, a fork-shaped piece or link, *j*, is attached by an axis, *j'*, and the front end of *j* is pivoted to the breech-block *k* at *k'*. The breech-block is therefore held between the prongs of the forked link *j*. When the crank-shaft *c* is rotated in the direction of the arrow, the pivot *j'* describes the arc of a circle in the same direction. As the breech-block *k* can only slide back and forth in the direction of the arrow, it is evident that this rotation of the shaft *e*, as described, will pull the breech-block backward, with reference to the barrel and frame, along the guides *m'*, and at the same time the surface *l* of the fork-shaped link *j* will describe the arc of a circle around *k'*, and will, consequently, move down along the rear curved surface of the breech-block.

350. The Maxim Automatic Machine Gun—The Breech-block and Carrier.

THE BREECH-BLOCK AND CARRIER.—The breech-block consists of the part *k* (Figs. 366, 367), which moves backward and forward parallel to the axis of the barrel, the bearing *m*, sliding in the groove *m′* in the inner frame *b*; and of the part *n*, called the carrier, which is attached to the front part of the breech-block *k*, and moves backward and forward with it, but has also a vertical sliding motion along its front. This carrier, *n*, extracts the loaded cartridges from the belt which carries them, feeds them to the barrel, and extracts the empty case after firing. Its action is as follows: When in the firing position (Fig. 366) the

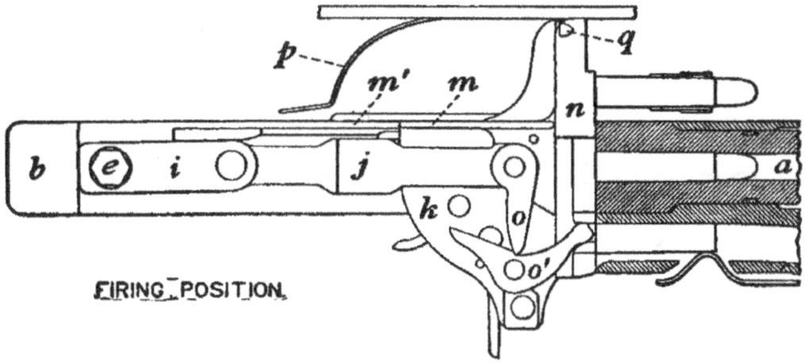

FIG. 366.

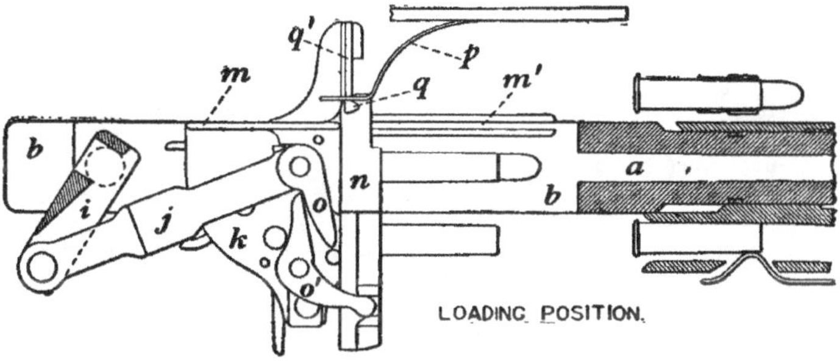

FIG. 367.

forked link *j* is nearly horizontal, and the carrier *n* is at its extreme upward position. After firing, when *j* rotates downward (Fig. 367), the cam *o*, which is a part of *j*, is no longer in contact with the second cam *o′*, pivoted to *k*, and working *n*. The carrier *n* is then held up by two guides on the sides of the inner frame, shown in Fig. 371, upon which the arms, *q*, rest, and is pressed down as the block slides backward, by two springs, *p*, attached to the fixed casing, and acting on the two projecting arms, *q*, on the carrier. When the block moves forward again, the forked link *j* rises, and the cam *o*, acting on *o'*, raises the carrier at the end of the forward movement of the block. Two grooves in the rear face of *n*, slide on bearings, *q'*, on the front of the breech-block *k*. The front surface of the carrier *n* is grooved, so that the heads of the cartridges will fit in it. As it rises, these grooves engage the rim of the cartridge in the chamber, and that of a second cartridge above this in the feed-belt. When the piece is fired, the backward motion of the breech-block causes the carrier *n* to draw the loaded cartridge out of the feed-belt, and the empty case out of the chamber. The carrier *n* is then forced down by the springs *p*, as explained, which brings the loaded cartridge opposite the chamber, and the empty shell opposite the ejector-tube. A forward motion of the breech-block then deposits the cartridge in the chamber, and ejects the empty case.

351. The Maxim Automatic Machine Gun—The Firing Mechanism —The Feed.

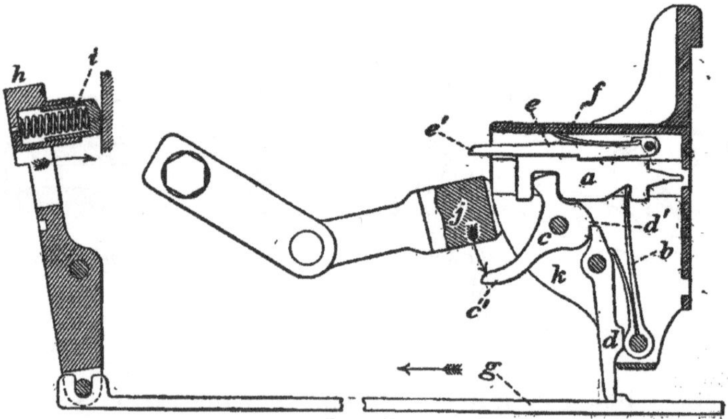

FIG. 368.

THE FIRING MECHANISM. — This is contained in the interior of the breech-block *k*. It consists (Fig. 368) of the firing-pin *a*, the main spring *b*, which acts also as a sear-spring, the tumbler *c*, sear *d*, safety-sear *e*, and its spring *f*.

The action is as follows: When the forked link *j* (see also previous figures) moves downward, it strikes the projecting end *c′* of the tumbler *c*, and causes the latter to rotate in the direction of the arrow. The upper part of the tumbler, bearing in a notch in the firing-pin *a*, draws back the latter, compressing the mainspring *b*, till the sear *d* catches under the notch *d′* of the tumbler. At the same time, the safety-sear *e* drops into a notch on the upper side of the firing-pin *a*. If the firing is to be continuous, the trigger-rod *g* is kept constantly pulled backward in the direction of the arrow, by pressing with the thumbs on the lever *h*. If the pressure upon *h* is relieved, the spring *i* forces the trigger-rod *g* forward, and the firing ceases, the trigger being no longer pulled.

Supposing the trigger-rod to be kept pulled, the sear *d*, striking against the projection on the trigger-rod *g* as the lock moves forward, is disengaged from the notch *d′* in the tumbler, and the firing-pin is held back by the safety-sear *c* alone. As the forked link *j* rises in closing the breech, it strikes the projecting end *e′* of the safety-sear, just as the breech is closed, disengaging *c* from its notch in the firing-pin, which then moves forward and fires the cartridge. In this case the firing is automatic and continuous.

If the firing is to be by single shots, the trigger-rod *g* is not kept in the pulled position. In this case the forked link *j* rises and disengages the safety sear *e*, as before. The firing-pin is now held back only by the sear *d* and tumbler *c*. Pulling the trigger-rod disengages *d*, and fires the cartridge.

THE FEED.—The cartridges are contained in belts, made by uniting two strips of canvas, with intervals between them to hold the former. These belts, with their cartridges, are contained in a box placed below the gun. Over the rear end of the barrel is a box-shaped feed attached to the casing (Fig. 369).

This feed contains a slide, *a*, having a pin, *b*, acted on by a lever, *c*. The slide has two spring-pawls, *d*; and two other spring-pawls, *e*, are fixed to the feed-box, but not attached to the slide. The belt containing the cartridges is passed into the feed-box. till the first cartridge is caught by the spring-pawls *d*. As the barrel recoils, a projection on the inner movable frame, strikes the lower end of the lever at *c'*, and causes the slide *a* to move to the left, by the action of this lever on the pin *b*. The spring-pawls *d*, moving with the slide *a*, push the cartridges and the belt to the left, till a cartridge is in position to be caught by the groove in the carrier as it rises. The lower spring-pawls *e*, being pivoted to the feed-box, do not slide, and hence hold the

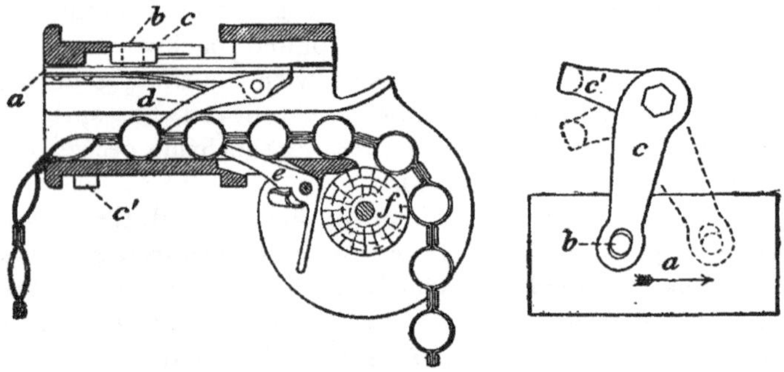

FIG. 369.

belt and cartridges in place, while the slide *a*, with the pawls *d*, moves back again to the right to engage over another cartridge. *f* is a wooden roller, over which the belt passes, and the mouth of the feed-box has guides for directing the motion of the belt.

352. The Maxim Automatic Machine Gun—Action of the Mechanism. The Figs. 370 and 371 show the assembled gun and mechanism. On the exterior right side of the fixed casing, is a curved arm, *a*, a stop, *b*, and a buffer-spring, *c*.

When the gun is fired, supposing the trigger to remain pulled, the barrel, inner frame, and breech mechanism recoil together for a short distance.

At the end of this recoil the curved arm *d* of the bent lever strikes against the curved arm *a*, fixed to the outer case. This causes a rotation of the crank-shaft *e*, and, as previously explained, the breech-block *k* is drawn back from the chamber, thus opening the breech, and at the same time drawing a loaded cartridge out of the belt, and extracting the empty case from the chamber. As long as the rotation of the crank-shaft *e* continues, the breech-block *k* moves backward.

During the last part of its motion, the carrier *n* is forced downward by the springs *p*, Fig. 371, and thus the loaded

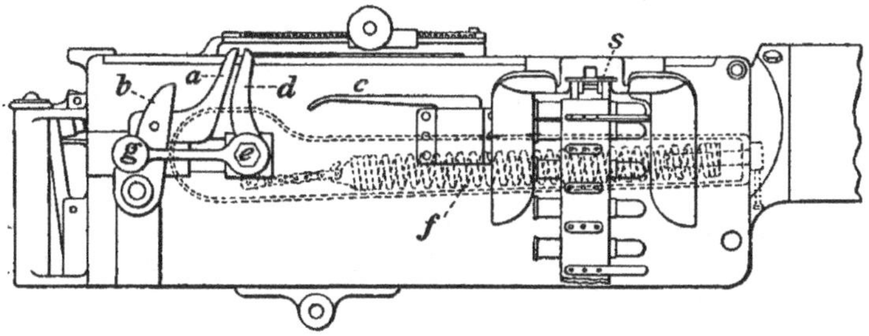

FIG. 370.

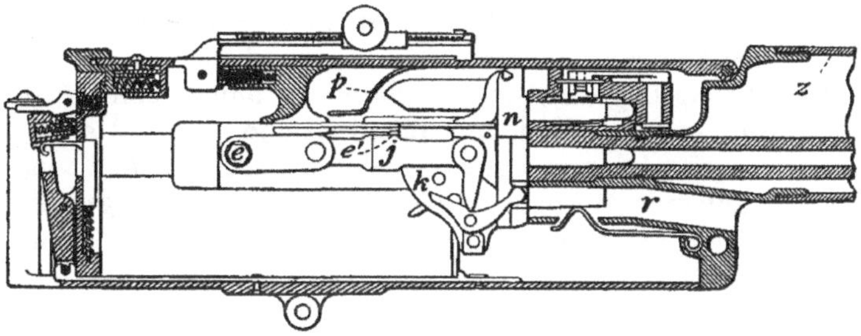

FIG. 371.

cartridge is brought in line with the chamber, and the empty case with the ejector-tube *r*.

During this time also, the firing-pin has been cocked, and the strong spiral spring *f* extended by this rotation of the crank-shaft *e*, as explained. The rotation of the crank-shaft *e* continues, till the outer arm *g* of the bent lever, strikes the buffer-spring *c*, fixed to the casing. The reaction of this spring, and the tension of the spiral spring *f*, now cause the crank-shaft *e* to rotate in the opposite direction, and the spiral spring *f* also forces the barrel and frame forward.

As the rotation of *e* continues, the breech-block *k* moves forward, the loaded cartridge is thrust into the barrel, and the empty shell out through the ejector-tube *r*. As the breech closes, the carrier *n* rises, and grasps another cartridge which has been fed forward by the slide *s* in the feed-box, as explained, and when the breech is completely closed the forked link *j* strikes the safety sear *e'*, and fires the cartridge.

The barrel is surrounded by a bronze casing, *z*, which is filled with water for keeping the barrel cool during the rapid firing, and a provision is made for the escape of the steam if the water is heated to the boiling-point.

353. The Hotchkiss Revolving Cannon—General Features—Rotating Mechanism—Loading Mechanism—Extracting Mechanism —Action.

GENERAL FEATURES.—This machine gun differs from the others in its weight and calibre, being much heavier, and firing a projectile weighing about one pound, which may be either shell or canister. It resembles in some respects the Gatling gun, already described, and is composed of a group of five barrels assembled around a central shaft, the whole revolving in front of a heavy breech, which contains all the mechanism. It differs from the Gatling gun in having only one loading, one firing, and one extracting apparatus for the five barrels.

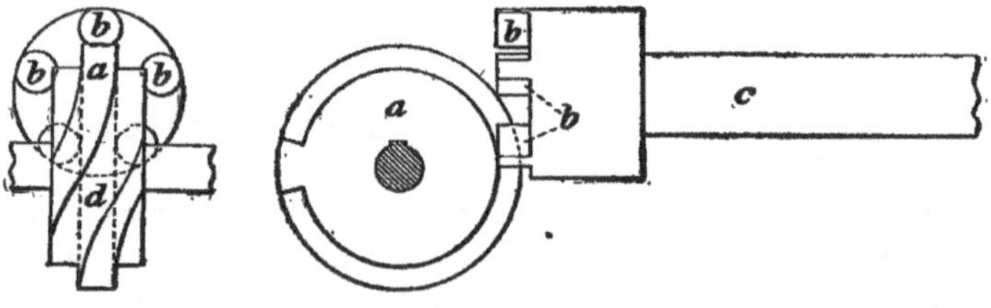

FIG. 372.

ROTATING MECHANISM.—This consists (Fig. 372) of a cam-wheel, *a*, operated by a crank, and working against a series of studs, *b*, on the rear end of the central shaft *c*.

This cam-wheel is mounted in a recess in the breech,

and its peculiar feature is that the grooves *d*, in its surface, are partly screw-threads, and partly planes at right angles to the axis of the cam-wheel shaft. The object of this arrangement will be explained.

LOADING MECHANISM. — On the left side of the solid breech is situated a loading piston. This moves back and forward in a recess, parallel to the axis of the barrels. The piston itself (*b*, Fig. 373) is cylindrical, and has an arm, *a*, which is flat, and carries a toothed rack.

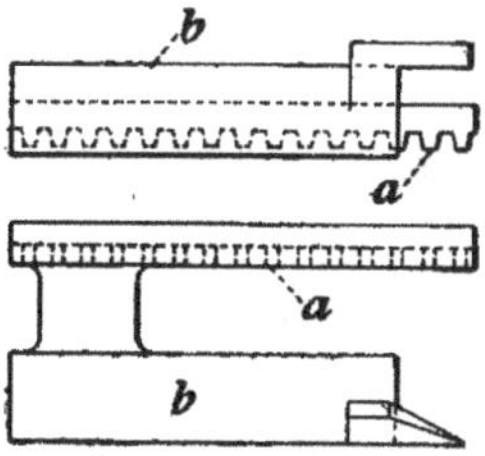

FIG. 373.

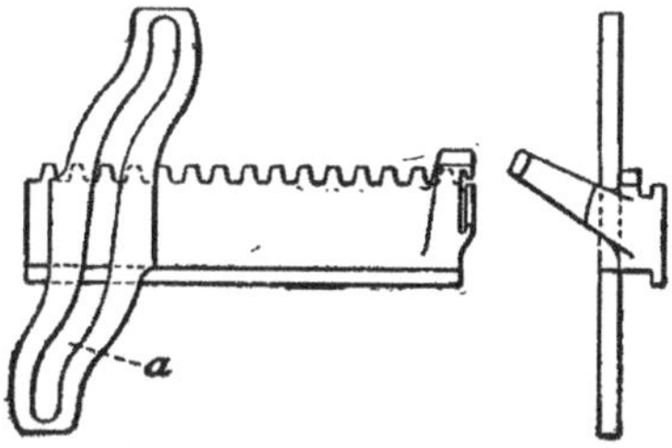

FIG. 374.

EXTRACTING MECHANISM.—The extractor is also situated in a recess on the left side of the breech, and below and parallel to the loading-piston. Its shape is shown in Fig. 374, the front end carrying a hook-shaped extractor, the rear end having an arm with a slot, *a*, and the upper edge forming a toothed rack.

ACTION OF LOADING AND EXTRACTING MECHANISM.—When the cam-wheel is rotated by the crank, if the spiral part of its grooves, *d*, Fig. 372, is bearing against one of the studs, *b*, the central shaft and barrels rotate. If, however, the plane grooves act against these studs. the barrels do not move, and are held in position by the binding of these plane surfaces against the studs.

In Fig. 375, *a* represents the axis of the cam-wheel, *b* and *c* the loading-piston and extractor respectively, in their relative positions when assembled.

The crank *d* is attached to the axis *a*, and the toothed wheel *e* is mounted on an independent axis on the left side of the breech, gearing into *b* and *c*. When the axis *a* rotates, a pin, *g*, on the end of the crank *d*, engaging in the slot *f*, at the end of the extractor-arm, draws back the extractor. This occurs while the plane grooves of the cam-wheel are bearing against the studs on the central shaft, and, consequently, there is no rotation of the barrels.

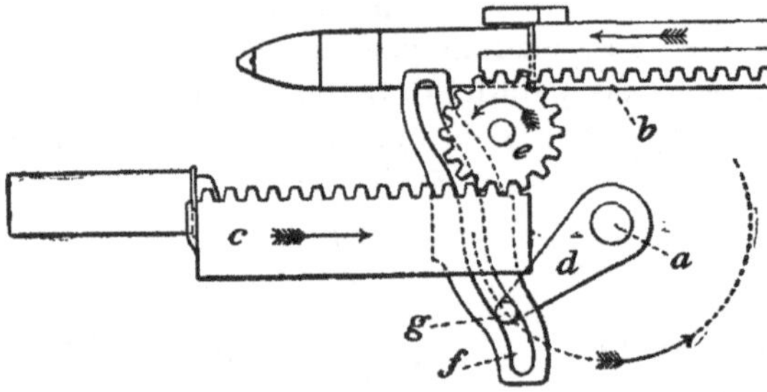

FIG. 375.

As the extractor moves backward, it withdraws the empty case from the barrel last fired. By the backward motion of the extractor, the wheel *e* is caused to rotate, and, acting on the rack of the loading-piston, it causes the latter to move forward, thus pushing the loaded cartridge in front of it into the chamber. The loading and extracting are performed while the barrels stand still. As the rotation of the cam-wheel still continues, the pin *g* in the slot *f* reaches a part of this slot which is concentric with *a*. At this time the extractor and loading-piston stand still, while the barrels rotate.

Continued rotation of the cam-wheel beyond this point of rest, reverses the motion of the loading-piston and extractor, pushing the latter forward, and drawing the former backward, and so on.

354. The Hotchkiss Revolving Cannon — The Feed — The Firing Mechanism.

THE FEED.—The cartridges are contained in zinc or tin cases—ten in a case. These cases are inserted in a feed-tray, *a*, Fig. 376, mounted on the left side of the breech; the act of inserting the case causing it to open and allow the cartridges to enter the tray. Resting against the top of the loading-piston *b*, is a hinged lid, *c*, attached to the breech. As the loading-piston moves forward, pushing the cartridge in front of it into the barrel, this hinged lid prevents the entrance of a second cartridge. When the loading-piston moves back in rear of the lid, the latter drops

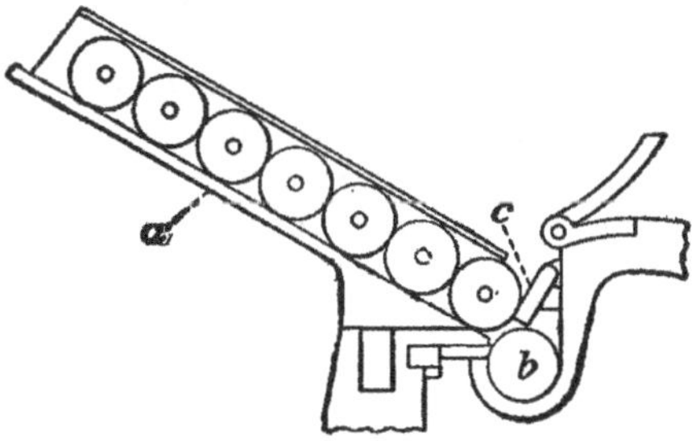

FIG. 376.

down by its own weight and that of the cartridges resting against it, and allows a fresh cartridge to drop in front of the loading-piston. The piston immediately moves forward, raising the lid, and keeping back the other cartridges.

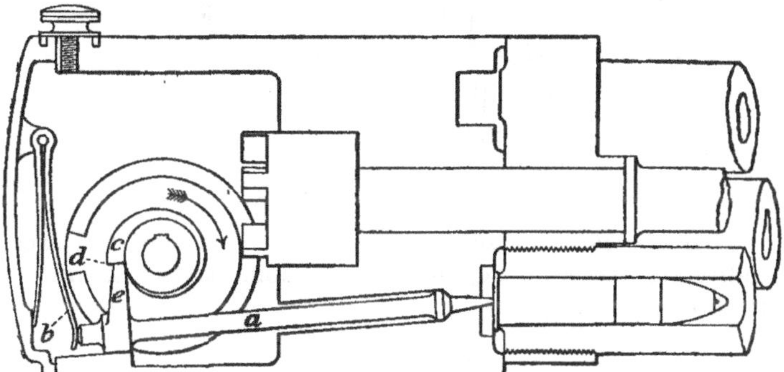

FIG. 377.

THE FIRING MECHANISM.—This is arranged as follows, (Fig. 377): On the right hand lower side of the breech is a strong firing-pin, *a*, moving in a recess, and acted on by the main spring *b*. On the shaft of the cam-wheel is a spiral cam, *c*, which is cut off abruptly at *d*. The arm *e* of the firing-pin, bears against this spiral cam, and the pin is constantly pressed forward by the main spring. As the cam-wheel rotates, the spiral cam, acting on the arm *e*, gradually draws back the firing-pin, compressing the main spring *b*. When this spring is compressed to its full extent, the spiral ends abruptly at *d*, and at this time the barrels are standing still. The firing-pin then moves forward, driven by the main spring, and fires the cartridge. Hence the loading, firing, and extraction in this gun are performed while the barrels are stationary. The recoil is borne by the heavy breech, which is made of cast-iron, and is recessed to receive the various parts of the mechanism.

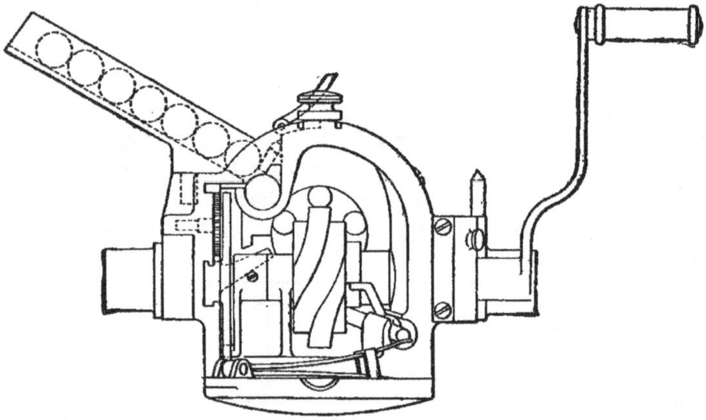

FIG. 378.

These parts are strong and not liable to break, and are readily accessible for repair. The various parts assembled are shown in Fig. 378.

RAPID-FIRE GUNS.

355. Characteristics of Rapid-fire Guns—Object—History.

CHARACTERISTICS.—A rapid-fire gun is distinguished from a machine gun by having a larger calibre, loading by hand, having generally one barrel, and an artificial means of checking recoil and returning the gun to the firing position. It uses metallic ammunition, and rapidity of fire is obtained by the use of a simple breech mechanism, which works quickly, cocking the firing-pin, and extracting the empty case, in the act of opening.

OBJECT.—The object of rapid-fire guns is to defend naval vessels against the attack of swift torpedo boats, by delivering a rapid and easily directed fire of projectiles having sufficient energy to penetrate the plates of these boats; and also for piercing the lighter-armored parts of large ships. In the land service their use is not so well defined. In order to utilize the rapid fire of which they are capable, it is necessary that the aim shall be maintained upon a given object, and not altered by the recoil of piece or carriage. In the naval service, owing to the character of the mounting, this object is very readily attained, the gun being mounted on an elastic or spring-return carriage, fixed to the vessel, by which arrangement the gun is brought back to the firing position after discharge without derangement of the aim. In the land service similar mountings have been provided, attached to wheel-carriages, but in general the shock of recoil alters the direction of the piece, due to the mobility of the carriage, and hence it must be redirected after each fire. In the latest mounting for the land service, spring-return devices have been abandoned, and a rigid carriage adopted. This carriage is provided with a spade at the end of the trail, which is forced into the ground by the recoil, and, when fixed, holds the gun and carriage in place.

HISTORY.—The Hotchkiss revolving cannon, already described, was first used when torpedo boats were adopted, but it was found impossible to give sufficient velocity to the projectiles from this gun to pierce their plates, or to increase the calibre much beyond 1.75 inches, owing to the great weight of the group of barrels. Under these circumstances Hotchkiss invented a single-barrelled rapid-fire gun, with a sliding breech-block, using metallic ammunition, and firing a much heavier projectile than the revolving cannon, with a muzzle-velocity of 1800 ft.-seconds. From this time these guns have rapidly developed in power, and many different systems of breech-closing have been devised.

The principal systems are the Hotchkiss, Nordenfelt, Driggs-Schroeder, Maxim, Gruson, Krupp, and Armstrong, and a few of these will be described as types.

356. The Hotchkiss Rapid-fire Gun—The Gun—The Breech Mechanism—Action.

THE GUN.—The body of the gun consists of a tube and a jacket, united by shrinkage.

The jacket extends to the rear of the tube, and is slotted vertically to receive the breech-block.

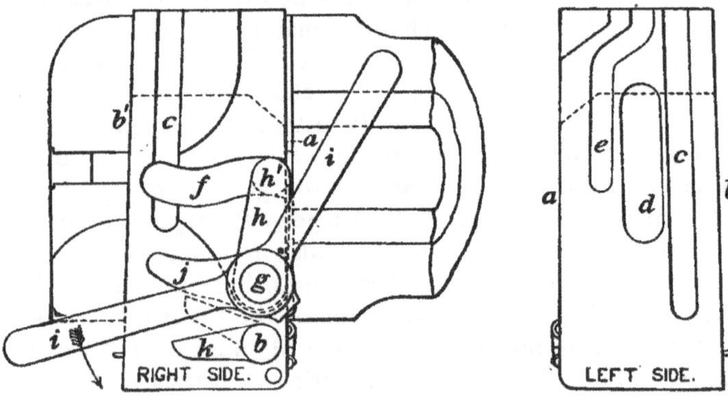

FIG. 379.

BREECH MECHANISM.—The breech mechanism consists (Fig. 379) of a wedge-shaped block, which rises and falls in a vertical direction in the slot at the rear end of the jacket, instead of moving horizontally, as in the Krupp. Its front surface, *a*, is perpendicular to the axis of the bore, and its rear surface, *b'*, is inclined to that axis, so that the block, in rising, gradually moves forward towards the barrel. *c c* are guide-grooves parallel to the rear face *b'*, and corresponding projections in the breech-slot fit into these grooves, and guide the block in its motion. On the left side of the block is the stop-groove *d*. A bolt passes through the left side of the breech, and entering this groove, prevents the block from falling out of its recess when the breech is opened. In front of this groove, and on the same side of the block, is the extractor-groove *e*. The extractor is exactly similar to that already described in the Hotchkiss mountain gun, except that it works in a recess on the left side of the breech, instead of on the top. Its lug bears in the groove *e*, and as the block falls, the extractor is moved back very slowly at first, extracting the empty case from the chamber, and then very quickly, owing to the sudden change of form of the groove *e*, ejecting the case from the gun. On the right-hand side of the block is a groove, *f*, called the studway. A crank-shaft, *g*, passes through the right-hand side of the breech, projecting into the breech slot, and to this inner projection is attached the crank *h*, with a stud, *h'*, at its extremity, working in the groove or studway *f*. The crank-shaft *g* is operated by two handles, *i*, attached to it on the outside.

ACTION.—When the crank-shaft *g* is turned by the handles *i* in the direction of the arrow, the stud *h'* moves at first in a part of the groove *f*, concentric with *g*, and hence no motion of the block occurs. During this time the hammer is cocked, as will be explained.

The stud *h'* now enters the eccentric part of the groove *f*, and causes the block to descend. As soon as it is started, it will fall by its own weight, till arrested by the stop-bolt bearing against the top of the stop-groove *d*. A reversal of the rotation of the crank-shaft *g*, after loading, causes the block to rise, and the block in rising, forces the projectile home.

The opening of the breech-block in firing is prevented as follows: When the breech is closed, the weight of the block is supported by the stud *h'*. At this time the vertical plane through the centre of *h'* is in front of that through the centre of the crank-shaft *g*. Hence the weight of the block acting vertically tends to keep the latter closed.

357. The Hotchkiss Rapid-fire Gun—Firing Mechanism—Action—Remarks.

FIRING MECHANISM.—The firing mechanism (Fig. 380) is contained in recesses in the front portion of the block, and consists of a hammer, *a*, mounted on a rocking-shaft *b* (see also Fig. 379), a main spring, *c*, a sear, *d*, and a sear-spring, *e*.

The rocking-shaft *b* passes through the lower front corner of the breech-block, and projects beyond the right side, carrying on its right extremity, a curved arm or cam, *k*, Fig. 379, called the cocking-toe. The crank-shaft *g*, which is

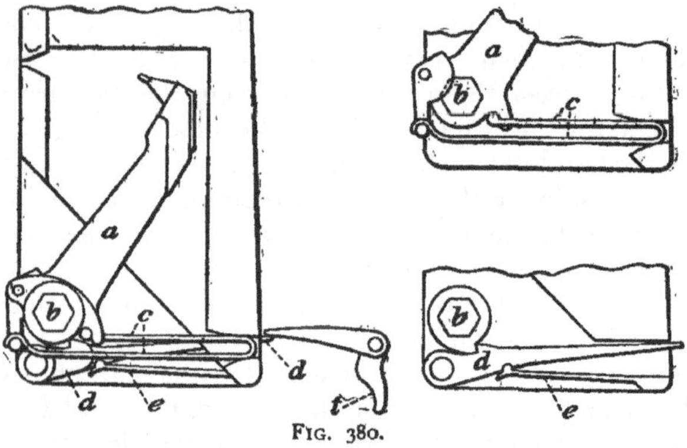

FIG. 380.

fixed to the breech (Fig. 379), carries a cam, *j*, called the cocking-cam, which is just above *k*.

ACTION.—When the handles, *i*, Fig. 379, are rotated in the direction of the arrow, the cam *j* comes in contact with the cocking-toe *k*, on the rocking-shaft *b*, Figs. 379 and 380, drawing back the hammer *a*, and compressing the main-spring *c*. At this time there is no motion of the breech-block, because the stud *h'*, Fig. 379, is moving in the concentric part of its studway, *f*, as explained.

The rotation of the handles, *i*, continues, till the hammer *a* is cocked, the sear *d* catching in a notch on the rocking-shaft *b*, the block all the while remaining motionless. As soon as the cocking is accomplished, the breech-block falls and opens the breech. The end of the sear *d* projects beyond the rear surface of the breech-block, and when the latter is home, the sear is in contact with the trigger *t*, and, pulling it, fires the charge. The piece cannot be fired before the breech is closed, 1st, because the firing-pin of the hammer is not opposite the primer; 2d, the trigger will not touch the sear; 3d, the cam *j* on the shaft *g* will catch the cocking-toe *k* before the firing-pin can reach the primer.

The main spring *e* is so connected with the rocking-shaft *b*, on which the hammer is mounted, that its lower leaf acts downward and its upper leaf upward, and hence the pressure and friction of the rocking-shaft in its bearings are very much diminished.

REMARKS.—All parts of the mechanism are readily accessible and easily dismounted. For aiming, a stock, *a*, Fig. 381, is bolted to the left side of the piece if the gun is on a

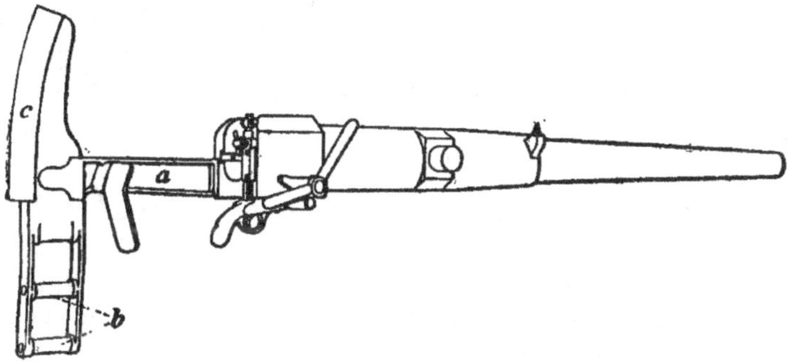

FIG. 381.

rigid carriage, or to the left side of the carriage if the gun recoils. This stock has handles, *b*, for grasping with the left hand at different elevations, and a rubber tube, *c*, against which the left shoulder rests in firing. For the naval service the gun is generally mounted with its trunnions resting in a fork. This fork turns in azimuth in a heavy socket, and this, combined with the vertical motion of the gun around the axis of the trunnions gives a motion in any direction.

358. The Nordenfelt Rapid-fire Gun—The Breech Mechanism—Action.

THE BREECH MECHANISM.—The breech mechanism in this system combines the sliding and rotating motions, so as to avoid guillotining the cartridge in forcing it home.

The breech-block (Figs. 382, 383, and 384) consists of two parts, the first part, or block proper, *B*, and the rear part, or wedge, *W*. The front part *B* rotates around the shaft *S*, which passes through the breech, while the rear part or wedge, *W*, has at first a vertical downward sliding motion along the back of *B*, till it reaches the position shown in Fig. 383, the two upper surfaces *r r*, then forming one continuous cylindrical surface, at which time both parts rotate backward together around the shaft *S*, Fig. 384. *C* is a

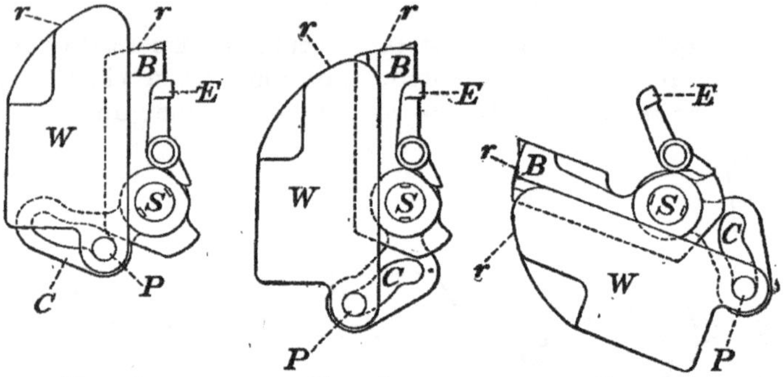

FIG. 382. FIG. 383. FIG. 384.

cam, fastened upon the shaft *S*, and having a slot in it, in which the pin *P*, attached to the wedge, works. *E* is the extractor and ejector.

ACTION.—A lever-handle is attached to the right extremity of the shaft *S*, outside the breech. After firing, the parts of the block are in the position shown in Fig. 382. As the lever-handle on the shaft *S* is rotated, it causes the cam *C* attached to *S* to rotate downwards. The slot in the cam, bearing on the pin *P*, forces the wedge *W* downwards, till it stands in the position shown in Fig. 383. The pin *P* is now at the end of the slot in the cam *C*, and a continuation of the rotation of the lever-handle and shaft *S*, causes both parts of the block to rotate together to the rear, opening the breech (Fig. 384). At the beginning of this rotation, the extractor *E* is moved slowly backward, withdrawing the empty case. This motion afterwards becomes more rapid, ejecting the case from the gun.

After the loaded cartridge is inserted in the gun, the rotation of the lever on the shaft *S* is reversed. This causes both parts of the block to rotate to the front, till the front part *B* comes in contact with the breech. The wedge *W* is then forced vertically upward by the action of the cam *C*, till it occupies the position shown in Fig. 382, completely closing the breech. The parts are held in position by the pin *P*, at the last moment of closing, entering a concentric part of the slot in the cam *C*, which supports the wedge in position.

359. The Nordenfelt Rapid-fire Gun—The Firing Mechanism—Action—Remarks.

THE FIRING MECHANISM.—This is arranged as follows (Fig. 385): The firing-pin *a* passes through the front part, *B*, of the breech-block. In rear, it has projecting lugs, *b*, one

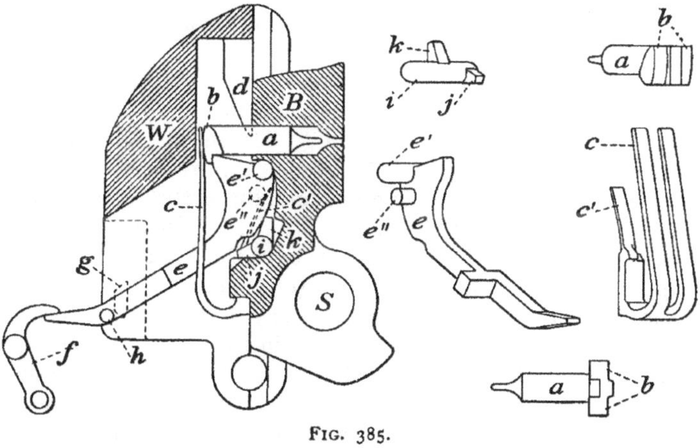

FIG. 385.

on each side. The middle part of the sliding wedge *W* is hollowed out, to receive the parts of the mechanism. The main spring *c* is secured at one end against the rear face of the block *B*, while the other end bears against the rear faces of the lugs *b*, of the firing-pin *a*, and urges the latter forward.

Two wedge-shaped lugs *d*, on the sliding part *W* of the block, one of which is shown, act upon the lugs *b* to cock the firing-pin, when the sliding part of the block descends.

The sear *e* works on a pivot, *e'*, resting against the rear face of the front part of the block, and catches under the head of the firing-pin as shown, when the latter is cocked. Its lower end projects beyond the rear face of the block, and is in contact with the trigger *f* when in the firing position. *g* is a safety lug on the wedge, and *h* the corresponding lug on the sear. *i* is a tappet-trigger, so called, consisting of a shaft working in a bearing on the rear face of the front part *B* of the block, and having upon it two arms or cranks *j* and *k*, set at an angle with each other. There is also a vertical groove on the front of the sliding wedge, not shown in the figure, which ends abruptly in a shoulder, and which works the tappet-trigger. This may be called the tappet-trigger groove.

Action.—In the position shown in the figure, the parts are ready for firing. On pulling the trigger *f*, the firing-pin *a* moves forward, under the action of the main spring *c*, exploding the cartridge. The lever on the main shaft *S* is now rotated, and the sliding wedge descends. As it moves down, the wedge-shaped lugs *d*, engage in front of the projecting lugs *b* of the firing-pin, and force the latter back against the action of the main spring *c*. When the firing-pin is forced back to its full extent, the shoulder of the tappet-trigger groove on the wedge *W*, acting on the arm *j*, causes the shaft of the tappet-trigger *i* to rotate, and forces the inner crank-arm *k* backward against the sear *e*, retaining the latter in the cocked position. The cartridge is now inserted, the lever-handle reversed, and the breech closed. As the wedge *W* rises into its closed position, the wedge-shaped lugs *d* rise above the firing-pin, and the projecting part of the sear *e* comes in contact with the trigger *f*. The firing-pin is now held back by the sear *e*, which in turn is held by the arm *c*′, of the main spring *c*, bearing on the lug *e*′′. A pull on the trigger *f* depresses the projecting end of the sear *e*, disengages the sear from the firing-pin, and allows the latter to move forward and fire the cartridge.

REMARKS.—The cartridge cannot be fired before the breech is completely closed, for the following reasons:

1st. While the wedge is rising, the lugs, *d*, are in front of the projections, *b*, on the firing-pin. Hence, if the firing-pin moves forward at this time, the projections, *b*, will strike against the lugs, *d*, and prevent the firing-pin from reaching the cartridge.

2d. The lugs, *d*, clear the projections, *b*, on the firing-pin before the breech is completely closed. At this instant, however, the safety lug *g* on the wedge comes just in front of the corresponding safety lug *h* on the sear, so that the sear cannot be moved till the breech is completely closed, at which time *g* rises above *h*.

360. The Driggs-Schroeder Rapid-fire Gun—The Breech Mechanism —Action.

This gun is an American invention, and its breech system combines the rotating and sliding movements, as in the Nordenfelt, but its distinguishing feature is that the breech-slot does not extend through the top of the breech, and the block rests in grooves on the top and sides of this slot. This gives greater strength, protects all the working parts, and enables the weight of these parts to be reduced, thereby facilitating the operations of opening and closing the breech.

THE BREECH MECHANISM.—The breech mechanism consists of a block, *a*, Fig. 386, having grooves, *b*, and projections, *c*, cut upon its top and sides, which fit into corresponding recesses in the top and sides of the breech recess. *d* is a shaft passing through the sides of the breech, and through the block *a*, about which motion of the latter takes place. *e* is a cam attached to the shaft *d*, and rotating with it; *ff*′ a surface in the interior of the block of the shape shown, which is in contact with the cam *e*, the rear part of *f* being inclined backward and upward, and the front part *f′* being circular and concentric with the axis of *d*. The rear surface *f* ends in a cylindrical pin, *g*. *h* is an inclined surface on the lower rear end of the block, against which the cam *e* acts at a certain period of its rotation. *i* is a slot in

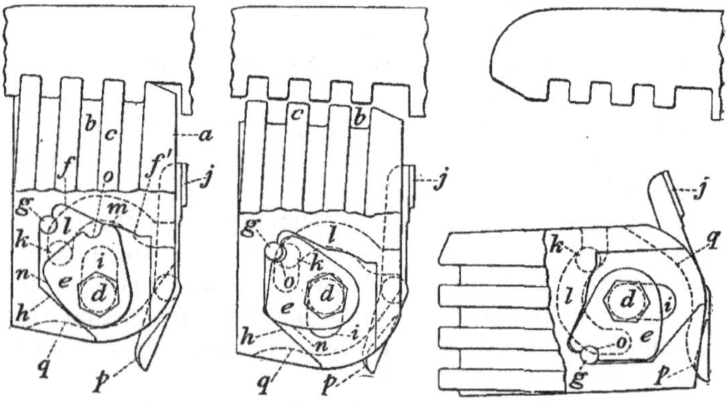

FIG. 386. FIG. 387. FIG. 388

the breech-block, which allows the block to slide in a direction at right angles to the axis of *d*. *j* is the extractor, there being two of these; *k* a guide-bolt screwed through the breech-casing, and fitting into the guide-groove *l* on the side of the block. There are two of these guide-bolts, one on each side.

ACTION.—In Fig. 386 the block is shown in the firing position, in 387 in the partly opened, and in 388 in the fully opened position. When the shaft *d* is rotated to the rear by hand after firing, the toe *m* of the cam *e* passes along the cam surface *ff′*, permitting the block to descend. Should the block not drop freely, the lower face *n* of the cam, acting on the inclined surface *h*, forces the block downward, the slot *i* in the block allowing this motion with reference to the shaft *d*. The guide-groove *l* is also so shaped as to allow this motion with reference to the guide-bolts, *k*.

This vertical motion of the block continues until the block has descended a distance sufficient to disengage the projections, *c*, from the corresponding recesses in the top and sides of the breech. The block is now resting on the toe *m* of the cam, and the guide-bolt, as shown in Fig. 387, is just entering the curved portion of the guide-groove. The block now has a double movement, downward by virtue of the toe continuing to move along the surface *ff'*, and rotary owing to the shape of the guide-grooves. This motion continues till the notch *o* in the cam comes to a full bearing against the cylindrical pin *g*, when the block will rotate around *d* as an axis till the breech is open. As the block rotates to the rear, the extractor-groove *q* strikes the tail *p* of the extractor *j*, and rotates it backward, slowly at first, and then more rapidly, owing to the shape of the abutting surfaces; extracting and ejecting the empty case. The charge is now inserted, and the rotation of *d* reversed. This causes the block to rotate around *d* till its projections, *c*, are ready to enter their recesses in the breech, the notch *o* of the cam *e*, bearing against the pin *g* of the block.

When the rotation of the block is finished, the pressure of the cam on the surface *f*, causes the block to rise, and seat itself in the recesses in the breech. The toe *m* of the cam then passes to the concentric surface *f'*, and supports the weight of the block during firing and keeps it in its seat in the breech. The movement along the concentric surface *f'* is continued until the toe *m* of the cam passes over the centre of rotation *d*, and prevents the downward thrust of the block from having any tendency to turn the cam backward, and it is therefore held rigidly and securely in place.

361. The Driggs-Schroeder Rapid-fire Gun—The Firing Mechanism—Action—Remarks.

THE FIRING MECHANISM.—This consists (Fig. 389) of a firing-pin, *a*, working in a recess in the block, and having a shoulder, *b*, in front.

A strong spiral main spring, *c*, bears against this shoulder, its rear end resting against a corresponding shoulder in the block. *g* is the full-cock notch; *h*, the sear, acting vertically in a recess at the rear end of the block; *i*, the sear-spring, seated in a recess in the block, and acting against the lower end of the sear to force the latter vertically upward. *k* is a projecting lug on the firing-pin *a*, which bears in a circular recess, *j*, in the front upper surface of the cam *e*.

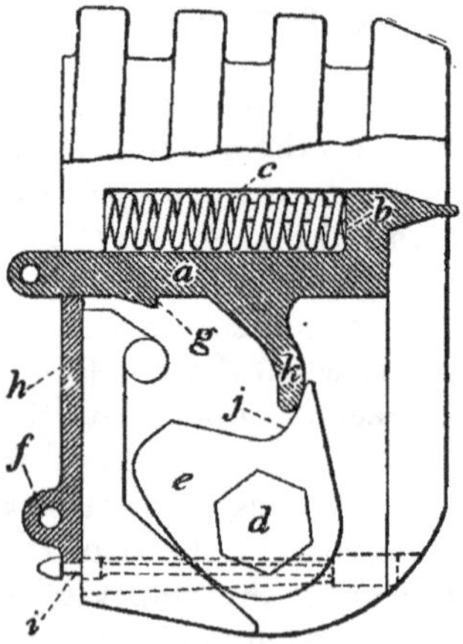

FIG. 389.

ACTION.—When the cam *e* is rotated to the rear by the shaft *d*, the surface of the circular recess *j*, acting against the projecting lug *k* on the firing-pin, withdraws the point of the latter through its hole in the block, so that it will allow the block to descend freely. As the rotation of the cam *e* continues, the firing-pin is drawn back still further, compressing the spiral main spring *c*, till the firing-pin is fully cocked, at which time the sear *h*, acted on by its spring *i*, rises, and engages in front of the full-cock notch *g*, and the firing-pin is thus retained in its cocked position. When the breech is closed, the cam *e* is rotated forward, and the lug *k* is no longer in contact with the surface of the groove *j*. The sear *h*, being pulled vertically downward by a lanyard attached at *f*, is disengaged from the full-cock notch, and the firing-pin *a* moves forward, firing the cartridge.

REMARKS.—The cartridge cannot be fired before the breech is completely closed, because until it is closed the groove *j* in the cam *e* is in such a position as to catch the lug *k* of the firing-pin if the latter should move forward, and thus prevent the firing-pin from striking the cartridge.

362. The Maxim Semi-automatic Rapid-fire Gun—The Breech Mechanism—Action.

This gun differs from those previously described in being semi-automatic; that is, the firing of the cartridge causes the barrel and breech mechanism to recoil together, opens the breech, cocks the firing-pin, and thus prepares the gun for the insertion of a fresh cartridge. The act of inserting the cartridge closes the breech, and if the trigger be kept pulled, fires the piece.

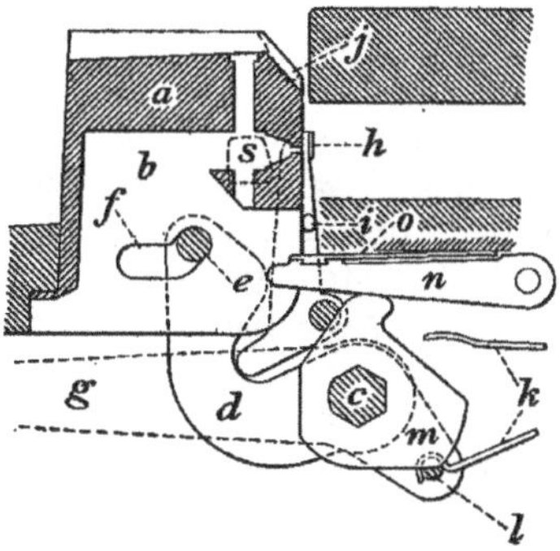

FIG. 390.

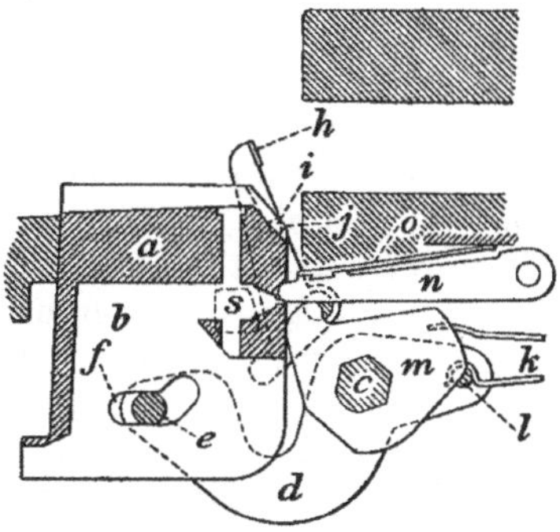

FIG. 391.

THE BREECH MECHANISM.—This consists (Figs. 390 and 391) of a breech-block *a*, hollowed out at *b* to receive the firing mechanism. This block moves vertically upward and downward in a slot in the rear part of the barrel.

Attached to a projection on the lower side of the barrel is the shaft *c*, about which motion of the breech-block takes place. Two arms *d*, one on each side, are attached to the shaft *c* and rotate with it. These arms are connected in rear at their upper ends, by a pin *e*, which passes through the breech-block *a*, and works in a slot *f* in the block. This slot is at first concentric with the axis of *c*, and is afterwards eccentric to that axis. *g* is a handle attached to the shaft *c*, outside the gun, for the purpose of starting the mechanism, and can be readily detached. *h* is the extractor and ejector, having two projections *i*, one on each side, which fit in corresponding recesses *j* in the breech-block. *k* is a strong spring, one end of which bears against the pin *l* on the short arm of the lever *d*, and which presses the long arm of *d* constantly upward. *m* is a cam fastened to the shaft *c*, and *n* is a catch pivoted to the jacket, and having no motion in the direction of recoil. It is constantly pressed downward by the spring, *o*.

ACTION OF THE BREECH MECHANISM.—The parts as represented in Fig. 390 are in the firing position, the breech being closed. Before firing, the operating handle *g* is removed from the shaft *c*. When the piece is fired, all the parts except the catch *n*, with its spring *o*, move to the rear in recoil, the barrel sliding in the jacket, which remains fixed.

As they move to the rear, the cam *m* slides along the fixed catch *n*, till the projecting corner of *m* has passed beyond the end of *n*, when the spring *o* forces *n* downward. When the end of the recoil is reached, a strong spring, not shown in the figure, acts to draw all the parts back to their former position. As the parts move forward, the projecting toe or corner of the cam *m*, strikes against the end of the catch *n*, and as the forward motion of the parts continues, *m* is forced to rotate backwards.

This causes the arms *d* to rotate downward, carrying with them the pin *e*, which passes through the slot *f* in the breech-block. As the front part of this slot is concentric with reference to *c*, no motion of the breech-block occurs at first, but at this time the firing-pin is cocked.

As the forward motion of the parts continues, the toe or corner of the cam *m* is freed from the catch *n* by the upward motion of *m*, which strikes against *n*. When the pin *e* enters the eccentric part of the slot *f* the breech-block descends. In its descent it forces the extractor backward, slowly at first, and afterwards more rapidly, by the action of the studs *s* on the sides of the block which bear on the tail of the extractor. This downward motion of the block, and backward motion of the extractor, continues, till the lugs *i* on the extractor, catch in the recesses *j*, of the block. The breech is now open (Fig. 391) and ready for loading. When the long arm of *d* rotates downward, the short arm moves upward, and compresses the spring *k*. This spring tends to raise the breech-block by its action on the short arm of *d*, but the block cannot rise, because it is held by the projections *i* of the extractor bearing in the recesses *j* of the block. When the cartridge is inserted, its rim strikes against the extractor, and frees the lugs *i* from the recesses *j*, and the pressure of the spring *k* on the short arm of the lever *d* causes the block to rise, thus completely closing the breech.

363. The Maxim Semi-automatic Rapid-fire Gun—The Firing Mechanism—Action—Remarks.

THE FIRING MECHANISM.—This consists (Figs. 392 and 393) of a hammer *E*, rotating around a shaft *a*, which passes

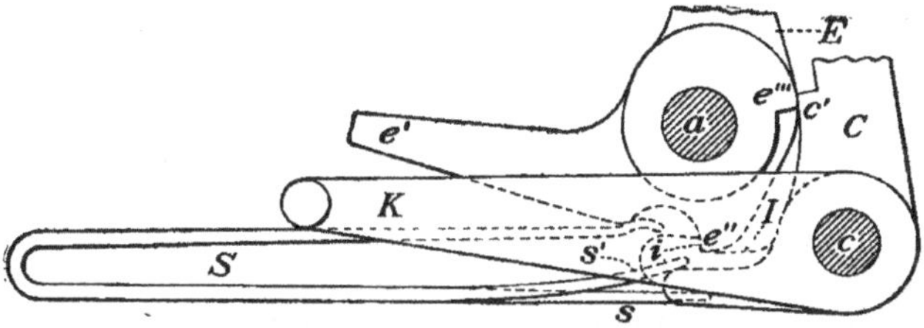

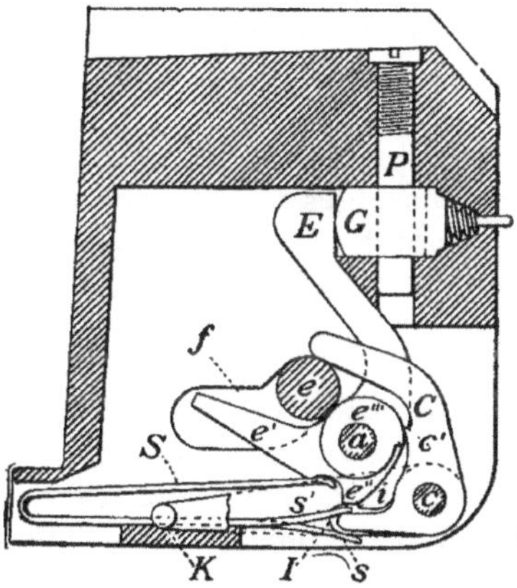

FIG. 392.

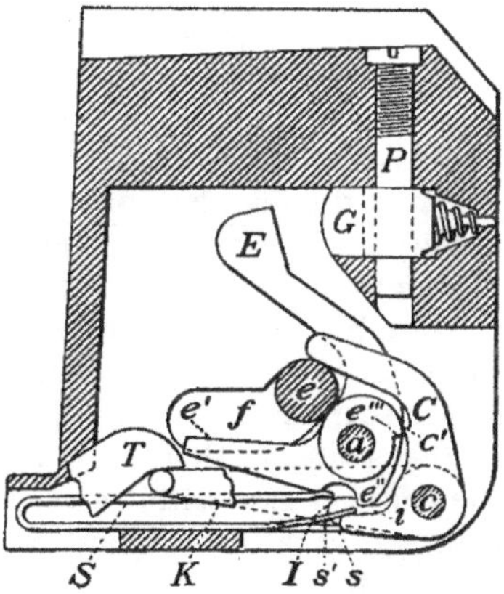

FIG. 393.

through the breech-block. This hammer is pressed constantly forward by the main spring *S*, the upper branch of which acts against the lower bent end *e'* of the hammer. The firing-pin *G* is a separate piece, which is held in place by the bolt *P* passing through a slot in it, this slot allowing the pin to move backwards and forwards for a short distance. In front, the firing-pin is acted on by a spiral spring, which forces it backward as soon as the pressure of the hammer *E* is removed. The sear *I* rotates around a shaft *c* in the breech-block, and is constantly pressed upward by the spring *s*, which is a fork of the lower branch of the main spring *S*. The safety-sear *C* also rotates around the same shaft *c*, and is constantly pressed downward by the spring *s'*, which is also a fork of the lower branch of the main spring *S*. The hammer *E* has two notches *e''* and *e'''* which engage in the corresponding notches *i* of the sear *I* and *c'* of the safety-sear, *C* respectively.

The sear *I* is attached to a lever *K*, which is pivoted around *c*, and extends to the rear, where it comes in contact with a trigger, *T*, Fig. 393.

ACTION OF FIRING MECHANISM.—As represented in Fig. 392, the charge has just been fired. When counter-recoil begins, the pin *e* which passes through the slot *f* in the breech-block, and which connects the upper extremities of the arms attached to the main shaft, begins to move along the concentric portion of the groove *f*. During this time there is no vertical motion of the block, as explained, but the pin *e* strikes against the lower part, *e'*, of the hammer, and depresses it, thus drawing back the upper part, *E*, and compressing the main spring *S*. This continues till the hammer is fully cocked, at which time the notch *e''* on the hammer *E* is in front of the corresponding notch *i* of the sear *I*, and the notch *e'''* on the hammer is engaged in the notch *c'* of the safety-sear *C*. The hammer is now held back by the safety-sear *C*, this latter having been lowered by the action of the spring *s'* till it is in front of the slot *f*. As soon as the hammer *E* is withdrawn from the firing-pin *G*, the latter is forced back by its spiral spring, and the block is now free to descend.

When the cartridge is inserted the breech-block rises.

As it does so, the pin *e* moves up along the slot *f*, and strikes the end of the safety-sear *C*, which projects in front of the slot *f*, and is therefore in the path of the pin *e*. This disengages the notch c' of the safety-sear from the notch e''' of the hammer E. The notch e'' of the hammer E then engages with the notch i of the sear, and the hammer is now held back only by the sear I. Pulling the trigger disengages the notch i of this sear from the notch e'' of the hammer, and fires the piece. If the trigger be kept pulled, it is evident that the piece will fire on the closing of the breech, because the hammer E is then held back only by the safety-sear C, and the pin e strikes this safety-sear in closing, as explained, and frees its notch c' from the notch e''' of the hammer.

REMARKS.—The piece cannot be fired before the breech is completely closed, 1st, because the firing-pin is not opposite the primer; and 2d, because firing cannot take place till the safety-sear is disengaged, and this is done only by the act of closing, and cannot be accomplished till that time.

364. Ammunition for Rapid-fire Guns—Projectiles—Cases.

The ammunition for the different systems of rapid-fire guns is similar in almost every respect, differing only in the kinds of fuze used, each system having its own, but all being of the same type. All these guns use metallic ammunition.

PROJECTILES.—The projectiles are of four kinds: common shell (Fig. 394), made of cast iron; steel shell (Fig. 395); shrapnel (Fig. 396); and canister (Fig. 397). The two kinds of shell are similar in construction except that the base of the steel shell is an independent piece screwed in, and having its inner surface concave, so that it will tend to form a gas-check when the bursting-charge explodes, and retain the gases till they acquire sufficient pressure to rupture the walls. The point of the steel shell is made sharp for armor-piercing, while that of the common shell is cut off, and the two may always be distinguished by this difference in their points. The fuzes are placed in the base, leaving the point and head solid. Many of the steel projectiles are made by electro-welding.

The shrapnel (Fig. 396) is composed of a body, a head, and a base. The body is made of a steel tube weakened longitudinally by six cuts, in order that it may rupture readily. The head is of brass, fitted with a combination fuze, and the upper end of the body is crimped into a recess

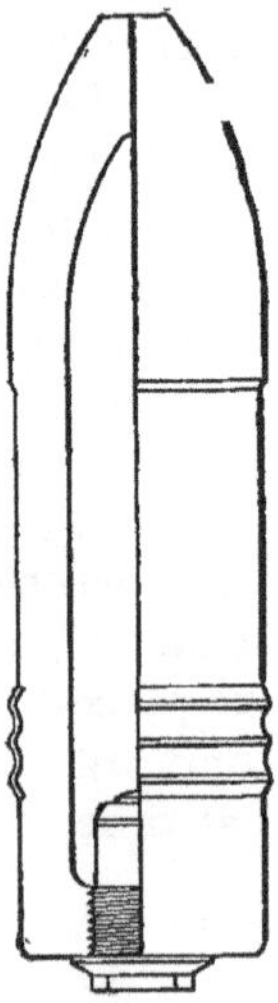

FIG. 394.

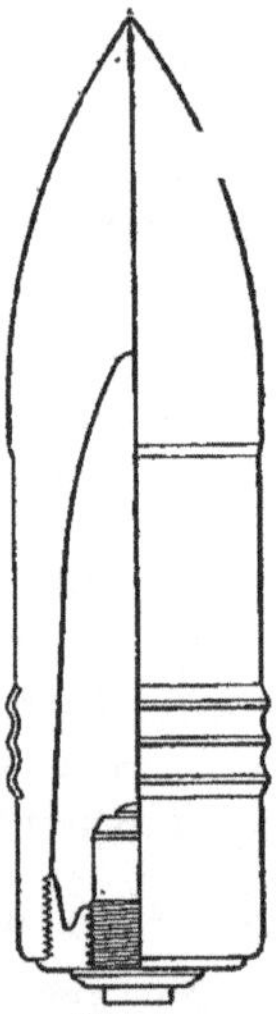

FIG. 395.

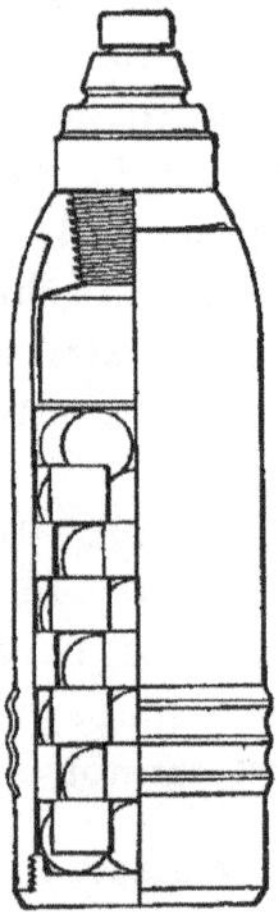

FIG. 396.

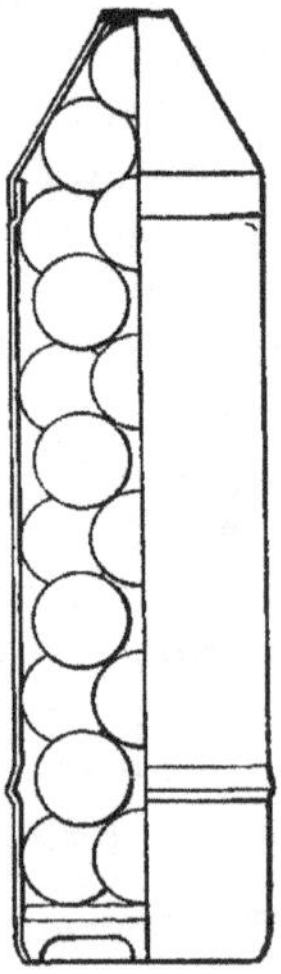

FIG. 397.

FIG. 398.

FIG. 399.

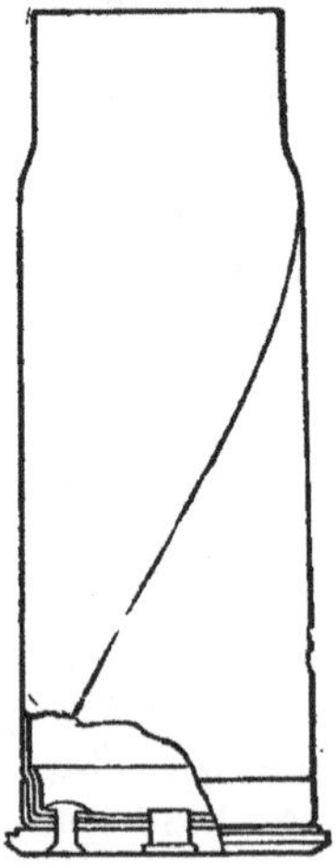

FIG. 400.

in the head. The base is a steel plug forced into the body under pressure, or screwed in. The interior is filled with bullets, the layers being separated by cast-iron disks, as explained in the U. S. shrapnel for field-guns. The bursting-charge is contained in a tin cup in front.

The case-shot (Fig. 397) consists of a thin sheet-brass case, with a conical head, and a bottom of soft brass acting as a rotating band. The bottom is strengthened on the inside with a loose plate of sheet iron. The case is filled with hardened lead balls packed in sawdust.

CASES.—These are of three kinds:

1. The solid drawn case (Fig. 398), made as described under small-arm ammunition.

2. The built-up case (Fig. 399), consisting of a drawn tube of brass, bent inward at the head, and furnished with an inner and an outer cup of brass, which are riveted to a sheet-iron disk on the outside to strengthen the construction. In some cases the sheet-iron disk is omitted, and the outer cup used.

3. The wrapped case (Fig. 400), consisting of a sheet of brass of trapezoidal shape wrapped into a cylinder, and the head formed as in a built-up cartridge.

The latter case is now abandoned except for the Hotchkiss revolving cannon. All the cartridges are centre-primed. The projectile, fuze, charge, and case are assembled, forming a complete cartridge, the only limit in size being the weight readily handled by one man. On this account, for the larger calibres the projectile and case are separated.

The assembled cartridge is shown in Fig. 401.

FIG. 401.

INDEX.

A

B

C

D

E

F

G

H

J

K

L

M

N

O

P

Q

R

S

T

U

V

W

X

Z

Spotted something that isn't quite right?

Let us know and we'll fix it for future reprints. And as a thank you for your diligence, we'll send you a free Sothis book of your choice.

publisher@sothis.press

This page intentionally left blank

Made in the USA
Coppell, TX
05 November 2026

11307508R00402